Food Engineering Series

For further volumes:
http://www.springer.com/series/5996

M. Anandha Rao

Rheology of Fluid, Semisolid, and Solid Foods

Principles and Applications

Third Edition

Springer

M. Anandha Rao
Department of Food Science
Cornell University
Geneva, NY, USA

ISSN 1571-0297
ISBN 978-1-4614-9229-0 ISBN 978-1-4614-9230-6 (eBook)
DOI 10.1007/978-1-4614-9230-6
Springer New York Heidelberg Dordrecht London

Library of Congress Control Number: 2013951523

Printed on acid-free paper

Springer is part of Springer Science+Business Media (www.springer.com)

Preface to the Third Edition

Following the very good acceptance of the second edition of *Rheology of Fluid and Semisolid Foods: Principles and Applications*, it is an honor and a pleasure to present the third edition.

Again, the book is divided into eight chapters: Chap. 1—Introduction: Food Rheology and Structure, Chap. 2—Flow and Functional Models for Rheological Properties of Fluid Foods, Chap. 3—Measurement of Flow and Viscoelastic Properties, Chap. 4—Rheology of Food Gum and Starch Dispersions, Chap. 5—Rheological Behavior of Processed Fluid and Semisolid Foods, Chap. 6—Rheological Behavior of Food Gels, Chap. 7—Role of Rheological Behavior in Sensory Assessment of Foods and Swallowing, and Chap. 8—Application of Rheology to Fluid Food Handling and Processing.

Several changes have been incorporated in this third edition and these revisions should help readers better appreciate the important role that rheological properties play in food science, as well as to utilize them to characterize foods. The new topics covered in the second edition have been listed in the preface of that edition. In the third edition, the contents of the second edition have been left intact. Additional topics covered in the 3rd edition include: In Chap. 1, a section on microstructure in general and on nanometer-scale milk protein fibrils together with their quantitative characterization in terms of persistence and contour lengths. In Chap. 2, a section on the phase diagram of a colloidal glass of hard spheres and its relationship to milk protein dispersions were added. Additions in Chap. 3 include, sections on Microrheology, including detailed descriptions of single particle and multiparticle micro rheological measurements, and Diffusive Wave Spectroscopy. In Chap. 4, a short section on starch spherulites was added. Because the Bostwick consistometer is used extensively for quality control of pureed foods, e.g., baby foods, in Chap. 5, a section on correlation of Bostwick consistometer data in terms of property-based dimensionless groups was added. Chapter 6 contains a new section on the effect of calcium on the morphology and functionality of whey protein nanometer-scale fibrils. In Chap. 7, it was pointed out that the roles of tribology and rheology are considered for the sensory perception of foods.

My association with the Riddet Institute, Massey University, NZ, helped shape the changes I made in this third edition. I thank Prof. Harjinder Singh for inviting me to work at the Riddet Institute, 2007–2011. My special thanks to Dr. Simon Loveday for valuable collaboration. Again, the mistakes in this book are mine and I hope that many of them can be corrected in a future edition.

Department of Food Science
Cornell University
Geneva, NY, USA
May 18, 2013

M. Anandha Rao, PhD

Contents

Chapter 1
Introduction: Food Rheology and Structure

By definition, rheology is the study of deformation and flow of matter. The science of rheology grew considerably due to research work done on synthetic polymers and their solutions in different solvents that, in turn, was necessary due to the many uses of the polymers ("plastics") in day-to-day and industrial applications. Nevertheless, because of the biological nature of foods, food rheology offers many unique opportunities of study and there exists a large body of food rheology literature. Many foods are composed mainly of biopolymers and aqueous solutions containing dissolved sugars and ions. The former are large molecules, often called macromolecules, such as proteins, polysaccharides, and lipids from a wide range of plant and animal sources. In addition, water is an important component in many foods and plays a major role in the creation of edible structures and their storage stability (Rao 2003). Processed foods may be viewed as soft, edible, structures that are created as a result of the responses of proteins, polysaccharides, and lipids in aqueous media to different processing methods, such as thermal processing, homogenization, and other physical treatments. Most, if not all, of those responses are physical in nature. The measured rheological responses are those at the macroscopic level. However, they are directly affected by the changes and properties at the microscopic level (Rao 2006). Thus, it would be helpful to understand the role of structure of foods on their rheological behavior.

Rheological properties are based on flow and deformation responses of foods when subjected to normal and tangential stresses. A thorough study of the rheology of fluid foods requires knowledge of tensors and the basic principles of fluid flow, such as the equation of continuity (conservation of mass) and the equation of motion (conservation of momentum). The necessary basic equations of fluid flow can be derived (Bird et al. 1960) by conducting a balance of either mass or momentum on a stationary volume element of finite dimensions (e.g., Δx, Δy, Δz in rectangular coordinates) through which the fluid is flowing.

$$\text{Input} + \text{Generation} = \text{Output} + \text{Accmulatio} \tag{1.1}$$

Because mass is not generated, in the derivation of the equation of continuity, the second term on the left-hand side of Eq. 1.1 can be omitted. The applicable partial

M. A. Rao, *Rheology of Fluid, Semisolid, and Solid Foods*, Food Engineering Series, DOI 10.1007/978-1-4614-9230-6_1,

differential equations are obtained in the limit as the dimensions of the control volume tend to zero. The primary objective here is to point out to readers the origins of useful relationships and the assumptions made in deriving them, and present the equation of continuity, and the equation of motion in Cartesian, cylindrical, and spherical coordinates in Appendixes 1–A; for the actual derivation of the transport equations the aforementioned references should be consulted. The use of some of these equations in deriving equations for specific measurement geometries will be illustrated in Chap. 3 and in processing applications, in Chap. 8. In vector and tensor form, the transport equations are quite concise (Appendix 1–A).

Foods can be classified in different manners, for example, as solids, gels, homogeneous liquids, suspensions of solids in liquids, and emulsions. Fluid foods are those that do not retain their shape but take the shape of their container. Fluid foods that contain significant amounts of dissolved high molecular weight compounds (polymers) and/or suspended solids exhibit non-Newtonian behavior. Many non-Newtonian foods also exhibit both viscous and elastic properties, that is, they exhibit viscoelastic behavior.

Fluid and semisolid foods exhibit a wide variety of rheological behavior ranging from Newtonian to time dependent and viscoelastic. Fluid foods containing relatively large amounts of dissolved low molecular weight compounds (e.g., sugars) and no significant amount of a polymer or insoluble solids can be expected to exhibit Newtonian behavior. A small amount (~1 %) of a dissolved polymer can substantially increase the viscosity and also alter the flow characteristics from Newtonian of water to non-Newtonian of the aqueous dispersion. It is interesting to note that whereas the rheological properties are altered substantially, magnitudes of the thermal properties (e.g., density and thermal conductivity) of the dispersion remain relatively close to those of water. Further, with one or two exceptions, such as that of food polymer dispersions, it is difficult to predict precisely the magnitudes of the viscosity of fluid foods mainly because foods are complex mixtures of biochemical compounds that exhibit a wide variation in composition and structure. Therefore, studies on rheological properties of foods are useful and important for applications that include handling and processing, quality control, and sensory assessment of foods. The latter is an important field of study to which food scientists have made significant contributions.

The classification of rheological behavior and the measurement of rheological properties of fluid foods were reviewed among others by Sherman (1970), Ross-Murphy (1984), Rao (1977a, b, 2005), Rao and Steffe (1992), Steffe (1996) and others. In addition to these references, one must consult books on synthetic polymer rheology for valuable information on viscoelastic behavior, measurement techniques, and on the role of fundamental properties such as molecular weight on viscoelastic behavior. In particular, the texts by Bird et al. (1977a, b), Ferry (1980), and Tschoegl (1989) have much useful information on viscoelasticity. Barnes et al. (1989) discussed rheology in a lucid manner.

Techniques for measuring rheological properties, especially flow properties, were well covered by Van Wazer et al. (1963) and those of viscoelastic properties by Walters (1975), Whorlow (1980), Dealy (1982), and Macosko (1994). An examination of the early experimental efforts covered in Van Wazer et al. (1963)

would be educational for the innovations in experimental techniques and developments in interpretation of rheological behavior. In this respect, a review (Markovitz 1985) of the studies conducted soon after the Society of Rheology was formed also provides a fascinating picture of the development of the science of rheology. While one should appreciate the ease with which many rheological measurements can be performed today primarily due to the availability of powerful desktop computers, it is essential that food rheologists understand the underlying principles in rheological measurements and interpretation of results. In addition, low-friction compressed-air bearings, optical deformation measurement systems, and other developments also have contributed to measurements of foods and other low viscosity materials.

Stress and Strain Tensors

While scalar quantities have magnitudes, vectors are defined in terms of both direction and magnitude and have three components. Tensors are second-order vectors and have nine components. In a Cartesian system of coordinates, a stress tensor ($\tilde{\tilde{\tau}}$) of force imposed on unit surface area of a test material can be resolved in terms of nine components (σ_{ij}), three normal and six tangential stresses. In simple shearing (viscometric) flow that is encountered in the flow geometries: capillary, Couette, cone-and-plate, and parallel plate, $\sigma_{12}=\sigma_{21}$ while the other four tangential components, σ_{13}, σ_{23}, σ_{31}, and σ_{32} are equal to zero, so that the stress tensor may be written as:

$$\tilde{\tilde{\tau}} = \begin{Vmatrix} \sigma_{11} & \sigma_{12} & 0 \\ \sigma_{21} & \sigma_{21} & 0 \\ 0 & 0 & \sigma_{33} \end{Vmatrix} \tag{1.2}$$

The stresses, σ_{11}, σ_{22}, and σ_{33} are normal stresses that are equal to zero for Newtonian fluids and may be of appreciable magnitudes for some foods, such as doughs. The components of the deformation tensor, $\tilde{\tilde{e}}$, in viscometric flows are:

$$\tilde{\tilde{e}} = \begin{Vmatrix} 0 & \dot{\gamma} & 0 \\ \dot{\gamma} & 0 & 0 \\ 0 & 0 & 0 \end{Vmatrix} \tag{1.3}$$

where, $\dot{\gamma}$ is the shear rate.

Rheological properties of food materials over a wide range of phase behavior can be expressed in terms of viscous (viscometric), elastic and viscoelastic functions which relate some components of the stress tensor to specific components of the strain or shear rate response. In terms of fluid and solid phases, viscometric functions are generally used to relate stress with shear rate in liquid systems, while elastic functions are related to the appropriate stress function to strain in solids. Viscoelastic properties cover the combination where a material exhibits both viscous and elastic properties.

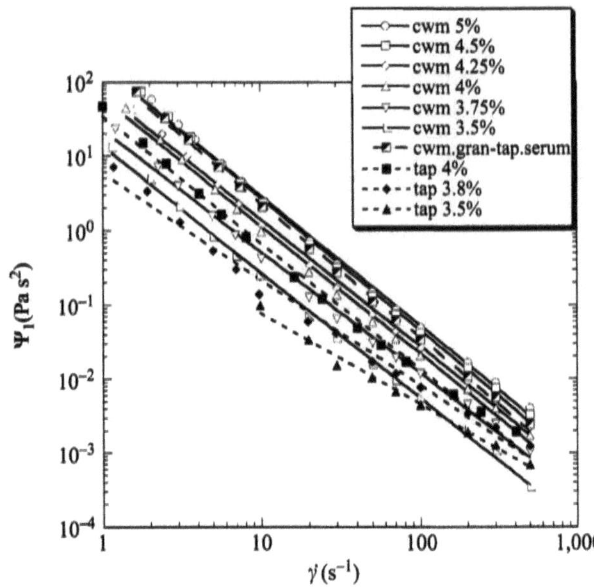

Fig. 1.1 First normal stress coefficient data of starch dispersions with different concentrations as a function of shear rate. *cwm* cross-linked waxy maize, *tap* tapioca, *gran* granule. (Genovese and Rao 2003)

Viscometric Properties

Three material functions: the viscosity function η and the first and second normal stress coefficients Ψ_1 and Ψ_2 can be evaluated from the stress tensor using Eqs. (1.3–1.5) which relate specific stress components σ_{ij} to the shear rate $\dot{\gamma}$:

$$\sigma_{21} = \eta(\dot{\gamma}) \tag{1.4}$$

$$\sigma_{11} - \sigma_{22} = \psi_1(\dot{\gamma})^2 \tag{1.5}$$

$$\sigma_{22} - \sigma_{33} = \psi_2(\dot{\gamma})^2 \tag{1.6}$$

Steady state experiments employing rotational, capillary and tube flow viscometers are commonly used to generate data for evaluation of these material properties. Reported values of the viscometric coefficients for many materials, especially synthetic polymers, indicate that η and Ψ_1 are large and positive, whereas Ψ_2 is small and negative (Macosko 1994). While the first normal stress difference is known to be responsible for the climbing fluid film phenomenon often referred to as the "Weissenberg effect," extrudate swell, and normal force pump, the second normal stress difference determines where the free surface of a fluid flowing down a trough would be convex (Barnes et al. 1989; Macosko 1994). Magnitudes of $\psi_1(\dot{\gamma})$ of starch dispersions shown in Fig. 1.1 illustrate typical values and role of shear rate (Genovese and Rao 2003). In Chap. 3, the relationship between some of the rheological parameters from steady shear and oscillatory shear, as well as the first normal stress coefficient in dynamic shear, Ψ_1 (ω), will be discussed.

Most rheological studies have been concentrated on the viscosity function and dynamic viscoelastic properties, and much less on normal stress differences (Rao 1992). The former has been shown to depend on the molecular weight and also plays an important role in handling of foods, for example, flow and heat transfer applications. Beginning about the mid 1960s, there has been a gradual paradigm shift from normal stresses to dynamic rheological properties as measures of viscoelastic properties. One reason for the shift appears to be the difficulties in obtaining reliable magnitudes of normal stresses with cumbersome rheometers and, compared to dynamic rheological properties, their limited practical application. Another reason is that with the availability of automated rheometers, measurement of dynamic rheological parameters has become very popular and manageable. Therefore, one does not encounter much data on normal stress functions of foods.

The relationships between stress and strain, and the influence of time on them are generally described by constitutive equations or rheological equations of state (Ferry 1980). When the strains are relatively small, that is, in the linear range, the constitutive equations are relatively simple. Equations 1.5 and 1.6 are constitutive equations of viscoelastic behavior under continuous shear (nonlinear range). In the limit of low strains, relationships exist between the above material functions and those obtained from dynamic and creep experiments. Constitutive equations will be discussed further in Chap. 4.

Shear Stress–Shear Rate Relationships

To understand rheological properties of foods in a systematic manner, it is convenient to first study the principles of rheological behavior of fluid and solid foods. Viscosity data must be obtained with well-designed viscometers and expressed in fundamental units. Viscosity data in units that are specific to a particular brand viscometer, such as the time to flow out of a specific cup or the manufacturer's arbitrary viscosity, will not be applicable universally, and the viscosity values are of limited value. Viscosity is best determined using geometries in which the shear rate can be calculated from the dimensions of the measuring system and experimental data, such as the velocity or volumetric flow rate of a fluid or the rotational speed of a rotating cylinder, cone or plate. In general, the determination of stress corresponding to the shear rates is not very difficult. However, calculation of applicable shear rate requires effort. Unfortunately, there have been too many instances in which such effort has not been expended. In addition to the dynamic viscosity determined in shear fields, one can also determine viscosity in extensional shear fields (Padmanabhan 1995) as discussed later.

Units in Rheological Measurement

Shear Rate, denoted by the symbol, $\dot{\gamma}$, is the velocity gradient established in a fluid as a result of an applied shear stress. It is expressed in units of reciprocal seconds, s^{-1}. Shear Stress is the stress component applied tangentially. It is equal to

Table 1.1 Conversion factors for viscosity, kinematic viscosity, and activation energy of flow

The SI unit for viscosity is Pa s
1 Pa s = 1000 cP = 1 N s m^{-2} = 1 kg m^{-1} s^{-1} = 0.67197 Ibm ft^{-1} s^{-1}
1 poise (P) = 1 dyne s cm^{-2} = 0.1 Pa s
1 cP = 0.001 Pa s = 1 mPa s = 0.01 P
1 Ibm ft^{-1} s^{-1} = 1.4882 kg m^{-1} s^{-1} = 1.4882 Pa s = 1488.2 cP
Kinematic viscosity (cSt) = viscosity (cP)/density (g cm^{-3})
1 St = 100 cSt = 0.0001 m^2 s^{-1}
1 calorie = 4.1868 JOUle (J)
The gas constant, R = 8.314 (J mol^{-1} K^{-1})
Boltzmann constant, $k = 1.38 \times 10^{-23}$ N m K^{-1}

the force vector (a vector has both magnitude and direction) divided by the area of application and is expressed in units of force per unit area (Pa). "The nomenclature committee of the Society of Rheology σ be used to denote shear stress." However, the symbol τ that was used to denote shear stress for a long time can be still encountered in rheology literature.

Viscosity, is the internal friction of a fluid or its tendency to resist flow. It is denoted by the symbol η for Newtonian fluids, whose viscosity does not depend on the shear rate, and for non-Newtonian fluids to indicate shear rate dependence by η_a. Depending on the flow system and choice of shear rate and shear stress, there are several equations to calculate. Here, it is defined by the equation:

$$\eta_a = \frac{\text{shear stress}}{\text{shear rate}} = \frac{\sigma}{\dot{\gamma}} \tag{1.7}$$

The preferred units of viscosity are Pa s or mPa s. Some of the older units and their relation to the SI unit are given in Table 1.1. It is clear that the shear rate employed in the calculation must be specified when the magnitude of apparent viscosity is discussed. Apparent viscosity has many useful applications in characterizing a fluid food; in particular, in the characterization of shear-thinning fluids, the apparent viscosity at low shear rates, called the zero shear rate viscosity (η_0), is a useful parameter. Because much work was done with viscosity expressed in poise in sensory assessment of viscosity, the same tradition will be continued in Chap. 7 on sensory assessment of viscosity.

Types of Fluid Flow Behavior

The major types of fluid flow behavior can be described by means of basic shear diagram of shear rate versus shear stress, such as Figs. 1.2 and 1.3. In Fig. 1.2, the shear stresses are plotted against the shear rates (independent variable) which is the conventional method. However, some authors plot shear rates against the shear stresses (independent variable) as shown in Fig. 1.3. With the introduction of controlled-stress rheometers, the use of shear stress as the independent variable is often desirable.

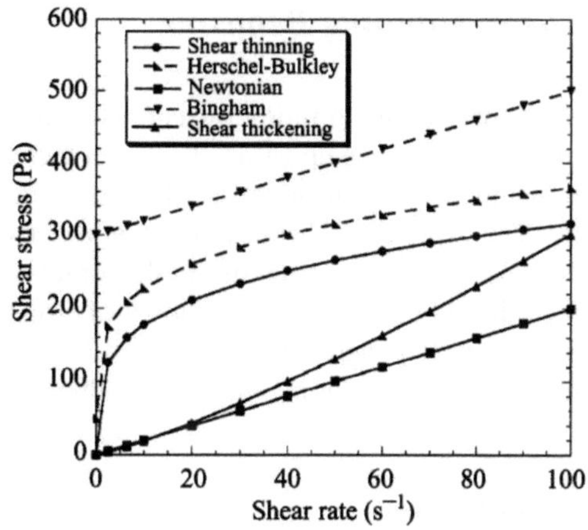

Fig. 1.2 Basic shear diagram of shear rate versus shear stress for classification of time-independent flow behavior of fluid foods: Newtonian, shear thinning, and shear thickening. Also, some foods have yield stress that must be exceeded for flow to occur: Bingham and Herschel–Bulkley (H–B)

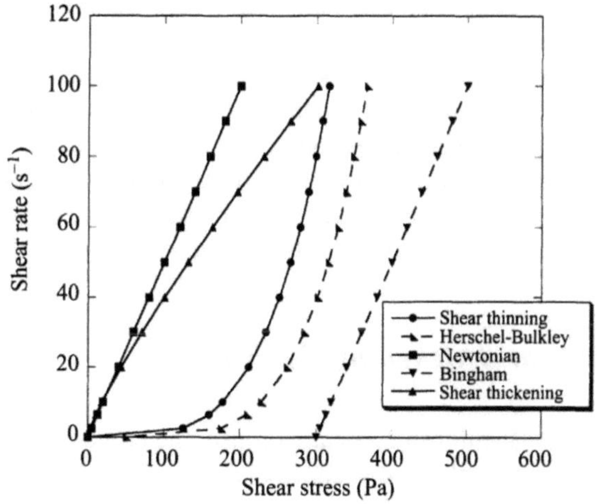

Fig. 1.3 Basic shear diagram similar to Fig. 1.2, except that shear stress is plotted as the independent variable: Newtonian, shear thickening, Bingham, and Herschel–Bulkley

Newtonian Behavior

With Newtonian fluids, the shear rate is directly proportional to the shear stress and the plot begins at the origin. Typical Newtonian foods are those containing compounds of low molecular weight (e.g., sugars) and that do not contain large concentrations of either dissolved polymers (e.g., pectins, proteins, starches) or insoluble solids. Examples of Newtonian foods include water, sugar syrups, most honeys, most carbonated beverages, edible oils, filtered juices and milk.

All other types of fluid foods are non-Newtonian which means that either the shear stress–shear rate plot is not linear and/or the plot does not begin at the origin, or the material exhibits time-dependent rheological behavior as a result of structural changes. Flow behavior may depend only on shear rate and not on the duration of shear (time-independent) or may depend also on the duration of shear (time-dependent). Several types of time-independent flow behavior of foods have been encountered.

Shear-Thinning Behavior

With shear-thinning fluids, the curve begins at the origin of the shear stress–shear rate plot but is concave upwards, that is, an increasing shear rate gives a less than proportional increase in shear stress. Shear-thinning fluids are popularly called pseudoplastic. The expression shear thinning is preferred compared to pseudoplastic because it is an accurate description of the shear rate–shear stress curve. Shear thinning may be thought of being due to breakdown of structural units in a food due to the hydrodynamic forces generated during shear. Most non-Newtonian foods exhibit shear thinning behavior, including many salad dressings and some concentrated fruit juices.

Yield Stress

The flow of some materials may not commence until a threshold value of stress, the yield stress (σ_0) (see Figs. 1.2 and 1.3), is exceeded. Although the concept of yield stress was questioned recently (Barnes and Walters 1985), within the time scales of most food processes the concept of yield stress is useful in food process design, sensory assessment, and modeling. Shear thinning with yield stress behavior is exhibited by foods, such as tomato concentrates, tomato ketchup, mustard, and mayonnaise. In the event, the shear rate–shear stress data follow a straight line with a yield stress, the food is said to follow the Bingham plastic model.

Shear-Thickening Behavior

In shear-thickening behavior also, the curve begins at the origin of the shear stress–shear rate plot and is concave downwards, that is, an increasing shear stress gives a less than proportional increase in shear rate. This type of flow has been encountered in partially gelatinized starch dispersions. The expression dilatant is popularly and incorrectly used to describe shear thickening. However, because dilatancy implies an increase in the volume of the sample during the test, it is incorrect to use it to describe shear-thickening rheological behavior. Strictly speaking, shear thickening should be due to increase in the size of the structural units as a result of shear.

Time-Dependent Behavior

Foods that exhibit time-dependent shear-thinning behavior are said to exhibit thixotropic flow behavior. Most of the foods that exhibit thixotropic behavior are heterogeneous systems containing a dispersed phase that is often very fine. At rest, the particles or molecules in the food are linked together by weak forces. When the hydrodynamic forces during shear are sufficiently high, the interparticle linkages are broken resulting in reduction in the size of the structural units that in turn offer lower resistance to flow during shear (Mewis 1979). This type of behavior is common to foods such as salad dressings and soft cheeses where the structural adjustments take place in the food due to shear until an equilibrium is reached. Time-dependent shear-thickening behavior is called antithixotropic behavior. Formerly, it was called rheopectic behavior, but the Society of Rheology recommended the expression antithixotropic. This type of behavior, although rare, is being detected in many foods due to the development of sensitive automated rheometers. When shear rate versus shear stress data are obtained first in ascending order of shear rate and immediately afterwards in descending order, the two curves will not coincide and values of the latter will be lower than the former for thixotropic foods (Fig. 1.4). Repetition of the experiments will result in an equilibrium hysteresis loop.

In antithixotropic behavior, the shear stress values in descending order of shear rates are higher than those in ascending order (Fig. 1.5). The characterization of these types of flow is an important research problem at this time. The primary difficulty in obtaining reliable thixotropic or antithixotropic data is that often loading of the test sample in to a measuring geometry induces structural changes that cannot be either controlled or expressed quantitatively. Therefore, considerable caution should be exercised in using the area of hysteresis loop as a measure of thixotropic behavior and a characteristic of the food sample because the magnitude of the area will depend on the shear to which the sample is subjected to prior to and during loading of the sample into a viscometer.

Apparent Viscosity

Additional information can be obtained from a basic shear (shear stress–shear rate) diagram (see Fig. 1.2) and the equations that describe the data in it. The yield stress σ_0 (if present) can be read off the diagram. One can calculate the apparent viscosity at any given shear rate. For example, the apparent viscosity at a shear rate of 50 s^{-1} is:

$$\eta_{a,50} = \frac{\sigma_{50}}{\dot{\gamma}_{50}} \tag{1.8}$$

where, σ_{50} is the shear stress corresponding to a shear rate of 50 s^{-1}. The zero shear viscosity η_0, is an important parameter in the study of shear-thinning fluids. It is the limiting viscosity at very low shear rates. The important role of η_0 in the study of biopolymer dispersions will be discussed in Chaps. 2 and 4.

Intrinsic Viscosity

Many foods contain high molecular weight polymers, such as proteins, pectins, and others. Often, they contribute significantly to the structure and viscosity of foods. In dilute solutions, the polymer chains are separate and the intrinsic viscosity, denoted

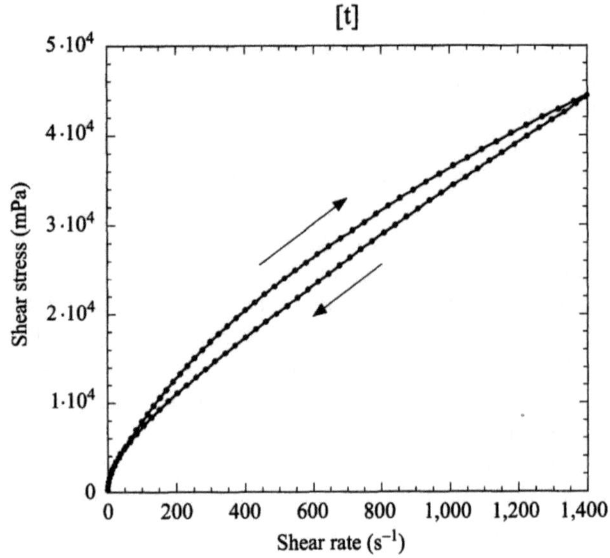

Fig. 1.4 Time-dependent shear-thinning (thixotropic) behavior of a Tapioca starch dispersion heated at 67 °C for 5 min, data of Tattiyakul (1997)

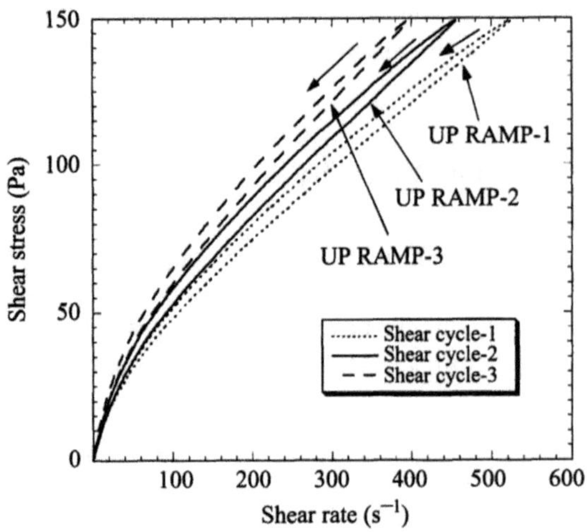

Fig. 1.5 Time-dependent shear-thickening (antithixotropic) behavior of a cross-linked waxy maize starch dispersion heated at 120 °C for 30 min, data of Chamberlain (1996). (Note that the decreasing shear curves of the samples had higher stress values than the increasing shear curves)

as $[\eta]$, of a polymer in solution depends only on the dimensions of the polymer chain. Because $[\eta]$ indicates the hydrodynamic volume of the polymer molecule and is related to the molecular weight and to the radius of gyration, it reflects important molecular characteristics of a biopolymer. The concentrations of polymers used should be such that the relative viscosities (η/η_s) of the dispersions are from about 1.2 to 2.0 to assure good accuracy and linearity of extrapolation to zero concentration (Morris and Ross-Murphy 1981; da Silva and Rao 1992). Intrinsic viscosity can be determined from dilute solution viscosity data as the zero concentration limit of specific viscosity (η_{sp}) divided by concentration (c):

$$[\eta] = \lim_{c \to 0} (\eta_{sp}/c) \tag{1.9}$$

where, $\eta_{sp} = [(\eta - \eta)/\eta_s]$ and η and η_s are the viscosities of the solution and the solvent, respectively. When a dilute solution exhibits shear-thinning behavior, its zero shear viscosity, η_0, at very low shear rates, sometimes referred to as "vanishing shear rates," may be used in place of the Newtonian viscosity η. We note that there are several ways of determining the intrinsic viscosity $[\eta]$ from experimental dilute solution viscosity data (Tanglertpaibul and Rao 1987). The two equations commonly employed for determining $[\eta]$ of food gums are those of Huggins (Eq. 1.10) and Kraemer (Eq. 1.11):

$$\frac{\eta_{sp}}{c} = [\eta] + k_1 [\eta]^2 c \tag{1.10}$$

$$\mathrm{In}(\eta_r)/c = [\eta] + k_2[\eta]^2 c \tag{1.11}$$

In Eq. (1.11), η_r is the relative viscosity (η/η_s). Equations 1.10 and 1.11 imply that plots of η_{sp}/c versus c and $\ln(\eta_r)/c$ versus c would result in straight lines, respectively. The extrapolations are usually done for relative viscosity values between 1.2 and 2.0 when the corresponding specific viscosities are between about 0.2 and 1.0. The Huggins constant k_1 is considered to be an index of the polymer–polymer interaction with a large number of the reported values being between 0.3 in good solvents and 1.0 in theta solvents; the higher values are attributed to the existence of association between the macromolecules. The Huggins and Kraemer constants (k_1 and k_2) are theoretically related by the equation:

$$k_1 = k_2 + 0.5 \tag{1.12}$$

The Huggins equation has been used much more extensively and in Fig. 1.6 intrinsic viscosity determination of mesquite seed gum, a galactose–mannose polymer, based on the equation is illustrated (Yoo et al. 1994).

For polyelectrolytes (charged polymers), a plot of η_{sp}/c versus c may be a curve. An alternate expression of Fuoss and Strauss (1948) can be used (Chamberlain and Rao 2000):

$$\frac{\eta_{sp}}{c} = \frac{[\eta]}{(1 + Bc^{1/2})} \tag{1.13}$$

When (c/η_{sp}) is plotted against $c^{1/2}$, a straight line is obtained with an intercept of $1/[\eta]$ and a slope of $B/[\eta]$.

The Mark–Houwink empirical equation relates intrinsic viscosity and the average molecular weight *(M):*

$$[\eta] = K(M)^a \tag{1.14}$$

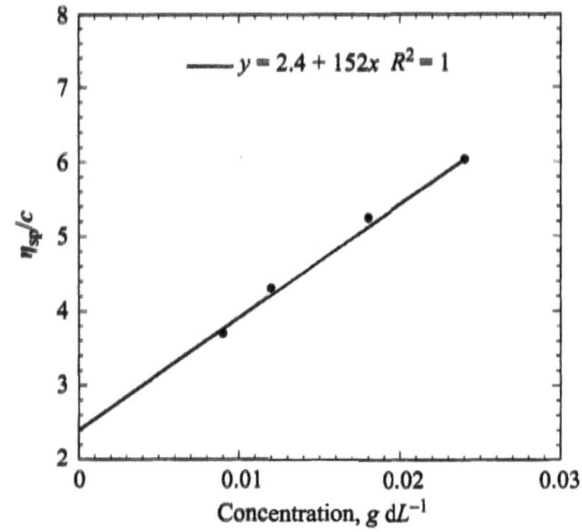

Fig. 1.6 Intrinsic viscosity determination of mesquite seed gum, a Galactose–Mannose Polymer, based on the Huggins equation

where K and a are constants at a specific temperature for a given polymer-solvent system. The magnitude of the constant a is generally in the range 0.5–0.8 when conformation of the polymer is random coil (Launay et al. 1986), for polymers that assume compact conformations, the magnitudes of the exponent a are lower than 0.5. Magnitudes of higher than 0.8 are encountered when hydrodynamic interactions are absent (free draining random coil), that is, when the solvent can drain freely between the macromolecules, as is the case when the conformations are elongated and rigid (rod like polymers). A large number of models have been used to deduce intrinsic viscosity–molecular weight relationships (Launay et al. 1986), including the equivalent sphere model (Flory–Fox equation), the random flight model for flexible chain molecules, the worm-like chain model, and rigid elongated molecules. Rigid elongated conformations have been associated with large values of the Mark–Houwink exponent with values of 1.7–1.8 being reported for some polyelectrolytes at low ionic strength and high degree of disassociation. The magnitudes of intrinsic viscosity and of molecular weight of several biopolymers are given in Table 1.2 (da Silva and Rao 1992). It is seen that significant differences in the magnitudes are encountered depending on the structural characteristics, such as molecular mass (length of the chain) and different degrees of substitution. For polymers with charges, $[\eta]$ and rheological behavior are also affected by the pH and the ionic strength of the solution; therefore, their values should be controlled in studies of $[\eta]$ and their rheological behavior.

Intrinsic viscosity is very useful as a measure of the hydrodynamic volume of food polymers and has been used for studying the role of polymer concentration on zero shear viscosity of several food polymers (Chap. 4).

Table 1.2 Examples of magnitudes of intrinsic viscosity [η] and average molecular weight (M) of some biopolymers

Biopolymer	Solvent	M	[η] (dl/g)	Reference
Apple pectin				
DE[a]—72.9%	NaCl 0.155 M	155,000	6.13	Michel (1982)
DE—60.5%		113,000		
Xanthan gum	Water		168	Launay et al. (1984)
	NaCl 2%		36.7	
Sodium alginate	Ionic force			
	0.01	200,000	5.99	
		500,000	17.2	Smidsrød (1970)
	1.0	200,000	3.72	
		500,000	8.26	
Locust bean gum	Water	1,600,000	11.2	Doublier (1975)
		1,100,000	7.70	
Gyar gum	Water	440,000	4.5	Robinson et al. (1982)
		1,650,000	12.5	

[a] DE stands for degree of esterification

Stress–Strain Behavior of Solid Foods

Equations 1.2–1.4 represent material functions under large deformations (e.g., continuous shear of a fluid). One may recall a simple experiment in an introductory physics course where a stress (σ) is applied to a rod of length L in a tension mode and that results in a small deformation ΔL. The linear relationship between stress (σ) and strain (γ) (also relative deformation, $\gamma = \Delta L/L$) is used to define the Young's modulus of elasticity E (Pa):

$$\sigma = E\gamma \tag{1.15}$$

If the solid does not shows time-dependent behavior, that is, it deforms instantaneously, one has an ideal elastic body or a Hookean solid. The symbol E for the modulus is used when the applied strain is extension or compression, while the symbol G is used when the modulus is determined using shear strain. The conduct of experiment such that a linear relationship is obtained between stress and strain should be noted. In addition, for an ideal Hookean solid, the deformation is instantaneous. In contrast, all real materials are either viscoplastic or viscoelastic in nature and, in particular, the latter exhibit time-dependent deformations. The rheological behavior of many foods may be described as viscoplastic and the applicable equations are discussed in Chap. 2.

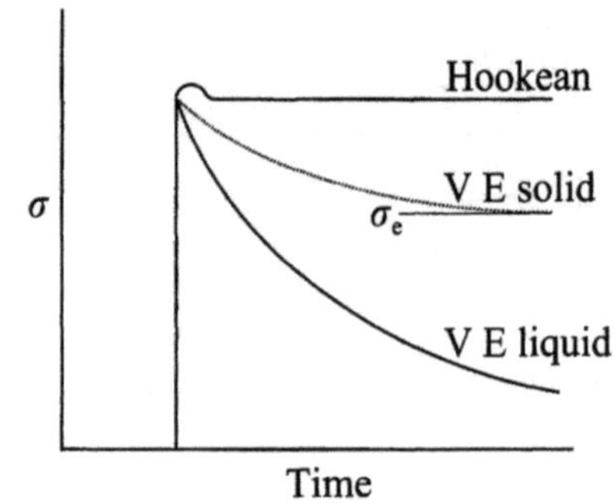

Fig. 1.7 Stress relaxation response of a fluid, ideal elastic solid, and a viscoelastic material. The experimental response is not shown separately because it is the same as that of the ideal elastic solid are identical

Linear Viscoelasticity

One convenient manner of studying viscoelasticity is by stress relaxation where the time-dependent shear stress is studied for step increase in strain. In Fig. 1.7, the stress relaxation of a Hookean solid, and a viscoelastic solid and liquid are shown when subjected to a strain instantaneously and held constant. The relaxation modulus can be calculated as:

$$G(t,\gamma) = \frac{\sigma(t,\gamma)}{\gamma} \tag{1.16}$$

Because a Hookean solid deforms instantaneously, the imposed strain has a time-dependent profile similar to that of the stress. For relatively small strains, shear normal stresses and the type of deformation (e.g., linear or shear) will not be important. Further, the extensional and the shear relaxation moduli are related as:

$$E(t) = 3G(t) \tag{1.17}$$

Boltzmann suggested that when the changes in stress and strain are small, one can express the change in stress as:

$$\mathrm{d}\sigma = \gamma \mathrm{d}G \tag{1.18}$$

Defining the memory function as: $M(t) = -dG(t)/\mathrm{d}t$, one can express a large linear deformation as the sum of all the small linear deformations that in turn can be written as the integral over all past times:

$$\sigma = -\int_{-\infty}^{t} M(t-t')\gamma(t')\mathrm{d}t' \tag{1.19}$$

where, t' is the past time variable in the range $-\infty$ to the present time t. The elapsed time is denoted as $s = (t - t')$, and Eq. 1.17 can be written as:

$$\sigma = -\int_{-\infty}^{t} M(s)\gamma(t-s)\mathrm{d}s \tag{1.20}$$

The above equation is a one-dimensional model of linear viscoelastic behavior. It can be also written in terms of the relaxation modulus after noting that:

$$\mathrm{d}\sigma = G\mathrm{d}\sigma = G\frac{\mathrm{d}\gamma}{\mathrm{d}t}\mathrm{d}t = G\dot{\gamma}\,\mathrm{d}t \tag{1.21}$$

$$\sigma = -\int_{-\infty}^{t} G(t-t')\dot{\gamma}\,\mathrm{d}t' \tag{1.22}$$

Therefore, the stress is an integral over past time of the product of the relaxation modulus and the rate of strain. The relaxation modulus can be expressed in terms of a series of relaxation times, λ_n, and the constants, G_n:

$$G(t) = \sum_{n=1}^{N} G_n \exp(-t/\lambda_n) \tag{1.23}$$

In the unlikely event of $G(t)$ being described by a single exponential term:

$$G(t) = G_0 \exp(-t/\lambda) \tag{1.24}$$

From the above equation, one obtains the simple Maxwell model:

$$\sigma = -\int_{-\infty}^{t} G_0 e^{-(t-t')/\lambda}\dot{\gamma}\mathrm{d}t' \tag{1.25}$$

One can express linear viscoelasticity using the relaxation spectrum $H(\lambda)$, that is, using the relaxation time λ. The relationship between the relaxation modulus and the spectra is:

$$G(s) = \int_{0}^{\infty} \frac{H(\lambda)}{\lambda} e^{-s/\lambda}\,\mathrm{d}\lambda \tag{1.26}$$

Linear Viscoelasticity in Differential Form

Three equations are basic to viscoelasticity: (1) Newton's law of viscosity, $\sigma = \eta\dot{\gamma}$, (2) Hooke's law of elasticity, Eq. 1.15, and (3) Newton's second law of motion, $F=ma$, where m is the mass and a is the acceleration. One can combine the three equations to obtain a basic differential equation. In linear viscoelasticity, the conditions are such that the contributions of the viscous, elastic, and the inertial elements are additive. The Maxwell model is:

$$\sigma + \lambda\frac{\mathrm{d}\sigma}{\mathrm{d}t} = \eta\dot{\gamma} = \lambda G_0\dot{\gamma} \tag{1.27}$$

A convenient manner of interpreting viscoelastic behavior of foods is in terms of a spring that has a modulus E and a dashpot that represents a Newtonian fluid with viscosity η that can be arranged either in series (Maxwell model) (Fig. 1.8, left) or in parallel (Kelvin–Voigt model). In the Maxwell model, for slow motions the dashpot, that is, Newtonian behavior dominates. For rapidly changing stresses, that is, at short times, the model approaches elastic behavior. The use of models such as the Maxwell and the Kelvin–Voigt (Fig. 1.8, right) and their combinations is valid only when the experimental data are obtained within the linear viscoelastic range. Many studies on foods have reported measurements in the linear viscoelastic range so that the Maxwell and the Kelvin-Voigt models and their combinations were used in the interpretation of results of such studies (Chaps. 3 and 5).

The relaxation time for a Maxwell element (Fig. 1.8 left) is defined as:

$$\tau_1 = \frac{\eta}{E} \tag{1.28}$$

From data in a stress relaxation experiment (Chap. 3), where the strain is constant and stress is measured as a function of time, $\sigma(t)$, the relaxation time may be estimated from the time necessary for $[\sigma(t)/\sigma(0)]$ to become $(1/e)=0.368$. Typically, several Maxwell elements are used to fit experimental data, $\sigma(t)$. For the Kelvin–Voigt element (Fig. 1.8 right) under stress, the equation is:

$$\sigma = \eta\frac{\mathrm{d}\gamma}{\mathrm{d}t} + E\gamma \tag{1.29}$$

For constant stress, the above equation can be integrated to describe the deformation:

$$\lambda = \frac{\sigma}{E}[1 - \exp(-E/\eta)t] \tag{1.30}$$

In terms of the retardation time, τ_2, the above equation is:

$$\gamma = \frac{\sigma}{E}[1 - \exp(-t/\tau_2)] \tag{1.31}$$

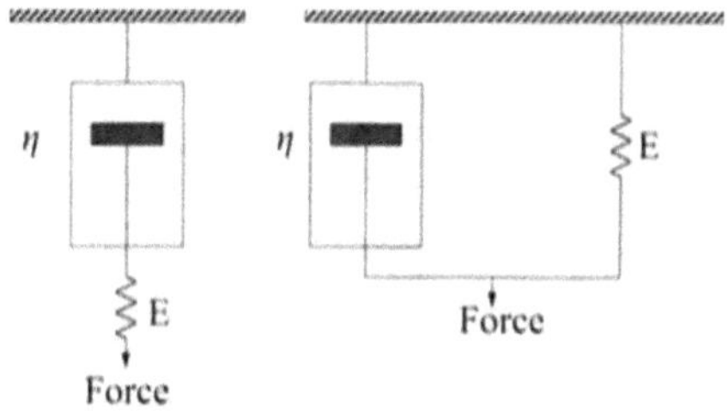

Fig. 1.8 Maxwell Model (*left*) and Kelvin–Voigt Model (*right*) illustrate mechanical analogs of viscoelastic behavior

The Kelvin–Voigt elements are used to describe data from a creep experiment and the retardation time (τ_2) is the time required for the spring and the dashpot to deform to (1–1/*e*), or 63.21 % of the total creep. In contrast, the relaxation time is that required for the spring and dashpot to stress relax to 1/*e* or 0.368 of $\sigma(0)$ at constant strain. To a first approximation, both τ_1 and τ_2 indicate a measure of the time to complete about half of the physical or chemical phenomenon being investigated (Sperling 1986).

Length Scale of Food Molecules and Foods

When studying the rheological behavior of a food, knowledge of the composition of the food, especially the important structuring components (e.g., dissolved polymers, suspended solids), the structure of the food itself (e.g., homogeneous or phase-separated gel, emulsion), and the processing and storage conditions would be helpful as they all often affect the behavior. The structure of a food is the result of specific and nonspecific interactions at levels ranging from the molecular (<1–100 nm) to the supramolecular ($2\times10^3-10^7$ nm) (Clark and Ross-Murphy 1987; Aguilera and Stanley 1999). Specific interactions at the molecular level are between distinct atoms that result in covalent bonds, hydrogen bonding, enzyme–substrate coupling, as well as hydrophobic interactions. Much of the work in understanding foods, especially biopolymer gels, has been based on studies at the molecular level from which the structural details at the supramolecular level have been inferred.

In dispersions, however, the structure of the food particles plays a major role in defining the rheological behavior. Either large molecules or finely subdivided bulk matter could be considered to be colloidal matter that is, the particles are in the range $10^{-9}-10^{-6}$ m in dimension. Natural colloidal systems include milk, cloudy fruit juice, and egg white. With a colloidal particle, the surface area of the particle is so much greater than its volume that some unusual behavior is observed, for example, the particle does not settle out by gravity (i.e., they neither float nor sink). Therefore, principles of colloidal science are also useful in understanding the rheological behavior. For example, colloids can be classified as lyophilic (solvent loving) or lyophobic (solvent fearing) and when the solvent is water they are called hydrophilic and hydrophobic, respectively.

At the molecular length scales, different spectroscopic methods are suitable for studies. For supramolecular structures, that have length scales >2,000 nm, various techniques, such as microscopic, light scattering, and laser diffraction, have become routine tools to study quantify/understand microstructure and they have been reviewed in various chapters of this book and elsewhere (Aguilera and Stanley 1999). In addition to microscopic and size distribution data, fractal dimension has been used to characterize food particles. Fractal dimension indicates the degree to which an image or object outline deviates from smoothness and regularity. The term fractal was coined by Mandelbrot (1982) who introduced dimensions "between" the conventional Euclidean dimensions of 1, 2, and 3, in order to describe structures that are not Euclidean lines, surfaces, or solids. One characteristic of fractal objects is their "self-similarity," the attribute of having the same appearance at all magnifications. However, real materials or "natural fractals" are self-similar only over a limited range of scales (Marangoni and Rousseau 1996). A fractal dimension from 1 to 2 describes the area filling capacity of a convoluted line and a fractal dimension from 2 to 3 describes the volume filling capacity of a highly rugged surface (Barret and Peleg 1995). Based on this definition, smooth surfaces are associated with a value of surface fractal dimension, $D_f = 2.0$, while extremely convoluted surfaces have values approaching 3.0 (Nagai and Yano 1990).

The fractal dimension can be estimated by several techniques, including structured walk (Richardson's plot), bulk density–particle diameter relation, sorption behavior of gases, pore size distribution, and viscoelastic behavior. The fractal dimension obtained by each method has its own physical meaning (Rahman 1997).

A nanometer-scale or nano-scale particle may be defined as a substance with two or three dimensions between 1 and 100 nm (http://www.nano.gov/). Self-assembly of molecules occurs in many biological materials. One widespread example of biological self-assembly is the folding of proteins into their compact three-dimensional structures. However, a wide variety of pathological conditions arise when proteins fail to fold or to remain folded correctly (Dobson 2003). The deposition of protein aggregates with fibrillar structures has been implicated in amyloidosis, a generic term for a subset of protein misfolding diseases (e.g., Alzheimer's and Creutzfeldt-Jakob) and in amyloid-related disorders (e.g., Huntington's disease).

Because of the high ratio of length (micrometer) to width (nanometer) of fibrils, their solutions can form mesophases such as nematic phases, providing opportunities for the development of new protein-based functional foods. The mechanism of self-assembly varies and appears to be specific for each protein (Dobson 2003). Loveday et al. (2009) reviewed the general characteristics of nanometer-scale fibrils from proteins.

Polymer theory deals with three types of fibrils, called filaments in biology, characterized by two length scales: the persistence length, l_p, and the contour length L_c. The persistence length, defined as: $l_p = \kappa/(k_B T)$, is the typical length at which thermal fluctuations begin to bend the polymer filament in different directions; it is used to characterize the flexibility or rigidity of a filament. Increase in filament rigidity results in decrease in shear modulus of a gel because of reduced entanglement. The contour length, L_c, of a filament is its length at maximum extension. A filament is considered flexible when $l_p << L_c$, and rigid in the opposite situation: $l_p >> L_c$; many biological filaments are in a third intermediate category: semi flexible filaments with l_p and L_c are of comparable magnitude (Loveday et al. 2009).

Phase Transitions in Foods

The expression transition refers to a change in physical state and, in a food, the transition of concern is often either from liquid to solid, solid to liquid, or solid to solid. It is caused primarily by a change in temperature (heating and/or cooling) or pressure (Roos 1998). However, auxiliary conditions, such as pH and presence of divalent ions, as well as enzymatic action aid liquid to solid transitions. For example, gels can be created from Casein either by enzymatic action followed by precipitation with Ca^{2+} or by acid coagulation.

The thermodynamic definition of a phase transitions is based on changes occurring in Gibbs free energy, G, and chemical potential, μ, at the transition temperature (Sperling 1986, 2001). A first-order transition is defined as one in which the first derivatives of G and μ with respect to temperature exhibit discontinuities at the transition temperature. Concomitantly, a step change occurs in enthalpy, entropy, and volume at the transition temperature. Important first-order transitions in foods include, crystallization, melting, protein denaturation, and starch gelatinization. Invariably, in a food many compounds are present, so that a transition may occur over a range of temperatures instead of a fixed temperature (Rao 2003). Starch gelatinization will be covered in Chap. 4.

A second-order transition is defined as one in which the second derivatives of G and μ with respect to temperature exhibit discontinuities at the transition temperature. Although glass transition of amorphous foods has the properties of a second-order transition, there are no well-defined second-order transitions in foods (Roos 1998).

Knowledge of the magnitudes of the temperatures over which the transition takes place is useful in understanding the role of various components. The differential scanning calorimeter (DSC) is used extensively to determine first-order and transition temperatures, more so than other techniques. A DSC measures the rate of heat flow into or out of a cell containing a sample in comparison to a reference cell in which no thermal events occur. It maintains a programmed sample cell temperature by adjusting heat flow rates. Data obtained with a DSC depends to some extent on the heating/cooling rate that should be specified when discussing the data; common heating rates are 5 °C min^{-1} and 10 °C min^{-1}. The heat flow versus temperature diagrams are known as thermograms.

Glass Transition in Foods

At the glass transition temperature, T_g, the amorphous portions of a polymer soften and flow. Due to coordinated molecular motion, in the glass transition region, the polymer softens and the modulus decreases by about three orders of magnitude. Figure 1.9 illustrates a DSC curve for an idealized glass transition in which T_g can be taken as the temperature at which one-half of the change in heat capacity, Δc_p, has occurred (Sperling 1986). Besides DSC, other experimental techniques that have been used to determine T_g include dilatometry, static and dynamic mechanical measurements, as well as dielectric and magnetic measurements (Sperling 1986).

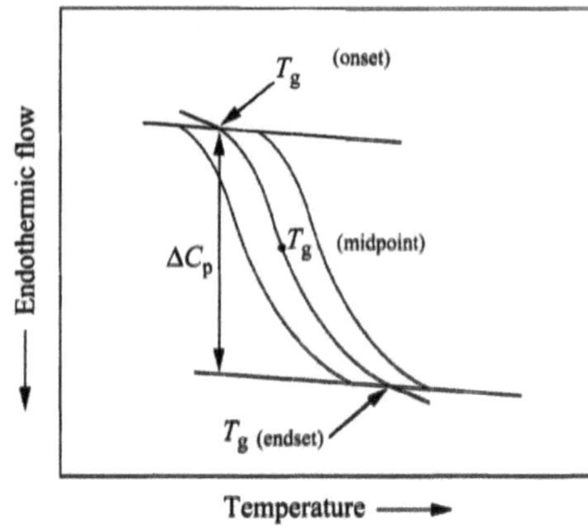

Fig. 1.9 A DSC curve for an idealized glass transition in which T_g can be taken as the temperature at which one-half of the change in heat capacity, ΔC_p, has occurred

Roos (1995) noted that the decrease in viscosity above T_g is responsible for various changes, such as stickiness and collapse of dried foods, agglomeration, and crystallization of food components (e.g., lactose in dried milk). In addition, the crispness of various low moisture foods is lost above T_g. Determination of T_g values as a function of solids or water content and water activity can be used to establish state diagrams, that may be used to predict the physical state of food materials under various conditions; they may also be used to show relationships between composition and temperature that are necessary to achieve changes in food processing and for maintaining food quality in processing and storage.

Figure 1.10, adapted from Rao (2003), is an idealized phase diagram for a frozen food, in which the temperature of the food is plotted against the solids fraction in the food. Below and to the right of the T_g curve, the product is in a glassy state. Besides the T_g of water, commonly accepted as 136 K and of the anhydrous solute, the glass transition temperature of unfrozen solute–water phase, T_g', and the corresponding solids weight fraction, X_g' are shown in the figure. Acceptance of the important role of glassy and rubbery states, and glass transition to better understand processing, storage, and stability of low moisture and frozen foods was largely due to the efforts of Levine and Slade; much useful information can be found in their reviews (e.g., Levine and Slade 1992; Slade and Levine 1998).

Glass transition temperature and molecular weight. Data on glass transition temperatures and the molecular weights of monodisperse polymers may be described by a linear relationship linking the T_G value to the inverse of the molecular weight of a polymer:

$$T_G = T_{G\infty} - \frac{A}{M} \tag{1.32}$$

where, A is a constant, M is the molecular weight of the polymer, and $T_{G\infty}$ is the T_G for infinite molecular weight. For maltodextrins, the values of the parameters $T_{G\infty}$ and the constant A were found to be 520 K and 53,003 Kg mol^{-1}, respectively. Values of $T_{G\infty}$ reported for various food polymers are in the range: 500–520 K (Rao et al. 2012).

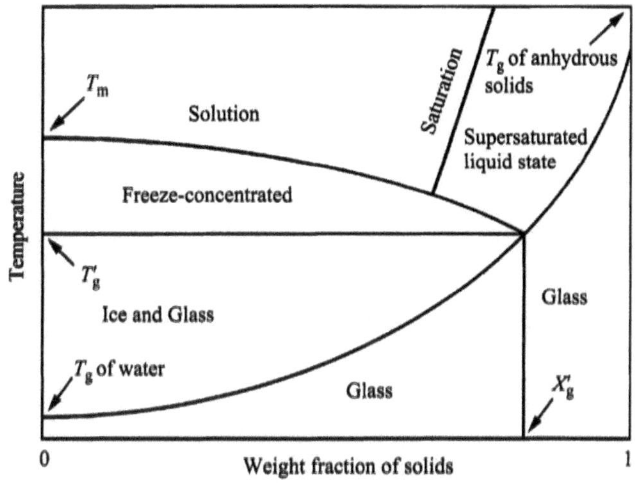

Fig. 1.10 State diagram illustrating glass transition, T_g, and melting, T_m, temperatures, and the different phases: Glass transition temperature of unfrozen solute–water phase, T_g' and the corresponding solids weight fraction, X_g'. (Adapted from Rao 2003)

Appendix 1-A

Momentum and Heat Transport Equations for Incompressible Fluids

Transport Equations in Vector Notation

The equations of continuity, motion, and energy in vector notation are given here:

$$\nabla \cdot \mathrm{v} = 0$$

$$\rho \frac{\mathrm{DV}}{\mathrm{Dt}} = -\nabla p - [\nabla \cdot \tau] + \rho\, g$$

$$k\nabla^2 T = -[\nabla \cdot q] - [\tau : \nabla \mathrm{V}]$$

The Equation of Continuity (Bird et al. 1960)

Cartesian coordinates (x, y, z):

$$\frac{\partial \rho}{\partial t} + \frac{\partial}{\partial x}(\rho v_x) + \frac{\partial}{\partial y}(\rho v_y) + \frac{\partial}{\partial z}(\rho v_z) = 0$$

Source: Bird, R. B., Stewart, W. E., and Lightfort, E. N. 1960. *Transport Phenomena,* John Wiley and sons, New York.

Cylindrical coordinates (r, θ, z):

$$\frac{\partial \rho}{\partial t}+\frac{1}{r}\frac{\partial}{\partial r}(\rho r v_r)+\frac{1}{r}\frac{\partial}{\partial \theta}(\rho v_\theta)+\frac{\partial}{\partial z}(\rho v_z)=0$$

Spherical coordinates (r, θ, ϕ):

$$\frac{\partial \rho}{\partial t}+\frac{1}{r^2}\frac{\partial}{\partial r}\left(\rho r^2 v_r\right)+\frac{1}{r\,\mathrm{Sin}\,\theta}\frac{\partial}{\partial \theta}(\rho v_\theta\,\mathrm{Sin}\,\theta)+\frac{1}{r\,\mathrm{Sin}\,\theta}\frac{\partial}{\partial \phi}(\rho v_\phi)=0$$

Equation of Motion in Rectangular Coordinates (*x*, *y*, *z*) in terms of σ (Bird et al. 1960)

x-component:

$$\rho\left(\frac{\partial v_x}{\partial t}+v_x\frac{\partial v_x}{\partial x}+v_y\frac{\partial v_x}{\partial y}+v_z\frac{\partial v_x}{\partial z}\right)=-\frac{\partial p}{\partial x}+\left(\frac{\partial \sigma_{xx}}{\partial x}+\frac{\partial \sigma_{yx}}{\partial y}+\frac{\partial \sigma_{zx}}{\partial z}\right)+\rho g_x \quad \text{(A)}$$

y-component:

$$\rho\left(\frac{\partial v_y}{\partial t}+v_x\frac{\partial v_y}{\partial x}+v_y\frac{\partial v_y}{\partial y}+v_z\frac{\partial v_y}{\partial z}\right)=-\frac{\partial p}{\partial y}+\left(\frac{\partial \sigma_{xy}}{\partial x}+\frac{\partial \sigma_{yy}}{\partial y}+\frac{\partial \sigma_{zy}}{\partial z}\right)+\rho g_y \quad \text{(B)}$$

z-component:

$$\rho\left(\frac{\partial v_z}{\partial t}+v_x\frac{\partial v_z}{\partial x}+v_y\frac{\partial v_z}{\partial y}+v_z\frac{\partial v_z}{\partial z}\right)=-\frac{\partial p}{\partial z}+\left(\frac{\partial \sigma_{xz}}{\partial x}+\frac{\partial \sigma_{yz}}{\partial y}+\frac{\partial \sigma_{zz}}{\partial z}\right)+\rho g_y \quad \text{(C)}$$

The Equation of Motion in Spherical Coordinates (*r*, *θ*, *ϕ*) in terms of σ (Bird et al. 1960)

r-component:

$$\rho\left(\frac{\partial v_r}{\partial t}+v_r\frac{\partial v_r}{\partial r}+\frac{v_\theta}{r}\frac{\partial v_r}{\partial \theta}+\frac{v_\phi}{r\sin\theta}\frac{\partial v_r}{\partial \phi}-\frac{v_\theta^2+v_\phi^2}{r}\right) \quad \text{(A)}$$

$$=-\frac{\partial p}{\partial r}+\left(\frac{1}{r^2}\frac{\partial}{\partial r}\left(r\sigma^2{}_{rr}\right)+\frac{1}{r\sin\theta}\frac{\partial}{\partial \theta}(\sigma_{r\theta}\sin\theta)+\frac{1}{r\sin\theta}\frac{\partial \sigma_{r\phi}}{\partial \phi}-\frac{\sigma_{\theta\theta}+\sigma_{\phi\phi}}{r}\right)+\rho g_r$$

θ-component:

$$\rho\left(\frac{\partial v_\theta}{\partial t}+v_r\frac{\partial v_\theta}{\partial r}+\frac{v_\theta}{r}\frac{\partial v_\theta}{\partial \theta}+\frac{v_\phi}{r\sin\theta}\frac{\partial v_\theta}{\partial \phi}+\frac{v_r v_\theta}{r}-\frac{v^2\phi\cot\theta}{r}\right)$$
$$=-\frac{1}{r}\frac{\partial p}{\partial \theta}+\left(\frac{1}{r^2}\frac{\partial}{\partial r}\left(r^2\sigma_{r\theta}\right)+\frac{1}{r\sin\theta}\frac{\partial}{\partial \theta}\right)(\sigma_{\theta\theta}\sin\theta) \tag{B}$$
$$+\frac{1}{r\sin\theta}\frac{\partial\sigma_{\theta\phi}}{\partial\phi}+\frac{\sigma_{r\theta}}{r}-\frac{\cot\theta}{r}\sigma_{\phi\phi}\Bigg)+\rho g_\theta$$

ϕ-component:

$$\rho\left(\frac{\partial v_\phi}{\partial t}+v_r\frac{\partial v_\phi}{\partial r}+\frac{v_\theta}{r}\frac{\partial v_\phi}{\partial \theta}+\frac{v_\phi}{r\sin\theta}\frac{\partial v_\phi}{\partial \phi}+\frac{v_\phi v_r}{r}+\frac{v_\theta v_\phi}{r}\cot\theta\right)$$
$$=-\frac{1}{r\sin\theta}\frac{\partial p}{\partial \phi}+\left(\frac{1}{r^2}\frac{\partial}{\partial r}(r^2\sigma_{r\phi})+\frac{1}{r}\frac{\partial\sigma_{\theta\phi}}{\partial\theta}+\frac{1}{r\sin\theta}\frac{\partial\sigma_{\phi\phi}}{\partial\phi}\right. \tag{C}$$
$$\left.+\frac{\sigma_{r\phi}}{r}+\frac{2\cot\theta}{r}\sigma_{\theta\phi}\right)+\rho g_\phi$$

Equation of Motion in Cylindrical Coordinates (r, θ, z) in terms of σ (Bird et al. 1960)

r-component:

$$\rho\left(\frac{\partial v_r}{\partial t}+v_r\frac{\partial v_r}{\partial r}+\frac{v_\theta}{r}\frac{\partial v_r}{\partial \theta}-\frac{v_\theta^2}{r}+v_z\frac{\partial v_r}{\partial z}\right)$$
$$=-\frac{\partial p}{\partial r}+\left(\frac{1}{r}\frac{\partial}{\partial r}(r\sigma_{rr})+\frac{1}{r}\frac{\partial\sigma_{r\theta}}{\partial\theta}-\frac{\sigma_{\theta\theta}}{r}+\frac{\partial\sigma_{rz}}{\partial z}\right)+\rho g_r \tag{A}$$

θ-component:

$$\rho\left(\frac{\partial v_\theta}{\partial t}+v_r\frac{\partial v_\theta}{\partial r}+\frac{v_\theta}{r}\frac{\partial v_\theta}{\partial \theta}+\frac{v_r v_\theta}{r}+v_z\frac{\partial v_\theta}{\partial z}\right)$$
$$=-\frac{1}{r}\frac{\partial p}{\partial \theta}+\left(\frac{1}{r^2}\frac{\partial}{\partial r}(r^2\sigma_{r\theta})+\frac{1}{r}\frac{\partial\sigma_{\theta\theta}}{\partial\theta}+\frac{\partial\sigma_{\theta z}}{\partial z}\right)+\rho g_\theta \tag{B}$$

z-component:

$$\rho\left(\frac{\partial v_z}{\partial t}+v_r\frac{\partial v_z}{\partial r}+\frac{v_\theta}{r}\frac{\partial v_z}{\partial \theta}+v_z\frac{\partial v_z}{\partial z}\right)$$
$$=-\frac{\partial p}{\partial z}+\left(\frac{1}{r}\frac{\partial}{\partial r}(r\sigma_{rz})+\frac{1}{r}\frac{\partial\sigma_{\theta z}}{\partial\theta}+\frac{\partial\sigma_{zz}}{\partial z}\right)+\rho g_z \tag{C}$$

References

Aguilera, J. M. and Stanley, D. W. 1999. *Microstructral Principles of Food Processing and Engineering*, Aspen Publishers, Gaithersburg, Maryland, USA.
Barnes, H. A. and Walters, K. 1985. The yield stress myth? *Rheol. Acta* 24: 323–326.
Barnes, H. A., Hutton, J. F., and Walters, K. 1989. *An Introduction to Rheology*, Elsevier Science Publishers B.V., Amsterdam, The Netherlands.
Bird, R. B., Stewart, W. E., and Lightfoot, E. N. 1960. *Transport Phenomena*, John Wiley & Sons, New York.
Bird, R. B., Armstrong, R. C. and Hassager, O. 1977a. *Dynamics of Polymeric Liquids–Fluid Mechanics*, John Wiley and Sons, New York.
Bird, R. B., Hassager, O., Armstrong, R. C. and Curtiss, C. F. 1977b. *Dynamics of Polymeric Liquids–Kinetic Theory*, John Wiley and Sons, New York.
Chamberlain, E. K. 1996. Characterization of heated and thermally processed cross-linked waxy maize starch utilizing particle size analysis, microscopy and rheology. M. S. Thesis, Cornell University, Ithaca, NY.
Chamberlain, E. K. and Rao, M. A. 2000. Concentration dependence of viscosity of acid-hydrolyzed amylopectin solutions. Food Hydrocolloids 14: 163–171.
Clark, A. H. and Ross-Murphy, S. B. 1987. Structural and mechanical properties of biopolymer gels. *Adv. Polym. Sci.*, 83: 57–192.
Dobson, C. M. 2003. Protein folding and misfolding. Nature 426(18): 884–890.
Doublier, J. L. 1975. Propriétés rheologiques et caracteristiques macromoleculaires de solutions aqueuses de galactomannanes. Doctoral thesis, Universite Paris VI, France.
Fuoss, R. M. and Strauss, U. P. 1948. Polyelectrolytes. II. Poly-4-vinylpyridonium chloride and poly-4-vinyl-N-n-butylpyridonium bromide. *J. Polym. Sci.*, 3(2): 246–263.
Genovese, D. B. and Rao, M. A. 2003. Apparent viscosity and first normal stress of starch dispersions: role of continuous and dispersed phases, and prediction with the Goddard-Miller model. *Appl. Rheol.* 13(4): 183–190.
Launay, B., Cuvelier, G. and Martinez-Reyes, S. 1984. Xanthan gum in various solvent conditions: intrinsic viscosity and flow properties. In *Gums and Stabilisers for the Food Industry 2*, ed. G.O. Phillips, D. J. Wedlock, and P. A. Williams, pp. 79–98, Pergamon Press, London.
Launay, B., Doublier, J. L., and Cuvelier, Ci. 1986. Flow properties of aqueous solutions and dispersions of polysaccharides, in *Functional Properties of Food Macromolecules*, eds. J. R. Mitchell and D. A. Ledward, Ch. 1, pp. 1–78, Elsevier Applied Science Publishers, London.
Levine, H. and Slade, L. 1992. Glass transition in foods, in *Physical Chemistry of Foods*, H. G. Schwartzberg and R. W. Hartel, pp. 83–221, Marcel Dekker, New York.
Loveday S. M., Rao, M. A., Creamer, L. K. and Singh, H. 2009. Factors affecting rheological characteristics of fibril gels: The case of β-lactoglobulin and α-lactalbumin. *J Food Sci.* 74(3): R47–R55.
Mandelbrot, B. B. 1982. *The Fractal Geometry of Nature*, W. H. Freeman, New York.
Marangoni, A. G. and Rousseau, D. 1996. Is plastic fat governed by the fractal nature of the fat crystals? *J. Am. Oil Chem. Soc.* 73(8): 991–994.
Markovitz, H. 1985. Rheology: in the beginning. *J. Rheol.* 29: 777–798.
Mewis, J. 1979. Thixotropy—a general review. *J. Non-Newtonian Fluid Mech.* 6: 1–20.
Michel, F. 1982. Etude du comportement potentiometrique et viscosimetrique de pectines hautement methylees en presence de saccharose. Doctoral thesis, Universite de Dijon, France.
Morris, E. R. and Ross-Murphy, S. B. 1981. Chain flexibility of polysaccharides and glicoproteins from viscosity measurements, in *Techniques in Carbohydrate Metabolism*, B310, ed. D. H. Northcote, pp. 1–46, Elsevier Science, Amsterdam.
Nagai, T. and Yano, T. 1990. Fractal structure of deformed potato starch and its sorption characteristics. *J. Food Sci.* 55(5): 1334–1337.
Padmanabhan, M. 1995. Measurement of extensional viscosity of viscoelastic liquid foods. *J. Food Eng.* 25: 311–327.

Rahman, M. S. 1997. Physical meaning and interpretation of fractal dimensions of fine particles measured by different methods. *J. Food Eng.* 32: 447–456.

Rao, M. A. 1977a. Rheology of liquid foods—a review. *J. Texture Stud.* 8: 135–168.

Rao, M. A. 1977b. Measurement of flow properties of fluid foods—developments limitations, and interpretation of phenomena. *J. Texture Stud.* 8: 257–282.

Rao, M. A. 1992. Measurement of Viscoelastic Properties of Fluid and Semisolid Foods, in "Viscoelastic Properties of Food," ed. M. A. Rao and J. F. Steffe, pp. 207–232, Elsevier Applied Science Publishers, New York.

Rao, M. A. 2003. Phase transitions, food texture and structure, in *Texture in Food, Volume 1: Semi-Solid Foods*, ed. B. M. McKenna, pp. 36–62, Woodhead Publishing Ltd., Cambridge, UK.

Rao, M. A. 2005. Rheological properties of fluid foods, in *Engineering Properties of Foods*, eds. M. A. Rao, S. S. H. Rizvi, and A. K. Datta, 3rd ed., pp. 41–99, CRC Press, Boca Raton, FL.

Rao, M. A. 2006. Influence of food microstructure on food rheology, in *Understanding and Controlling the Microstructure of Complex Foods*, ed. D. J. McClements, Woodhead Publishing Ltd., Cambridge, UK. (In Press).

Rao, M. A. and Steffe, J. F. 1992. *Viscoelastic Properties of Foods*, pp. 1–444. Elsevier Applied Science Publishers, New York.

Rao M. A., Tattiyakul J., and Liao H. J. 2012. Rheological and thermal properties of starch and starch based biopolymers, Chapter 5, in Starch-Based Polymeric Materials and Nanocomposites: Chemistry, Processing, and Applications, eds: J. Ahmed, B. Tiwari, S. H. Imam, and M. A. Rao, Taylor and Francis Group, Boca Raton, FL.

Robinson, G., Ross-Murphy, S.B., and Morris, E.R. 1982. Viscosity-molecular weight relationships, intrinsic chain flexibility and dynamic solution properties of guar galactomannan. *Carbohydrate Research* 107: 17–32.

Roos, Y. 1995. Characterization of food polymers using state diagrams. *J. Food Eng.* 24(3): 339–360.

Roos, Y. H. 1998. Role of water in phase-transition phenomena in foods, in *Phase/State Transitions in Foods*, eds. M. A. Rao and R. W. Härtel, pp. 57–86, New York, Marcel Dekker.

Ross-Murphy, S. B. 1984. Rheological Methods, in *Biophysical Methods in Food Research*, ed. H. W.-S. Chan, pp. 138–199, Blackwell Scientific Publications, Oxford, U.K.

Sherman, P. 1970. *Industrial Rheology*, Academic Press, New York.

Slade, L. and Levine, H. 1998. Selected aspects of glass transition phenomena in baked goods, in *Phase/State Transitions in Foods*, eds. M. A. Rao and R. W. Härtel, pp. 87–94, Marcel Dekker, New York.

Smidsrød, 0. 1970. Solution properties of alginate. *Carbohydrate Research*, 13: 359.

Sperling, L. H. 1986. *Introduction to Physical Polymer Science*, 1st ed., New York, Wiley.

Sperling, L. H. 2001. *Introduction to physical polymer science*, 3rd ed., New York, John Wiley.

Steffe, J. F. 1996. *Rheological Methods in Food Process Engineering*, 2nd ed., Freeman Press, East Lansing, MI.

Tanglertpaibul, T. and Rao, M. A. 1987. Intrinsic viscosity of tomato serum as affected by methods of determination and methods of processing concentrates. *J. Food Sci.* 52: 1642–1645 & 1688.

Tattiyakul, J. 1997. Studies on granule growth kinetics and characteristics of tapioca starch dispersions using particle size analysis and rheological methods. M. S. Thesis, Cornell University, Ithaca, NY.

Tschoegl, N. W. 1989. *The Phenomenological Theory of Linear Viscoelastic Behavior—An Introduction*, Springer-Verlag, New York.

Van Wazer, J. R., Lyons, J. W., Kim, K. Y., and Colwell, R. E. 1963. *Viscosity and Flow Measurement A Handbook of Rheology*, Interscience, New York.

Walters, K. 1975. *Rheometry*, Chapman and Hall, London.

Whorlow, R. W. 1980. *Rheological Techniques*, Halsted Press, New York.

Yoo, B., Figueiredo, A. A., and Rao, M. A. 1994. Rheological properties of mesquite seed gum in steady and dynamic shear. *Lebens. Wissen, und Techno!* 27: 151–157.

Suggested Reading[1]

Bames, H.A., Hutton, J. F., and Walters, K. 1989. *An Introduction to Rheology*, Elsevier Science Publishers B.V., Amsterdam, The Netherlands.

Bird, R. B., Armstrong, R. C., and Hassager, O. 1977a. *Dynamics of Polymeric Liquids—Fluid Mechanics*, John Wiley and Sons, New York.

Bird, R. B., Hassager, O., Armstrong, R. C., and Curtiss, C. F. 1977b. *Dynamics of Polymeric Liquids—Kinetic Theory*, John Wiley and Sons, New York.

Dealy, J. M. 1982. *Rheometers for Molten Polymers*, Van Nostrand Reinhold Co., New York.

Ferry, J. D. 1980. *Viscoelastic Properties of Polymers*, John Wiley, New York.

Macosko, C. W. 1994. *Rheology: Principles, Measurements, and Applications*, VCH Publishers, New York.

Sperling, L. H. 1986. *Introduction to Physical Polymer Science*, John Wiley, New York.

[1] The following books contain much useful information on the science of rheology that should be useful to food professionals.

Chapter 2
Flow and Functional Models for Rheological Properties of Fluid Foods

A flow model may be considered to be a mathematical equation that can describe rheological data, such as shear rate versus shear stress, in a basic shear diagram, and that provides a convenient and concise manner of describing the data. Occasionally, such as for the viscosity versus temperature data during starch gelatinization, more than one equation may be necessary to describe the rheological data. In addition to mathematical convenience, it is important to quantify how magnitudes of model parameters are affected by state variables, such as temperature, and the effect of structure/composition (e.g., concentration of solids) of foods and establish widely applicable relationships that may be called functional models.

Rheological models may be grouped under the categories: (1) empirical, (2) theoretical, and (3) structural. Obviously, an empirical model, such as the power law (Eq. 2.3), is deduced from examination of experimental data. A theoretical model is derived from fundamental concepts and it provides valuable guidelines on understanding the role of structure. It indicates the factors that influence a rheological parameter. The Krieger–Dougherty model (Krieger 1985) (Eq. 2.26) for relative viscosity is one such model. Another theoretical model is that of Shih et al. (1990) that relates the modulus to the fractal dimension of a gel.

A structural model is derived from considerations of the structure and often kinetics of changes in it. It may be used, together with experimental data, to estimate values of parameters that help characterize the rheological behavior of a food sample. One such model is that of Casson (Eq. 2.6) that has been used extensively to characterize the characteristics of foods that exhibit yield stress. Another structural model is that of Cross (1965) (Eq. 2.14) that has been used to characterize flow behavior of polymer dispersions and other shear-thinning fluids. While application of structure-based models to rheological data provides useful information, structure-based analysis can provide valuable insight in to the role of the structure of a dispersed system. For example, as discussed in Chap. 5, it allows for estimating the contributions of interparticle bonding and network of particles of dispersed systems.

Flow models have been used also to derive expressions for velocity profiles and volumetric flow rates in tube and channel flows, and in the analysis of heat transfer phenomenon. Numerous flow models can be encountered in the rheology literature

M. A. Rao, *Rheology of Fluid, Semisolid, and Solid Foods,* Food Engineering Series,
DOI 10.1007/978-1-4614-9230-6_2,

Table 2.1 Some two- and three-parameter flow models for describing shear rate ($\dot{\gamma}$) versus shear stress (σ) data

$\sigma = \eta\dot{\gamma}$	Newtonian model*
$\sigma = \dfrac{\dot{\gamma}}{\left[\dfrac{1}{\eta_0} + K_E(\sigma)^{(1/n_E)-1}\right]}$	Ellis model for low-shear rate data containing η_0 (Brodkey 1967)
$\sigma = \left[\eta_\infty\dot{\gamma} + K_s\dot{\gamma}^{n_s}\right]$	Sisko model for high-shear rate data containing η_∞ (Brodkey 1967)
$\eta_a = \eta_\infty + \dfrac{\eta_0 - \eta_\infty}{1 + (\alpha_c\dot{\gamma})^m}$	Cross model for data over a wide range of shear rates*
$\eta_a = \eta_\infty + \dfrac{\eta_0 - \eta_\infty}{[1 + (\lambda_c\dot{\gamma})^2]^N}$	Carreau model for data over a wide range of shear rates*
$\sigma = K\dot{\gamma}^n$	Power law model used extensively in handling applications*
$\sigma - \sigma_0 = \eta'\dot{\gamma}$	Bingham model*
$\sigma - \sigma_{0H} = K_K\dot{\gamma}^{n_H}$	Herschel–Bulkley model*
$\sigma^{0.5} = K_{0c} + K_c(\dot{\gamma})^{0.5}$	Casson model used especially in treating data on chocolates*
$\sigma^{0.5} - \sigma_{0M} = K_M\dot{\gamma}^{n_M}$	Mizrahi and Berk (1972) model is a modification of the Casson model
$\sigma^{n_1} = \sigma_0^{n_1} + \eta_\infty(\dot{\gamma})^{n_2}$	Generalized model of Ofoli et al. (1987)*
$\sigma = [(\sigma_{oV})^{1/n_V} + K_V\dot{\gamma}]^{n_V}$	Vocadlo (Vocadlo and Moo Young 1969) model

*Discussed in text

and some from the food rheology literature are listed in Table 2.1. Also, here those models that have found extensive use in the analysis of the flow behavior of fluid foods are discussed. Models that account for yield stress are known as viscoplastic models (Bird et al. 1982). For convenience, the flow models can be divided in to those for time-independent and for time-dependent flow behavior.

Time-Independent Flow Behavior

Newtonian Model

The model for a Newtonian fluid is described by the equation:

$$\sigma = \eta\gamma \tag{2.1}$$

As per the definition of a Newtonian fluid, the shear stress, σ, and the shear rate, $\dot{\gamma}$, are proportional to each other, and a single parameter, η, the viscosity, characterizes the data. For a Bingham plastic fluid that exhibits a yield stress, σ_0, the model is:

$$\sigma - \sigma_0 = \eta'\gamma \tag{2.2}$$

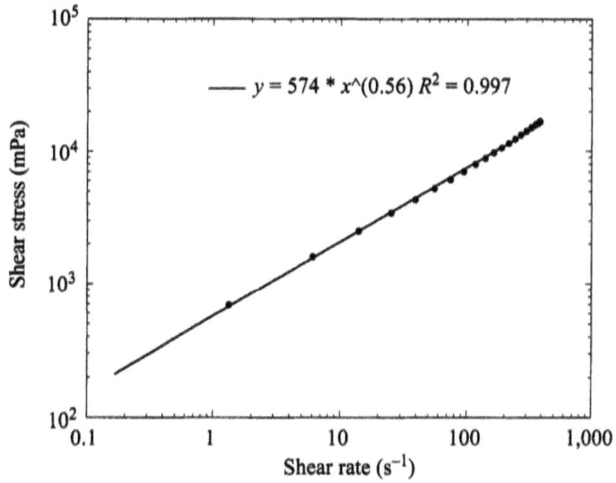

Fig. 2.1 Plot of log shear rate ($\dot{\gamma}$) versus log shear stress (σ) for a 2.6% Tapioca starch dispersion heated at 67 °C for 5 min (Tattiyakul 1997) to illustrate applicability of the power law model

where, η' is called the Bingham plastic viscosity.

As shown in Fig. 2.1, the Newtonian model and the Bingham plastic model can be described by straight lines in terms of shear rate and shear stress, and the former can be described by one parameter η and the latter by two parameters: η' and σ_0, respectively. However, the shear rate–shear stress data of shear-thinning and shear-thickening fluids are curves that require more than one parameter to describe their data. Given that the equation of a straight line is simple, it is easy to understand attempts to transform shear rate–shear stress data in to such lines. An additional advantage of a straight line is that it can be described by just two parameters: the slope and the intercept.

Power Law Model

Shear stress–shear rate plots of many fluids become linear when plotted on double logarithmic coordinates and the power law model describes the data of shear-thinning and shear thickening fluids:

$$\sigma = K\dot{\gamma}^n \tag{2.3}$$

where, K the consistency coefficient with the units: Pa s^n is the shear stress at a shear rate of 1.0 s^{-1} and the exponent n, the flow behavior index, is dimensionless that reflects the closeness to Newtonian flow. The parameter K is sometimes referred to as consistency index. For the special case of a Newtonian fluid ($n=1$), the consistency index K is identically equal to the viscosity of the fluid, η. When the magnitude of $n<1$ the fluid is shear-thinning and when $n>1$ the fluid is shear-thickening in nature. Taking logarithms of both sides of Eq. 2.3:

$$\log\sigma = \log K + n\log\dot{\gamma} \tag{2.4}$$

The parameters K and n are determined from a plot of log σ versus log $\dot{\gamma}$, and the resulting straight line's intercept is log K and the slope is n. If a large number of σ versus $\dot{\gamma}$ data points, for example, >15 (it is easy to obtain large number of points with automated viscometers) are available, linear regression of log $\dot{\gamma}$ versus log σ will provide statistically best values of K and n. Nevertheless, a plot of experimental and predicted values of log $\dot{\gamma}$ and log σ is useful for observing trends in data and ability of the model to follow the data. Figure 2.1 illustrates applicability of the power law model to a 2.6% tapioca starch dispersion heated at 67 °C for 5 min. Linear regression techniques also can be used for determination of the parameters of the Herschel–Bulkley (when the magnitude of the yield stress is known) and the Casson models discussed later in this chapter.

Because it contains only two parameters (K and n) that can describe shear rate–shear stress data, the power law model has been used extensively to characterize fluid foods. It is also the most used model in studies on handling of foods and heating/cooling of foods. Extensive compilations of the magnitudes of power law parameters can be found in Holdsworth (1971, 1993). Because it is convenient to group foods in to commodities, a compilation of magnitudes of power law parameters of several food commodities are given in Chap. 5. In addition, the influence of temperature in quantitative terms of activation energies, and the effect of concentration of soluble and insoluble solids on the consistency index are given.

Although the power law model is popular and useful, its empirical nature should be noted. One reason for its popularity appears to be due to its applicability over the shear rate range: 10^1-10^4 s^{-1} that can be obtained with many commercial viscometers. Often, the magnitudes of the consistency and the flow behavior indexes of a food sample depend on the specific shear rate range being used so that when comparing the properties of different samples an attempt should be made to determine them over a specific range of shear rates. One drawback of the power law model is that it does not describe the low-shear and high-shear rate constant-viscosity data of shear-thinning foods.

Herschel–Bulkley Model

When yield stress of a food is measurable, it can be included in the power law model and the model is known as the Herschel–Bulkley model:

$$\sigma - \sigma_{0\mathrm{H}} = K_{\mathrm{H}} \dot{\gamma}^{n_{\mathrm{H}}} \tag{2.5}$$

where, $\dot{\gamma}$ is shear rate (s^{-1}), σ is shear stress (Pa), n_{H} is the flow behavior index, K_{H} is the consistency index, and $\sigma_{0\mathrm{H}}$ is yield stress. It is noted here that the concept of yield stress has been challenged (Barnes and Walters 1989) because a fluid may deform minutely at stress values lower than the yield stress. Nevertheless, yield stress may be considered to be an engineering reality and plays an important role in many food products.

If the yield stress of a sample is known from an independent experiment, K_{H} and n_{H} can be determined from linear regression of log σ—$\sigma_{0\mathrm{H}}$ versus log($\dot{\gamma}$) as

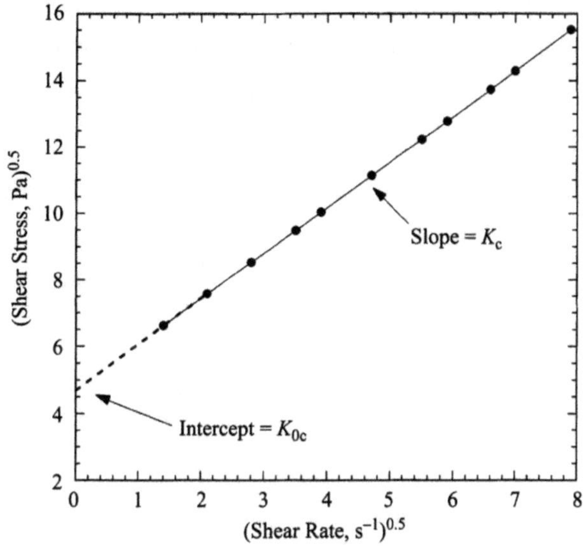

Fig. 2.2 Plot of $(\dot{\gamma})^{0.5}$ versus $(\sigma)^{0.5}$ for a food that follows the Casson model. The square of the intercept is the yield stress and that of slope is the Casson plastic viscosity

the intercept and slope, respectively. Alternatively, nonlinear regression technique was used to estimate σ_{0H}, K_H, and n_H (Rao and Cooley 1983). However, estimated values of yield stress and other rheological parameters should be used only when experimentally determined values are not available. In addition, unless values of the parameters are constrained a priori, nonlinear regression provides values that are the best in a least squares sense and may not reflect the true nature of the test sample.

Casson Model

The Casson model (Eq. 2.6) is a structure-based model (Casson 1959) that, although was developed for characterizing printing inks originally, has been used to characterize a number of food dispersions:

$$\sigma^{0.5} = K_{0c} + K_c(\dot{\gamma})^{0.5} \tag{2.6}$$

For a food whose flow behavior follows the Casson model, a straight line results when the square root of shear rate, $(\dot{\gamma})^{0.5}$, is plotted against the square root of shear stress, $(\sigma)^{0.5}$, with slope K_c and intercept K_{0c} (Fig. 2.2). The Casson yield stress is calculated as the square of the intercept, $\sigma_{0c} = (K_{0c})^2$ and the Casson plastic viscosity as the square of the slope, $\eta_{ca} = (K_c)^2$. The data in Fig. 2.2 are of Steiner (1958) on a chocolate sample. The International Office of Cocoa and Chocolate has adopted the Casson model as the official method for interpretation of flow data on chocolates. However, it was suggested that the vane yield stress would be a more reliable measure of the yield stress of chocolate and cocoa products (Servais et al. 2004).

The Casson plastic viscosity can be used as the infinite shear viscosity, η_∞, (Metz et al. 1979) of dispersions by considering the limiting viscosity at infinite shear rate:

$$\left(\frac{d\sigma}{d\dot{\gamma}}\right)_{\dot{\gamma}\to\infty} = \left(\frac{d(\sqrt{\sigma})}{d\dot{\gamma}}\frac{d\sigma}{d(\sqrt{\sigma})}\right)_{\dot{\gamma}\to\infty} \tag{2.7}$$

Using the Casson equation the two terms in the right-hand side bracket can be written as:

$$\frac{d(\sqrt{\sigma})}{d\dot{\gamma}} = \frac{K_C}{2\sqrt{\dot{\gamma}}} \tag{2.8}$$

and

$$\frac{d\sigma}{d(\sqrt{\sigma})} = 2\sqrt{\sigma} \tag{2.9}$$

Combining the above two equations,

$$\eta_\infty = \eta_{Ca} = (K_C)^2 \tag{2.10}$$

Quemada Model

Quemada et al. (1985) proposed a viscosity equation for dispersed systems based on zero-shear, η_0, and infinite-shear, η_∞, viscosities, and a structural parameter, λ, dependent on the shear rate, that may be written as:

$$\frac{\eta}{\eta_\infty} = \frac{1}{\left\{1-\left[1-\left(\frac{\eta_\infty}{\eta_0}\right)^{0.5}\right]\lambda\right\}^2} \tag{2.11}$$

where,

$$\lambda = \frac{1}{\left[1+(t_C\dot{\gamma})^{0.5}\right]} \tag{2.12}$$

The time constant t_c is related to the rate of aggregation of particles due to Brownian motion. For highly concentrated dispersed systems, η_∞ will be much lower than η_0, so that $(\eta_\infty/\eta_0) \ll 1$ and the dispersion may have a yield stress, and Eq. (2.11) reduces to the Casson model (Eq. 2.6) (Tiu et al. 1992) with the Casson yield stress, $\sigma_{0c} = (\eta_\infty/t_c)$. Thus the Casson–Quemada models can be used to examine dispersions whose rheological behaviors range from only shear-thinning to shear thinning with yield stress. The Casson–Quemada models were used to study the role of cocoa sol-

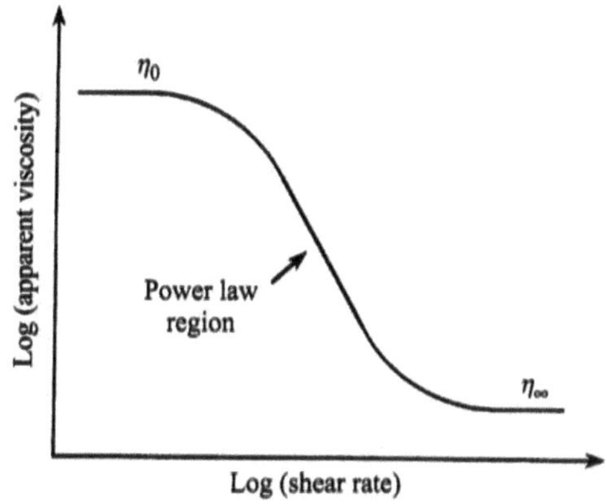

Fig. 2.3 Plot of shear rate versus apparent viscosity for shear thinning foods identifying three separate regions: a zero-shear viscosity at low shear rates, a power law region at intermediate shear rates, and an infinite-shear viscosity at high-shear rates. Often, only data in the power law region are obtained

ids and cocoa butter on cocoa dispersions (Fang et al. 1996, 1997) to be discussed in Chap. 5.

A general model for shear rate–shear stress data that under specific assumptions reduces to the Herschel–Bulkley, the Casson, and other models was presented by Ofoli et al. (1987):

$$\sigma^{n_1} = \sigma_0^{\,n_1} + \eta_\infty (\dot{\gamma})^{n_2} \tag{2.13}$$

where, n_1 and n_2 are constants, and η_∞ is the infinite shear viscosity. It is important to note that one model may be applicable at low-shear rates and another at high-shear rates (Dervisoglu and Kokini 1986). While applicability of the flow models themselves may be interesting, it is much more important to study the role of food composition on a model's parameters and apply the model to better understand the nature of foods.

Apparent Viscosity—Shear Rate Relationships of Shear-Thinning Foods

At sufficiently high polymer concentrations, most shear-thinning biopolymer (also called a gum or a hydrocolloid) dispersions exhibit similar three-stage viscous response when sheared over a wide shear rate range (Fig. 2.3): (1) at low-shear rates, they show Newtonian properties with a constant zero-shear viscosity (η_0) over a limited shear range that is followed by, (2) a shear-thinning range where solution viscosity decreases in accordance with the power law relationship; the reciprocal of the shear rate at which the transition from Newtonian to pseudoplastic behavior occurs is the characteristic time or the time constant, and (3) attains a limiting and constant infinite-shear-viscosity (η_∞). The three regions may be thought of being due rearrangement in the conformation of the biopolymer molecules in the dispersion due to shearing. In stage 1 when the magnitude of $\dot{\gamma}$ is low, there is little rearrangement of the polymer chains, while in stage 2 the chains undergo gradual rearrangement with $\dot{\gamma}$ resulting in a power law behavior. In stage 3, the shear rate is sufficiently high that the polymer chains do not undergo much rearrangement.

Cross and Carreau Models

The apparent viscosity (η_a) of the solution can be correlated with shear rate ($\dot{\gamma}$) using the Cross (Eq. 2.14) or the Carreau (Eq. 2.15) equations, respectively.

$$\eta_a = \eta_\infty + \frac{\eta_0 - \eta_\infty}{1+(\alpha_c \dot{\gamma})^m} \tag{2.14}$$

$$\eta_a = \eta_\infty + \frac{\eta_0 - \eta_\infty}{[1+(\lambda_c \dot{\gamma})^2]^N} \tag{2.15}$$

where, α_c and λ_c are time constants related to the relaxation times of the polymer in solution and m and N are dimensionless exponents. Because magnitudes of $\eta\infty$ of food polymer dispersions with concentrations of practical interest are usually very low in magnitude, they are difficult to determine experimentally. Therefore, to avoid consequent errors in estimation of the other rheological parameters in Eqs. 2.14 and 2.15, often η_∞ has been neglected (Abdel-Khalik et al. 1974; Lopes da Silva et al. 1992). The Cross and Carreau models described well the shear dependence of aqueous dispersions of high methoxyl pectins and locust bean gum (Lopes da Silva et al. 1992), konjac flour gum (Jacon et al. 1993), and mesquite gum solution (Yoo et al. 1995), and other gums (Launay et al. 1986). In general, the model of Cross has been used in studies in Europe and that of Carreau in North America. In Chap. 4, the applicability of the Cross and Carreau models to locust bean gum dispersions will be discussed in more detail.

For small values of η_∞, the Cross exponent m tends to a value $(1-n)$, where n is the power law flow behavior index (Launay et al. 1986; Giboreau et al. 1994). For the shear rate, $\dot{\gamma}_c$ where $\eta_{ap}=(\eta_0+\eta_\infty)/2$, the Cross time constant $\alpha_c=1/\dot{\gamma}_c$. Generally, $\dot{\gamma}_c$ gives an order of magnitude of the critical shear rate marking the end of the zero shear rate Newtonian plateau or the onset of the shear-thinning region. It is therefore important to recognize the shear rate dependence of the rheological behavior of polysaccharide polymers in solution and the difficulty involved in obtaining experimental data over the applicable shear rate range of $10^{-6}-10^4$ s^{-1} (Barnes et al. 1989). The low-shear rate region of about $10^{-3}-10^0$ is often used for the characterization and differentiation of structures in polysaccharide systems through the use of stress controlled creep and non destructive oscillatory tests. The shear rate range of about 10^1-10^4 s^{-1} falls within the operational domain of most commercial rheometers, so that the range of $10^{-3}-10^{4}\ s^{-1}$ can sometimes be effectively covered by a combination of measuring procedures and instruments.

Both the Carreau and the Cross models can be modified to include a term due to yield stress. For example, the Carreau model with a yield term given in Eq. (2.16) was employed in the study of the rheological behavior of glass-filled polymers (Poslinski et al. 1988):

$$\eta_a = \sigma_0 \dot{\gamma}^{-1} + \eta_p \left[1 + (\lambda_p \dot{\gamma})^2\right]^{-N} \tag{2.16}$$

where, σ_0 is the yield stress, η_p is the plateau viscosity, and λ_p and N are constants to be determined from experimental data. Rayment et al. (1998) interpreted the rheological behavior of guar gum dispersions containing raw rice starch in terms of the Cross model with yield stress (Eq. 2.17). We note that, when yield stress is exhibited, the term plateau viscosity is used instead of zero-shear viscosity:

$$\eta_a = \sigma_0 \dot{\gamma}^{-1} + \eta_P \left[1 + (\alpha_c \dot{\gamma})\right]^{-m} \tag{2.17}$$

Models for Time-Dependent Flow Behavior

Considerable care should be exercised in determining reliable time-dependent rheological data because of the often unavoidable modification in structure due to sample handling and during loading the sample in a viscometer or rheometer measuring geometry. Nevertheless, with careful attention to details, such as allowing a sample to relax in the rheometer measuring geometry, rheological data can be obtained to characterize time-dependent rheological behavior.

Weltman Model

The Weltman (1943) model has been used to characterize thixotropic (Paredes et al. 1988) behavior and of antithixotropic behavior (da Silva et al. 1997) of foods:

$$\sigma = A - B \log t \tag{2.18}$$

where, σ is shear stress (Pa), t is time (s), and A (value of stress at $t = 1$ s) and B are constants. A plot of σ versus log time should result in a straight line. In thixotropic behavior B takes negative values and in antithixotropic behavior it takes positive values. Table 2.2 shows typical magnitudes of the constants A and B for cross-linked waxy maize starch dispersions.

Table 2.2 Weltman Equation parameters for cross-linked waxy maize gelatinized starch dispersions. (Da Silva et al. 1997)

		Shear rate (s^{-1})			
Weltman Parameter	Conc. (%)	50	100	200	300
A	3	6.97×10^{-2}	5.76×10^{-2}	4.22×10^{-2}	
	4	2.97×10^{-1}	2.08×10^{-1}	1.72×10^{-1}	9.54×10^{-2}
	5		5.39×10^{-1}	3.88×10^{-1}	3.55×10^{-1}
B	3	1.58×10^{-3}	5.71×10^{-4}	1.66×10^{-5}	
	4	3.15×10^{-3}	2.07×10^{-3}	5.29×10^{-3}	7.83×10^{-3}
	5		8.30×10^{-5}	1.06×10^{-2}	8.79×10^{-3}
Correlation coef.	3	1.00	1.00	1.00	
	4	1.00	1.00	1.00	1.00
	5		1.00	1.00	1.00

Tiu–Boger Model

A model to study thixotropic behavior of foods exhibiting yield stress was devised by Tiu and Boger (1974) who studied the time-dependent rheological behavior of mayonnaise by means of a modified Herschel–Bulkley model:

$$\sigma = \lambda\left[\sigma_{0\mathrm{H}} + K_{\mathrm{H}}(\dot{\gamma})^{n\mathrm{H}}\right] \tag{2.19}$$

where, σ is the shear stress (Pa), $\dot{\gamma}$ is the shear rate (s^{-1}), λ is a time-dependent structural parameter that ranges from an initial value of unity to an equilibrium value λ_e, σ_{0H} is the yield stress (Pa), K_H is the consistency index (Pa s^n), and n_H is the flow behavior index. The decay of the structural parameter with time was assumed to obey a second-order equation:

$$\frac{d\lambda}{dt} = -k_1(\lambda - \lambda_e)^2 \tag{2.20}$$

where, the constant k_1 is a function of shear rate to be determined experimentally. While the determination of σ_{0H}, and n_H is straight forward, estimation of k_1 and λ_e requires the use of values of apparent viscosities (η_a) (Tiu and Boger 1974):

$$\lambda = \frac{\eta_a \dot{\gamma}}{\sigma_{0\mathrm{H}} + K_{\mathrm{H}}\dot{\gamma}^{n\mathrm{H}}} \tag{2.21}$$

and

$$\frac{d\eta_a}{dt} = -a_1\left(\eta_a - \eta_e\right)^2 \tag{2.22}$$

where,

$$a_1(\dot{\gamma}) = \frac{k_1\dot{\gamma}}{\sigma_{0H} + K_H\dot{\gamma}^{nH}} \tag{2.23}$$

It can be shown (Tiu and Boger 1974), that a plot of $1/(\eta_a - \eta_e)$ versus time will yield a straight line of slope a_1 and repeating the procedure at other shear rates will establish the relationship between a_1 and $\dot{\gamma}$, and hence k_1 and $\dot{\gamma}$ from Eq. 2.20. For a commercial mayonnaise sample, values of the different parameters were: $\lambda_e = 0.63$, $\sigma_{0H} = 7.0$ Pa, $K_H = 28.5$ Pa s^n, $n_H = 0.32$, and k_1 was a weak function of shear rate; specifically, $k_1 = 0.012\ \dot{\gamma}^{0.13}$ (Tiu and Boger 1974).

Role of Solids Fraction in Rheology of Dispersions

Following Einstein's work (Einstein 1906, 1911) (Eq. 2.24) on dilute rigid sphere dispersions, models for estimating viscosity of concentrated nonfood dispersions of solids are based on volume fraction (ϕ) of the suspended granules and the relative viscosity of the dispersion, $\eta_r = (\eta/\eta_s)$, where η is the viscosity of the dispersion η_s is the viscosity of the continuous phase (Jinescu 1974; Metzner 1985).

$$\eta_r = 1 + 2.5\,\phi \tag{2.24}$$

Metzner (1985) pointed out that at high solids concentration levels, the theoretical equation (Eq. 2.25) of Frankel and Acrivos appears to do a good job of portraying experimental data of rigid solids dispersed in polymer melts.

$$\eta_r = \frac{9}{8}\left[\frac{(\phi/\phi_m)^{1/3}}{1 - (\phi/\phi_m)^{1/3}}\right] \tag{2.25}$$

Wildemuth and Williams (1984) modeled the η_r of rigid sphere suspensions with a shear-dependent maximum volume fraction (ϕ_m). The applicability of a shear dependent ϕ_m (Wildemuth and Williams 1984) to food dispersions has not been tested.

The Krieger–Dougherty (1959) relationship (Eq. 2.26) is based on the assumption that an equilibrium exists between individual spherical particles and dumbbells that continuously form and dissociate:

$$\eta_r = \left(1 - \frac{\phi}{\phi_m}\right)^{-[\eta]\phi_m} \tag{2.26}$$

where, $[\eta]$ and ϕ_m are the intrinsic viscosity and maximum packing fraction of solids. Theoretically, $[\eta]$ should be 2.5 for rigid spheres and ϕ_m should be about 0.62 if the spheres are of uniform diameter (Krieger 1985), but Choi and Krieger (1986) found it necessary to use values of $[\eta]$ of 2.65–3.19 to fit Eq. 2.25 to viscosity-

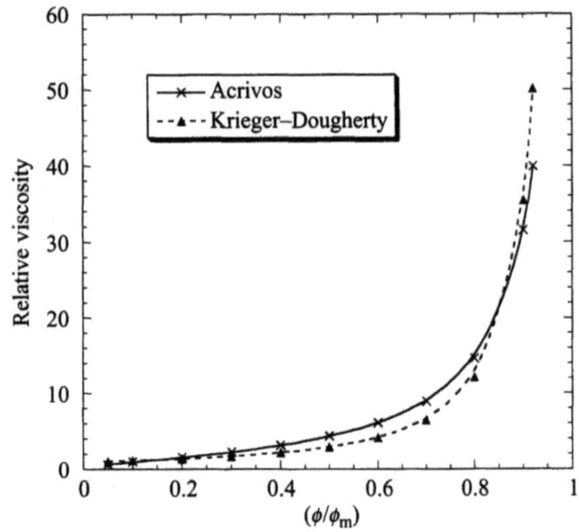

Fig. 2.4 Relative viscosity versus volume fraction ratio ($\O/\O_m$) predicted by models of Frankel–Acrivos and Krieger–Dougherty $\O_m=0.62$ and $[\eta]=2.5$

volume fraction data on sterically stabilized poly methylmethacrylate spheres. For calculating values of % according to the Krieger–Dougherty equation, a value of $\phi_m=0.62$ was assumed (Choi and Krieger 1986). Both Eqs. 2.24 and 2.25 contain the ratio of volume fraction of solids in a dispersion to the maximum volume fraction and the values of η_r predicted (Fig. 2.4) are close to each other. In Fig. 2.4 for the Krieger–Dougherty model, ϕ_m was taken to be 0.62 and $[\eta]$ to be 2.5. It is emphasized that the above equations were derived for rigid solids; because of the polydisperse and deformable nature of gelatinized starch dispersions, it is not surprising that attempts to predict their viscosity with Eq. 2.25 (Noel et al. 1993; Ellis et al. 1989) were not successful.

Most food particles are not spherical in shape so that the empirical equation (Eq. 2.25) that described well (Kitano et al. 1981; Metzner 1985) the relative viscosity versus concentration behavior of suspensions of spheres and fibers with aspect (L/D) ratios ≤30 in polymer melts is of interest:

$$\eta_r = [1 - (\phi/A)]^{-2} \tag{2.27}$$

Equation 2.27 is an empirical equation that is a simple modification of the Maron–Pierce equation for dispersions of spherical rigid solids:

$$\eta_r = [1 - (\phi/\phi_m)]^{-2} \tag{2.28}$$

For fluids that obey the power law model (Eq. 2.3), Metzner (1985) suggested that the viscosities of the suspension and of the continuous phase be evaluated at the same shearing stress. For rigid particles, the value of A decreases as aspect ratio of suspended particles increases; for example, when the aspect ratio is 1.0 (smooth sphere) the magnitude of A is 0.68, and when the aspect ratio is 6–8 (rough crystal)

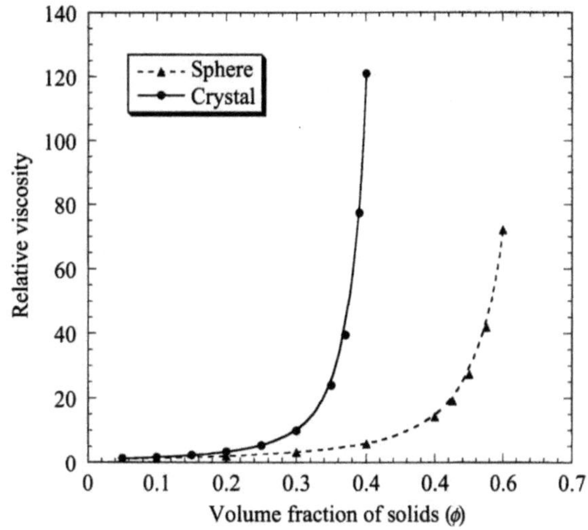

Fig. 2.5 Relative viscosity, η_r, versus volume fraction of solids predicted by model of Kitano et al. (1981) for rigid spheres (A=0.68) and crystal-like solids (A=0.44)

A is 0.44 (Metzner 1985). The shape of many food particles is not spherical and may be considered to be closer to a rough crystal. Figure 2.5 illustrates predictions of η_r by Eq. 2.25 for dispersions of spherical and rough crystal-like rigid particles. For the dispersion of rough crystal-like rigid particles, high values of η_r are attained at solids loading much lower than for rigid spherical particles (Fig. 2.5).

Because of the compressible nature of food dispersions, the direct determination of the magnitude of ϕ is not easy as it depends on the centrifugal force employed in the separation of the phases. Therefore, rheological properties of plant food dispersions, such tomato concentrate and concentrated orange juice, are based on the mass of pulp. In starch dispersions, they are based on the mass fraction of starch granules, denoted as cQ, as described in Chap. 4.

Applicability of Eq (2.27) for suspensions of tomato pulp of narrow size distribution was shown by Yoo and Rao (1994). Tomato pulp particles with average particle diameters 0.71 and 0.34 mm retained on the two sieves (sieve no. 40 and 60, respectively) were produced (Fig. 2.6). The apparent average diameters were calculated as in Kimball and Kertesz (1952). Because the Casson viscosity was shown to be equal to infinite shear viscosity (Metz et al. 1979), it is less arbitrary and has a theoretical foundation. Considering the values of the single parameter A of tomato puree samples (Fig. 2.7), the particle shape appears to be close to rough or irregular spherical shape (Yoo and Rao 1994). Also, the magnitude of A of the TP6 sample was higher than that of the TP4 sample because of its lower aspect ratio.

Quemada et al. (1985) proposed a model similar to that given by Eq. 2.27, but with A=0.5 and the structural parameter k=1.80:

$$\eta_r = \left(1 - \frac{1}{2}k\phi\right)^{-2} \tag{2.29}$$

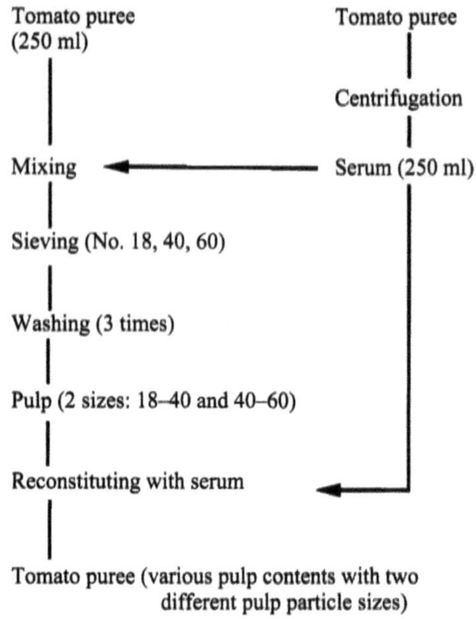

Fig. 2.6 Procedure for producing tomato pulp particles of narrow particle size distribution. (Yoo and Rao 1994)

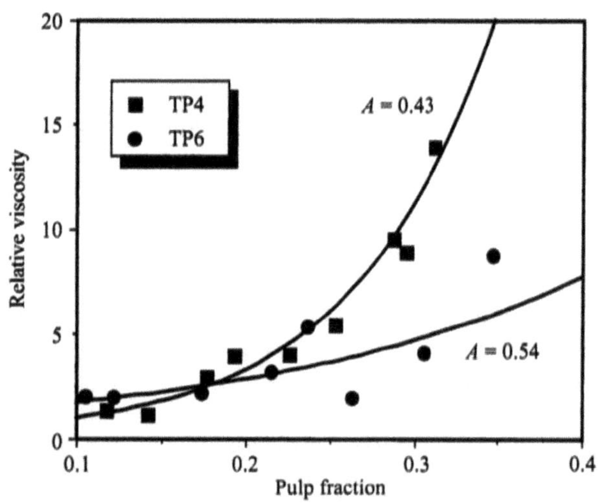

Fig. 2.7 Applicability of the model of Kitano et al. (1981) (Fig. 2.5) to tomato particles. (Yoo and Rao 1994)

They reported values of k in the range 2.50–3.82 for dispersions of rigid solids, and 1.70–1.85 for red blood suspensions. For gelatinized 2.6 % tapioca STDs, the relative viscosity was calculated using Eq. 2.30:

$$\eta_{r_\infty} = \frac{\eta_\infty}{\eta_s} \tag{2.30}$$

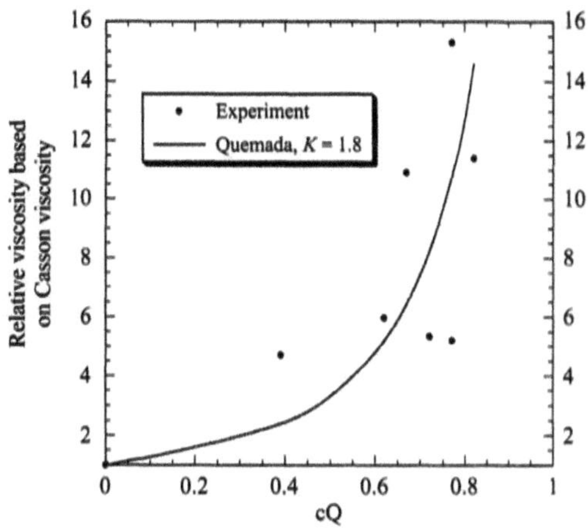

Fig. 2.8 The relative viscosity (η_r) values of tapioca starch dispersions strongly depended on their volume fraction. The line in the figure represents values predicted by the model of Quemada et al. (1985) with the structural parameter $k=1.8$. (Tattiyakul et al. 2009)

where, $\eta_{r\infty}$ is the relative viscosity based on the Casson viscosity, η_∞ (mPa s) at 20 °C, and η_s 178 is the viscosity (mPa s) of the supernatant determined at 20 °C.

The model of Quemada et al. (1985) with the structural parameter $k=1.80$ gave a satisfactory fit (Fig. 2.8). Because heated tapioca starch granules are deformable, the value $k=1.80$ of red blood cells was selected for the data on tapioca STDs. Given that the Krieger–Dougherty (1959) model predictions were substantially different than experimental η_r values of starch dispersions (Ellis et al. 1989; Noel et al. 1993; Tattiyakul 1997), the reasonable applicability of Eq. 2.29 (Quemada et al. 1985) to tapioca STDs is noteworthy. However, the model of Quemada et al. (1985) is a phenomenological model, while that of Krieger–Dougherty (1959) is based on intrinsic viscosity of a single sphere. Saunders (1961) and Parkinson et al. (1970) found that the viscosity of a suspension increased with decrease in the average particle size. The particle size dependence can be explained by recognizing that as the particle size decreases, the number of particles in a given volume increases, resulting in a decrease in the mean distance between the particles. Another result of decreasing the particle size is to increase the potential for particle–particle interaction (Agarwala et al. 1992).

The important role of volume fraction on the structure of rigid sphere dispersions has been uncovered recently; as the volume fraction of hard spheres is increased, the equilibrium phase changes from a disordered fluid to coexistence with a crystalline phase ($0.494<\phi<0.545$), then to fully crystalline ($\phi=0.545$), and finally to a glass ($\phi=0.58$) (Pham et al. 2002).

Colloidal Glass

Studies using colloidal hard spheres (HS) have led to valuable insights in to their phase and rheological behavior: first, a disordered fluid to crystal transition at $\phi=0.494$ and coexistence of crystal and liquid domains for $0.494\leq\phi\leq0.545$. Fur-

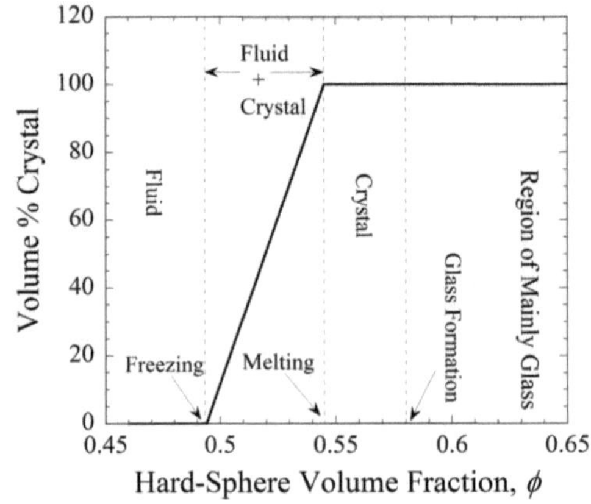

Fig. 2.9. Phase behavior of colloidal hard sphere dispersions, from Pusey and van Megen (1986)

ther, as the volume fraction of HS is increased, transition to fully crystalline state, and finally to a colloidal glassy state. (Pusey and van Megen 1986; Phan et al. 1996). The expression "colloidal glass" is used to differentiate it from the well- and long-known temperature-driven glassy state (Loveday et al. 2007).

The phase behavior of colloidal HS dispersions is summarized in Fig. 2.9 and it is discussed next. As the solids concentration is increased gradually, at a solids concentration, $\phi = 0.494$, crystals (clusters of particles) appear that coexist with the liquid. Thus, the coexistence of colloidal fluid and crystal phases is analogous to that of a simple liquid and solid at a first-order phase transition. Further, in the coexistence region: $0.494 \leq \phi \leq 0.545$, we find a linear dependence with ϕ which, when extrapolated to 0 and 100 %, provides the "freezing" and "melting" concentrations. The liquid to crystal transition at $\phi = 0.494$ is referred to as the beginning of freezing. These observations have been verified by others whose investigations were based on computer simulation, theory, and three-dimensional microscopy (e.g., Phan et al. 1996; Weeks et al. 2004).

As the volume fraction of solids, ϕ, is increased beyond $\phi = 0.545$, the particles are increasingly caged by others, and at a critical value, ϕ_G, the caging stops all long-range particle motion, and the system is considered to be glassy (Pham et al. 2002). Pusey and van Megen (1986) observed that the highly concentrated ($\phi > 0.58$), viscous, samples exhibited only partial heterogeneous crystallization even when left undisturbed for several months. The concentration of particles was sufficiently high that particle diffusion was hindered to the point where crystals did not form on that time scale and the suspensions remained in the metastable amorphous phase created by the earlier sample tumbling. Thus, for $\phi \approx 0.58$ hard-sphere colloidal dispersions form glasses and the glassy state is present over a range of solids concentration. The effective volume fraction of the most concentrated glassy sample was close to $\phi = 0.637$ expected for the random close-packed HS (so called Bernal) glass. The discovery of a glass composed of equal-sized spheres is especially interesting since, although there has been considerable theoretical work and computer simulations on such model glasses earlier, real glasses composed of spherical units were not identified experimentally.

Rheology of Protein Dispersions

Many protein particles are crystal shaped (Loveday et al. 2007). For example, bovine serum albumin (BSA) is a heart shaped globular protein molecule that can be approximated as an equilateral triangle with sides of 8 nm and a depth of 3 nm, and it has an equivalent hydrodynamic radius of 3.7 nm in the range of pH 4 and 8 (Ferrer et al. 2001). The β-lactoglobulin (β-LG) dimer is not spherical, but can be approximated as a prolate ellipsoid with length = 6.9 nm and width = 3.6 nm. From literature, it is known that the value of $\phi_{max} = \sim 0.71 - 0.74$ for prolate ellipsoid particles. In addition, in the Krieger-Dougherty model, one can use a value $[\eta] = 3.6$ for ellipsoid particles.

In milk, the casein is aggregated with calcium phosphate as casein micelles, with a mean size of about 300 nm. After the calcium phosphate is removed, the resulting sodium caseinate exists in solution mainly as a mixture of casein monomers and casein nanometer-scale particles (10–20 nm); further, gels may be produced from dispersions of sodium caseinate by heating, acidification, and high-pressure processing (Dickinson 2006). In sodium caseinate, different caseins interact with each other to form associated structures, which exist as a dynamic system of casein monomers, casein complexes, and aggregates (Lucey et al. 2000). The average radius of gyration of caseinate aggregates has been shown to be in the range 22–48 nm; the aggregates have been shown to be not spherical but highly elongated structures. The extent of aggregation of sodium caseinate depends on the relative proportions of the different monomeric caseins and also on the temperature, pH, ionic strength, and calcium ion concentration (Dickinson 2006).

Farrer and Lips (1999) obtained zero-shear viscosity data on dispersions of sodium caseinate (pH 6.8, 0.1 M NaCl) over the concentration range 3–28 % w/v. Their values of relative viscosity were calculated using solvent viscosity of 1 mPa s. Panouille et al. (2005) obtained zero-shear viscosity data for dispersions of phosphocaseinate (pH 6.0, polyphosphate 2 % w/v); the phosphocaseinate was obtained after the colloidal calcium phosphate had been removed from the casein micelles. An empirical model, based on concentration, *C*, (instead of volume fraction) was used to fit the relative viscosity data of the phosphocaseinate dispersions up to a concentration of about 10 % w/v:

$$\eta_r = \left(1 - \frac{C}{C_c}\right)^{-2} \tag{2.31}$$

In Fig. 2.10, the relative viscosity data on sodium caseinate dispersions from different laboratories are shown. As expected, the viscosity increases gradually with concentration up to about 10 % and then more steeply at higher concentrations. Pitkowski et al. (2008) noted that this behavior was also found in multiarm star polymers and polymeric micelles. The high-end sodium caseinate dispersions were of higher concentration than in previous studies (Farrer and Lips 1999; Panouille et al. 2005) and the viscosity data were in good agreement with the extrapolated data of Farrer and Lips (1999) (Loveday et al. 2010).

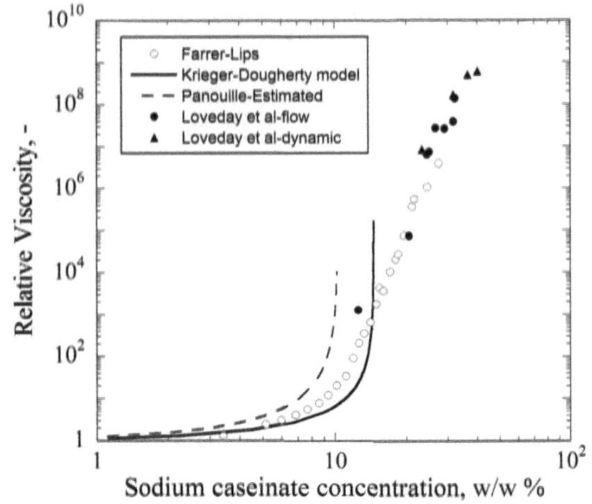

Fig. 2.10 Relative viscosity vs. concentration of Na caseinate dispersions. Data of Farrer-Lips (1999) on dispersions of Na caseinate pH=6.8, 0.1 M NaCl (Loveday et al. 2010). Solid line: values using the Krieger and Dougherty (1959) model for hard spheres with $[\eta]=2.5$ and $\phi_{max}=0.65$. Dotted line: values calculated using equation that fit data on dispersions of phospho-caseinate (pH 6.0, polyphosphate 2% w/v) in water by Panouille et al. (2005)

The dashed line in Fig. 2.10 represents relative viscosity values of the phospho-caseinate dispersions predicted by the empirical model (Eq. 2–31). Also shown in this figure as a solid line are values of relative viscosity predicted by the Krieger-Dougherty model (Eq. 2–26) with $[\eta]=2.5$ and $\phi_{max}=0.65$. It is interesting to note that the values predicted by the model were lower than the experimental data up to a concentration of about 14% w/w, but increased more steeply than the data at higher concentrations. Because of uncertainties in the determination of the sodium caseinate particle volume fraction, the polydispersity of the particles, and their softness, the Krieger-Dougherty rheological model developed for hard-sphere dispersions predicted the trends accurately but not the absolute values of viscosity.

Modulus of Gels of Fractal Floes

In addition to the volume fraction of solids, their fractal nature also affects rheological properties. Shih et al. (1990) developed a scaling relationship for the elastic properties of colloidal gels by considering the structure of the gel network to be a collection of close packed fractal floes of colloidal particles. They defined two separate rheological regimes depending on the strength of the interfloc links relative to that of the floes themselves: (1) the strong-link regime is observed at low particle concentrations, allowing the floes to grow to be very large, so that they can be considered weak springs. Therefore, the links between floes have a higher elastic constant than the floes themselves, and (2) the weak-link regime is observed at high particle concentrations, where the small floes are stronger springs, and the links between floes have a lower elastic constant than the floes themselves. The weak-link regime should be applicable to gels that are well above the gelation threshold (Shih et al. 1990) where the elastic modulus, G', is related to the particle volume fraction (ϕ) by the following relationship:

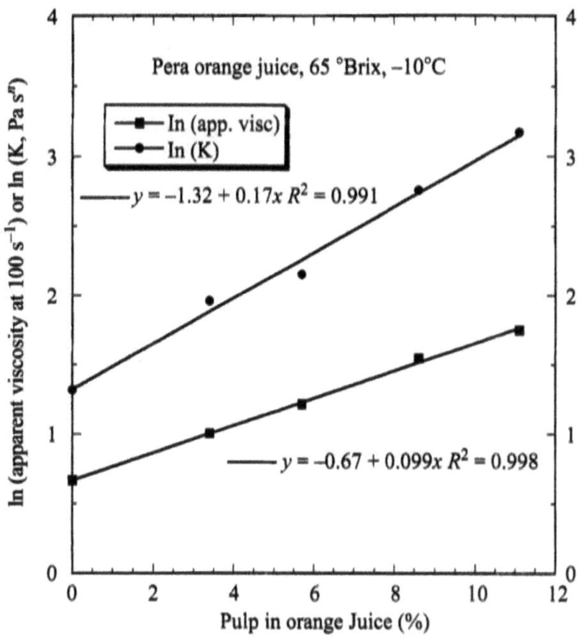

Fig. 2.11 Plateau modulus of starch dispersion plotted against volume fraction of starch granules. Fractal dimensions of the starch granules were calculated from the slopes of the lines

Table 2.3 Fractal dimension of selected foods based on rheological data. (Rao 2007)

Network of Particles	Fractal Dimension, D_f
Palm oil or lard fat	2.82–2.88
Cocoa butter	2.37
Salatrim®	2.90
Milk fat and/canola oil blends	1.97–1.99
Whey protein isolate + $CaCl_2$ gels	2.30–2.60
Soy protein isolate gels, pH 3.8 and 0.2 M NaCl	2.30
Starch gels	2.79–2.81
Egg white protein gel, pH 3.7	1.90–2.10

$$G' \sim \phi^{(d-2)/(d-D_f)} \tag{2.32}$$

where, D_f is the fractal dimension of the colloidal floe and d is the Euclidean dimension of the network—usually three. The power relationship between the modulus and the volume fraction of solids implied in Eq. 2.31 is illustrated in Fig. 2.11 for two gelatinized starch dispersions (Genovese and Rao 2003). Such plots have been utilized to determine the fractal dimension of several food gels (Table 2.3) and values of D_f between about 1.9 and 2.9 have been reported (Rao 2007). Wu and Morbidelli (2001) extended the above model to include gels that are intermediate between the strong-link and the weak-link regimes.

Effect of Soluble and Insoluble Solids Concentration on Apparent Viscosity of Foods

The effect of concentration on the zero-shear viscosity of biopolymer dispersions can be expressed in terms of the coil overlap parameter, $c[\eta]$, and the zero-shear specific viscosity as described in Chap. 4 in connection with food gum dispersions.

Unlike biopolymer dispersions where the intrinsic viscosity is known and the polymer concentration can be chosen a priori, often for fluid foods the concentration of soluble (e.g., pectins in fruit juices) and insoluble solids can be determined only posteriori, and the determination of their zero-shear viscosities is also difficult due to instrument limitation and due to the existence of yield stress. However, in many foods, it may be possible to identify the components, called key components, that play an important role in the rheological properties.

The effect of concentration (c) of soluble or insoluble solids on either apparent viscosity (η_a) or the consistency index of the power law model (K) can be described by either exponential or power law relationships:

$$\eta_a \propto \exp(ac) \tag{2.33}$$

$$\eta_a \propto c^b \tag{2.34}$$

$$K \propto \exp(a'c) \tag{2.35}$$

$$K \propto c^{b'} \tag{2.36}$$

where, a, a', b, and b′ are constants to be determined from experimental data. Other models that are applicable to specific foods are discussed under the flow properties of specific foods in Chaps. 4 and 5.

We consider the viscosity data on Pera concentrated orange juice (COJ) of Vitali (Vitali and Rao 1984a, b) to illustrate the exponential model for the effect of soluble solids (°Brix) and insoluble solids (% Pulp) (Tables 2.3 and 2.4). The influence of soluble solids on apparent viscosity at a shear rate of 100 s^{-1} (η_a, 100) and on K shown in Fig. 2.12 and that of insoluble solids on the same rheological parameters shown in Fig. 2.13 can be described by exponential relationships.

Table 2.4 Effect of °Brix on apparent viscosity and consistency index of the power law model, Pera orange juice, 5.7 % pulp, −10 °C. (Vitali and Rao 1984a, b)

°Brix	($\eta_{a,100}$ Pa s	ln, $\eta_{a,100}$	K, Pa s^n	*ln* K
50.8	0.29	−1.2379	0.89	−0.1165
56.5	0.65	−0.4308	2.18	0.7793
57.9	0.83	−0.1863	2.56	0.9400
61.6	1.62	0.4824	5.61	1.7246
65.3	3.38	1.2179	23.58	3.1604

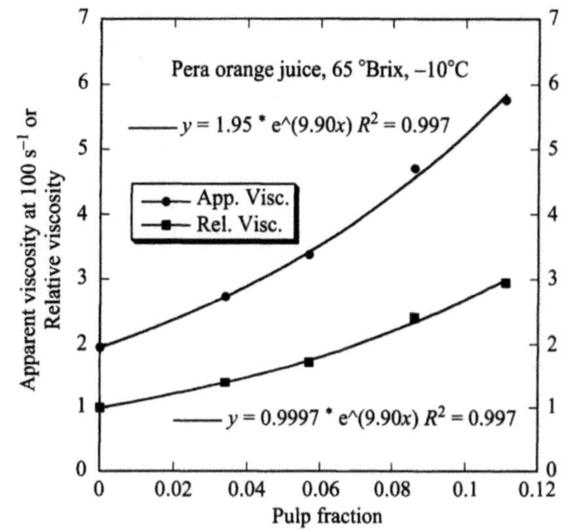

Fig. 2.12 The influence of insoluble solids on apparent viscosity at a shear rate of 100 s^{-1}(η_a, 100) and on the consistency coefficient, *K*, of pera concentrated orange juice, data of Vitali and Rao (1984b)

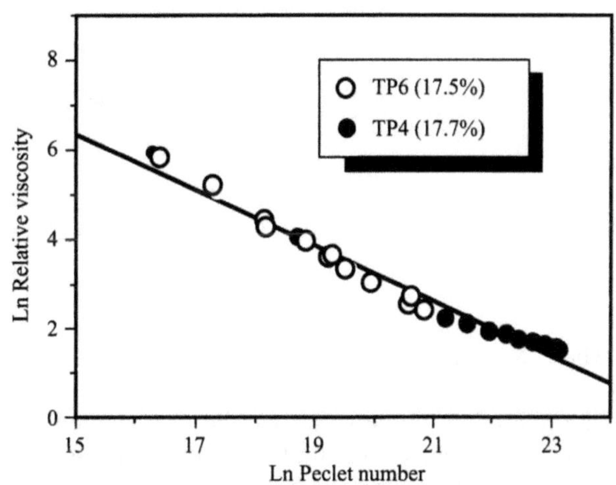

Fig. 2.13 The influence of soluble solids on apparent viscosity at a shear rate of 100 s^{-1} (η_a, 100) and on the consistency coefficient, *K*, of Pera concentrated orange juice, Data of Vitali and Rao (1984a)

The effect of concentration (*c*) of soluble solids (°Brix) and insoluble solids (Pulp) on either apparent viscosity or the consistency index of the power law model of FCOJ can be described by exponential relationships (Vitali and Rao 1984a, b). Equations 2.36 and 2.37 are applicable to the consistency coefficient (*K*) of the power law model. In the case of FCO J, it should be noted that insoluble solids are expressed in terms of pulp content determined on a 12 °Brix sample by centrifugation at 360 × g for 10 min (Table 2.5).

$$K = K^{c} \exp(B_{K}^{c}\ {}^{\circ}\mathrm{Brix}) \tag{2.37}$$

$$K = K^{p} \exp(B_{K}^{P}\ {}^{\circ}\mathrm{Brix}) \tag{2.37}$$

Table 2.5 Effect of Pulp content on apparent viscosity and consistency index of the power law model, pera orange juice, 65 °Brix – 10 °C. (Vitali and Rao 1984a, b)

Pulp %	$\eta_{a,100}$ Pa s	$ln, \eta_{a,100}$	K, Pa s^n	ln K
0	1.95	0.6678	3.74	1.3191
3.4	2.73	1.0043	7.11	1.9615
5.7	3.38	1.2179	8.63	2.1552
8.6	4.70	1.5476	15.77	2.7581
11.1	5.76	1.7509	23.91	3.1743

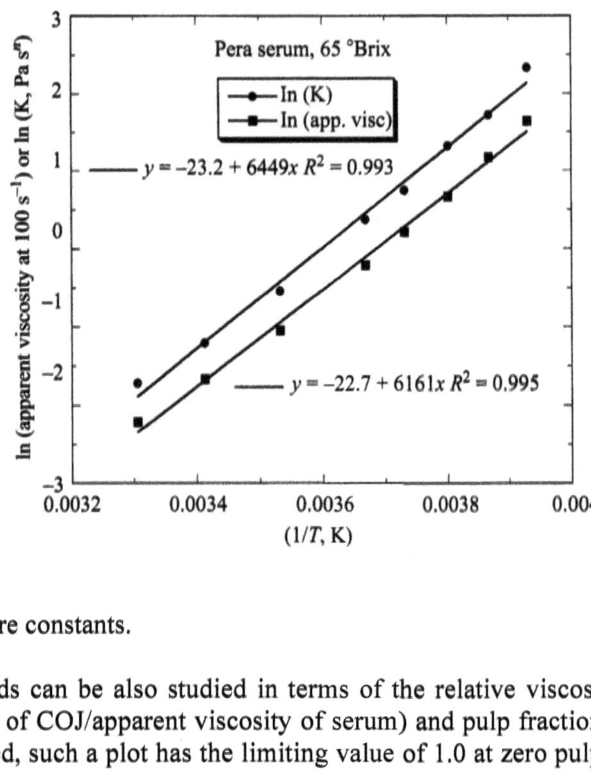

Fig. 2.14 The strong influence of pulp fraction on the viscosity of concentrated orange juice (COJ) and the relative viscosity (η_r) of COJ

where, K^c, K^p, B_K^C and B_K^P are constants.

The role of insoluble solids can be also studied in terms of the relative viscosity (η_r=apparent viscosity of COJ/apparent viscosity of serum) and pulp fraction (Fig. 2.12) and, as expected, such a plot has the limiting value of 1.0 at zero pulp fraction (serum). The curve in Fig. 2.14 illustrates the strong influence of pulp fraction on the viscosity of COJ. The values of η_a,100 also, as expected, follow a profile similar to that of η_r (Fig. 2.14). The two curves are described by the equations:

$$\eta_r = \exp(9.90 \times \text{pulp fraction}) \quad (2.39)$$

$$\eta_{a\ 100} = 1{,}95 \times \exp(9.90 \times \text{pulp fraction}) \quad (2.40)$$

Peclet Number of Dispersions

Yoo and Rao (1996) studied the influence of two different sizes of tomato pulp particles (TP4 and TP6) at a pulp weight fraction of 17 % in terms of Peclet number and

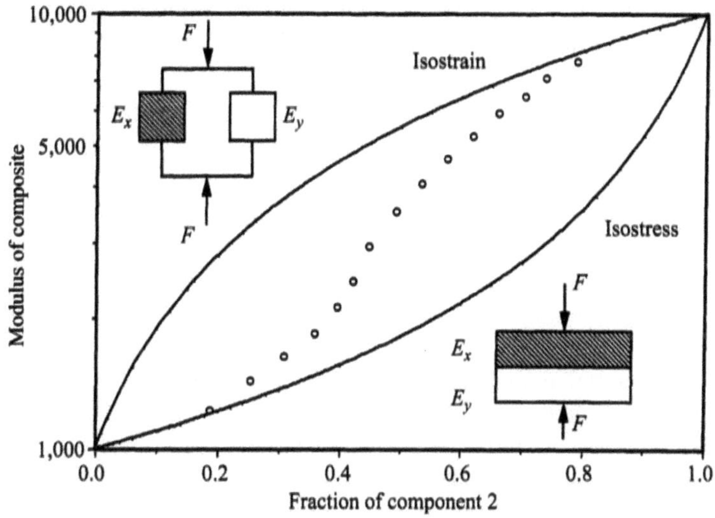

Fig. 2.15 Peclet number versus relative viscosity of tomato pulp particles. At equal values of pulp weight fraction, a TP6 sample with small diameter particles was more viscous than a TP4 sample with large diameter particles

relative viscosity (Krieger 1985; Tsai and Zammouri 1988). The Peclet numbers (P_e) were calculated using the equation:

$$P_e = \frac{\eta_0 r_0^{\,3} \dot{\gamma}}{kT} \tag{2.41}$$

where, η_0 is the viscosity of the suspending liquid (serum), r_0 is the particle radius, k is the Boltzmann constant (1.38×10^{-23} N m K^{-1}), and T is the absolute temperature.

The Peclet number compares the effect of imposed shear with the effect of diffusion of the particles. When $P_e \gg 1$, hydrodynamic effects dominate and a dispersion of spherical particles exhibits shear-thinning behavior. In contrast, when $P_e \ll 1$, the distribution of particles is only slightly altered by the flow (Hiemenz and Rajagopalan 1997). As shown in Fig. 2.15, at equal values of pulp weight fraction, a TP6 sample with small diameter particles was more viscous than a TP4 sample with large diameter particles. A nearly linear relationship exists between the relative viscosity of TP sample and the Peclet number. Similar linear relationship was found for glass bead suspensions (Tsai and Zammouri 1988). However, the slope of the linear relationship of TP6 sample is slightly higher than that of TP4 sample, indicating that at high-shear rates, the TP6 aggregates with small particles are more sensitive to shear than those with large particles. The different aspect ratios of the TP4 and TP6 particles discussed above is another reason for the deviation from a single linear relationship. From these observations, it can be concluded that the effect of particle size on the relative viscosity of tomato puree sample can be correlated with the Peclet number of the particle.

Emulsions

Many foods are oil-in-water or water-in-oil emulsions (o/w), with dispersed particle size range of 0.01 – 10 μm. Many of the equations discussed for food suspensions are also applicable to emulsions. In a dilute emulsion, the particles are far apart and the interparticle interactions are relatively weak. Skim milk is an example of a dilute emulsion with the concentration of fat droplets (dispersed phase) < 1 %. However, in a concentrated emulsion, the particles are close to each other and there are strong interparticle interactions. Mayonnaise has a dispersed phase (oil) concentration of 70 – 80 %. As with foods containing insoluble solids, the volume fraction of the dispersed phase in emulsions is important.

The separation of phases in an emulsion occurs spontaneously in the direction of decreasing Gibbs free energy. Thus it can be said that there is more surface energy in an emulsion when the dispersed droplets are in a highly subdivided state than when they are in a coarser state of subdivision. To be stable, an o/w emulsion should exhibit yield stress, and the forces applied to the continuous phase by the dispersed phase due to the applicable forces (e.g., gravity and buoyancy) should be below the emulsion's yield stress.

The words stable and unstable are often used to describe emulsions that are better understood by examining the underlying processes. The coarsening of a thermodynamically unstable emulsion is called coalescence or aggregation; coalescence is a process by which two or more particles fuse together to form a single larger particle; aggregation is a process by which small particles lump together to form aggregates (Hiemenz and Rajagopalan 1997). The van der Waals force between particles in a dispersion is usually attractive and is strong at short distances between particles. Therefore, the emulsion will be unstable and coagulate unless there are repulsive interactions between particles. Two methods are used to overcome van der Waals attraction: (1) electrostatic stability in which the electrostatic force resulting from overlapping electrical double layers of two particles, and (2) polymer-induced or steric stability in which a suitable polymer that is adsorbed on the particle surfaces may be used. In food emulsions, polymer-induced stability is encountered commonly.

Very often, the microstructure and the macroscopic states of dispersions are determined by kinetic and thermodynamic considerations. While thermodynamics dictates what the equilibrium state will be, kinetics determine how fast that equilibrium state will be determined. While in thermodynamics the initial and final states must be determined, in kinetics the path and any energy barriers are important. The electrostatic and the electrical double-layer (the two charged portions of an interfacial region) play important roles in food emulsion stability. The Derjaguin-Landau-

Verwey-Overbeek (DLVO) theory of colloidal stability has been used to examine the factors affecting colloidal stability.

An emulsifier may be thought of as a single chemical component or mixture of components that have the capacity to promote emulsion formation and stabilization by interfacial action; in contrast, a stabilizer confers long-term stability on an emulsion involving often adsorption or another mechanism. Polymers, such as gum Arabic, egg albumin, casein, and gelatin have been used for a long time to stabilize food emulsions. Other stabilizers include xanthan gum, guar gum, and whey protein isolate. Because some of these are charged polymers (polyelectrolytes), their stabilizing influence is due to both electrostatic and polymeric effects, that is, electrosteric stabilization. Lecithins (e.g., from egg yolk and soybean) are the most important food emulsifiers derived from natural sources without chemical reaction (Dickinson 1993). They consist of mixtures of many phospholipid components with phosphatidylcholine (PC) and phosphatidylethanolamine (PE) being present in large proportions.

A simplified view of some of the possible effects of polymer molecules on a dispersion include (Hiemenz and Rajagopal 1997): (1) at low polymer concentrations bridging flocculation, where a polymer chain forms bridges by adsorbing more than one particle, (2) at higher polymer concentrations, "brush-like" layers can form on the solid particles that can mask the influence of van der Waal's attraction between the particles, thereby imparting stability, called steric stabilization, (3) at moderate to high polymer concentrations, the polymer chains may be excluded in the region between the particles resulting in depletion flocculation, and (4) at high polymer concentrations, polymer-depleted regions may be created by demixing the polymer resulting in depletion stabilization. Generally, polymers containing only one kind of repeat unit (homopolymers) are not good for steric stabilization. One requirement is that the polymer be adsorbed at the oil–water interface and the polymer is considered to reside partially at surface sites and partially in loops or tails in the solution. Under the right circumstances, an adsorbed polymer layer stabilizes a dispersion against aggregation. It may be said that the approaching particles in an aggregation step experience an increase in free energy of repulsion, ΔG_R, that can be divided into enthalpic (ΔH_R) and entropie (Δ, S_R) contributions:

$$\Delta G_R = \Delta H_R - T\Delta S_R \tag{2.42}$$

In Eq. 2.39, the terms describe changes in enthalpy and entropy in the overlapping region of the adsorbed layers of two particles. Because ΔH_R and ΔS_R can be either positive or negative, it is possible for the value of ΔG_R to change sign with change in temperature and the critical temperature for the onset of flocculation is known as the critical flocculation temperature (Hiemenz and Rajagopalan 1997).

Because of the many factors affecting the rheology of food emulsions, the rheological properties cannot be predicted a priori so that experimental studies on model o/w and food emulsions are necessary. Whey proteins, especially β-lactoglobulin, possess good emulsification properties and whey protein stabilized emulsions can be converted into emulsion gels by thermal treatment (Aguilera and Kessler 1989; Dickinson and Yamamoto 1996a). In general, the incorporation of protein-coated emulsion droplets into a heat-set globular protein network resulted in an increase in gel strength. Further, a self-supporting heat-set emulsion gel can be was formed at a protein content much lower than that required under similar thermal processing conditions (Dickinson and Yamamoto 1996b). Addition of pure egg yolk L-α- phosphatidylcholine after emulsification caused an increase in strength of a heat-set β-lactoglobulin emulsion gels (Dickinson and Yamamoto 1996b). The positive influence of pure egg lecithin added after emulsification on the elastic modulus of the whey protein concentrate emulsion gel was attributed to lecithin-protein complexation. Crude egg lecithin also gave a broadly similar increase in gel strength. However, pure soybean lecithin was not as effective in reinforcing the network and crude soybean lecithin was ineffective (Dickinson and Yamamoto 1996a). These results support the general hypothesis that filler particles (e.g., globular protein-coated droplets) that interact with a gel matrix (a network of aggregated denatured whey protein) tend to reinforce the network and increase the gel strength (Dickinson and Yamamoto 1996b).

Other relevant studies on physical properties of emulsions are those of McClements et al. (1993), Dickinson and Pawlowsky (1996), Dickinson et al. (1996), and Demetriades et al. (1997). The application of nuclear magnetic resonance (NMR) technique (Simoneau et al. 1993) and ultrasonic spectroscopy (Demetriades et al. 1996) to study the stability of emulsions were discussed. The rheological behavior of salad dressings and mayonnaises that are emulsions will be discussed in Chap. 5.

Effect of Temperature on Viscosity

A wide range of temperatures are encountered during processing and storage of fluid foods, so that the effect of temperature on rheological properties needs to be documented. The effect of temperature on either apparent viscosity at a specified shear rate (Eq. 2.42) or the consistency index, K, of the power law model (Eq. 2.43) of a fluid can be described often by the Arrhenius relationship. The effect of temperature on apparent viscosity can be described by the Arrhenius relationship:

$$\eta_{\mathrm{a}} = \eta_{\infty\mathrm{A}} \exp(E_{\mathrm{a}}/RT) \tag{2.43}$$

where, η_{a} is the apparent viscosity at a specific shear rate, $\eta_{\infty\mathrm{A}}$ is the frequency factor, E_a is the activation energy (J mol^{-1}), R is the gas constant (J mol^{-1}K^{-1}), and T is temperature (K).

Although the name of Arrhenius is associated with Eq. 2.42, Moore (1972) credited J. de Guzman Carrancio for first pointing out this relationship in 1913. The quantity E_a is the energy barrier that must be overcome before the elementary flow process can occur. The term: exp $(-E_a/RT)$ may be explained as a Boltzmann fac-

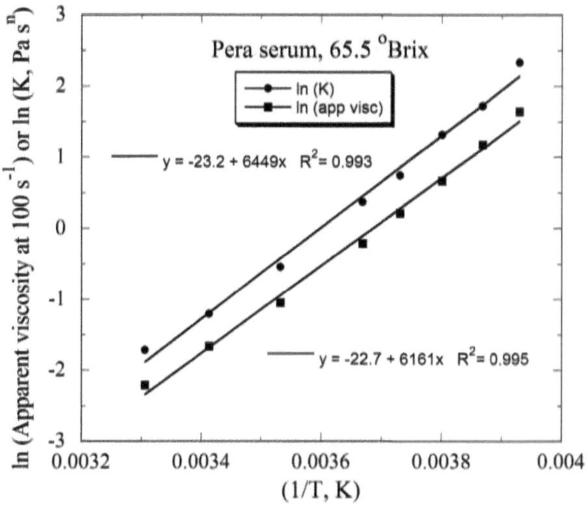

Fig. 2.16 Applicability of the Arrhenius Model to the apparent viscosity versus temperature data on a concentrated orange juice serum sample (Vitali and Rao 1984b) is Shown

tor that gives the fraction of the molecules having the requisite energy to surmount the barrier. Hence, E_a is the activation energy for viscous flow. From a plot of ln η_a (ordinate) versus $(1/T)$ (abscissa), E_a equals (slope × R) and $\eta_{\infty A}$ is exponential of the intercept.

The Arrhenius equation for the consistency coefficient is:

$$K = K_{\infty} \exp\left(E_{aK} / RT\right) \tag{2.44}$$

where, K_{∞} is the frequency factor, E_{ak} is the activation energy (J/mol), R is the gas constant, and T is temperature (K). A plot of ln K (ordinate) versus $(1/T)$ (abscissa) results in a straight line, and $E_{ak} = (\text{slope} \times R)$, and K_{∞}. is exponential of the intercept. The activation energy should be expressed in joules (J), but in the earlier literature it has been expressed in calories (1 calorie = 4.1868 J). The applicability of the Arrhenius model to the apparent viscosity versus temperature data on a concentrated orange juice serum sample (Vitali and Rao 1984a, b) is shown in Fig. 2.16.

The Arrhenius equation did not describe very well the influence of temperature on viscosity data of concentrated apple and grape juices in the range 60–68 °Brix (Rao et al. 1984, 1986). From nonlinear regression analysis, it was determined that the empirical Fulcher equation (see Ferry 1980, p. 289, Soesanto and Williams 1981) described the viscosity versus temperature data on those juice samples better than the Arrhenius model (Rao et al. 1986):

$$\log \eta = A' + \frac{B'}{T - T_{\infty}} \tag{2.45}$$

Table 2.6 Magnitudes of the parameters of the Arrhenius Equation for the effect of temperature on concentrated Apple and Grape juices. (Rao et al. 1986)

Sample	η_∞	E_a, $kJ\,mol^{-1}$	SSQ	R^2
Apple 68.3 °Brix	1.366E-15	79	0.10260	1.00
Apple 64.9 °Brix	1.671E-15	76	0.02870	1.00
Apple 60.1 °Brix	2.610E-13	62	0.00110	1.00
Apple 55.0 °Brix	5.050E-11	48	0.00100	0.99
Grape 68.3 °Brix	2.150E-14	73	0.31770	1.00
Grape 64.5 °Brix	9.810E-13	63	0.03950	1.00
Grape 59.9 °Brix	2.940E-12	58	0.00620	1.00
Grape 54.0 °Brix	7.610E-11	49	0.00130	1.00

Table 2.7 Magnitudes of the parameters of the Fulcher Equation for the effect of temperature on concentrated Apple and Grape juices. (Rao et al. 1986)

Sample	A′	B′	T_∞	SSQ	R^2
Apple 68.3 °Brix	−7.26	1,049	132.8	0.01356	1.00
Apple 64.9 °Brix	−7.56	1,035	130.9	0.01360	1.00
Apple 60.1 °Brix	−6.87	911	124.3	0.00208	1.00
Apple 55.0 °Brix	−6.07	896	94.0	0.00048	1.00
Grape 68.3 °Brix	−6.75	1,039	125.7	0.14110	1.00
Grape 64.5 °Brix	−6.65	1,041	115.1	0.02150	1.00
Grape 59.9 °Brix	−6.81	1,048	107.0	0.00080	1.00
Grape 54.0 °Brix	−6.56	1,085	84.5	0.00080	1.00

The magnitudes of the parameters of the Arrhenius and the Fulcher equations for the studied concentrated apple and grape juices are given in Tables 2.6 and 2.7, respectively. The physical interpretation of the three constants in the Fulcher equation is ambiguous, but by translating them in terms of the WLF parameters their significance can be clarified and it is functionally equivalent to the WLF equation (Ferry 1980; Soesanto and Williams 1981):

$$\log\left(\frac{\eta}{\rho T}\bigg/\frac{\eta_0}{\rho_0 T_0}\right) = -\frac{c_1^0\left(T-T_0\right)}{c_2^0+\left(T-T_0\right)} \tag{2.46}$$

Specifically, T_∞ and B' are related to c_1^0, c_2^0 and a reference temperature T_0, often called the glass transition temperature, by the following two equations:

$$T_\infty = T_0 - c_2^0 \tag{2.47}$$

$$B' = \frac{c_1^0}{\left(T_0-T_\infty\right)} \tag{2.48}$$

However, it should be pointed out that Soesanto and Williams (1981) also determined the values of the glass transition temperature of the very high concentration sugar solutions (91.9–97.6% by weight) by regression techniques and not experimentally.

The WLF equation was also used to correlate viscosity versus temperature data on honeys (Al-Malah et al. 2001; Sopade et al. 2003). Because of the empirical nature of the Fulcher equation and the empirical origin of the WLF equation, their use with viscosity data of relatively dilute fruit juices serves mainly the objective of obtaining a useful correlation.

Combined Effect of Concentration and Temperature

The flow behavior index (n) is assumed to be relatively constant with temperature and concentration, and the combined effects of temperature and concentration on the power law consistency index, K, are described by:

$$K = Ac^{b1} \exp(E_a/RT) \tag{2.49}$$

Tomato concentrates and concentrated milk samples are examples of foods in this category. Alternatively, the combined effects of temperature and concentration on the power law consistency index, K, are described by:

$$K = A' \exp\left(\frac{E_{\mathrm{a}}}{RT} + bc\right) \tag{2.50}$$

Mixing Rules for two Component Blends

Earlier, the role of suspended particles in fluid media in increasing the viscosity of a suspension and different equations relating the volume fraction of solids were discussed. For general two component or polymer composites, the Takayanagi models (Ross-Murphy 1984; Sperling 1986) provide means for calculating the upper (so called isostrain) and lower (so called isostress) limits of values for the shear modulus G_{c} of a composite formed from components x and y with shear moduli G_x and G_y, respectively. In the former, the polymers are arranged in parallel with respect to deformation, while in the latter they are arranged in series. In the parallel arrangement, the deformation of the weaker component is limited by the modulus of the stronger material and both components are deformed to the same extent (isostrain). In the series arrangement, the strength of the weaker component limits the force transmitted to the stronger material; therefore, both are subjected to the same stress (isostress).

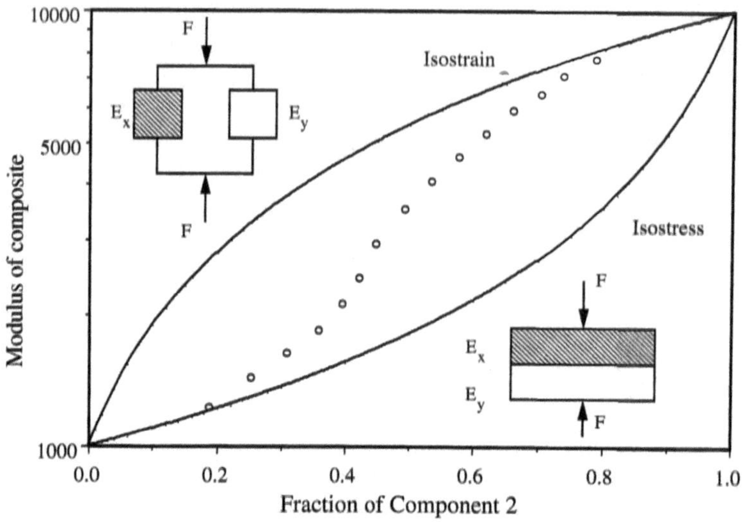

Fig. 2.17 Modulus of the composite gel (G_c) Plotted against volume fraction of component y ($G=10{,}000$). When the weaker component (x, $G=1{,}000$ Pa) dominates and becomes the continuous phase, G_c follows the lower bound isostress limit, with increasing fraction of y, there will be a phase inversion and G_c reaches the upper bound limit, path indicated by open circles

If ϕ_x and $\phi_y = 1-\phi_x$ are the respective volume fractions of the two components, the equations for the upper and lower bound values are:

$$G_c = \phi_x G_x + \phi_y G_y \quad \text{(upper bound, isostrain)} \tag{2.51}$$

$$(1/G_c) = (\phi_x/G_x) + (\phi_y/G_y) \quad \text{(lower bound, isostress)} \tag{2.52}$$

Values of G_c predicted by the above two models are shown in Fig. 2.17. Implicit in the above models is that the experimental value of G_c lies between those of G_x and G_y. For a simple phase-separated system, the lower bound form (isostress) should predict G_c prior to phase inversion when the supporting phase (x) is the weaker one. It will then show a transition to the upper bound behavior. More complex models can be found in Sperling (1986).

The Takayanagi models have been used to better understand the rheological behavior of a starch-soyprotein system (Chap. 4), tomato paste (Chap. 5), and mixed gel systems (Chap. 6). However, given that most foods contain several major components ($\tilde{c}$), the number and distribution of phases (P) is much more complex as seen from the phase rule:

$$\text{P} = (\tilde{c} - \tilde{f} + 2) \tag{2.53}$$

where, $\tilde{f}$ is the number of degrees of freedom.

Treatment of Rheological Data Using Models

Prior to using a model for description of rheological data, it would be desirable tocritically examine a number of issues, such as:

1. How reliable are the data? Are the data reproducible? Are the measurement techniques reliable? Are the test samples reproducible, reliable, and suitable for rheological measurements?
2. Plot the data (σ—$\dot{\gamma}$)-look for trends in data and artifacts of rheometer. Sometimes, most, but not all, of the data can be used for subsequent analysis. If the data are not reliable, why fit models?
3. Rheological model: (a) Is it appropriate for the experimental data? (b) How reliable is the model parameter estimation software? Has the reliability of the software been checked? (c) What are you looking for in the data (e.g., effect of temp.)? (d) Compare experimental data with a model's predictions because R^2 values often do not indicate the appropriateness of the model for the selected data!

References

Abdel-Khalik, S. I., Hassager, O., and Bird, R. B. 1974. Prediction of melt elasticity from viscosity data. *Polymer Eng. and Sci.* 14: 859–867.

Agarwala, M. K., Patterson, B. R., and Clark, P. E. 1992. Rheological behavior of powder injection molding model slurries. *J. Rheol.* 36: 319–334.

Aguilera, J. M. and Kessler, H. G. 1989. Properties of mixed and filled dairy gels. *J. Food Sei.* 54: 1213–1217, 1221.

Al-Malah, K.-I.-M., Abu-Jdayil, B., Zaitoun, S., and Al-Majeed-Ghzawi, A. 2001. Application of WLF and Arrhenius kinetics to rheology of selected dark-colored honey. *J. Food Process Eng.* 24(5): 341–357.

Barnes, H. A. and Walters, K. 1989. The yield stress myth? *Rheol. Acta* 24: 323–326.

Barnes, H. A., Hutton, J. F., and Walters, K. 1989. An Introduction to Rheology, Elsevier Science Publishers B.V., Amsterdam, The Netherlands.

Bird, R. B., Dai, G. C., and Yarusso, B. J. 1982. The rheology and flow of viscoplastic materials. *Rev. Chem. Eng.* 1: 1–70.

Brodkey, R. S. 1967. The Phenomena of Fluid Motions, Addison-Wesley, Reading, MA.

Casson, N. 1959. A flow equation for pigment-oil suspensions of the printing ink type, in Rheology of Disperse Systems, ed. C. C. Mill, pp. 82–104, Pergamon Press, New York.

Choi, G. R. and Krieger, I. M. 1986. Rheological studies on sterically stabilized model dispersions of uniform colloidal spheres II. Steady-shear viscosity. *J. Colloid Interface Sei.* 113: 101–113.

Cross, M. M. 1965. Rheology of non-Newtonian fluids: a new flow equation for pseudoplastic systems. *J. Colloid Sci.* 20: 417–437.

Da Silva, P. M. S., Oliveira, J. C., and Rao, M. A. 1997. The effect of granule size distribution on the rheological behavior of heated modified and unmodified maize starch dispersions. *J. Texture Stud.* 28: 123–138.

Demetriades, K., Coupland, J., and McClements, D. J. 1996. Investigation of emulsion stability using ultrasonic spectroscopy, in 1996 IFT Annual Meeting Book of Abstracts, pp. 109–110, Institute of Food Technologists, Chicago, IL.

Demetriades, K., Coupland, J., and McClements, D. J. 1997. Physical properties of whey protein stabilized emulsions as related to pH and NaCI. *J. Food Sei.* 62: 342–347.

Dervisoglu, M. and Kokini, J. L. 1986. Steady shear rheology and fluid mechanics of four semi-solid foods. *J. Food Sei.* 51: 541–546, 625.

Dickinson, E. 1993. Towards more natural emulsifiers. Trends Food *Sei.* Technol. 4: 330–334.

Dickinson, E. 2006. Structure formation in casein-based gels, foams, and emulsions. *Colloids Surf A* 288:3–11.

Dickinson, E. and Pawlowsky, K. 1996. Effect of high-pressure treatment of protein on the rheology of flocculated emulsions containing protein and polysaccharide. *J. Agric. Food Chem.* 44: 2992–3000.

Dickinson, E. and Yamamoto, Y. 1996a. Viscoelastic properties of heat-set whey protein-stabilized emulsion gels with added lecithin. *J. Food Sei.* 61:811–816.

Dickinson, E. and Yamamoto, Y. 1996b. Effect of lecithin on the viscoelastic properties of β-lactoglobulin-stabilized emulsion gels. *Food Hydrocollids* 10: 301–307.

Dickinson, E., Hong, S.-T., and Yamamoto, Y. 1996. Rheology of heat-set emulsion gels containing beta-lactoglobulin and small-molecule surfactants. *Neth. Milk Dairy J.* 50: 199–207.

Einstein, A. 1906. Eine neue bestimmung der molekuldimension. *Ann. Physik* 19: 289–306.

Einstein, A. 1911. Berichtigung zu meiner arbeit: Eine neue bestimmung der molekuldimension. *Ann. Physik* 34: 591–592.

Ellis, H. S., Ring, S. G., and Whittam, M. A. 1989. A comparison of the viscous behavior of wheat and maize starch pastes. *J. Cereal Sei.* 10: 33–44.

Fang, T. N., Tiu, C., Wu, X., and Dong, S. 1996. Rheological behaviour of cocoa dispersions. *J. Texture Stud.* 26: 203–215.

Fang, T., Zhang, H., Hsieh, T. T., and Tiu, C. 1997. Rheological behavior of cocoa dispersions with cocoa butter replacers. *J. Texture Stud.* 27: 11–26.

Farrer, D. and Lips, A. 1999. On the self-assembly of sodium caseinate. *Int Dairy J* 9:281–6.

Ferrer, M. L., Duchowicz, R., Carrasco, B., Torre, J. G., Acuña, A. U. 2001. The conformation of serum albumin in solution: A combined phosphorescence depolarization-hydrodynamic modeling study. *Biophysical Journal* 80:2422–2430.

Ferry, J. D. 1980. Viscoelastic Properties of Polymers, John Wiley, New York.

Genovese, D. B. and Rao, M. A. 2003. Role of starch granule characteristics (volume fraction, rigidity, and fractal dimension) on rheology of starch dispersions with and without amylose. *Cereal Chem.* 80: 350–355.

Giboreau, A., Cuvelier, G., and Launay, B. 1994. Rheological behavior of three biopolymer/water systems with emphasis on yield stress and viscoelastic properties. *J. Texture Stud.* 25: 119–137.

Hiemenz, R C. and Rjagopalan, R. 1997. Principles of Colloid and Surface Chemistry, 3rd ed., Marcel Dekker, Inc., New York.

Holdsworth, S. D. 1971. Applicability of rheological models to the interpretation of flow and processing behavior of fluid food products. *J. Texture Stud.* 4: 393–418.

Holdsworth, S. D. 1993. Rheological models used for the prediction of the flow properties of food products: a literature review. *Trans. Inst. Chem. Engineers* 71, Part C: 139–179.

Jacon, S. A., Rao, M. A., Cooley, H. J., and Walter, R. H. 1993. The isolation and characterization of a water extract of konjac flour gum. *Carbohydr. Polym.* 20: 35–41.

Jinescu, V. V. 1974. The rheology of suspensions. *Int. Chem. Eng.* 143: 397–420.

Kimball, L. B. and Kertesz, Z. I. 1952. Practical determination of size distribution of suspended particles in macerated tomato products. *Food Technol.* 6: 68–71.

Kitano, T., Kataoka, T., and Shirota, T. 1981. An empirical equation of the relative viscosity of polymer melts filled with various inorganic fillers. *Rheol. Acta* 20: 207–209.

Krieger, I. J. 1985. Rheology of polymer colloids, in Polymer Colloids, eds. R. Buscall, T. Corner, and J. F. Stageman, pp. 219–246, Elsevier Applied Science, New York.

Krieger, I. M. and Dougherty, T. J. 1959. A mechanism for non-Newtonian flow in suspensions of rigid spheres. *Trans. Soc. Rheol.* 3: 137–152.

Launay, B., Doublier, J. L., and Cuvelier, G. 1986. Flow properties of aqueous solutions and dispersions of polysaccharides, in Functional Properties of Food Macromolecules, eds. J. R. Mitchell and D. A. Ledward, pp. 1–78, Elsevier Applied Science Publishers, London.

Lopes da Silva, J. A. L., Gonpalves, M. P., and Rao, M. A. 1992. Rheological properties of high-methoxyl pectin and locust bean gum solutions in steady shear. *J. Food Sei.* 57: 443–448.

Loveday, S. M., Creamer, L. K., Singh, H. and Rao, M. A. 2007. Phase and rheological behavior of high-concentration colloidal hard-sphere and protein dispersions. *J Food Sci.* 72(7): R101–107.

Loveday, S. M., Rao, M. A., Creamer, L. K. and Singh, H. 2010. Rheological behavior of high-concentration sodium caseinate dispersions. *J Food Sci.* 74:N30–N35.

Lucey, J. A., Srinivasan, M., Singh, H. and Munro, P. A. 2000. Characterisation of commercial and experimental sodium caseinates by multi-angle laser light scattering and size-exclusion chromatography. *J Agric Food Chem* 48:1610–6.

McClements, D. J., Monaham, F. J., and Kinsella, J. E. 1993. Effect of emulsion droplets on the rheology of whey protein isolate gels. *J. Texture Stud.* 24: 411–422.

Metz, B., Kossen, N. W. F., and van Suijdam, J. C. 1979. The rheology of mould suspensions, in Advances in Biochemical Engineering, eds. T. K. Ghose, A. Fiechter and N. Blakebrough, Vol. 2, p. 103, Springer Verlag, New York.

Metzner, A. B. 1985. Rheology of suspensions in polymeric liquids. *J. Rheol.* 29: 739–775.

Mizrahi, S. and Berk, Z. 1972. Flow behaviour of concentrated orange juice: mathematical treatment. *J. Texture Stud.* 3: 69–79.

Moore, W. J. 1972. Physical Chemistry, 4th ed., Prentice Hall, Inc., Englewood Cliffs, New Jersey.

Noel, T. R., Ring, S. G., and Whittam, M. A. 1993. Physical properties of starch products: structure and function, in Food Colloids and Polymers: Stability and Mechanical Properties, eds. E. Dickinson and P. Wolstra, pp. 126–137. Royal Soc. Chem., Cambridge, UK.

Ofoli, R. Y., Morgan, R. G., and Steffe, J. F. 1987. A generalized rheological model for inelastic fluid foods. *J. Texture Stud.* 18: 213–230.

Panouille, M., Benyahia, L., Durand, D. and Nicolai, T. 2005. Dynamic mechanical properties of suspensions of micellar casein particles. *J Colloid Interface Sci* 287:468–75.

Paredes, M. D. C., Rao, M. A., and Bourne, M. C. 1988. Rheological characterization of salad dressings. 1. Steady shear, thixotropy and effect of temperature. *J. Texture Stud.* 19: 247–258.

Parkinson, C., Matsumoto, S., and Sherman, P. 1970. The influence of particle-size distribution on the apparent viscosity of non-Newtonian dispersed system. *J. Colloid Interface Sei.* 33: 150–160.

Pham, K. N., Puertas, A. M., Bergenholtz, J., Egelhaaf, S. U., Moussaid, A., Pusey, P. N., Schofield, A. B., and Cates, M. E. 2002. Multiple glassy states in a simple model system. *Science* 296: 104–106.

Phan, S-E., Russel, W. B., Cheng, Z., Zhu, J., Chaikin, P. M., Dunsmuir, J. H. and Ottewill, R. H. 1996. Phase transition, equation of state, and limiting shear viscosities of hard sphere dispersions. *Physical Review E* 54:6633–6645.

Pitkowski, A., Durand, D. and Nicolai, T. 2008. Structure and dynamical properties of suspensions of sodium caseinate. *J. Colloid Interface Sci.* 326:96–102.

Poslinski, A. J., Ryan, M. E., Gupta, R. K., Seshadri, S. G., and Frechette, F. J. 1988. Rheological behavior of filled polymeric systems I. Yield stress and shear-thinning effects. *J. Rheol.* 32: 703–735.

Pusey, P. N, and van Megen, W. 1986. Phase behaviour of concentrated suspensions of nearly hard colloidal spheres. *Nature* 320:340–342.

Quemada, D., Fland, P., and Jezequel, P. H. 1985. Rheological properties and flow of concentrated diperse media. *Chem. Eng. Comm.* 32:61–83.

Rao, M. A. 2007. Influence of food microstructure on food rheology, in Understanding And Controlling the Microstructure of Complex Foods, ed. D. J. McClements, Woodhead Publishing Ltd., Cambridge, UK.
Rao, M. A. and Cooley, H. J. 1983. Applicability of flow models with yield for tomato concentrates. *J. Food Process Eng.* 6:159–173.
Rao, M. A., Cooley, H. J., and Vitali, A. A. 1984. Flow properties of concentrated fruit juices at low temperatures. *Food Technology* 38(3): 113–119.
Rao, M. A., Shallenberger, R. S., and Cooley, H. J. 1986. Effect of temperature on viscosity of fluid foods with high sugar content, in Engineering and Food, eds. M. LeMaguer and P. Jelen, Vol. 1, pp. 23–31, Elsevier Applied Science Publishers, New York.
Rayment, P., Ross-Murphy, S. B., and Ellis, P. R. 1998. Rheological properties of guar galactomannan and rice starch mixtures. II. Creep measurements. *Carbohydr. Polym.* 35: 55–63.
Ross-Murphy, S.B. 1984. Rheological methods. In Biophysical Methods In Food Research, pp. 138–199, ed. H.W. Chan, Blackwell Scientific Publications, London.
Saunders, F. L. 1961. Rheological properties of monodisperse latex systems I. Concentration dependence of relative viscosity. *J. Colloid Sei.* 16: 13–22.
Servais, C., Ranc, H., and Roberts, I. D. 2004. Determination of chocolate viscosity. *J. Texture Stud.* 34(5–6): 467–498.
Shih, W.-H., Shih, W. Y., Kim, S.-I., Liu, J., and Aksay, I. A. 1990. Scaling behavior of the elastic properties of colloidal gels. *Phys. Rev.* A 42(8): 4772–4779.
Simoneau, C., McCarthy, M. J., and German, J. B. 1993. Magnetic resonance imaging and spectroscopy for food systems. *Food Res. Intern.* 26: 387–398.
Soesanto, T. and Williams, M. C. 1981. Volumetric interpretation of viscosity for concentrated and dilute sugar solutions. *J. Phys. Chem.* 85: 3338–3341.
Sopade, P.-A., Halley, P., Bhandari, B., D'Arcy, B., Doebler, C., and Caffin, N. 2003. Application of the Williams-Landel-Ferry model to the viscosity-temperature relationship of Australian honeys. *J. Food Eng.* 56(1): 67–75.
Sperling, L. H. 1986. Introduction to Physical Polymer Science, John Wiley, New York.
Steiner, E. H. 1958. A new rheological relationship to express the flow properties of melted chocolate. Rev. *Internationale de la Choeolatiere.* 13: 290–295.
Tattiyakul, J. 1997. Studies on granule growth kinetics and characteristics of tapioca starch dispersion during gelatinization using particle size analysis and rheological methods. M. S. thesis, Cornell University, Ithaca, NY.
Tattiyakul, J., Liao, H-J. and Rao, M.A. 2009. Role of structure in the measurement of flow properties of food and starch dispersions: a review. *International Journal of Food Properties* 12(1):2–10.
Tiu, C. and Boger, D. V. 1974. Complete rheological characterization of time-dependent food products. *J. Texture Stud.* 5: 329–338.
Tiu, C., Podolsak, A. K., Fang, T. N., and Watkins, J. B. 1992. Rheological behavior of water-creosote and creosote-water emulsions. *Rheol. Acta* 31: 381–389.
Tsai, S. C. and Zammouri, K. 1988. Role of interparticular van der Waals force in rheology of concentrated suspensions. *J. Rheol.* 32: 737–750.
Vitali, A. A. and Rao, M. A. 1984a. Flow properties of low-pulp concentrated orange juice: effect of temperature and concentration. *J. Food Sei.* 49: 882–888.
Vitali, A. A. and Rao, M. A. 1984b. Flow properties of low-pulp concentrated orange juice: Serum viscosity and effect of pulp content. *J. Food Sei.* 49: 876–881.
Vocadlo, J. J. and Moo Young, M. 1969. Rheological properties of some commercially available fats. *Can. Inst. Food Technol. J.* 2: 137–140.
Weltman, R.N. 1943. Breakdown of thixotropic structure as a function of time. *J. Appl. Phys.* 14: 343–350.
Wildemuth, C. R. and Williams, M. C. 1984. Viscosity of suspensions modeled with a shear-dependent maximum packing fraction. *Rheol. Acta* 23: 627–635.

Wu, H. and Morbidelli, M. 2001. A model relating structure of colloidal gels to their elastic properties. *LangmuirM*: 1030–1036.

Yoo, B. and Rao, M. A. 1994. Effect of unimodal particle size and pulp content on rheological properties of tomato puree. *J. Texture Stud.* 25: 421–436.

Yoo, B. and Rao, M. A. 1996. Creep and dynamic rheological behavior of tomato concentrates: effect of concentration and finisher screen size. *J. Texture Stud.* 27: 451–459.

Yoo, B., Rao, M. A., and Steffe, J. F. 1995. Yield stress of food suspensions with the vane method at controlled shear rate and shear stress. *J. Texture Stud.* 26: 1–10.

Chapter 3
Measurement of Flow and Viscoelastic Properties

Techniques for measuring rheological properties of polymeric materials have been well described previously by others (e.g., Whorlow 1980; Macosko 1994). The text by Van Wazer et al. (1963) is still a valuable reference that explains in detail many facets of earlier attempts to measure rheological properties of polymeric materials as well as basic equations of viscometric flows. The unique nature of fluid foods prompted this author to review both the rheological properties of fluid foods and their measurement about 30 years ago (Rao 1977a, b). Subsequent efforts on rheology of foods include those of Rao (1992, 2005) and Steffe (1996).

The study of the Newtonian and non-Newtonian flow behavior of foods requires considerable care and adequate instrumentation. Poorly designed instruments can provide data that can be misleading and of little value. A viscometer must be capable of providing readings that can be converted to shear rate ($\dot{\gamma}$) and shear stress (σ) in the proper units of s^{-1} and Pa, respectively. Furthermore, a well-designed instrument should allow for the recording of the readings so that time-dependent flow behavior can be studied. The shear stresses that need to be measured in the case of low-viscosity foods are low in magnitudes so that instruments that minimize friction by the use of gas bearings are very useful for such foods.

Ideally, for viscosity measurement, the flow in the selected geometry should be steady, laminar and fully developed, and the temperature of the test fluid should be maintained uniform. For obtaining data under isothermal conditions, the temperature of the sample should be maintained constant at the desired level. For foods that exhibit Newtonian behavior, viscometers that operate at a single shear rate (e.g., glass capillary) are acceptable. For foods that exhibit non-Newtonian behavior, data should be obtained at several shear rates and the commonly used viscometric flow geometries in rheological studies on foods include: (1) concentric cylinder, (2) plate and cone (cone-plate), (3) parallel disk (also called parallel plate), (4) capillary/tube/pipe, and (5) slit flow. The equations of continuity and motion can be applied to the aforementioned well-defined geometries. As shown later, a single cylinder rotating in an infinite fluid can also be used (Rao 1995), if the experiments are conducted properly.

The concepts of laminar and fully developed flow are frequently used in studies on flow and heat transfer. Both are best explained using flow in a straight tube as an

M. A. Rao, *Rheology of Fluid, Semisolid, and Solid Foods,* Food Engineering Series,
DOI 10.1007/978-1-4614-9230-6_3, © Springer Science+Business Media New York 2014

example. In steady fully developed flow, the radial velocity profile does not change along the length of the tube. Laminar flow means that all fluid elements are flowing parallel to each other and that there is no bulk mixing between adjacent elements. The criterion for laminar flow of a Newtonian fluid, with density ρ and viscosity η, and average velocity $\bar{v}$ in a tube is that the Reynolds number (Eq. 3.1), named after Osborne Reynolds, is <2,000

$$\frac{D\bar{v}\rho}{\eta} < 2{,}000 \tag{3.1}$$

Viscosity measurements are done under laminar flow conditions. Under conditions of turbulent flow of Newtonian fluids, the measured viscosity will be higher. In general, because non-Newtonian fluid foods are highly viscous, usually laminar flows are encountered.

A priori estimation of applicable shear rates for food processing application is also not easy because they depend upon both the properties of the food and the shearing conditions of a specific operation. Often, order of magnitude estimates of the shear rates are made assuming Newtonian flow behavior and the process equipment is designed conservatively. Sestak et al. (1983) presented characteristic ranges of shear rates for a few operations that should be, especially the very high shear rate values, used with considerable caution: coagulation tanks with agitators, 2–20 s^{-1}; helical channel extruder, 1,000–3,000 s^{-1}; and filling machine nozzle, 30–3,000 s^{-1}. For flow in pipes of circular cross-section, assuming laminar flow far from fittings, the shear rates for Newtonian fluids can be estimated from the expression: $(32Q/\pi D^3)$, where Q is the volumetric flow rate ($m^3\,s^{-1}$) of the food and D is the internal diameter of the tube (m). However, if the flow properties are being measured to correlate with human assessment of quality, Shama and Sherman (1973) showed that viscous foods are manipulated in the mouth at shear rates of about 10–30 s^{-1}, while low-viscosity foods are subjected to a shear stress of about 10 Pa that results in shear rates of about 100–200 s^{-1}.

The relationships for shear stress in capillary/tube and slit geometries can be deduced relatively easily from the pressure drop over a fixed length after fully developed flow has been achieved. Similarly, in rotational viscometer geometries (concentric cylinder, cone-plate, parallel disk), shear stress can be calculated from the measured torque and the dimensions of the test geometry being used. In contrast to shear stress, the derivation of expressions for the shear rate requires solution of the continuity and momentum equations with the applicable boundary conditions. The derivation of equations for the shear rates in capillary, concentric cylinder, and cone-plate geometries is outlined in Appendix 3-A through 3-C. It will be shown that in a properly designed cone-plate geometry, the shear rate depends only on the rotational speed and not on the geometrical characteristics. In all other flow geometries, the dimensions of the measuring geometry (capillary, concentric cylinder, parallel disks) play important roles.

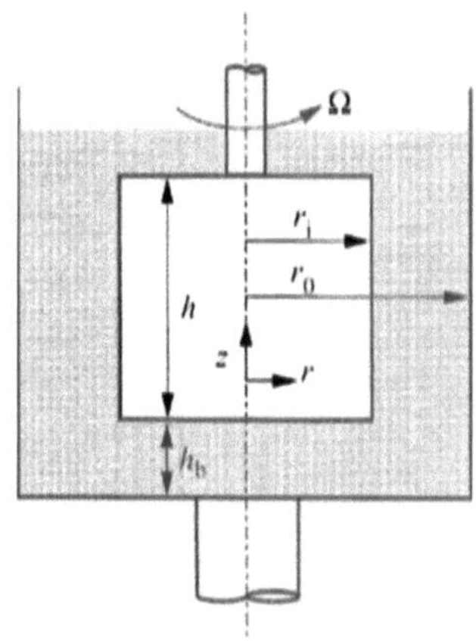

Fig. 3.1 Schematic diagram of a concentric cylinder geometry

Rotational Viscometers

In a rotational viscometer, the shear rate is derived from the rotational speed of a cylinder or a cone. Often, considerable effort is spent on obtaining low shear viscosity data, such as zero shear viscosity, that are difficult to obtain in view of the low shear rates needed, for example, $<0.01\ s^{-1}$. The maximum rotational speed that is used is left to the discretion of the operator. If the flow properties of the food are needed to design for processing operations, it is necessary to use shear rates that span the range expected to be used in the process and these may be quite high.

Concentric Cylinder Viscometer

In this geometry, a cylinder (bob) is placed concentrically (coaxially) inside a cup containing a selected volume of the test fluid (Fig. 3.1). It is easy to imagine that the process of loading a sample into the cup and insertion of the bob will either destroy or alter the structure of the test fluid. Also, it may require considerable effort to insert a bob into a cup containing a very high-viscosity material, for example, a dough. The bob is rotated and the drag of the fluid on the bob is measured by means of a torque sensor. In the Couette geometry, the cup is rotated and the main advantage is that higher shear rates can be obtained prior to the onset of turbulence due to Taylor vortices. By changing the rotational speed (shear rate) and measuring the resulting shear stress, it is possible to obtain viscosity data over a wide range of shearing conditions. In automated viscometers, the bob is programmed for an increase-in-speed ramp up to a predetermined shear rate and a decrease-in-speed ramp back to zero speed. In controlled stress instruments, the variable controlled is the stress (torque) and the rotational speed is the response measured. Today, many rotational viscometers are available commercially whose data collection function, and calculation of shear rate–shear stress data and parameters of rheological models (e.g., power law model) are performed by dedicated computers using software provided by the viscometer manufacturers. Nevertheless, it is instructive to know the applicable equations for the different geometries.

In a concentric cylinder geometry, the shear stress can be determined from the total torque (M)

$$\sigma = \frac{M}{2\pi r_{\mathrm{i}}^2 h} \tag{3.2}$$

where r_{i} is the radius of the rotating bob.

Newtonian Shear Rate

The Newtonian shear rate in a concentric cylinder geometry, $\dot{\gamma}_{\mathrm{N}}$, can be calculated exactly from the expression (see Appendix 3-A)

$$\dot{\gamma}_{\mathrm{N}} = \frac{2\Omega}{[1-(r_{\mathrm{i}}/r_{\mathrm{o}})^2]} \tag{3.3}$$

where Ω is the angular velocity of the rotating bob, r_{i} is the radius of the bob, and r_{o} is the radius of the cup. Recalling that viscosity is obtained by dividing the magnitude of shear stress by the corresponding shear rate, the viscosity of a Newtonian fluid from concentric cylinder flow data is given by the Margules equation

$$\eta = \left(\frac{M}{4\pi h\Omega}\right)\left(\frac{1}{r_i^2} - \frac{1}{r_0^2}\right) \tag{3.4}$$

where η is viscosity, M is the torque on the bob, Ω is the angular velocity of the rotating bob, h is the length of the bob in contact with the fluid, r_{i} is the radius of the bob, and r_{o} is the radius of the cup. For a given instrument with a given geometry, this equation reduces to

$$\eta = CM/\Omega \tag{3.5}$$

where C is the instrument constant $= 1/[(4\pi h)\{(1/r_{\mathrm{i}}^2)-(1/r_0^2)\}]$.

The instrument constant provided by manufacturers is usually determined using Newtonian fluids. Although viscometer manufacturers usually provide conversion factors to calculate the torque values for each bob that accompanies their instrument, their validity should be checked (e.g., by suspending precision weights from a bob positioned horizontally) periodically due to changes in response of torque transducers.

Krieger Method

In contrast to a cone and plate geometry to be discussed next, the shear rate of non-Newtonian foods cannot be determined from a simple expression involving the angular velocity and often one must use a suitable relationship between rotational speed and shear stress to correct for non-Newtonian behavior. More complex equations are needed to describe the flow of non-Newtonian fluids in concentric cylinder geometry. For example, for fluids that can be described by the power law model, an expression presented by Krieger and Elrod (Van Wazer et al. 1963) has been used extensively in the literature

$$\dot{\gamma} = \frac{\Omega}{\ln \varepsilon}\left[1 + \ln \varepsilon \frac{d(\ln \Omega)}{d(\ln \sigma_i)} + \frac{(\ln \varepsilon)}{3\Omega}\frac{d^2\Omega}{d(\ln \sigma_i)^2}\right] \tag{3.6}$$

where Ω is the angular velocity of the rotating bob, $\dot{\gamma}$ is the shear rate at the bob, σ_i is the shear stress at the bob, and ε = (cup radius/bob radius).

Power Law Fluids

For fluids that can be described by the power law model, an expression that provides the influence of both the deviation from Newtonian behavior of the fluid and that of the cup to bob ratio is (McKelvey 1962; Brodkey 1967)

$$\dot{\gamma} = \frac{2\Omega r_i^2}{\left[1-\left(\frac{r_i}{r_0}\right)^2\right]}\left\{\frac{1-\left(\frac{r_i}{r_o}\right)}{n\left[1-\left(\frac{r_i}{r_0}\right)\right]^{2/n}}\right\} \tag{3.7}$$

From Eq. 3.7, as shown in Table 3.1, it can be deduced that for non-Newtonian foods the correction term to Newtonian shear rate depends on the extent to which a fluid deviates from Newtonian behavior and the size of gap between the inner and outer cylinders. To minimize errors in the calculated shear rates, it would be preferable to employ concentric cylinders with narrow gap between them. Therefore, some of the commercially available units have the ratio of the radii (r_i/r_o) of about 0.95. However, it is not obvious that the software provided by viscometer manufacturers contains corrections for non-Newtonian flow behavior, so that, whenever possible, it is advisable to use narrow gap concentric cylinder systems for fluids that deviate considerably from Newtonian behavior.

For a single cylinder rotating in an infinite fluid (e.g., large container), assuming the shear rate far from the rotating cylinder, $\dot{\gamma}$ is zero in Eq. 3A.12, the expression for the shear rate is

Table 3.1 Values of shear rate correction factors for concentric cylinder geometry

(rj/ro)	$XX = \left\{1-\left(\frac{r_i}{r_o}\right)^2\right\}$	Value of n	$YY = n\left\{1-\left(\frac{r_i}{r_o}\right)^{2/n}\right\}$	Correction Factor $\frac{XX}{YY}$
0.95	0.0975	0.20	0.0803	1.215
0.95	0.0975	0.50	0.0927	1.051
0.95	0.0975	1.00	0.0975	1.000
0.95	0.0975	1.20	0.0983	0.992
0.95	0.0975	2.00	0.1000	0.975
0.75	0.4375	0.20	0.1887	2.318
0.75	0.4375	0.50	0.3418	1.280
0.75	0.4375	1.00	0.4375	1.000
0.75	0.4375	1.20	0.4571	0.957
0.75	0.4375	2.00	0.5000	0.875
0.50	0.7500	0.20	0.1998	3.754
0.50	0.7500	0.50	0.4688	1.600
0.50	0.7500	1.00	0.7500	1.000
0.50	0.7500	1.20	0.8220	0.912
0.50	0.7500	2.00	1.0000	0.750

$$2\sigma_i \frac{d\Omega_i}{d\sigma_i} = \dot{\gamma}_i \tag{3.8}$$

Equation 3.8 can be used when only a single cylinder system is available instead of concentric cylinder system, but care should be exercised in making sure that the sample characteristics are not changed during collection of the experimental data, such as settling of solids of a dispersion. For a test sample, angular velocity versus shear stress data should be obtained from which the magnitude of $(d\Omega_i/d\sigma_i)$ is calculated and substituted in Eq. 3.8.

Other expressions for concentric cylinder geometry include that for Bingham plastic fluids where the yield stress must be taken into account which leads to the Reiner-Riwlin equation

$$\Omega = \frac{M}{4\pi h\eta'}\left(\frac{1}{r_i^2} - \frac{1}{r_0^2}\right) + \frac{\sigma_0}{\eta'}\ln\left(\frac{r_0}{r_i}\right) \tag{3.9}$$

where σ_0 is yield stress and η' is the plastic viscosity. Equations for the shear rate at the rotating bob can be also derived for fluids that follow other rheological models.

Steiner's Method

For foods, such as chocolates, that can be described by the Casson model (Eq. 2.5), Steiner (1958) chose not to develop an explicit expression for the non-Newtonian shear rate, but related $\dot{\gamma}_N$ to the Casson model parameters K_{0c}, K_c, and shear stress,

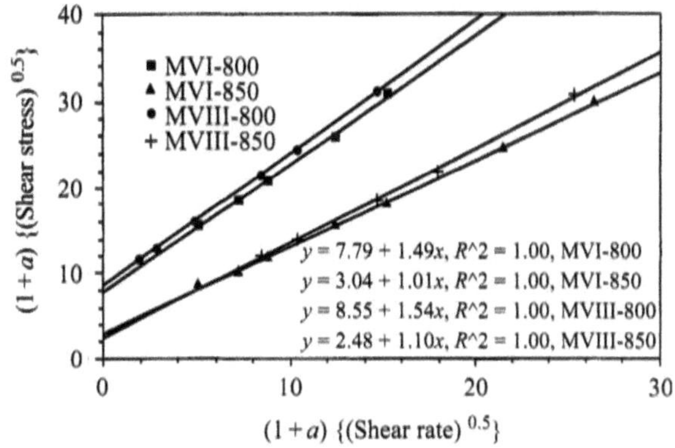

Fig. 3.2 Steiner plot for flow data on a chocolate sample

Table 3.2 Casson yield stress, $\sigma_{0c}=(K_{0c})^2$ and plastic viscosity, $\eta c_a=(K_c)^2$, values of chocolate samples with couvertures 800 and 850 determined with two rotovisco concentric cylinder systems MVI, $(r_i/r_o)=0.95$ and MVIII, $(r_i/r_o)=0.72$. (Steiner 1958)

Measuring system and sample	Intercept = $2K_{0c}$	σ_{OC} (Pa)	Slope = K_c	η_{Ca} (Pa s)
MVI, 800	7.79	15.2	1.49	2.22
MVI, 850	3.04	2.3	1.01	1.02
MVII, 800	8.55	18.3	1.54	2.37
MVIII, 850	2.48	1.5	1.10	1.21

σ. Steiner's approach is valid for values of $a=(r_i/r_o)$ between 0.5 and 0.9, and $\dot{\gamma}_N$ values greater than 0.1 s^{-1} when $a=0.9$ and 0.01 s^{-1} when $a=0.5$.

$$\left\{(1+a)\sqrt{\sigma}\right\} = K_c\left\{(1+a)\sqrt{\dot{\gamma}_N}\right\} + 2K_{0c} \tag{3.10}$$

The limited restrictions on the ranges of shear rate and the values of the ratio (r_i/r_o) for using Eq. 3.10 are due to retaining only the first term in the series expansion of $\ln(r_i/r_o)$ and neglecting the higher-order terms. From a plot $\{(1+a)\sqrt{\dot{\gamma}_N}\}$ (abscissa) versus $\{(1+a)\sqrt{\sigma}\}$ (ordinate; Fig. 3.2), where $a=(r_i/r_o)$, the Casson model parameters can be calculated from the slope (m) $K_c=m$ and intercept (b) $K_{0c}=b/2$. It will be recalled that the Casson yield stress is calculated as $\sigma_{0c}=(K_{0c})^2$ and the Casson plastic viscosity as $\eta c_a=(K_c)^2$. Values of σ_{0c} and ηc_a (Table 3.2) of two chocolate samples whose viscosity data were obtained by Steiner (1958) with two concentric cylinder systems: MVI, $(r_i/r_o)=0.95$ and MVIII, $(r_i/r_o)=0.72$ were determined by plotting $\{(1+a)\sqrt{\dot{\gamma}_N}\}$ versus $\{(1+a)\sqrt{\sigma}\}$ (Fig. 3.2). Magnitudes of the Casson parameters of the two chocolate samples determined with the two measuring systems are in reasonable agreement with each other. Steiner's approach to fluids that follow the Casson model is yet another example of the need to correct Newtonian shear rate data obtained with concentric cylinder systems.

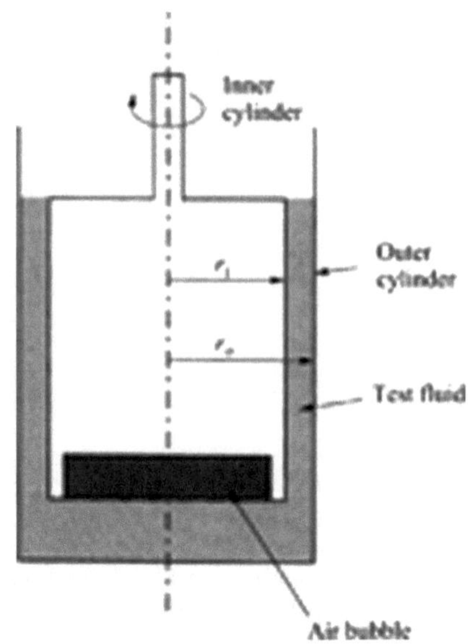

Fig. 3.3 Schematic of a concentric cylinder geometry air bubble to minimize shear at the bottom of a rotating cylinder

Data points shown were recalculated from original data of Steiner (1958). MVI and MVII indicate concentric cylinders with different dimensions, and 800 and 850 indicate different chocolate samples.

A number of concentric cylinder design factors affect the rheological data obtained and these are discussed next.

End Effects in Concentric Cylinder Systems

In Eq. 3.2, the measured torque should be only due to the lateral surface ($2\pi r_i h$) of a vertical cylinder. Therefore, any significant drag on the ends of the rotating cylinder should be either minimized by design, for example, by providing an interface of air (see Fig. 3.3), or corrected for the influence of the end. The correction for the drag on the bottom, the height equivalent to the end may be calculated by determination of the torque exerted by the rotating cylinder at different fill heights and extrapolation of the torque to zero height that is valid for a Newtonian fluid. Assuming torsional flow between parallel disks (Macosko 1994), the extra torque (M_{bb}) contributed by the end can be estimated

$$M_{bb} = \frac{2\pi r_i^3 (1/n)}{3+n} \left(\frac{\Omega_i}{h_b} \right)^n \tag{3.11}$$

where h_b is the gap between the bottom of the cylinder (Fig. 3.3) and the cup, and n is the power law flow behavior index that can be obtained from a plot of log Ω, versus log M. The total torque due to the lateral surface of the bob and the bob's end is $M + M_{bb}$.

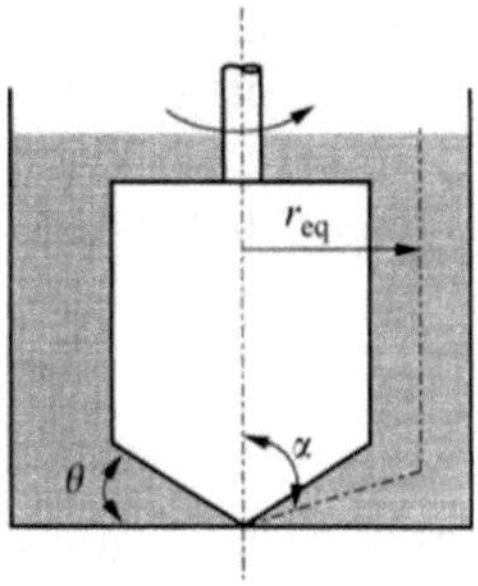

Fig. 3.4 Mooney–Ewart geometry to account for end effect in a concentric cylinder geometry

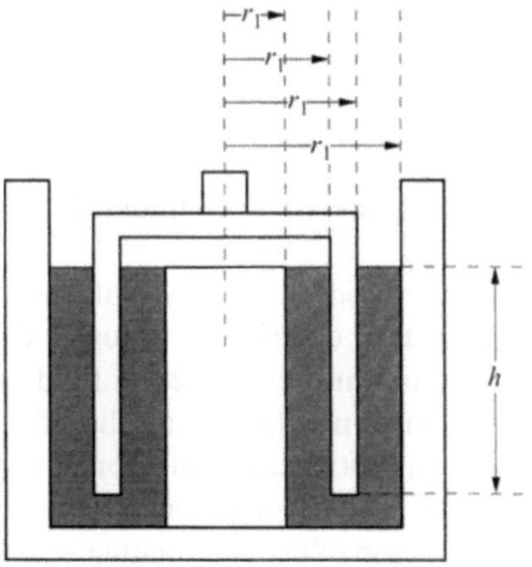

Fig. 3.5 Double concentric cylinder geometry for low-viscosity fluids

Another approach of accounting for end effect, called the Mooney–Ewart approach (Whorlow 1980), is to replace the end with a cone whose influence can be calculated (Fig. 3.4). Ignoring the discontinuity where the conical and cylindrical surfaces meet, the total torque is given in terms of the arithmetic mean stress, σ_m

$$M = 2\pi r_{eq}^2 \left(h + \frac{r_{eq}}{3\sin a}\right)\sigma m \tag{3.12}$$

The stress σ_m occurs at a radius r_{eq}

$$r_{eq}^2 = \frac{2r_i^2 r_0^2}{r_i^2 + r_0^2} \tag{3.13}$$

Double Concentric Cylinder System

For low-viscosity fluids, one manner of obtaining increasing the torque (stress) value is to increase the area of shear as shown in Fig. 3.5. Compared to the geometry shown in Fig. 3.1, in view of the thin wall of the inner component, the test sample is disturbed much less when the double concentric cylinder geometry is assembled for a rheological test and the end effects are negligible. The design is such that $(r_1/r_2)=(r_3/r_4)$. For the geometry shown in Fig. 3.5, the expression for the shear stress, analogous to Eq. 3.2 is

$$\sigma = \frac{M}{2\pi h(r_2^2 + r_3^2)} \tag{3.14}$$

Recognizing that the fluid is sheared on the surfaces with radii r_2 and r_3, one can write an equation analogous to Eq. 3.3

$$\dot{\gamma}_N = 2\Omega\left[\frac{1}{1-(r_1/r_2)^2} + \frac{1}{1-(r_3/r_4)^2}\right] \tag{3.15}$$

Slip in Concentric Cylinder Systems

In some foods, a thin layer of low-viscosity fluid forms at the solid–fluid interface that in turn contributes to lower viscosity values. The boundary condition that at the solid–fluid interface the fluid velocity is that of the wall is not satisfied. This phenomenon is known as slip effect. Mooney (1931) outlined the procedures for the quantitative determination of slip coefficients in capillary flow and in a Couette system. The development for the concentric cylinder system will be outlined here for the case of the bob rotating and details of the derivation can be found in Mooney (1931).

For a concentric cylinder system, the differential equation for angular velocity can be written as

$$\frac{d\omega}{dr} = \frac{T\varphi}{2\pi r^3} \tag{3.16}$$

Integration from the inner to the outer cylinder gives

$$\Omega = \frac{\beta(\sigma_1)T}{2\pi r_1^3} + \frac{\beta(\sigma_2)T}{2\pi r_2^3} + \int_{r_1}^{r_2}\left(\frac{T\varphi}{2\pi r^3}\right)dr \tag{3.17}$$

In the above equations, ω and Ω are the angular velocity of liquid and cylinder, respectively, β is the slip coefficient (defined as β = slip velocity/shearing stress), φ is the fluidity, and T is the torque per unit height (Mh^{-1}).

In a controlled shear rate viscometer such as the Haake Rotovisco, T is determined as a function of Ω, while in constant stress rheometers, such as the Carri-Med, Ω is determined as a function of T. In order to determine surface slip in a co-axial cylinder viscometer, one must vary the ratio r_1/r_2 as well as T and Ω (Mooney

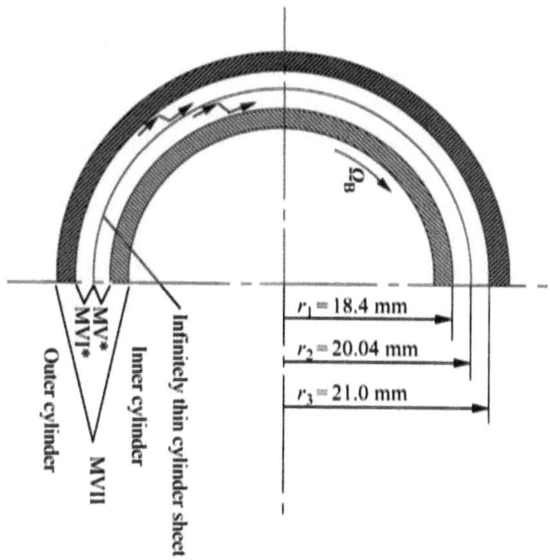

Fig. 3.6 Diameters of concentric cylinders used in a study of slip velocities using the method of Mooney. (Qiu and Rao 1989)

1931). Specifically, one can determine Ω at constant T using three combinations of cups and bobs of radii r_2/r_1, r_3/r_2, and r_3/r_1. The coefficient of slip is given by

$$\beta = \frac{2\pi r_2^3}{T}(N_{12} + N_{23} + N_{13}) \tag{3.18}$$

where N is the rotational speed for the appropriate concentric cylinder system. When the bob is rotating, as is the case in many rheometers, the slip velocities for the cup (v'_{cup}) and bob (v'_{bob}) are given by the equations

$$v'_{cup} = \beta\left[\frac{T}{2\pi r_{cup}^2 L}\right] \tag{3.19}$$

$$v'_{bob} = r_{bob}\Omega - \beta\left[\frac{T}{2\pi r_{bob}^2 L}\right] \tag{3.20}$$

The velocity difference across the test material is: $v'_{bob} - v'_{cup}$ and the ratio of this velocity to the velocity difference without slip $[(v'_{bob} - v'_{cup})/v_{bob}]$ is the factor used to correct the shear rate. It must be noted that Mooney (1931) also suggested that one may use only two concentric cylinder systems and this suggestion was interpreted in a quantitative manner by Yoshimura and Prud'homme (1988a, b) who derived the necessary equations for such systems.

To estimate slip velocities, the rotational speeds at specific values of torque for the three concentric cylinder units (Fig. 3.6) specified in the theory of Mooney (1931) were determined (Qiu and Rao 1989; Grikshtas and Rao 1993). The magnitudes of the radii of the three concentric cylinder units r_1, r_2, and r_3 were: 18.4,

20.04, and 21.0 mm, respectively. Experiments were conducted so that the rotational speeds N_{12}, N_{13}, and N_{23} were determined at the same magnitude of torque, where the subscripts 12, 13, and 23 denote the appropriate combination of cup and bob; for example, the subscript 12 denotes a bob of radius r_1 and a cup of radius r_2. The slip coefficient (β) and the slip velocity ratio (ν-ratio) were calculated using Eqs. 3.18, 3.19, and 3.20. The shear rates corrected for slip were calculated as the product of the ν-ratio and the uncorrected for slip shear rates calculated using Krieger and Elrod's equation (Eq. 3.6).

Grikshtas and Rao (1993) showed that Kiljanski's method for estimating slip velocities using two concentric cylinder units gave results comparable to that of Mooney. Yoshimura and Pru'dhomme (1988a, b) described procedures for estimating slip velocities in a parallel plate geometry.

Turbulent Flow

As stated earlier, rheological measurements should be conducted under laminar flow conditions and that turbulent flow conditions will result in higher viscosities. Thus, a sudden increase in viscosity with increase in rotational speed may be due to the onset of turbulent flow and may be erroneously attributed to shear-thickening behavior of the test fluid. In the design of concentric cylinder system, one has the option of rotating either the inner or the outer cylinder. In this respect, Taylor showed that when the Reynolds number reaches a certain value, vortices appear whose axes are located along the circumference and when the outer cylinder is rotated, the vortices appear at higher rotational speeds (Schlichting 1960). An equation for predicting the onset of vortex formation for Newtonian fluids credited to Ludwig Prandtl is (Schlichting 1960)

$$\frac{v_i w}{\upsilon}\sqrt{\frac{w}{r_i}} > 41.3 \tag{3.21}$$

where w is the width of the gap, v_i is the peripheral velocity of the inner cylinder, r_i is the radius of the inner cylinder, and $\upsilon = \eta\rho$ is the kinematic viscosity of the fluid.

Plate and Cone Geometry

The apex of a cone is brought into close proximity, but not into contact, of a horizontal plate (Fig. 3.7). Often, the apex is truncated slightly to eliminate a sharp point. The minimum gap between the cone and plate is usually of the order of 50 μm so that this geometry may not be suitable for dispersions containing larger diameter solids. The test fluid fills the gap between the cone and the plate, and because the gap is small, only a small volume (typically, 1–5 mL) of fluid is needed. The cone is rotated and the torque is measured at various speeds of rotation. A cone and plate viscometer can be used to obtain shear stress–shear rate curves and shear-stress

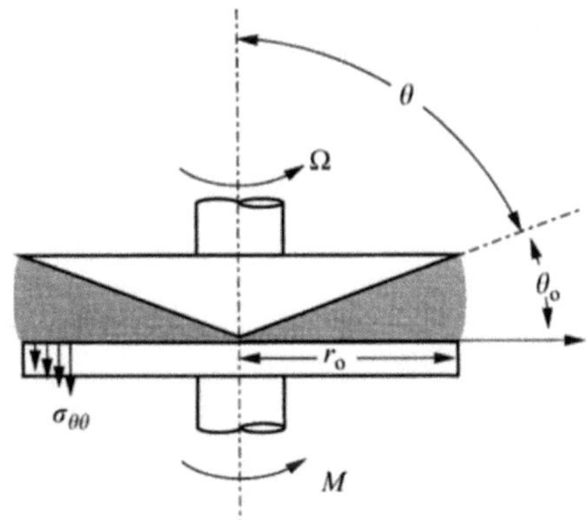

Fig. 3.7 Schematic diagram of a cone and plate geometry

versus time at constant shear rate curves as described earlier for concentric cylinder geometry. The shear rate and shear stress are given by the following equations:

$$\text{Shear stress}, \sigma_{\theta\phi} = \frac{3T_{\text{cn}}}{D} \tag{3.22}$$

$$\text{Shear rate}, \dot{\gamma} = \frac{\Omega}{\theta_0} \tag{3.23}$$

where T_{cn} is the torque per unit area, D is the diameter of the rotating cone or plate, Ω is the angular velocity, and θ_0 is the cone angle in radians. In general, θ_0 is usually quite small (2–4°). It is readily apparent from Eq. 3.22 that the magnitude of $\dot{\gamma}$ depends only on the angular velocity and the cone angle, so that it is uniform in a cone-plate geometry.

The concentric cylinder and the cone-plate geometries together with low-inertia rheometers are well suited for studying time-dependent viscous (e.g., thixotropic and antithixotropic) and viscoelastic (stress growth, creep, and dynamic) properties of foods. The alteration of the structure during sample loading is a concern in both geometries, but especially in the concentric cylinder system where typically the test sample is poured first into the cup and the rotating cylinder is inserted subsequently. At least, a reasonable period of time for recovery of sample structure should be given prior to rheological measurements. With some samples (e.g., gelatinized starch and other food gels), the rheological measurements can be conducted after the sample has been created in situ in the measuring system.

Parallel Disk Geometry

The parallel disk geometry, also called the parallel plate geometry, consists of two disks with radius r_0 separated by the gap h (Fig. 3.8). Assuming steady, laminar, and isothermal flow, the expression for shear rate is

Fig. 3.8 Schematic diagram of a parallel disk geometry

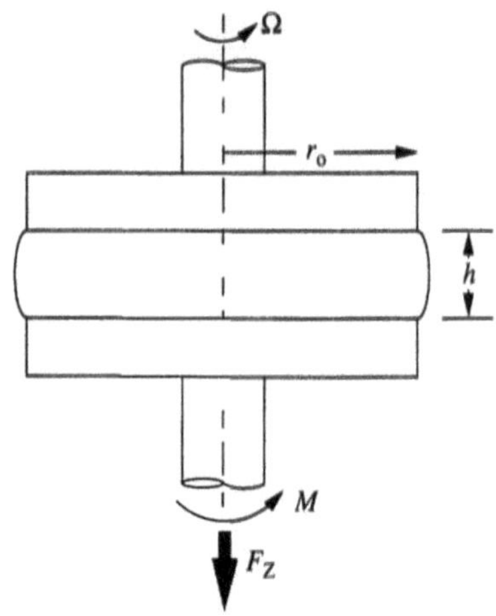

$$\dot{\gamma} = \frac{\Omega r_0}{h} \tag{3.24}$$

The shear stress can be determined from the measured torque, M

$$\sigma = \frac{3M}{2\pi r_0^3}\left[1 + \frac{1}{3}\frac{\mathrm{d}\ln M}{\mathrm{d}\ln \dot{\gamma}}\right] \tag{3.25}$$

where Ω is the angular velocity, r_0 is the disk radius, and h is the gap between the disks. Although the shear field is nonhomogeneous, the parallel disk geometry is useful in handling dispersions that contain relatively large-size particles, such as gelatinized starch dispersions (SDs). Yang and Rao (1998) used a parallel disk geometry to obtain rheological data during gelatinization of a corn SD. They found that the paraffin oil placed at the edge of the starch sample did not penetrate into the starch. The gap h between the disks can be selected to accommodate foods containing large particles and a rough guide for a suitable gap width is for $h \approx 10 \times$ particle diameter.

Mixer Viscometer

Some rotational viscometers employ a rotating disc, bar, paddle, or pin at a constant speed (or series of constant speeds). It is extremely difficult to obtain true shear stress, and the shear rate usually varies from point to point in the rotating member. In particular, the velocity field of a rotating disc geometry can be considerably distorted in viscoelastic fluids. Nevertheless, because they are simple to operate and give results easily, and their cost is low, they are widely used in the food industry. Although they may be useful for quality control purposes, especially Newtonian foods, the reliability of their values should be verified by comparison with data ob-

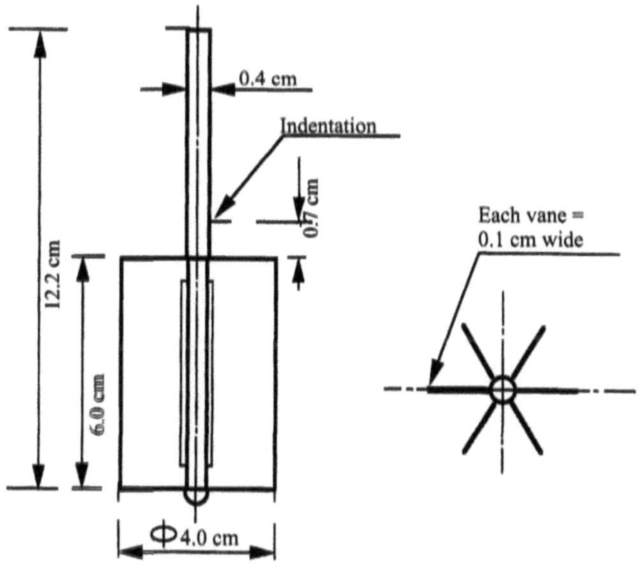

Fig. 3.9 Schematic diagram of a vane that can be used for obtaining yield stress, and shear rate versus shear stress data

tained with well-defined geometries (capillary/tube, concentric cylinder, and cone-plate).

Many foods are suspensions of solid matter in a continuous medium. In concentric cylinder and plate and cone viscometers, the solid matter may be larger than the gap between the stationary and the rotating parts, or it may settle due to agitation. In addition, a thin layer of the continuous phase may separate at the moving and stationary surfaces and this can induce errors in the measured values of shear rate. Foods that can pose problems during measurement with rotational viscometers include: mustard, tomato puree, apple sauce, and other suspensions. For these and other complex foods, approximate shear rate–shear stress data can be obtained by a technique using a rotating vane (Fig. 3.9) that minimizes settling and separation of the product. The technique was developed for the determination of power consumption during mixing of fluids (Metzner and Otto 1957) and adapted to studying rheology of fermentation broths (Bongenaar et al. 1973). It is based on the assumption that the average shear rate around the paddle is directly proportional to the rotational speed (N)

$$\dot{\gamma} = k_s N \tag{3.26}$$

where k_s is a constant that must be determined for each paddle (Rao 1975; Rao and Cooley 1984) and N is rotational speed (revolutions s^{-1}) of the paddle. The constant k_s can be determined from a semilogarithmic plot of ($1 - n$; abscissa) versus log[$P/(KN^{n+1}d^3)$] (ordinate; Rieger and Novak 1973), and the slope of the resulting line is equal to $-\log k_s$. For a given impeller, tests must be conducted such that the necessary data are obtained: P, the power (Nm s^{-1}); N, the rotational speed (s^{-1}); d the

Fig. 3.10 Plot of $(1-n;$ abscissa) versus log $[P/(KN^{n+1}d^{\beta})]$ for an eight-blade vane

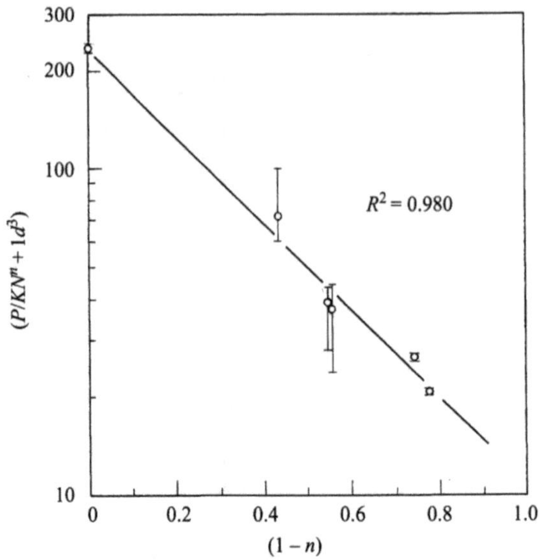

diameter of the impeller (m); and the power law parameters (K and n) of several test fluids so that a wide range of values of $(1-n)$ can be obtained. Figure 3.10 illustrates such a plot for an eight-blade impeller, $d=5.8$ cm and $h=4.4$ cm (not shown here). For geometries with large surface areas, the method of matching viscosities is less likely to contribute to errors. A value of k_s of about 19 was estimated from Figs. 3.9 and 3.10.

Magnitude of k_s can also be determined by matching the torque with a Newtonian fluid of a known viscosity to that with a non-Newtonian fluid (Metzner and Otto 1957; Rao and Cooley 1984). Wood and Goff (1973) applied the matching viscosity method to determine the effective shear rate in a Brabender Viscograph. The latter method was called (Rao and Cooley 1984) the Metzner-Otto-Wood-Goff (MOWG) method. Figure 3.11 illustrates the matching viscosity technique for the same impeller. Castell-Perez et al. (1987) and Briggs and Steffe (1996) employed the two techniques described earlier for small samples, while Lai et al. (2000) determined the value of k_s of the impeller–cup combination of the Rapid Visco Analyzer (RVA) to be 20.1.

Power Law Parameters Using a Mixer

For foods that follow the simple power law model, the flow behavior index is determined as the slope of a double logarithmic plot of log N (abscissa) versus log torque (ordinate). In order to determine the consistency index, magnitudes of torque must be determined in the same container for the test fluid and for a standard fluid whose rheological characteristics are known. Recognizing that the ratios of the torques will

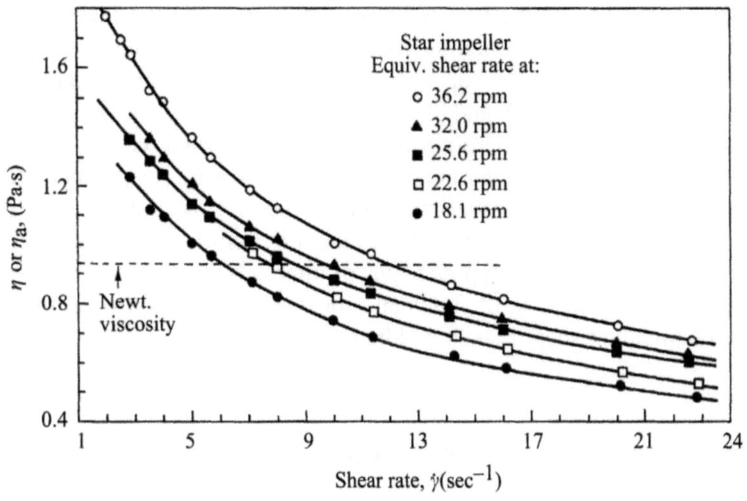

Fig. 3.11 Illustration of the matching viscosity technique for the eight-blade vane

be equal to the ratios of the shear stresses, one can write the algebraic expression for a test fluid (x) and the standard fluid (y)

$$\frac{T_x}{T_y} = \frac{\sigma_x}{\sigma_y} = \frac{K_x(k_s N_x)^{n_x}}{K_y(k_s N_y)^{n_y}} \tag{3.27}$$

where T_x and T_y are the magnitudes of torque measured with the test and the standard fluids, N_x and N_y are the rotational speeds employed with the test and standard fluids, K_x and K_y are the magnitudes of the consistency coefficient of the test and the standard fluids, and n_x and n_y are the magnitudes of the flow behavior index of the test and the standard fluids. Because all the terms other than K_x, the consistency index of the test fluid, are known, it can be calculated. For the special case of a Newtonian standard fluid, $N_y = 1$. Studies are needed to verify the assumption that the average shear rate of the paddle is directly proportional to the rotational speed. However, in the interim, for foods that cannot be studied with conventional viscometers, the careful use of vanes will enable the determination of the power law parameters that in turn can be used to calculate shear rates and shear stresses from the relationship

$$\sigma = K\dot{\gamma}^n \tag{2.3}$$

Flow Behavior Using a Vane

From numerical simulation of flow of power law fluids in a narrow-gap vane-in-cup geometry ($r_i = 8.5$ mm, $r_o = 9.5$ mm), Barnes and Carnali (1990) suggested that for shear-thinning fluids with the flow behavior index <0.5, the fluid within the blades is trapped and turns with the rotating vane. Thus, for a given rotational speed of the spindle, the shear stress at the cup wall is equal to that in conventional concentric cylinder viscometer geometry and identical flow curves would be obtained. Furthermore, significant and sudden drop in viscosity was observed with both geometries; compared with the concentric cylinder geometry, the shear rate at which the drop in viscosity occurred was higher with the vane geometry. It is emphasized that, at this time, the approximation to a concentric cylinder system is applicable to a vane-in-cup system and not for other types/shapes of impellers. In fact, some impellers are designed to promote mixing of fluids and for those impellers, the aforementioned method of equivalent shear rate should be used.

Yield Stress of Foods Using a Vane

Yield stress is an important rheological property in foods such as tomato catsup, tomato sauce, and melted chocolate. Experimental measurement of yield by the relaxation technique described by Van Wazer et al. (1963) is time consuming and the values are affected by prior shearing of sample. Yield stress values obtained by extrapolation of shear stress–shear rate data depend on the model used and prior shearing of the sample, so that different values of yield stress may be obtained for a given sample depending on the extent of alteration of structure. One manner of visualizing yield stress, shown in Fig. 3.12 (Keentok 1982), is that a sample with undisturbed structure has a high yield stress, called static yield stress. In contrast, a sample whose structure has been disturbed by shear has a lower magnitude of yield stress, called dynamic yield stress. The shearing of a sample may take place during loading of the sample into a measuring geometry and collection of rheological data.

A relatively simple technique using a vane with at least four blades was developed by Dzuy and Boger (1983, 1985). A six-blade vane (Fig. 3.9) was employed to examine yield stress of applesauce (Qiu and Rao 1988) and other pureed foods (Yoo et al. 1995). The studies of Keentok (1982) and James et al. (1987) also contain useful discussion on yield stress. Figure 3.13 illustrates the experimental setup used by Qiu and Rao (1988). In this technique, the maximum torque reading is determined at a low rotational speed (e.g., 0.4 rpm; Fig. 3.14). The vane yield stress, σ_{0v}, is then calculated from the equation

$$T_m = \frac{\pi D_V^3}{2}\left(\frac{H}{D_V}+\frac{1}{3}\right)\sigma_{0v} \tag{3.28}$$

where T_m is the maximum torque reading (N m), and D_v (m) is the diameter and H (m) the height of the paddle, respectively. The method is simple to use and consistent results were obtained for a wide range of food products, such as apple sauce

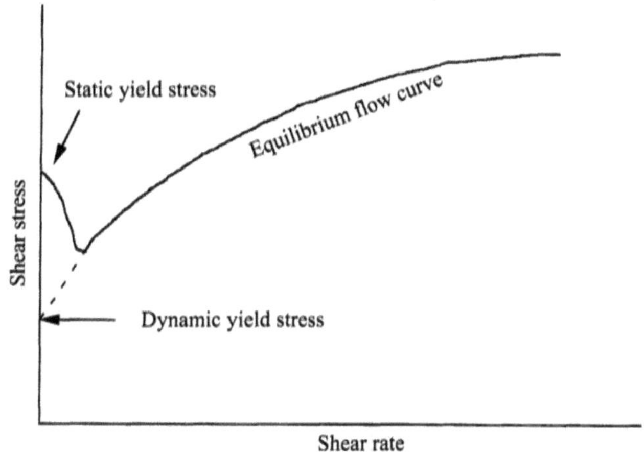

Fig. 3.12 Illustration of static and dynamic yield stress. (Keentok 1982)

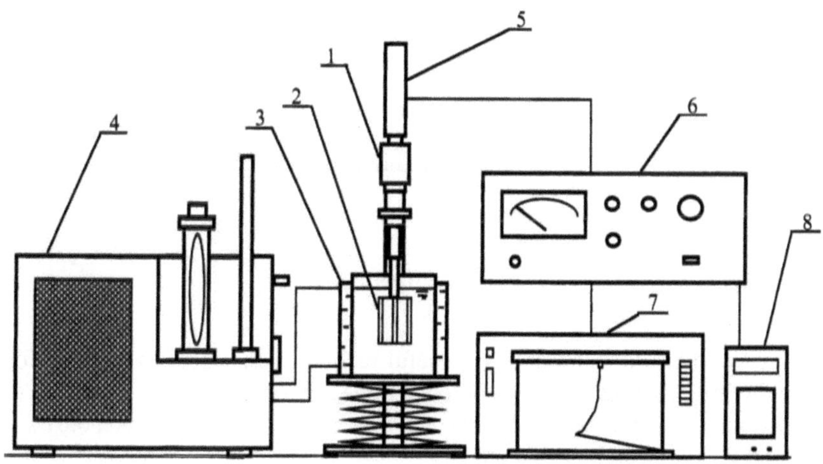

Fig. 3.13 Vane yield stress setup at controlled strain (Qiu and Rao 1988). The numbers denote separate pieces of equipment: 1—torque unit, 2—vane, 3—constant temperature vessel for sample, 4—constant temperature bath, 5—viscometer drive motor, 6—viscometer control panel, 7—chart recorder for torque output, and 8—voltmeter to monitor vane rpm signal.

(Qiu and Rao 1988), tomato purees (Yoo et al. 1995), and SDs (Genovese and Rao 2003a). In a study of the vane yield stress of tomato puree samples, a controlled shear rate viscometer and a controlled stress rheometer provided comparable values of yield stress (Yoo et al. 1995). However, in the controlled shear rate method, it is relatively easy to accurately determine the maximum torque; with the controlled stress rheometer, there is uncertainty in the determination of the representative T_m value (Fig. 3.15).

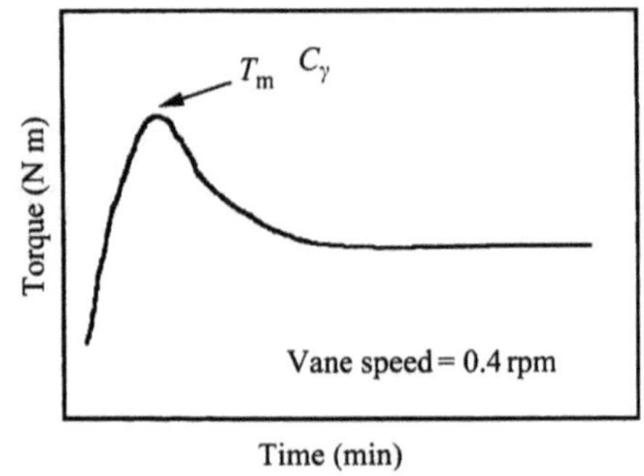

Fig. 3.14 Torque versus time plot with a vane in controlled strain operation where the maximum torque is used to calculate the yield stress

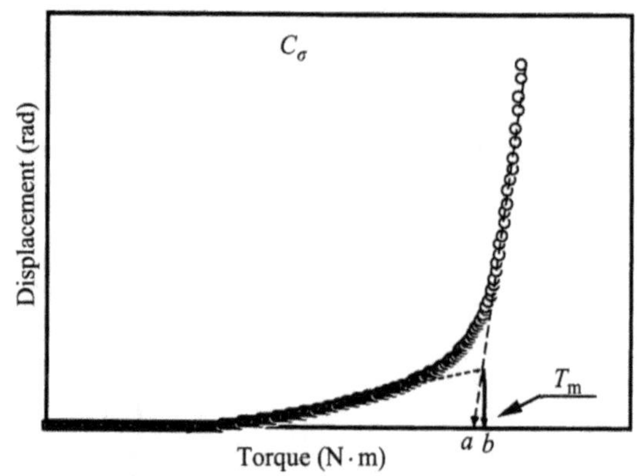

Fig. 3.15 Torque versus deformation plot with a vane in controlled stress operation

Structural Characteristics of Dispersions

The vane yield stress technique is a useful technique that applies small deformations in the initial stages and large deformations in the latter stages. From the initial linear portion of the torque–time curve at a low shear rate, for example, 0.01 s^{-1}, the shear modulus (G) can be calculated as

$$G = \sigma_e \dot{\gamma} t_e \tag{3.29}$$

where σ_e and t_e are the shear stress and time at the end of the initial linear portion of the plot, respectively. The product $\dot{\gamma} t_e$ is the strain corresponding to σ_e. Values of G obtained from vane data at $\dot{\gamma} = 0.01$ s^{-1} were of comparable magnitude to those of G' obtained from dynamic data in a frequency sweep (Genovese et al. 2004).

Michaels and Bolger (1962) and Metz et al. (1979) discussed a structural model for dispersions of particle floes that associate randomly to form weakly bonded aggregates and tenuous networks, giving rise to plastic properties. Genovese and Rao (2003a) showed that by applying an energy balance at the point of maximum deformation (yield point) in the vane test, the contributions of different structural components to the total yield stress can be estimated

$$\sigma_0 = \sigma_b + \sigma_V + \sigma_n \tag{3.30}$$

$$\sigma_0 = \sigma_{0_s} \tag{3.31}$$

$$\sigma_b = \sigma_{0_s} - \sigma_{0d} \tag{3.32}$$

$$\sigma_V = \eta_\infty \dot{\gamma} \tag{3.33}$$

where σ_b is the stress required to break the bonds between the floes, σ_v is the stress dissipated due to purely viscous drag, and σ_n is the stress required to break the aggregate network; σ_{0s} and σ_{0d} are the static and dynamic yield stresses of the samples with undisrupted and disrupted structure, respectively. Similar structural concepts, floes or clusters, have been used in recent studies on dispersions (Zhou et al. 1999; Maranzano and Wagner 2002); the stress decay following yield stress was associated with the breaking of bonds (James et al. 1987; Truong and Daubert 2000), and σ_n was associated with forces required to push aggregated particles away from each other and breaking their network (Zhou et al. 1999).

Genovese and Rao (2003a) suggested that first σ_{0s} of a sample with undisrupted structure be determined, followed by a flow test that serves the dual purposes of breaking down structure and also providing shear rate versus shear stress data, and finally determination of σ_{0d}. It appears that for most foods, the viscous component of yield stress will be small, so that the contributions of bonding and network components are of major interest. Another important practical use of σ_{0s} and σ_{0d} is creation of texture maps in which their values are plotted against corresponding values of deformations at yield, γ_{0s}, and γ_{0d}, respectively. Such a plot, shown for different SDs in Chap. 4, provides visual evidence of changes in structure of a product.

The concepts of interparticle bonding, net work structure, and viscous dissipation, as well as texture maps should be applicable to all structured dispersions, such as cosmetics and other consumer products. The vane yield stress test is a versatile test in which a fluid food is subjected to small deformations during the initial stages and large deformations during the latter stages of the experiment. From the former set of linear data, a shear modulus (G) of the sample can be estimated.

Torsion Gelometer for Solid Foods

In a torsion test, a capstan-shaped specimen is twisted in a viscometer, and the generated stress and strain are measured up to the point of material fracture. Torsion produces what is called a pure stress, a condition that maintains sample shape and volume during the test. The material can fail in shear, tension, compression, or in a combination mode, and the test does not dictate the mode of failure (Hamann 1983). The main disadvantages of torsion are: (1) specimen shaping and preparation are usually complex and tedious and (2) the technique is not applicable to soft or sticky materials. It was first used on fruits and vegetables (Diehl et al. 1979), but has been used on several other foods. Results of torsion gelometry and vane rheometry on cheddar, mozzarella, and processed cheeses were compared (Truong and Daubert 2001).

Pressure-Driven Flow Viscometers

Capillary/Tube Viscometer

Viscosity measurements can also be conducted using nonrotational pressure-driven flow viscometers. For example, the glass capillary viscometer in which flow is induced by gravitational force is used extensively for obtaining data on low-viscosity Newtonian fluids. In addition, small- and large-diameter tube viscometers can be constructed so that one can obtain shear rate–shear stress data of non-Newtonian foods (Fig. 3.16). The viscometers must be designed such that the pressure drop, Δp(Pa), over a known length of tube, L(m), and the corresponding volumetric flow rate, Q(m^3s^{-1}), of the food in laminar flow can be measured. Care must be exercised in designing the viscometer so that the Δp of food flow is measured over a section L of the tube free from the influence of fittings and the disturbances at the entrance and exit of the tube; the latter are known as end effects and are due to the higher pressure drop in flows that are not fully developed.

In the study of Vitali and Rao (1982), an entrance length of $100 \times D$ and an exit length of $40 \times D$ were used and the pressure drop was measured in fully developed flow; therefore, corrections for end (entrance and exit) effects were avoided. When it is not possible to measure Δp over a length of fully developed tube flow, the excess pressure drop due to end effects can be corrected using a Bagley plot obtained by using tubes with different (L/D) ratios and plotting pressure drop values at constant shear rate against those of (L/D; Fig. 3.17). The correction (ζ) to (L/D) is obtained by extrapolating to zero pressure drop so that the effective length at a specific shear rate for the specific food material is ($L+\zeta D$). Values of ζ from Fig. 3.17 were 1.10, 2.14, 2.59, 2.69, and 2.87 at the shear rates 40, 60, 90,120, and 250 s^{-1}. Another use for the data shown in Fig. 3.17 is to estimate the entrance pressure drop Δp_e.

Therefore, when pressure drop data over fully developed flow length cannot be obtained, the steps involved in obtaining shear rate–shear stress data can be summarized as: (1) Values of ($32Q/\pi D^3$) are plotted against the corresponding Δp for each tube. (2) At selected values of ($32Q/\pi D^3$), the corresponding pressures are obtained and plotted versus the (L/D) ratio of the tubes as in Fig. 3.17. (3) The correction factor ζ is obtained from Fig. 3.17 and the effective length at a specific shear rate is calculated as ($L+\zeta D$). (4) From the corrected shear stress and ($32Q/\pi D^3$), the shear rate–shear stress data can be obtained as discussed earlier. In the study of Jao et al. (1978) on soy dough rheology, the Bagley correction to end effects was used. Senouci and Smith (1988) found entrance pressure losses to be less than 10% of

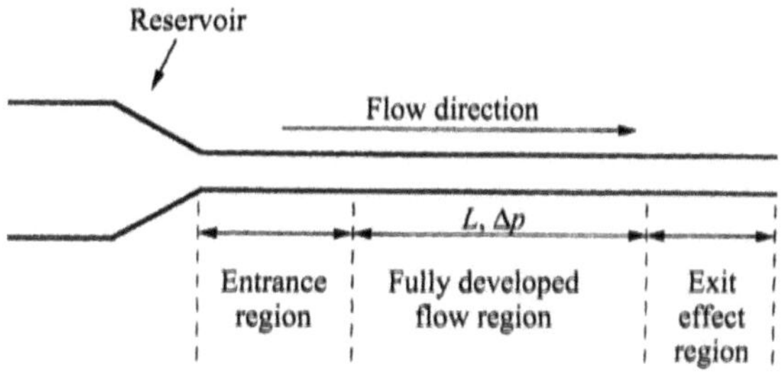

Fig. 3.16 Tube viscometer design considerations. (Vitali and Rao 1982)

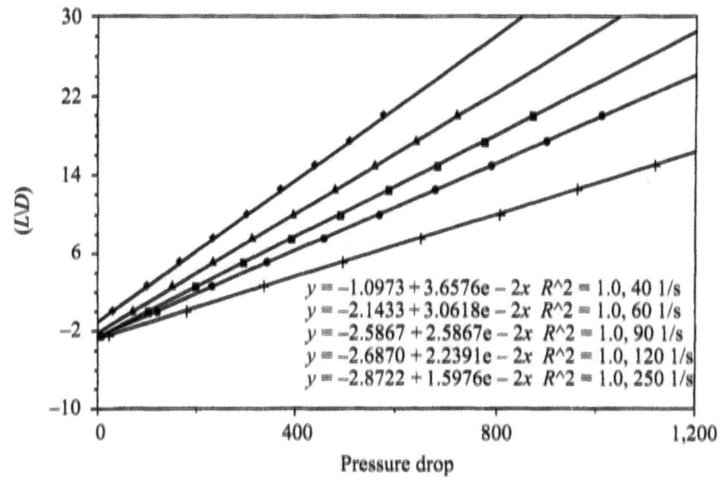

Fig. 3.17 Illustration of Bagley end correction technique. Pressure drop units in pounds per square inch as in Brodkey (1967)

the total pressure drop in a slit die viscometer for maize grits; in contrast, for potato flour, it was as large as 58 %.

An outline of the steps involved to derive the equations for shear rate and shear stress for fully developed flow in a tube is given in Appendix 2. The shear stress (σ_w) is given by Eq. 3.34 and the shear rate by Eq. 3.35, where the subscript w is used to emphasize that the values obtained are those at the pipe wall

$$\sigma_w = \frac{D\Delta p}{4L} \tag{3.34}$$

$$\dot{\gamma}_w = \frac{3}{4}\left(\frac{32Q}{\pi D^3}\right) + \frac{\sigma_w}{4}\frac{d(32Q/\pi D^3)}{d\sigma_w} \tag{3.35}$$

where D is the diameter of the capillary/tube and L is the length over which the pressure drop is measured. Equation 3.35 is known as the Weissenberg–

Table 3.3 Tube viscometer flow data on guava puree 13.6° Brix, 58 °C. (Data of Vitali and Rao 1982)

Test no.	Mercury column (cm)	Q ($cm^3\ s^{-1}$)	Shear stress (Pa)	Pseudoshear rate (s^{-1})	True shear rate (s^{-1})
1	11.3	3.31	29.1	285	394
2	16.4	7.95	42.3	684	945
3	20.2	12.58	52.1	1,080	1,500
4	23.0	20.28	59.3	1,750	2,410
5	27.2	28.58	70.2	2,460	3,400
6	30.1	38.04	77.6	3,270	4,520
7	33.0	47.39	85.1	4,080	5,630
8	34.8	55.54	89.8	4,780	6,600

Density of puree = 1,054 kg m^{-3}

Rabinowitsch–Mooney equation for laminar flow in a tube/pipe, in honor of the scientists who derived it, and is made up of a term for Newtonian shear rate, ($32Q/\pi D^3$), and a term that is a correction for non-Newtonian behavior. For fluid foods obeying the simple power law model, the flow behavior index, n, can be determined as the slope of a plot of log ($32Q/\pi D^3$; abscissa) and log (σ_w; ordinate), and it can be used to calculate the non-Newtonian shear rate, $\dot{\gamma}$, from the equation

$$\dot{\gamma} = \frac{3n+1}{4n}\left(\frac{32Q}{\pi D^3}\right) \tag{3.36}$$

One can now construct diagrams of shear rate versus shear stress for non-Newtonian foods. Vitali and Rao (1982) obtained volumetric flow versus pressure drop data given in Table 3.3 that were then converted to shear rate and shear stress values shown in Fig. 3.18 and also listed in the table.

Capillary and tube viscometers are relatively inexpensive. However, jackets for temperature control and large reservoirs for holding food samples can increase their cost and care must be exercised in analyzing the data obtained with them. As in the case of rotational viscometers, when food suspensions are being studied, the errors (slip effects) in the shear rates due to the formation of a thin layer of the continuous phase at the walls of the tube can be large. Much effort was spent on analyzing the potential errors in tube viscometers and at least the reviews (e.g., Van Wazer et al. 1963; Rao 1977b), if not the original references, should be consulted to appreciate the experiences of earlier workers. Capillary viscometers are not well suited for studying time-dependent rheological behavior because of the difficulty in measuring reliable values of time-dependent wall stresses.

Yield Stress Using a Capillary

If one can determine accurately the minimum pressure required, Δp_{min}, to cause flow in a horizontal tube, then the yield stress of the material can be estimated from

Fig. 3.18 Shear rate–shear stress data on guava puree obtained with a tube viscometer. (Vitali and Rao 1982). Both pseudo and true shear rates are shown.

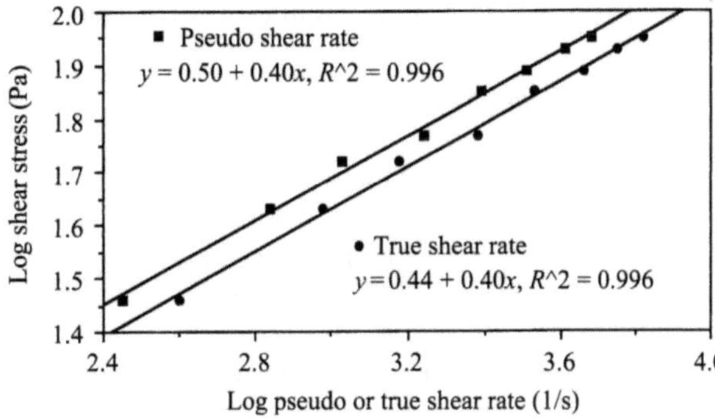

$$\sigma_0 = \frac{D\Delta p_{\min}}{4L} \tag{3.37}$$

The structure of the food sample would be disturbed considerably during the determination of $\Delta p_{\min}$, so that the measured yield stress would be closer to the dynamic yield stress than the static yield stress (Fig. 3.10). In contrast, in the vane method for determination of yield stress, both the static and dynamic yield stresses can be determined.

Die Swell or Jet Expansion

In a highly elastic material with a high first normal stress difference, $\sigma_{11} - \sigma_{22}$, the material expands to a larger diameter, known as die swell or jet expansion. The diameter of the expanded jet, D_l, may be estimated from an equation developed by Tanner (1988)

$$\frac{D_1}{D} = \left(1 + \frac{1}{8}\left\{\frac{\sigma_{11} - \sigma_{22}}{\sigma_w}\right\}^2\right)^{1/6} \tag{3.38}$$

The above equation provides approximate values because the effects of extensional viscosity and moisture evaporation are not considered.

Sample Calculation, Test No. 5

The following steps illustrate calculation of shear stress and shear rate. Note that centimeter-gram-second units were used as in the original work (Vitali and Rao 1982), but the final values are in SI units:

1. The pressure drop in centimeter of Hg was converted to Pa

$$\Delta p = \rho g \Delta h = 13.6\,\text{g cm}^{-3} \times 981\,\text{cm s}^{-2} \times 27.2\,\text{cm} = 3.63 \times 10^5\,\text{dyne cm}^{-2}$$

2. The shear stress at the wall

$$\sigma_w = \frac{D\Delta p}{4L} \frac{0.491 \times 3.63 \times 10^5}{4 \times 63.5} = 702\,\text{dyne cm}^{-2} = 70.2\,\text{Pa}$$

3. The pseudoshear rate was calculated using $\dot{\gamma}_N = (32Q/D^3)$

$$\dot{\gamma}_N = \frac{32 \times 28.58}{3.142 \times (0.491)^3} = 2.46 \times 10^3\,\text{s}^{-1}$$

4. The same procedures were used to calculate shear stresses and pseudoshear rates for all the data. A plot of log $\dot{\gamma}_N$ versus log σ_w shown in Fig. 3.16 and from linear regression of the data the slope (n') and the intercept (K') were calculated to be 0.40 and 3.17 Pa s^n. Because $n'=n$, the power law flow behavior index, the consistency index K was evaluated from

$$K = \frac{K'}{\left(\frac{3n+1}{4n}\right)^n} = \left[3.17/(1.38)^{0.4}\right] = 2.78\,\text{Pa}$$

Glass Capillary Viscometer

For low-viscosity Newtonian foods that can flow under normal gravitational force, glass capillary viscometers can be used. Recalling that the magnitude of viscosity can be obtained by dividing the equation for shear stress by that for shear rate, one can derive from the Hagen–Poiseuille equation

$$\eta = \left(\pi\ \Delta p\, r^4 t/8\, V L\right) \tag{3.39}$$

where η is the viscosity; Δp is the driving pressure for flow (usually product of fluid density; $\rho \times$ gravity; $g \times$ and mean height of fluid column, h, during flow, i.e., $\Delta p = \rho g h$); r the capillary radius; t the time of flow; V the volume of flow (note that the volumetric flow rate $Q = V/t$) and L the length of the capillary.

In glass capillary viscometers, both entrance and exit effects depend on the kinetic energy of the fluid stream in the capillary. Based on extensive experimental data, the kinematic viscosity, $\upsilon = \eta/\rho$, data with a correction term for kinetic energy is expressed

$$\upsilon = kt - K/t^m \tag{3.40}$$

where $k=\pi r^4 hg/8LQ$ and K and m are constants. The constant $m=2$ for capillaries ending in trumpet-shaped ends and $m=2/3$ for square ends. Typically, when $m=2$, the value of K is constant for a specific viscometer; for an Ubbelohde viscometer K was 60 (Van Wazer et al. 1963). More importantly, when viscosity measurements are conducted such that the efflux times are very long, the kinetic energy correction factor K/t^m may be omitted completely.

Therefore, when the time for a standard volume of fluid to pass through a length of capillary is measured for a standard viscosity (η_{st}) fluid (t_{st}) and a test fluid (t) and, if the densities of the test and standard fluids are nearly the same, the magnitude of viscosity of the test fluid can be calculated as

$$\eta = (t/t_{st})\eta_{st} \tag{3.41}$$

It is important that magnitudes of t and t_{st} are determined with care at a specific temperature, and that magnitude of viscosity of the standard fluid η_{st} is reliable. Glass capillary viscometers are not suitable for liquids that deviate substantially from Newtonian flow or contain large-size particles or a high concentration of suspended solids. In this case, the viscosity is directly proportional to the time of flow t.

Ubbelohde Viscometer

As stated in Chap. 1, for the determination of intrinsic viscosity, $[\eta]$, of a polymer, viscosity values of several dilute solutions, when the relative viscosities (η/η_s) of the dispersions are from about 1.2 to 2.0, are determined. To facilitate such measurements, the so-called Ubbelohde glass capillary viscometer is used that has a large reservoir to permit several successive dilutions of a polymer solution (Fig. 3.19). Because intrinsic viscosity measurement is important, the test procedure for using the Ubbelohde viscometer is outlined here in brief (Cannon Instrument Co. 1982).

(1) A measured volume of sample is charged directly from a pipette through tube G into the lower reservoir J. (2) The viscometer is placed in a constant temperature bath in a vertical position; it may take several minutes for the sample to reach the desired temperature. (3) Tube B is closed with a finger and suction is applied to tube C until the test fluid reaches the center of bulb C and suction on tube is removed. The finger from tube B is removed and immediately placed over tube A. (4) After the finger is removed from tube A, the efflux time for the liquid meniscus to pass from mark D to mark F is measured to the nearest 0.1 s. Essentially, the sample flows through the capillary H, so that care should be taken to keep it clean after a set of experiments. Steps 3 and 4 should be repeated to ensure that reproducible results can be obtained. The viscosity of the sample can be calculated from the efflux time by multiplying it with the viscometer constant. Alternatively, the values of times for flow of a standard volume of fluid of different polymer concentrations and of the solvent can be used to calculate the intrinsic viscosity. (5) The sample in reservoir is diluted by pipetting a measured volume of solvent and the viscosity of the diluted sample as described in steps 3 and 4. The intrinsic viscosity is determined by means

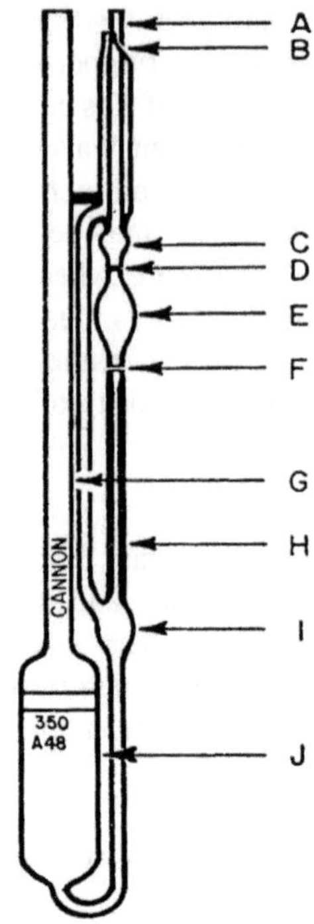

Fig. 3.19 Schematic diagram of an ubbelohde viscometer. (Courtesy of Cannon Instrument Co.)

of either the Higgins or Kraemer equation for non-polyelectrolytes, that is, plots of η_{sp}/c versus c or $\text{In}(\eta_r)/c$ versus c, respectively. For a polyelectrolyte, (c/η_{sp}) is plotted against $c^{1/2}$ from which a straight line is obtained with an intercept of $1/[\eta]$ and a slope of $B/[\eta]$, where B is a constant (see Eq. 1.13). Furthermore, determination of the time of flow of a polymer solution can be automated by detecting the position of the fluid's meniscus by optical means.

Slit (Channel) Rheometer

Slit or channel geometries have been used for studying primarily the rheological behavior of products processed in extruders, such as cooked cereal doughs. The applicable equations for shear rate and shear stress have been derived by Dealy (1982), Steffe (1996), and others, and their derivation is analogous to the corresponding equations in tube flow. For the fully developed flow of a Newtonian fluid in a slit of length L, width w, and height (thickness) h (Fig. 3.20), the relationship between the shear stress (σ) and the pressure gradient ($\Delta p/L$) is given by

$$\sigma = \left(\frac{\Delta p}{L}\right)\left(\frac{h}{2}\right) \tag{3.42}$$

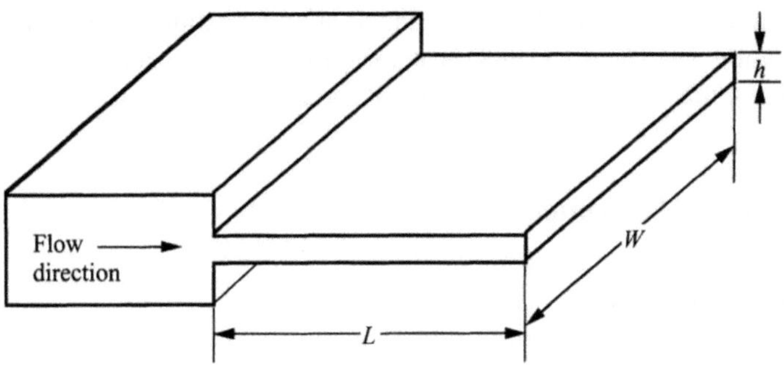

Fig. 3.20 Slit rheometer geometry. (Redrawn from Rao 1995)

where y is the coordinate in the direction of height. When the ratio of width to height should be greater than 10, edge effects can be neglected. The velocity profile for a Newtonian fluid is given by

$$v = \frac{3Q}{2hw}\left[1 - 4\left(\frac{y}{h}\right)^2\right] \tag{3.43}$$

At the wall ($y = h/2$), the shear stress and the shear rate of a Newtonian fluid are given by Eqs. 3.37 and 3.39, respectively

$$\dot{\gamma}\mathrm{N} = \frac{6Q}{h^2 w} \tag{3.44}$$

The shear rate for a non-Newtonian fluid can be obtained from

$$\dot{\gamma} = \left(\frac{6Q}{h^2 w}\right)\left(\frac{2+\beta}{3}\right) \tag{3.45}$$

where β equals

$$\beta = \frac{\mathrm{d}\ln(6Q/wh^2)}{\mathrm{d}\ln\left[\frac{h}{2}(\Delta p/L)\right]} = \frac{\mathrm{d}\ln\dot{\gamma}_\mathrm{N}}{\mathrm{d}\ln\sigma} \tag{3.46}$$

Equation 3.46 is analogous to the Weissenberg–Rabinowitsch–Mooney equation for laminar flow in a tube/pipe. As in pipe flow, the shear stress and the Newtonian shear rate are calculated first using Eqs. 3.42 and 3.45, respectively. Using the slope (β) of a plot of ln σ and ln $\dot{\gamma}$ in Eq. 3.45, the non-Newtonian shear rates ($\dot{\gamma}$) can be calculated. It is of interest to note that the slit rheometer has also been used for determining normal stress differences aimed at understanding extrudate swell. The study by Senouci and Smith (1988) is an example of the use of the slit rheometer for characterizing the rheological properties of foods.

Miscellaneous Viscometers

Falling Ball Viscometer

In the falling ball viscometer, the time for a standard sphere to fall a standard distance through a liquid at its limiting velocity is measured. The diameter of the ball is much smaller than the diameter of the container holding the fluid. This type of test is described by the Stokes equation

$$\eta = \frac{(2/9)(\rho_s - \rho_e)gR^2}{V} \tag{3.47}$$

where η is viscosity, ρ_s the density of the falling ball, ρ_e the density of the fluid, R is radius of the ball, g is gravity and V the limiting velocity. This is another simple instrument that is useful for clear Newtonian fluids. It has limited applicability for non-Newtonian fluids and cannot be used for opaque fluids unless a method for locating the ball at any time is part of the instrument.

Orifice Viscometer

The time for a standard volume of fluid to flow through an orifice is measured. The orifice may be regarded as a very short capillary. The viscometer is made up of a cup or bowl of standard dimensions with an accurately designed diameter orifice in the bottom. This type of viscometer is often used for quality control of fluid foods in food processing plants where a rapid, inexpensive method is required and extreme accuracy is not needed.

Other types of viscometers, such as an oscillation viscometer, that are useful for characterizing Newtonian foods are also available. However, their use for characterizing non-Newtonian fluid foods can be justified only if the complex flow fields can be analyzed and expressions are derived for the shear rate and shear stress.

Bostwick Consistometer

This instrument consists of a rectangular trough whose floor is graduated in millimeters. One end of the trough has a holding compartment 100 mL in capacity in which the test sample is held in place by a spring-loaded gate. It is used in the food industry to evaluate the consistency of fruit or vegetable purees off-line. The food product, such as a pureed baby food, is allowed to flow through a channel; the Bostwick consistency is the length of flow recorded in a specified time, typically 30 s. McCarthy and Seymour (1993) evaluated the flow observed in a Bostwick consistometer as a balance of gravitational and viscous forces dependent on the height of the fluid, fluid density, and fluid viscosity. For Newtonian fluids, they derived an expression showing the Bostwick measurement, L, to be a function of trough geometry and fluid properties

$$L = \xi N \left(\frac{gq^3}{3\upsilon} \right)^{0.2} t^{0.2} \tag{3.48}$$

where ξN is a similarity variable with a theoretical value of 1.411, g is the gravitational constant, q is fluid volume per unit width, υ is the kinematic viscosity, and t is time of flow. Experimental measurements with four Newtonian fluids (corn syrup and silicone oils of different viscosities) verified the theoretically predicted dependence of the extent of flow on kinematic viscosity (McCarthy and Seymour 1993).

Additional work (McCarthy and Seymour 1994) showed that the extent of flow at a given time was greater in a wider consistometer than in the standard consistometer. Another significant cause of deviation was that the flow analysis did not consider the inertial and gravitational flow regimes in the early stages of flow in the consistometer. After the experimental measurements for Newtonian and power law fluids were compared to theoretical predictions, it was suggested that the theory could be used to evaluate values for power law parameters. The Bostwick measurement length, L, after 30 s was found to be linearly related to $(\eta_a/\rho)^{-0.2}$.

Viscosity Measurement at High Temperatures

Many fluid foods are subjected to temperatures greater than 90 °C during pasteurization and thermal processing. Therefore, flow properties of fluid foods at these temperatures are useful in these processing applications. Relatively few studies have been conducted on rheological behavior of foods at high temperatures (~95 °C). Dail and Steffe (1990) used a pressurized tube (12.7 mm d) viscometer to study the flow behavior of 1.82 and 2.72 % waxy maize (WM) SDs at 121–143 °C. This system required a large mass (~ 160 kg) of test sample and, as with many tube/capillary viscometers, time-dependent rheological behavior at a fixed shear rate could not be detected. Abdelrahim et al. (1995) studied the flow behavior of WM SDs with an indirect-drive (magnetic coupling) pressurized (D100/300, Haake) concentric cylinder viscometer system. The uncertainty in rotational speeds due to slip of the magnetic coupling, friction in the mechanical bearings of the rotating cylinder, and the effect of the clutch mechanism were of concern. Reliable rheological data, in terms of the basic units of shear rate (s^{-1}) and shear stress (Pa), can be obtained on a specific food product at process temperatures using any one of the flow geometries (e.g., capillary, concentric cylinder, cone-plate).

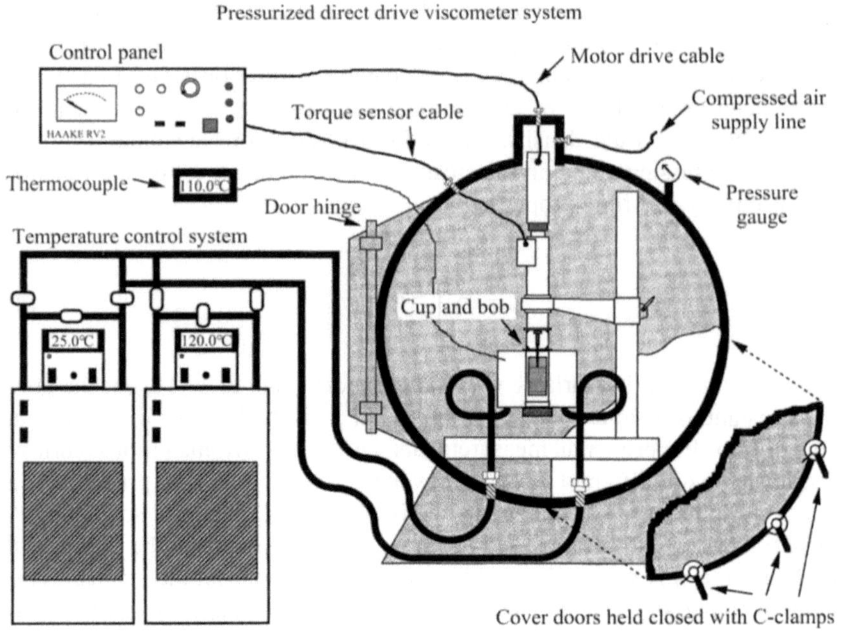

Fig. 3.21 Pressurized viscometer system for obtaining data at high temperatures. (from Liao 1998)

Fig. 3.22 Flow behavior (power law) data on tomato concentrates at 76–120 °C. (from Rao et al. 1999)

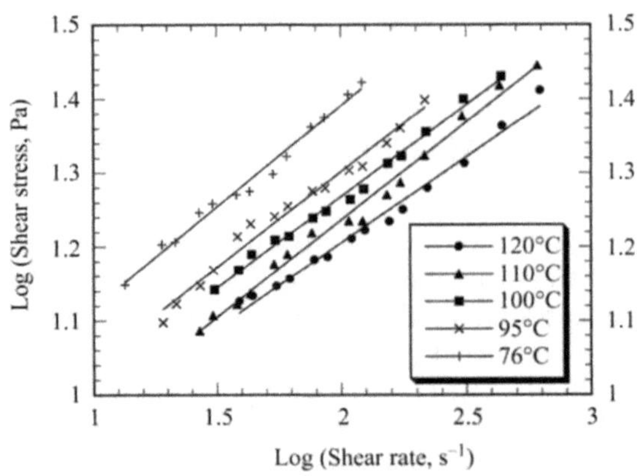

A pressurized direct-drive concentric cylinder viscometer system (Fig. 3.21) was used to obtain shear rate versus shear stress data on a tomato puree at several fixed temperatures between 76 and 120 °C (Fig. 3.22), and temperature versus apparent viscosity data at several shear rates on a 4 % waxy rice (WR) SD during gelatinization over the temperature range: 30–110 °C (Fig. 3.23; Rao et al. 1998). The drive motor, torque unit, and concentric cylinder unit and temperature control vessel of a

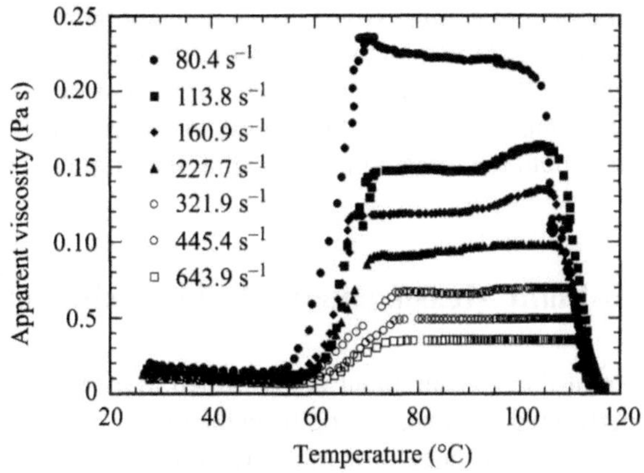

Fig. 3.23 Apparent viscosity versus temperature data on a starch dispersion as a function of shear rate (80–640 s^{-1}) and temperature. (35–118 °C; from Rao et al. 1999)

Haake RV2 viscometer system (Haake Inc.) were placed in a chamber (PRC) that could be pressurized to 0.2 MPa (two atmospheres). The temperature control vessel was insulated to minimize heat loss. A copper-constantan (36 gage wires) thermocouple placed in the well of the inner concentric cylinder measured the temperature of the test sample.

The control unit of the viscometer was kept outside the PRC so that the desired rotational speeds could be selected. The rotational speed and the corresponding torque data, as well as the temperatures sensed by the thermocouples were recorded on a laptop computer (Toshiba, T1200 XE) after analog to digital conversion via a data logger (Digistrip II, Kaye Instruments). The electrical lines to the drive motor and the torque unit inside the PRC were passed through airtight Swagelok fittings.

A high-temperature bath (Model 9500, Fisher Scientific) and a low-temperature bath (Model MH, Julabo, Germany), both containing silicone oil (Aldrich Chemical Co.) as the heating/cooling medium, were used to heat and cool the test sample, respectively. Insulated copper tubing was used for flow of the silicone oil between the oil baths and the viscometer temperature control vessel; the rigid tubing also helped in holding the viscometer firmly in the PRC. A set of three valves was used to either circulate the hot oil outside the PRC and keep it well agitated in the high-temperature bath or through the viscometer temperature control vessel. Another set of three valves was used with the low-temperature oil bath.

In a typical experiment, the temperatures of the high- and low-temperature baths were first set and allowed to reach equilibrium values while the silicone oil was circulated in the lines outside the PRC. Because of energy loss in the high-temperature lines, there was a substantial difference between the set temperature of the high-temperature bath and the sample temperature. Therefore, the temperature of the high-temperature bath was set about 20 °C higher than the desired sample temperature. The concentric cylinder system (MVI, r_i=20 mm, r_o=21 mm, h=60 mm) was carefully filled (~ 120 mL) with either tomato puree or SD, and secured in the temperature control vessel. To keep evaporation losses at a minimum, total experiment time at the high temperatures (95–120 °C) was kept close to 20 min. A Teflon® cover at the top of the concentric cylinder system and a piece of wet cloth wrapped above the concentric cylinder system helped considerably in reducing moisture loss from both the tomato puree and the SD samples.

In-Plant Measurement of Flow Behavior of Fluid Foods

An in-line measurement is performed in a process line; an on-line measurement is performed in a bypass loop from the main process line and the food may be returned to the main process line after measurement is performed. A near-line measurement is performed on a sample taken from a process line which is often discarded after measurement. Because foods are complex materials (e.g., suspensions, emulsions, gels), structural changes may take place during sampling (e.g., flow through a valve) for on-line and near-line measurements (Roberts 2003). Nevertheless, in principle, the previously described capillary flow, and rotational concentric cylinder, plate-cone, and mixer viscometers may be used for in-line, on-line, and near-line measurements. In this respect, Tamura et al. (1989) proposed a helical screw rheometer as an on-line viscometer. The empirical measurement methods described previously are used primarily in near-line measurements.

Roberts (2003) listed several requirements that both in-line and on-line measuring systems for foods should satisfy, including:

Free of hygiene risk: Must be constructed with a food-grade material of construction, permit standard clean-in-place procedures, and be free of dead flow zones.

Nondestructive: The system should not alter the quality of the product or perturb the production schedule and process.

Real-time operation: In order to minimize down time and waste or rework of a product, the response time should be short, typically seconds.

Physically robust and stable: In general, the system must require little maintenance and withstand the process operating conditions (e.g., temperature and pressure). The sensor signal must be unaffected by typical environment in a processing plant (e.g., mechanical vibration, electrical interference) and amenable to control operations.

Easy operation: It would be desirable that the sensor and system should be easy to operate, provide an acceptable signal for process control, and the results are not dependent on operator skills. However, determination of non-Newtonian rheological behavior also requires knowledge of the flow characteristics of the fluid food and its structure, as well as potential changes that can occur due to the shear rate and temperature prevalent in the measurement system. Thus, in addition to skilled operator, it would be desirable to have supervisors with a thorough knowledge of the rheological and physicochemical behavior of the food product being manufactured.

Tomographic Techniques

Some, if not all, of the requirements of in-line measurement techniques are satisfied in tomographic techniques that provide spatially and temporally resolved data. The techniques utilize the inherent properties of the food material and include those based on: magnetic, acoustic, optical, and electrical signals (Choi et al. 2002). These techniques have also been used in measurement of velocity profiles in tubes and rheology of stationary materials. Here, the emphasis is on determination of in-line measurement of rheological behavior of fluid foods using tube flow.

Magnetic resonance imaging (MRI) is a spectroscopic technique based on the interaction between nuclear magnetic moments and applied external magnetic fields. A sample is placed in a magnetic field within a radio frequency probe and the response of the test material: attenuation, frequency, and phase, to energy added in that frequency range is recorded. Two serious constraints of MRI are the need to include a nonmetallic and nonmagnetic test section in the flow system; in addition, the costs of setting up MRI systems in processing plants can be high.

Ultrasonic refers to sound waves with frequencies 20,000 Hz or greater which are beyond the range of human hearing. The sound waves are transmitted through the wall of a pipe and the reflections are analyzed. In principle, there are two different kinds of ultrasonic flow meters: transit time and Doppler flow meters. Both kinds measure primarily velocity. The primary advantages of ultrasonic Doppler velocity (UDV) meters over other types, such as turbine and conductivity meters are: no moving parts are involved, nonintrusive, low maintenance, hard to block, and work on nonconductive media. The UDV and MRI methods offer similar potential for rheological measurements under fully developed, steady, pressure-driven tube flow. In addition, the data processing techniques for MRI and UDV are somewhat similar.

Doppler meters measure the frequency shifts caused by liquid flow. The frequency shift is proportional to the liquid's velocity. Time of flight meters use the speed of the signal traveling between two transducers that increases or decreases with the direction of transmission and the velocity of the liquid being measured. Important parameters to consider when specifying ultrasonic flow meters include flow rate range, operating pressure, fluid temperature, and accuracy. One-beam Doppler flow meters are widely used, but multibeam profiling Doppler flow meters have been reported.

A schematic diagram of a UDV system is shown in Fig. 3.24. The relationship between fluid velocity, v, and UDV data is given by

$$v = \frac{c f_D}{2 f_0 \cos\theta} \tag{3.49}$$

where v is the velocity component along the axis of the ultrasound transducer, f_D is the Doppler shift frequency, f_0 is the frequency of the transmitted pulse, c is the speed of sound, and θ is the angle between the transducer and the flow direction, typically 45° (Dogan et al. 2003).

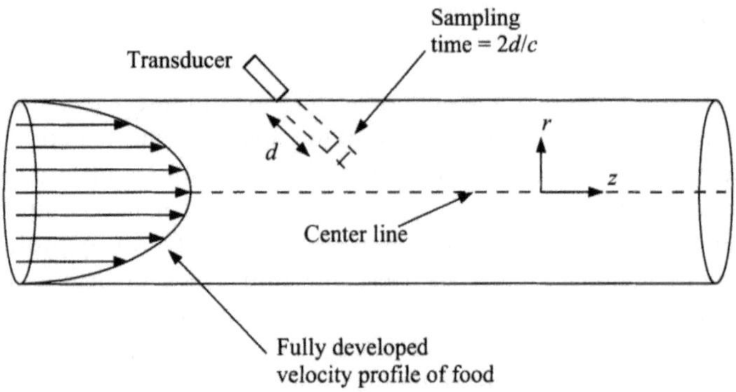

Fig. 3.24 A Schematic diagram of in-line measurement of flow properties with an ultrasonic Doppler velocity meter

The spatial location, *d,* of the velocity component in the above equation can be identified by a time-of-flight, Δt, measurement that relates the speed of the reflected wave to the distance traveled

$$d = \frac{c\Delta t}{2} \tag{3.50}$$

In turn, the values of *d* can be converted to the radial location in the pipe so that the velocity profile in a pipe can be obtained. The velocity profile is used to calculate velocity gradients (shear rates), (dv/dr), at specific locations using an even-order polynomial curve fit to the velocity data.

$$v(r) = a + br^2 + cr^4 + dr^6 + er^8 \tag{3.51}$$

Resolution of the velocity data and removal of data points near the center of the tube which are distorted by noise aid robustness of the curve fit; the polynomial curve fit introduced a systematic error when plug-like flow existed at radial positions smaller than 4 mm in a tube of 22 mm d. The curve fit method correctly fit the velocity data of Newtonian and shear-thinning behaviors but was unable to produce accurate results for shear-thickening fluids (Arola et al. 1999).

With velocimeter-based or point-wise rheological characterization, in addition to calculation of shear rate profile, the corresponding shear stress distribution is obtained by combining pressure drop measurements, and the linear relationship between the shear stress and the radial position in a pipe to be discussed later in this chapter.

UDV and pressure drop ΔP measurements were carried out on tomato concentrates with total solids: 8.75, 12.75, and 17.10 % (Dogan et al. 2003); the rheological parameters deduced from these data agreed well with those based on capillary flow data obtained at four different flow rates. Also, point-wise rheological characterization using MRI and UDV of 4.3° Brix tomato juice and 65.7 % corn syrup agreed well with off-line data using a rheometer (Choi et al. 2002). The UDV technique was also used on 1- to 3-mm diced tomatoes suspended in tomato juice and the yield stress of the suspension exhibited was characterized in terms of the Hershel–Bulkley model (Eq. 3.12) and the apparent wall slip region modeled as a Bingham fluid (Dogan et al. 2002).

For reliable characterization of a specific food by the UDV technique, extensive studies would be necessary to establish the operating parameters for that food and flow system. A change in the type of food or the composition of a specific food may necessitate a thorough evaluation of all operating parameters. Nevertheless, this technique may find a place in in-line characterization of rheological properties in food processing plants.

Vibrational Viscometers

Vibrational viscometers are robust, easy to install for in-line sensing of viscosity, offer minimal disruption of flow in a process line, operate over a wide range of temperatures (e.g., −40–400 °C), and provide real-time data. Vibrational viscometers actually measure kinematic viscosity (viscosity/density) and units are available capable of measuring kinematic viscosities ranging from 0.1 to 10^6 mPa s/g cm^{-3} (Centistokes). Typically, a vibrational viscometer employs a high-frequency (e.g., 650 Hz) torsional oscillation of a probe: sphere- or rod-shaped that undergoes damping by the fluid whose viscosity is of interest. The amplitude of oscillation is small, of the order of a micrometer, and the power consumed is converted to viscosity. When viscosity of a fluid changes, the power input to maintain constant oscillation amplitude is varied.

However, they may not indicate the true bulk viscosity of a suspension that forms a thin layer of the continuous phase (e.g., serum of tomato juice) around the immersed probe or when the probe is covered by a higher-viscosity gel due to fouling. Vibrational viscometers are suitable for measuring viscosities of Newtonian fluids, but not the shear-dependent rheological behavior of a non-Newtonian fluid (e.g., to calculate values of the power law parameters).

They may also be suitable for following gelation in near-line or laboratory experiments at a constant temperature. For example, a vibrational viscometer was used to determine the coagulation time of renneted milk at fixed temperatures (Sharma et al. 1989, 1992). However, in nonisothermal physical gelation, the elastic modulus depends on the temperature dependence of the resonant response so that precise correction for the influence of temperature must be known.

Extensional Flow Viscometry

Extensional or elongational flow is another type of deformation that can be used to obtain information regarding the rheological behavior of foods and it is different than the shear flows discussed previously. Any abrupt change in the flow geometry (e.g., sudden contraction, flow across an orifice) will generate a flow with an extensional shear component. In elongational flows, a velocity gradient along the direction of flow can stretch polymers far from equilibrium. In turn, extended polymers exert a force back on the solvent that leads to non-Newtonian behavior: viscosity enhancement and turbulent drag reduction. Applications of extensional flows to foods include atomization, swallowing, and dough sheeting (Padmanabhan 1995). Extensional flow data are very important because they differ considerably from shear data and the resistance to flow in extension can exceed that in shear by orders of magnitude (Sridhar et al. 1991). Three types of extensional flows have found wide acceptance: (1) uniaxial, (2) biaxial, and (3) planar. For any extensional flow, one can define the "stressing viscosities," η_1^+ and η_2^+

$$\eta_1^+ = \frac{\sigma_{11} - \sigma_{33}}{2(2+m)\dot{\varepsilon}} \tag{3.52}$$

$$\eta_2^+ = \frac{\sigma_{22} - \sigma_{33}}{2(1+2m)\dot{\varepsilon}} \tag{3.53}$$

For uniaxial extension $m=-1/2$ and only η_1^+ is defined, and for biaxial extension $m=1$ and $\eta_1^+ = \eta_2^+$ However, for planar extension $m=0$, and η_1^+ and η_2^+ are not equal (Macosko 1994).

Uniaxial Extension

In uniaxial extension, the material is stretched in one direction and compressed equally in the other two. For uniaxial extension or simple extension of a liquid sample of length, L, one meaningful measure of strain is (Dealy 1982)

$$\mathrm{d}\varepsilon = \mathrm{d}L/L \tag{3.54}$$

The strain rate is given by

$$\dot{\varepsilon} = \frac{\mathrm{d}\varepsilon}{\mathrm{d}t} = \frac{1}{L}\frac{\mathrm{d}L}{\mathrm{d}t} = \frac{\mathrm{d}\ln L}{\mathrm{d}t} \tag{3.55}$$

Because ($\mathrm{d}L/\mathrm{d}t$) is velocity, v, the above equation can be written as

$$\dot{\varepsilon} = \frac{v}{L} \tag{3.56}$$

When the strain rate, $\dot{\varepsilon}$, is maintained constant, the deformation obtained is called steady simple extension or steady uniaxial extension (Dealy 1982) and the extensional viscosity, η_E, is related to the normal stress difference

$$\eta_E(\dot{\varepsilon}) = \frac{\sigma_{11} - \sigma_{22}}{\dot{\varepsilon}} \tag{3.57}$$

One application of the above equation is to be able to estimate magnitudes of the normal stress difference: $(\sigma_{11} - \sigma_{22})$ as the product of $\dot{\varepsilon}$ and η_E. However, to obtain magnitudes of η_E from the above equation, one must have the means to measure both the normal stress difference and the strain rate, $\dot{\varepsilon}$. A fluid is said to be tension-thickening when η_E increases with increasing $\dot{\varepsilon}$, and tension-thinning when η_E decreases with increasing $\dot{\varepsilon}$. For the special case of a Newtonian fluid with viscosity, η

$$\eta_E = 3\eta \tag{3.58}$$

For a shear thinning fluid with a zero shear viscosity, η_0

$$\lim_{\dot{\varepsilon}\to 0}\left[\eta_E(\dot{\varepsilon})\right] = 3\eta_0 \tag{3.59}$$

The Trouton ratio, T_R, is defined as the ratio of the extensional viscosity to the shear viscosity

$$T_R = \frac{\eta_E(\dot{\varepsilon})}{\eta(\dot{\gamma})} \tag{3.60}$$

Often, it is not possible to reach a steady state in extension and it is convenient to define a transient extensional viscosity, $\bar{\eta}_E$ that is a function of time, *t,* and the extensional strain rate, $\dot{\varepsilon}$ (Barnes et al. 1989).

Even with solid foods, with which extensional viscosity measurement is relatively easy, the main difficulty in extensional viscosity measurement is in achieving steady state with respect to the applied rate of strain or stress. For example, Barnes et al. (1989) state that in the experiments conducted by Meissner, the Hencky strain (ln L/L_0, where L_0 is the original length and L is the length in deformation) of 7 corresponded to stretching the sample to 1,100 times its original length. Therefore, often the data obtained may be on the transient function, $\bar{\eta}_E(t, \dot{\varepsilon})$. Extensional flow viscosity measurement poses special challenge in low molecular weight polymers and the main problem is to achieve a continuous extensional flow field (Barnes et al. 1989) because the residence time, t_{res}, of polymer molecules in the velocity gradient is limited and the polydispersity of the molecules (Perkins et al. 1997). Although several techniques for measuring extensional stresses in polymer solutions have been proposed, rigorous interpretation of the results from these tests was only possible through the use of a rheological constitutive equation (Sridhar et al. 1991). This is because one generates a flow that is dominated by extension and the data obtained are then interpreted in terms of theologically meaningful material functions. Two such flows are called a spin-line or fiber-spin rheometer (Fig. 3.25) and the open-siphon technique (Fig. 3.26; Barnes et al. 1989). These two techniques, often suitably modified, have been employed by various authors for fluids and semisolids.

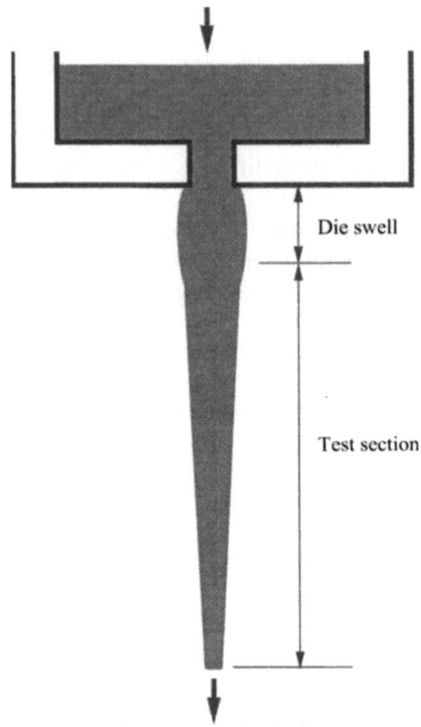

Fig. 3.25 A spin-line or fiber-spin rheometer for extensional viscosity measurement. (Redrawn from Barnes et al. 1989)

Filament Stretching

Sridhar et al. (1991) presented a filament stretching technique for measuring the extensional viscosity of polymer solutions (0.185 % polyisobutylene in kerosene with zero shear viscosity 36 Pa s at 20 °C) at a constant stretch rate. The sample to be tested was held between two teflon disks whose diameter ranged between 1 and 7 mm. The stationary top disk was attached to a microbalance system with a precision of 10^{-3} N. The output of the microbalance was recorded and converted to force on the filament. The bottom disc was attached to a central shaft of a material testing unit whose movement was controlled. A small amount of the test fluid in the vicinity of the disks remained undeformed. Therefore, the stretch rate was calculated from the decrease of filament radius with time, instead of increase in filament length with time. The diameter of the filament within the homogeneous portion of the filament during the stretching experiment was determined with an optics-based diameter measuring device. Several experiments were conducted to obtain a range of diameters as a function of time. As an alternative technique for measurement of the diameter, the authors suggested photographing the filament at up to four frames per second during the experiment and determination of the filament dimensions from the negatives. Two distinct stretching regions were observed: (1) The first local extension rate, $\dot{\varepsilon}_1$, was significantly greater than the overall extensional rate and

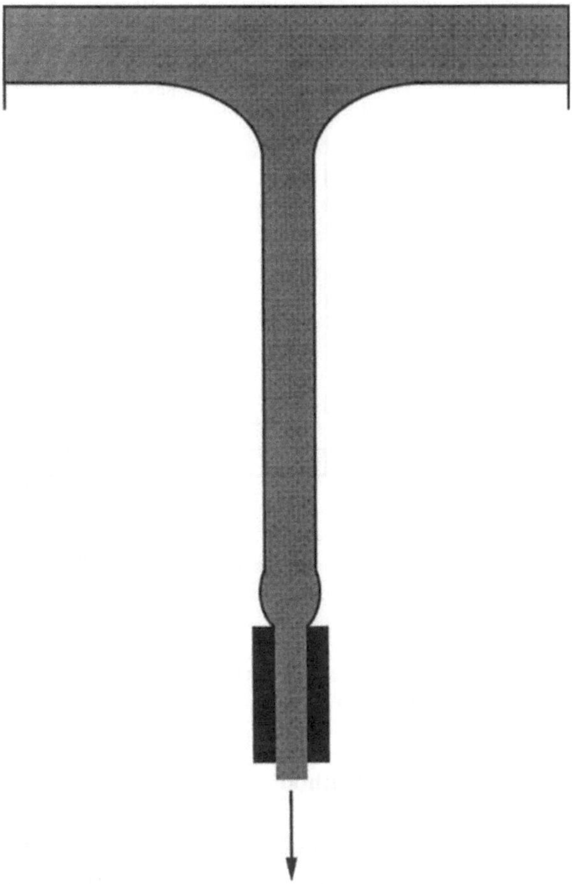

Fig. 3.26 Schematic diagram of the open-siphon technique for extensional viscosity measurement. (Redrawn from Barnes et al. 1989)

corresponded to significant necking of the fluid column when subjected to sudden stretching. This nonuniform stretching lasted up to half the time it took to stretch the fluid to the maximum length of about 160 mm. (2) The second extension rate region, $\dot{\varepsilon}$, corresponded to the homogeneous deformation of the fluid element when only a small amount of fluid remained next to the discs. Typical magnitudes of $\dot{\varepsilon}_1$ and $\dot{\varepsilon}_2$ obtained were: 0.72 s^{-1} and 0.33 s^{-1}, and 0.92 s^{-1} and 0.57 s^{-1}, respectively. Initially, when the extensional viscosity was independent of the stretch rate, the Trouton ratio increased slowly up to 3. Beyond this linear viscoelastic region, the polymer molecules extended considerably and the extensional viscosity increased dramatically to Trouton ratios $>10^3$. The flow rate and change in cross-sectional area of a fluid filament are used to determine the strain rate (Khagram et al. 1985) and such a device is called a fiber-spinning instrument.

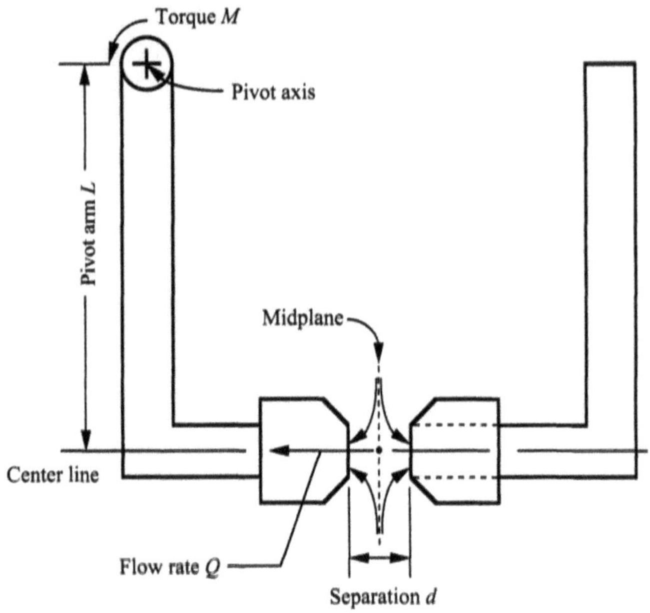

Fig. 3.27 Opposing jet devices can be used in which an area of intense extensional flow is created. (Redrawn from Macosko 1994)

Stagnation Point Method

In addition, stagnation point (it is a point at which the net magnitude of velocity is zero, such as at the midpoint of a cylinder or a sphere in cross-flow) devices, such as the dough roll mill, cross-slot flow, and opposing jet devices can be used in which an area of intense extensional flow is created without the need to sustain a continuous filament. The latter technique is illustrated in Fig. 3.27. Clark (1997) obtained extensional viscosity data on syrups and food gums dispersions using creation of a stagnation point technique. The ratio of extensional to shear viscosity was about 3.8 for a sugar syrup in reasonable agreement of 3.0 predicted by Trouton rule. The Trouton ratio for a 0.1 % xanthan gum dispersion in 0.1 % NaCI was about 4 at low and 16 at high strain rates, reflecting the limited and extensive extension of the molecule by the extensional flow field, respectively.

Biaxial Extension

Uniaxial extensional viscosity data are scarce, in part, because of difficulties in conducting uniaxial extension measurements in a reliable manner. In contrast, biaxial extension studies are relatively easy to conduct, and the availability of universal testing machines (UTM) in many food science and agricultural engineering

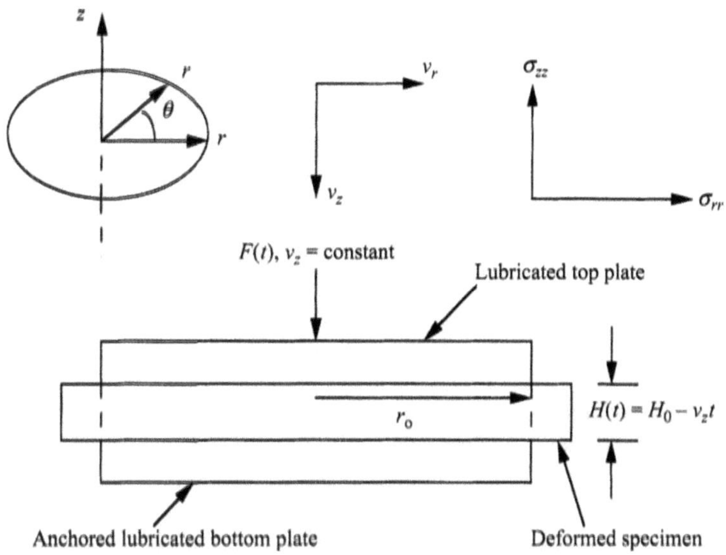

Fig. 3.28 Biaxial extensional flow test setup of Campanella et al. (1987)

departments that can be adapted for conducting biaxial extension measurements should help popularize them. In equal biaxial extension, the material is stretched equally in two directions and compressed in the other. It is convenient to work with circular disks that can be subjected to biaxial extension (Fig. 3.28) together with a UTM (Chatraei et al. 1981). This technique, called squeezing flow, was adapted to measurements on foods using lubrication between the food sample and the metal platens. Squeezing flow between two disks with lubricated surfaces can generate a homogeneous compression or equal biaxial extension in a high-viscosity polymer and other materials, such as cheeses and peanut butter (Chatraei et al. 1981; Casiraghi et al. 1985; Campanella et al. 1987). The extensional viscosity can be calculated assuming a homogeneous deformation, which implies that there is perfect slip at the wall. The requirement of perfect slip at the walls can be an asset in foods that tend to slip at walls in conventional viscometric flows, such as melted cheese and other foods. The pertinent equations for obtaining biaxial extensional viscosity are obtained assuming an incompressible material undergoing homogeneous deformation; the latter implies perfect slip at the wall (Chatraei et al. 1981; Campanella et al. 1987). For the system illustrated in Fig. 3.28, the velocity field is

$$v_z = \dot{\varepsilon}_T H(t); \quad v_r = -\dot{\varepsilon}_T \frac{r}{2}; \quad v_\theta = 0 \tag{3.61}$$

where v_z, v_r, and v_θ are the vertical, radial, and the angular velocity components; $H(t)$ is the instantaneous height at time t and r is the radial distance. In a UTM operated at a constant cross-head speed, the strain rate is given by

$$\dot{\varepsilon} = \frac{v_z}{H_0 - v_z t} \tag{3.62}$$

where H_0 is the initial height of the test specimen.

The experiment can be carried out at either constant or time-dependent force, and with the area of the plates being constant and neglecting edge effects, the normal stress difference is related to the applied force

$$\sigma_{zz} - \sigma_{rr} = \frac{F(t)}{\pi r_o^2} \tag{3.63}$$

The biaxial extensional viscosity (η_b) was defined as the ratio of the normal stress difference and the radial extension rate assumed to be half the strain rate

$$\eta_b = \frac{2F(t)(H_0 - v_z t)}{\pi r_o^2 \, v_z} \tag{3.64}$$

For an unvulcanized polydimethylsiloxane, the biaxial viscosity was approximately six times the shear viscosity over the biaxial extensional rates from 0.003 to 1.0 s^{-1} (Chatraei et al. 1981), a result expected for Newtonian fluids, that is, the relationship between the limiting value of biaxial extensional viscosity (η_b) at zero strain rate and the steady zero shear viscosity (η_0) of a non-Newtonian food is

$$\lim_{\dot{\varepsilon}_b \to 0} \left[\eta_b(\dot{\varepsilon}) \right] = 6\eta_0 \tag{3.65}$$

The lubricated squeezing flow technique was applied to melting American cheese where a UTM with lubricated platens was employed (Campanella et al. 1987) to obtain experimental data. This technique is particularly well suited for highly viscous semisolid foods prone to either slip or requiring considerable effort in pressure-driven methods, such as melted cheese, cheese spreads, corn massa, tomato pastes, mayonnaise, and peanut butter, and provides data at low values of strain rate of the order of about 0.01 s^{-1} Conversely, it is not easy to obtain high strain rates with UTMs. In addition, it should be ensured that the lubricant is not depleted and is in place during a biaxial strain experiment.

Planar Extension

In planar extension, the material is stretched in one direction, held to the same dimension in a second, and compressed in the third. The work of Perkins et al. (1997; coauthor Steven Chu is a 1997 physics Nobel laureate for studies on atomic physics) is an excellent example of the considerations in generating planar elongational flow for studying extensional viscosity of flourescently labeled DNA molecules. They tracked individual molecules in an imaging area 100 μμ 94 and measured the extension and the residence time of each DNA molecule. They classified the DNA molecular conformations as dumbbell, kinked, half-dumbbell, folded, uniform, kinked, coiled, or extended; the first three types were dominant at $\dot{\varepsilon} = 0.86\ \mathrm{s}^{-1}$ that were significantly greater than the inverse relaxation time ($\tau_r = 3.89$ s) determined at strains <0.3. For a single polymer coil, theory suggests that the polymer stretching occurs at a critical extensional strain rate, $\dot{\varepsilon}_c$

$$\dot{\varepsilon}_c \approx \frac{0.5}{\tau_1} \tag{3.66}$$

where τ_1 is the longest relaxation time of the polymer. For $\dot{\varepsilon} < \dot{\varepsilon}_c$ the molecules are in a coiled state and as $\dot{\varepsilon}$ is increased, the hydrodynamic force exerted across the polymer just exceeds the linear portion of the polymer's entropie elasticity (Perkins et al. 1997).

Cogswell's Treatment of Extensional Flow

One technique for measurement of extensional flow that has been used to study various doughs is that of Cogswell (1972, 1978) for entrance flows. The analysis is based on several assumptions (Padmanabhan and Bhattacharya 1993): (1) the flow is isothermal and creeping (negligible inertial effects), (2) the fluid is incompressible and has a pressure-independent viscosity, (3) the shear viscosity follows the power law model, $\eta_a = K\dot{\gamma}^{n-1}$, (4) there is no slip at the edge of the converging profile, and (5) that the entrance pressure drop (Δp_e) in converging flow from a circular barrel into a circular die can be considered to be made up of that due to shear ($\Delta p_{e,S}$) and extensional flow ($\Delta p_{e,E}$)

$$\Delta p_e = \Delta p_{e,E} + \Delta p_{e,S} \tag{3.67}$$

By applying force balances over a cone or wedge-shaped element, Cogswell (1972) developed relationships between pressure drop and stress. For a sudden planar contraction ($\theta = 90°$ in Fig. 3.29), the apparent planar extensional viscosity is (Padmanabhan and Bhattacharya 1993)

$$\eta_{p,E} = \frac{3(n+1)^2(\Delta p_{e,E})^2}{4\dot{\gamma}^2\eta_a} \tag{3.68}$$

and the apparent extension rate is

$$\dot{\varepsilon}_E = \frac{2\dot{\gamma}^2\eta_a}{3(n+1)\Delta p_{e,E}} \tag{3.69}$$

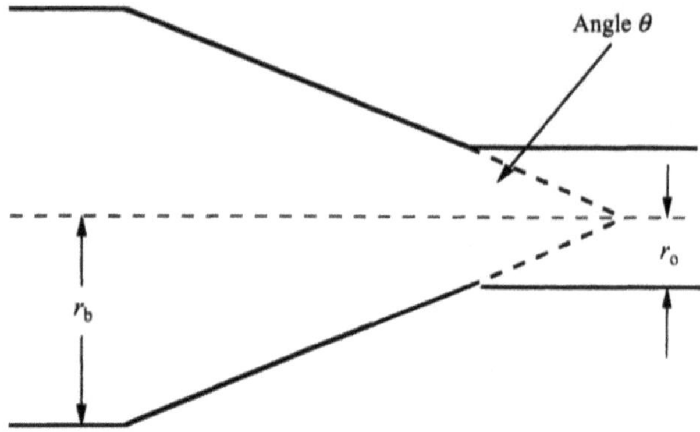

Fig. 3.29 Schematic diagram of extensional flow analysis geometry by Cogswell (1972)

The magnitude of $\Delta p_{e,E}$ is best obtained directly from experimental studies or application of Bagley's analysis. If it cannot be obtained from measurements as in Padmanabhan and Bhattacharya (1993), it can be estimated by: (1) first estimating the Δp_e using the procedure of Bagley discussed earlier, and (2) calculating $\Delta p_{e,E}$ from Eq. 3.64 after estimating $\Delta p_{e,S}$ from (Steffe 1996)

$$\Delta p_{e,S} = \left(\frac{4Q}{\pi r_o^3}\right)\left(\frac{3n+1}{4n}\right)^n \frac{2K}{3n\tan\theta}\left(1-\left(\frac{r_o}{r_b}\right)^{3n}\right) \tag{3.70}$$

The required general equations for calculating extensional viscosity can be derived assuming first that the average stress σ_E and the average extension rate ε_E are related by a power law relationship (Steffe 1996)

$$\sigma_E = K_E(\dot{\varepsilon}_E)^{n_E} \tag{3.71}$$

$$\Delta p_{e,E} = \left(\frac{4Q}{\pi r_o^3}\right)^{n_E}\left(\frac{2K_E}{3n_E}\right)\left(\frac{\tan\theta}{2}\right)^{n_E}\left(1-\left(\frac{r_o}{r_b}\right)^{3n_E}\right) \tag{3.72}$$

$$\dot{\varepsilon}_E = \left[\left(\frac{4Q}{\pi r_o^3}\right)\tan\theta\right]/2 \tag{3.73}$$

It should be noted that the above analysis requires knowledge of the actual flow angle θ of the test fluid that may not be the same as the one in the physical geometry.

Table 3.4 Viscoelastic parameters from shear, simple extension, and bulk compression

Parameter	Shear	Simple extension	Bulk compression
Stress-relaxation modulus	$G(t)$	$E(t)$	$K(t)$
Creep-compliance	$J(t)$	$D(t)$	$B(t)$
Storage modulus	$G'(\omega)$	$E'(\omega)$	$K'(\omega)$
Loss modulus	$G''(\omega)$	$E''(\omega)$	$K''(\omega)$
Complex modulus	$G^*(\omega)$	$E^*(\omega)$	$K^*(\omega)$
Dynamic viscosity	$\eta'(\omega)$	$\eta'_e(\omega)$	$\eta'_v(\omega)$
Complex viscosity	$\eta^*(\omega)$	$\eta^*_e(\omega)$	$\eta^*_v(\omega)$

Measurement of Viscoelastic Behavior of Fluid Foods

Although the normal stress and transient flow experiments provide useful data, particularly with respect to evaluation of constitutive equations, viscoelastic behavior of many foods has been studied by means of dynamic shear, creep-compliance, and stress-relaxation techniques. The paradigm shift from emphasis on normal stress measurements to dynamic rheological technique took place gradually beginning in the late 1950s. The application of each of these techniques to foods is considered in the following sections. It is customary to employ different symbols for the various rheological parameters in different types of deformation: shear, bulk, or simple extension. Following Ferry (1980), the symbols employed for the three types of deformation are given in Table 3.4.

Oscillatory Shear Flow

Small amplitude oscillatory shear (SAOS), also called dynamic rheological experiment, can be used to determine viscoelastic properties of foods. Here, both the expressions SAOS and dynamic rheological experiment will be used. The experiments can be conducted with the Couette, plate and cone, and parallel plate geometries. As noted earlier, uniform shear is achieved in the gap of a cone-plate geometry. However, the narrow gap may not be suitable for studying foods containing solids with dimension greater than about 100 μm. In a SAOS experiment, a sinusoidal oscillating stress or strain with a frequency ω is applied to the material and the phase difference between the oscillating stress and strain as well as the amplitude ratio are measured. The information obtained should be equivalent to data from a transient experiment at time $t=\omega^{-1}$.

In the SAOS tests, a food sample is subjected to a small sinusoidally oscillating strain or deformation $\gamma(t)$ at time t according to Eq. 3.74.

$$\gamma(t) = \gamma_0 \sin(\omega t) \tag{3.74}$$

where γ_0 is the strain amplitude and ω the angular frequency. The applied strain generates two stress components in the viscoelastic material: an elastic component in line with the strain and a 90° out of phase viscous component. Differentiation of Eq. 3.74 yields Eq. 3.75 which shows the strain rate $\dot{\gamma}(t)$ for evaluating the viscous component to be $\pi/2$ radians out of phase with the strain.

$$\dot{\gamma}(t) = \gamma_0\, \omega \cos(\omega t) \tag{3.75}$$

For deformation within the linear viscoelastic range, Eq. 3.76 expresses the generated stress (σ_0) in terms of an elastic or storage modulus G' and a viscous or loss modulus G''.

$$\sigma_0 = G'\gamma_0 \sin(\omega t) + G''\gamma_0 \cos(\omega t) \tag{3.76}$$

For a viscoelastic material, the resultant stress is also sinusoidal but shows a phase lag of δ radians when compared with the strain. The phase angle δ covers the range of 0 to $\pi/2$ as the viscous component increases. Equation 3.77 also expresses the sinusoidal variation of the resultant stress.

$$\sigma(t) = \sigma_0 \sin(\omega t + \delta) \tag{3.77}$$

The following expressions that define viscoelastic behavior can be derived from Eqs. 3.76 and 3.77:

$$G' = \left[\frac{\sigma_0}{\gamma_0}\right] \cos\delta \tag{3.78}$$

$$G'' = \left[\frac{\sigma_0}{\gamma_0}\right] \sin\delta \tag{3.79}$$

$$\tan\delta = \frac{G''}{G'} \tag{3.80}$$

where G'(Pa) is the storage modulus, G'' (Pa) is the loss modulus, and tan δ is the loss tangent.

The storage modulus G' expresses the magnitude of the energy that is stored in the material or recoverable per cycle of deformation. G'' is a measure of the energy which is lost as viscous dissipation per cycle of deformation. Therefore, for a perfectly elastic solid, all the energy is stored, that is, G'' is zero and the stress and the strain will be in phase (Fig. 3.30). In contrast, for a liquid with no elastic properties, all the energy is dissipated as heat, that is, G' is zero and the stress and the strain will be out of phase by 90° (Fig. 3.30).

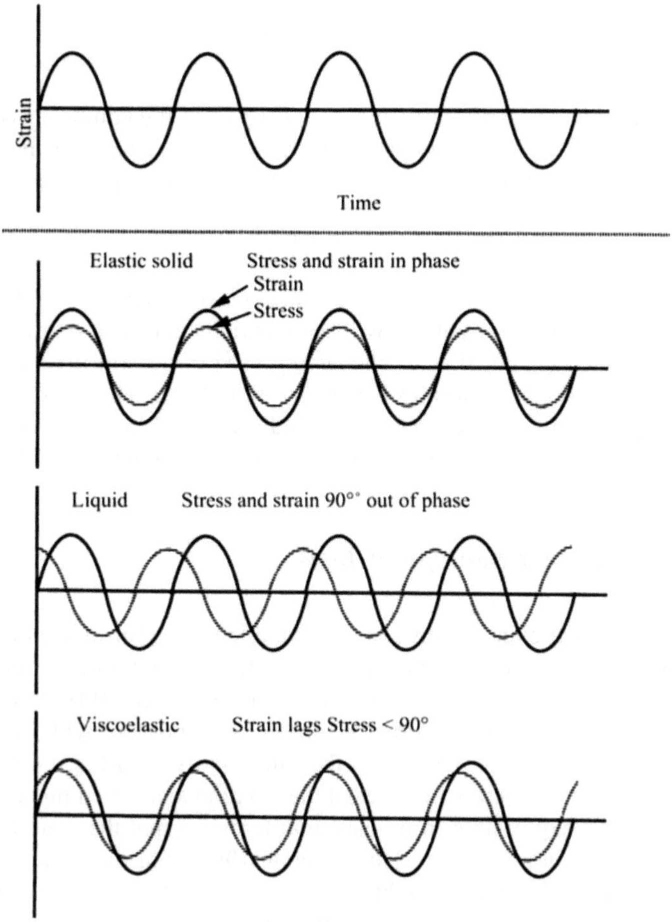

Fig. 3.30 Stress versus strain response of a Newtonian liquid, a viscoelastic liquid, and a perfectly elastic solid in dynamic tests. (Redrawn from Rao 1992)

For a specific food, magnitudes of G' and G'' are influenced by frequency, temperature, and strain. For strain values within the linear range of deformation, G' and G'' are independent of strain. The loss tangent is the ratio of the energy dissipated to that stored per cycle of deformation. These viscoelastic functions have been found to play important roles in the rheology of structured polysaccharides. One can also employ notation using complex variables and define a complex modulus $G^*(\omega)$

$$\left|G^*\right| = \sqrt{(G')^2 + (G'')^2} \tag{3.81}$$

We note that the dynamic viscosity and the dynamic rigidity are components of the complex dynamic viscosity, η^*

$$\eta^* = (G^* / \omega) = \eta' - i(G' / \omega) \tag{3.82}$$

where ω is the frequency of oscillation; the real part of the complex viscosity η' is equal to (G''/ω) and the imaginary part η'' is equal to (G'/ω). One can also determine the loss modulus G'', from oscillatory shear data using the expression

$$G'' = \omega\eta' \tag{3.83}$$

It should be noted that if G' is much greater than G'', the material will behave more like a solid, that is, the deformations will be essentially elastic or recoverable. However, if G'' is much greater than G' the energy used to deform the material is dissipated viscously and the material's behavior is liquid-like. The viscoelastic moduli determined over a range of frequencies will indicate transition zones corresponding to relaxation processes dependent on the material's structure (Ferry 1980).

Types of Dynamic Rheological Tests

Dynamic rheological experiments provide suitable means for monitoring the gelation process of many biopolymers and for obtaining insight into gel/food structure because they satisfy several conditions (Champenois et al. 1998; Lopes da Silva et al. 1998): (1) they are nondestructive and do not interfere with either gel formation or softening of a structure, (2) the time involved in the measurements is short relative to the characteristic times of the gelation and softening processes, and (3) the results are expressible in fundamental terms so that they can be related to the structure of the network (e.g., Durand et al. 1990). For the same reasons, they were used to follow the changes induced on potato cells by cellulase (Shomer et al. 1993), α-amylase (Champenois et al. 1998) on wheat starch, and can be used in studying other phase transitions in foods.

Once the linear viscoelastic region is established, three types of dynamic rheological tests can be conducted to obtain useful properties of viscoelastic foods, such as gels, and of gelation and melting. The linear viscoelastic region can be verified with a given sample by increasing the amplitude of oscillation and noting the magnitude of phase lag and the amplitude ratio. The limit of linearity can be detected when dynamic rheological properties (e.g., G' and G'') change rapidly from their almost constant values. They will be described briefly here and their results discussed extensively in Chap. 4 on food biopolymer dispersions and Chap. 6 on food gels. The three types of tests are:

(1) Frequency sweep studies in which G' and G'' are determined as a function of frequency (ω) at a fixed temperature. When properly conducted, frequency sweep tests provide data over a wide range of frequencies. However, if fundamental parameters are required, each test must be restricted to linear viscoelastic behavior. Figure 3.31 illustrates such data obtained on unmodified and acid-hydrolyzed amylopectin dissolved in 90% DMSO/10% water with a plate-cone geometry (Cham-

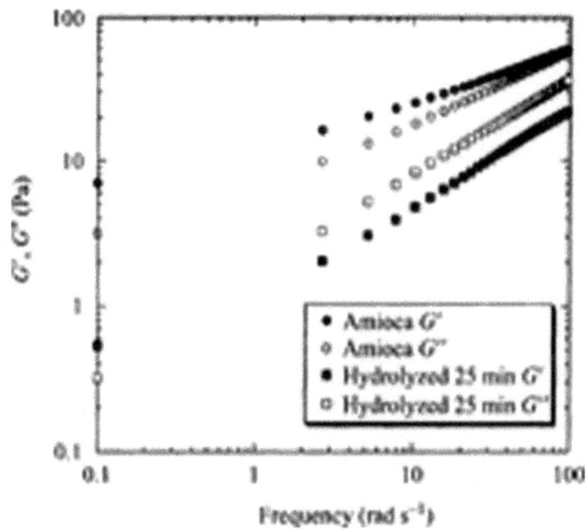

Fig. 3.31 Frequency sweep study: values of G' and G'' of 8 % acid converted waxy maize starches dissolved in 90 % DMSO/10 % water were determined as a function of frequency (ω) at 20 °C. (Chamberlain 1999)

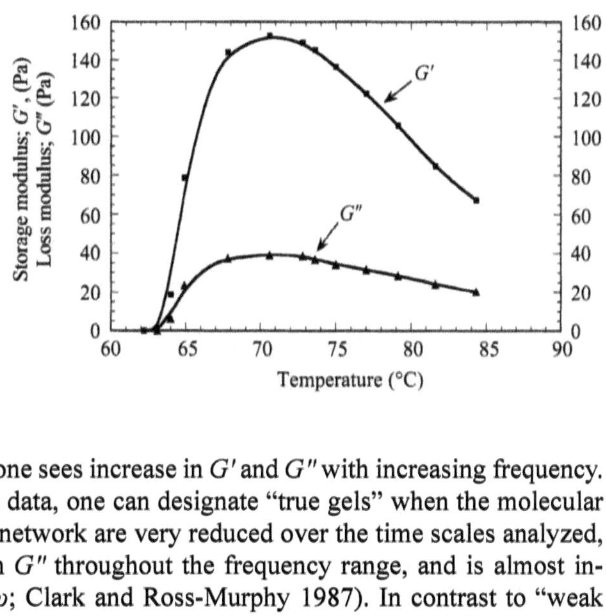

Fig. 3.32 Temperature sweep data in which G' and G'' at 6.28 rad s^{-1}, 3 % strain, were obtained on a 8 % tapioca starch dispersion heated from 60 to 85 °C at 4 °C min^{-1}. (Tattiyakul 1997)

berlain 1999). Frequently, one sees increase in G' and G'' with increasing frequency. Based on frequency sweep data, one can designate "true gels" when the molecular rearrangements within the network are very reduced over the time scales analyzed, such that G' is higher than G'' throughout the frequency range, and is almost independent of frequency (ω; Clark and Ross-Murphy 1987). In contrast to "weak gels," there is higher dependence on frequency for the dynamic moduli, suggesting the existence of relaxation processes occurring even at short time scales, and lower difference between moduli values.

(2) Temperature sweep studies in which G' and G'' are determined as a function of temperature at fixed ω. This test is well suited for studying gel formation during cooling of a heated dispersion (Rao and Cooley 1993), and gelatinization of a SD during heating (Fig. 3.32; Tattiyakul 1997) and gel formation of proteins (Owen et al. 1992). In Fig. 3.32, the gelatinization of starch was studied with a parallel plate (disc) geometry and considerable care was exercised to minimize moisture

Fig. 3.33 Values of G' and G'' during a gel cure experiment on low-methoxyl pectin + Ca^{2+} gels containing sucrose and sorbitol. (Grosso and Rao 1998)

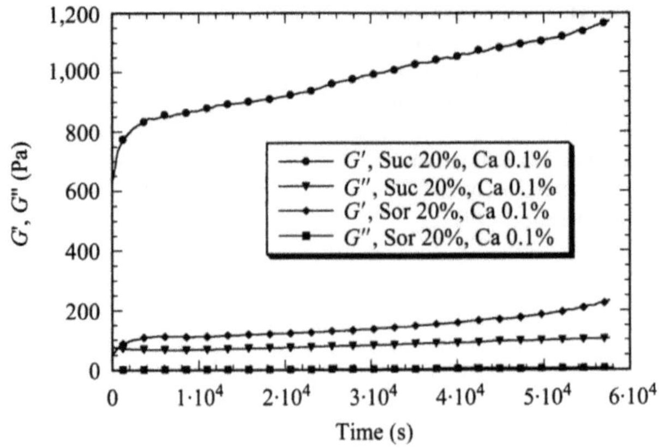

Fig. 3.34 Values of G' of 20 % sucrose-low-methoxyl pectin gels with 0.10 and 0.15 % calcium content, and of $\Delta G'$ due to increasing the calcium content

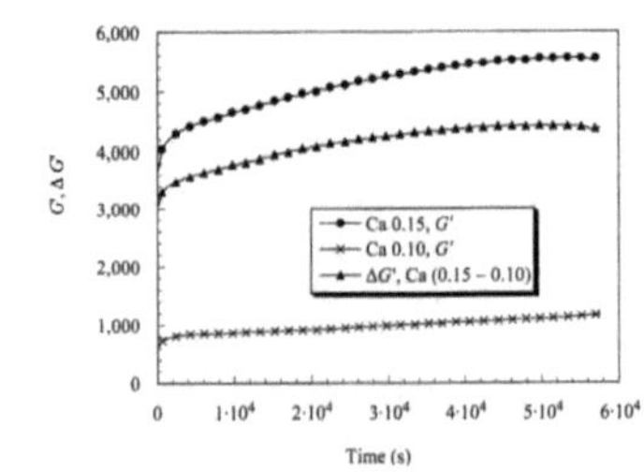

loss from the dispersions. The existence of the linear viscoelastic range can be verified in preliminary experiments conducted on a given product at several temperature steps (Owen et al. 1992).

(3) Time sweep study in which G' and G'' are determined as a function of time at fixed ω and temperature. This type of test, often called a gel cure experiment, is well suited for studying structure development in physical gels. Figure 3.33 is an example of a gel cure experiment on low-methoxyl pectin + Ca^{2+}gels containing sucrose with a cone-plate geometry (Grosso and Rao 1998).

Additional insight into gelation and melting phenomena can be obtained from the temperature and time sweep data by calculation of structure development rates (dG'/dt or $d\eta^*/dt$) or structure loss rates ($-dG'/dt$ or $d\eta^*/dt$) (Rao and Cooley 1993; Lopes da Silva et al. 1998). From the time sweep data, because G′ values of gels were recorded at the same time intervals, the influence of sugars and calcium concentrations on them could be examined quantitatively in terms of differences ($\Delta G'$) in G' values; Fig. 3.34 shows values of $\Delta G'$ of low-methoxyl pectin + Ca^{2+} gels containing 20 % sucrose due to increasing the calcium content from 0.10 to 0.15 % (Grosso and Rao 1998).

Dynamic rheological studies can also be used to study the characteristics of solid foods. Often, it is convenient to impose a sinusoidal compressive or elongational strain, so that storage and loss moduli obtained are denoted as E' and E'', respectively (Rao et al. 1995).

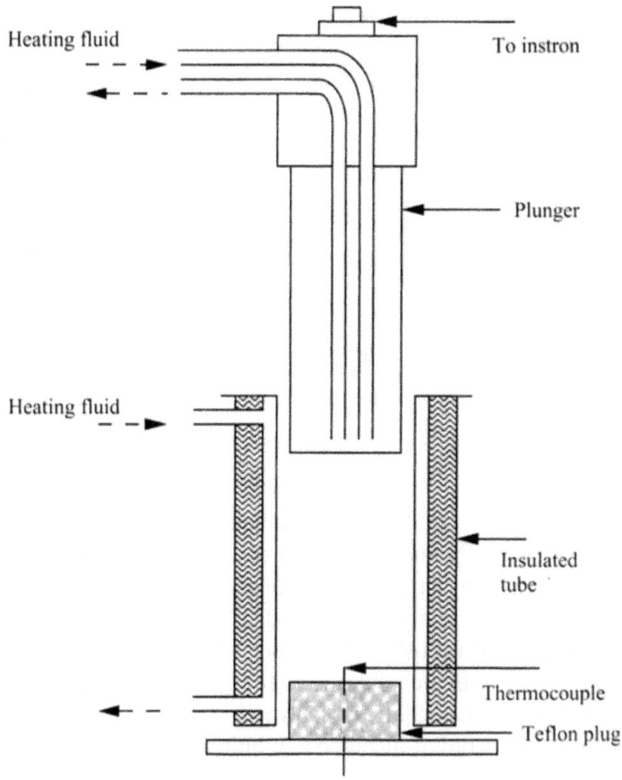

Fig. 3.35 Schematic diagram of a thermal scanning rigidity monitor. (Okechukwu et al. 1991)

Axial Motion between Coaxial Cylinders (Segel-Pochettino Geometry)

Axial motion between coaxial cylinders, the Segel-Pochettino geometry (Ferry 1980, p. 98), named a thermal scanning rigidity monitor (TSRM) was used to determine viscoelastic properties during sol–gel transition of foods (Wu et al. 1985a, 1985b; Hermann 1987; Okechukwu et al. 1991). A TSRM, shown schematically in Fig. 3.35, was used to assess the rheological changes that occur in cowpea and starch slurries as their temperatures were gradually raised from ambient to 90 °C (Okechukwu et al. 1991). The plunger was attached to a B load cell of an Instron UTM, (Instron Engineering Corp., Canton, MA) and the UTM was operated in an oscillatory mode with a cross-head speed of 0.05 cm min^{-1}, plunger displacement of 0.1 mm (Okechukwu et al. 1991). At the end of the test, the length of the plunger covered by the slurry was measured and used in estimating the apparent storage rigidity modulus, G'.

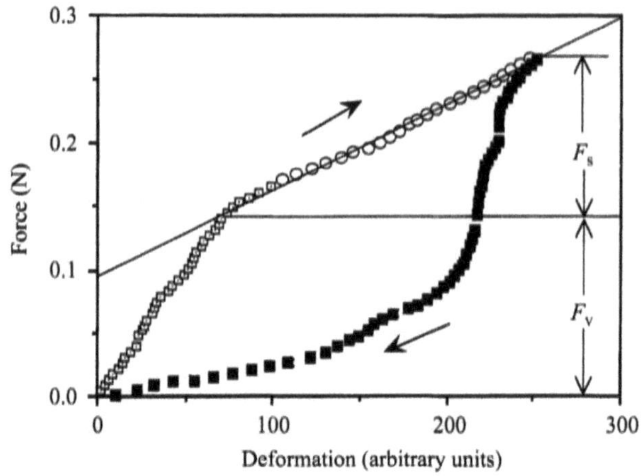

Fig. 3.36 Estimation of the storage (F_s) and viscous force (F_v) components of the peak force with a thermal scanning rigidity monitor. (Okechukwu et al. 1991)

Values of force were plotted against plunger displacement and delineated for estimates of the storage (F_s) and viscous force (F_v) components of the peak force (Fig. 3.36) that are considered typical components of a Kelvin–Voigt (Fig. 1.8) viscoelastic model. The areas under the force–displacement curves were numerically estimated using a subroutine (IMSL, Houston, TX). The storage force and the calculated areas were used to evaluate the storage rigidity modulus, G', and the energy loss, respectively (Wu et al. 1985a; Hamann 1987). G' was evaluated for a thin section of gel of length L, at radial distance r, within the annular region of the TSRM by Eq. 3.84

$$\frac{F_s}{2\pi Lr} = -G'\frac{dD}{dr} \tag{3.84}$$

where D is the plunger displacement and L is the length of the plunger covered by the slurry. Equation 3.80 can be solved for the physical boundaries of shear: at $r=r_1$, deformation $(D)=D_1$ and at $r=r_2$, $D=0$, where r_1 and r_2 are the radii of the plunger and tube, respectively, to give Eq. 3.85

$$G' = \frac{F_s}{2\pi L D_1} \ln\left(\frac{r_2}{r_1}\right) \tag{3.85}$$

To compensate for the additional compressive force experienced by the plunger at the flat surface, an additional term (Smith et al. 1949) is needed

$$G' = \frac{F_s}{2\pi L D_1}\left(\ln\frac{r_2}{r_1} - \frac{r_2^2 - r_1^2}{r_2^2 + r_1^2}\right) \tag{3.86}$$

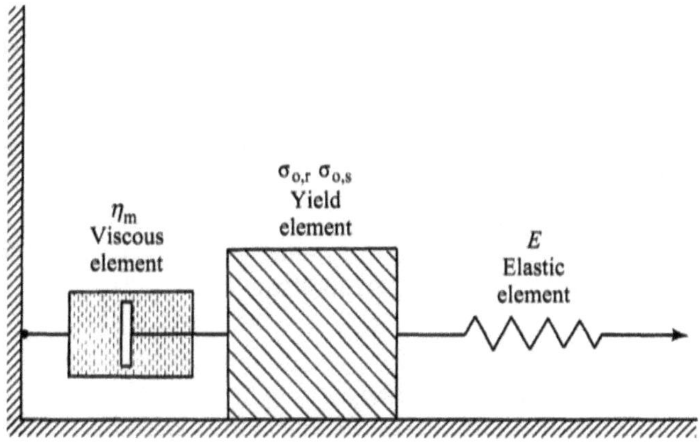

Fig. 3.37 Modified Bingham body to interpret results of dynamic rheological tests on salad dressings. (Elliott and Ganz 1977)

The % energy loss for each cycle ($\%E_{loss}$) was evaluated using Eq. 3.87

$$E_{loss} = \frac{\text{Area}_1 - \text{Area}_2}{\text{Area}_1} \times 100 \tag{3.87}$$

where Area_1 and Area_2 are the areas under the force–deformation plots for the positive and negative displacements of the plunger for a cycle, respectively. Because the TSRM can be used in conjunction with a UTM, viscoelastic behavior can be studied at several frequencies and deformations.

Vibrating Sphere Rheometer

A sphere vibrating at a specific frequency can be used to obtain magnitudes of G' and G'' at a specific frequency. Such an instrument was used to follow sol–gel transition of 5, 7, and 10% SDs (Hansen et al. 1990). However, such instruments seem to have a limited range of oscillation frequencies (e.g., 676–680 Hz). In addition, the reliability of the data obtained in comparison to data from dynamic rheological tests, in which cone-plate, parallel plate, and concentric cylinder geometries have been used, needs to be established.

Mechanical Models of Dynamic Rheological Data

The results of oscillatory shear data can be expressed in terms of mechanical models. Elliott and Ganz (1977) employed a modified Bingham body to interpret results of dynamic rheological tests on salad dressings. The model shown in Fig. 3.37 consists of an elastic element with modulus, E, a friction element having a rest yield stress $\sigma_{0,\,r}$, a sliding yield stress, $\sigma_{0,\,s}$, and a viscous element having a viscosity, η_m. This model was developed to deal with viscoelastic materials with yield stress. In dynamic rheological testing of these materials, when the amplitude of the imposed strain is sufficiently low so that the yield stress is not exceeded the stress response will be sinusoidal and in phase with the strain. For strains which result in stresses greater than the yield stress, the response will not be sinusoidal but somewhat square. In addition to salad dressings, this model was also employed for materials judged organoleptically to be unctuous (Elliott and Ganz 1971).

Time-Temperature Superposition

The method of reduced variables or time-temperature superposition (Ferry 1980, pp. 266–280) is based on the assumption that all relaxation times belonging to a given relaxation process have the same temperature dependence and that all contributions to a modulus (and compliance) should be proportional to ρT. Therefore, for example, changing the temperature from T to a reference temperature T_0 would cause a change in G' and G'' equivalent to multiplying them by the temperature-density correction factor, due to the effect of thermal expansion and to the entropy-spring nature of the stored elastic energy in the flexible chain theory: $(T_0\rho_0/T\rho)$. It was not possible to distinguish between $(T_0\rho_0/T\rho)$ and (T_0/T) because they differed by only about 20% for synthetic polymers (Ferry 1980). The difference between $(T_0\rho_0/T\rho)$ and (T_0/T) would be even less for aqueous food systems. In addition, multiplying the frequency values (ω) by the empirically determined shift factor a_T, that is, using ωa_T for the ordinate has the physical significance of the ratio of maximum relaxation times at different temperatures to that at T_0. Although the time-temperature superposition technique was developed empirically, there is much theoretical support. Expressions for a_T have been derived based on theories of dilute solution flexible random coils and undiluted polymers. Successful superposition means that at different temperatures the same molecular processes were taking place at different speeds (Lopes da Silva et al. 1994).

The time-temperature superposition method can be also applied to viscosity data (Ferry 1980). For any viscoelastic parameter, exact matching of the adjacent curves is an important criterion for the applicability of the method. In addition, when possible, the same values of a_T must superpose all the viscoelastic parameters and the temperature dependence of a_T should have a reasonable form based on experience. One advantage of the method is that the range of frequencies is extended beyond those available experimentally. The time-temperature method has been also referred to as thermorheological simplicity (Plazek 1996).

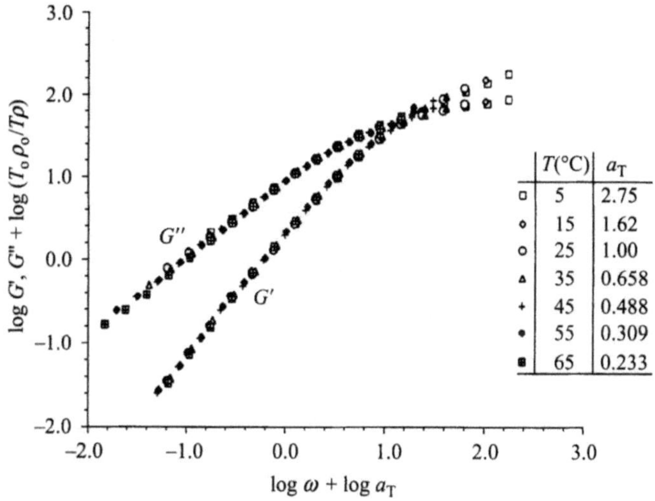

Fig. 3.38 Smooth master curves at T_0=25 °C were obtained for G' and G'' of a 1 % locust bean gum (LBG) dispersion by time-temperature superposition. (from Lopes da Silva et al. 1994)

Time-Temperature Superposition of Locust Bean (LB) Gum and Pectin Dispersions

Lopes da Silva et al. (1994) found that the frequency-temperature superposition, analogous to time-temperature superposition in transient rheological experiments, was applicable to a 1 % LB gum dispersion so that master curves at T_0 = 25°C were obtained for G' and G'' (Fig. 3.38). In contrast, smooth master curves could not be obtained for G' and G'' values of 3.5 % high-methoxyl pectin dispersions either separately or simultaneously (Fig. 3.39). The discrepancies were higher for 3.5 % low-methoxyl pectin dispersion. It was concluded that the time-temperature superposition technique was not applicable to the pectin dispersions due to their aggregated structure. For the studied samples, the vertical shift factor for the moduli ($T_0\rho_0/T\rho$) had a small effect on the master curve (Lopes da Silva et al. 1994).

Critical Stress and Strain

Although the dynamic experiments described earlier are to be conducted in the linear viscoelastic range, another experiment can be conducted in which the results obtained in the nonlinear range are useful. With a controlled stress rheometer, one can conduct an experiment in which the stress is increased continuously at a constant oscillatory frequency, say 1 Hz. Results obtained in such an experiment are

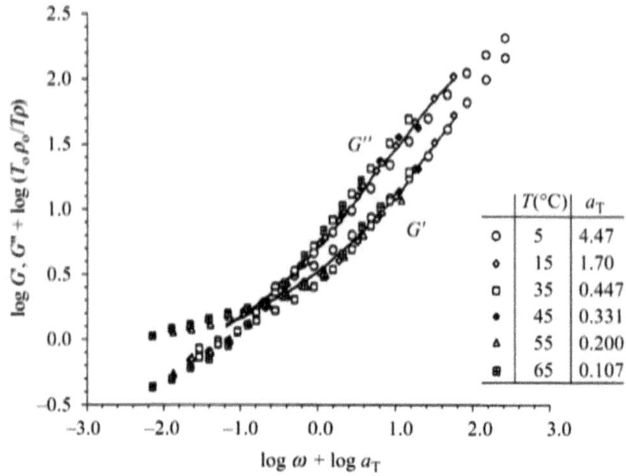

Fig. 3.39 Smooth master curves could not be obtained for G' and G'' values of 3.5 % high-methoxyl pectin dispersions by time-temperature superposition. (from Lopes da Silva et al. 1994)

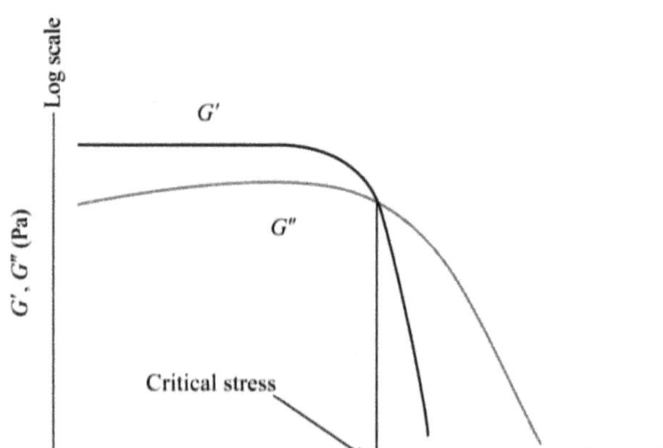

Fig. 3.40 Illustration of estimation of critical stress from a stress sweep at a fixed frequency dynamic rheological experiment. Alternatively, as described in the text, one may conduct a strain sweep experiment

shown schematically in Fig. 3.40. As the stress is increased continuously, initially, G' and G'' remain relatively constant until at a critical value of stress, σ_c, the magnitude of G' decreases sharply and that of G'' also decreases not as sharply after a slight increase. One may also use the value of the applied stress at which the curves of G' and G'' intersect as the critical stress at which structure of the sample breaks up and that at values of stress greater than the critical stress the food flows.

With a controlled strain rheometer, instead of using the stress, one may use the applied strain as the independent variable and work with values of G' and G''. The value of G' corresponding to the critical strain, γ_c, at which G'' reached its maximum

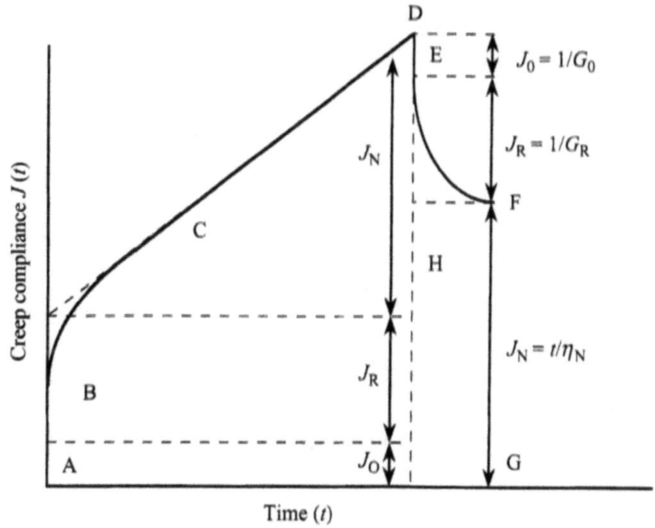

Fig. 3.41 Illustration of a creep-compliance curve. (Redrawn from Rao 1992)

value was used as an apparent yield stress of tomato concentrates (Yoo and Rao 1995). However, such yield values were lower than dynamic yield stress values determined using a vane. Additionally, as shown in Chap. 6, γ_c has been used in fractal models of gels. The critical stress, σ_c, or strain, γ_c, may be viewed as the point where the weakest bonds in the gel/food break.

Creep-Compliance Behavior

Creep-compliance studies conducted in the linear viscoelastic range also provide valuable information on the viscoelastic behavior of foods (Sherman 1970; Rao 1992). The existence of linear viscoelastic range may also be determined from torque-sweep dynamic rheological experiments. The creep-compliance curves obtained at all values of applied stresses in linear viscoelastic range should superimpose on each other. In a creep experiment, an undeformed sample is suddenly subjected to a constant shearing stress, σ_c. As shown in Fig. 3.41, the strain (γ) will increase with time and approach a steady state where the strain rate is constant. The data are analyzed in terms of creep-compliance, defined by the relation

$$J(t, \sigma_c) = \gamma/\sigma_c \tag{3.88}$$

A typical creep-compliance curve can be divided into three principal regions (Sherman 1970)

A region (A–B) of instantaneous compliance J_0 in which the bonds between the different structural units are stretched elastically. In this region, if the stress was removed, the structure of the sample would recover completely. The compliance is related to the corresponding modulus as

$$J_0 = 1/G_0 \tag{3.89}$$

where G_0 is the instantaneous elastic modulus. Magnitudes of instantaneous elastic modulus, G_0, and of storage modulus G' at 1 rad s^{-1} were found to be of the same order of magnitude for a starch paste (Giboreau et al. 1994) and tomato concentrates (Yoo and Rao 1996).

Region B–C corresponds to a time-dependent retarded elastic region with a compliance J_R. In this region, the bonds break and reform, but all of them do not break and reform at the same rate. The equation for this part using mean values for the parameters is

$$J_R = J_m\left[1-\exp(-t/\tau_m)\right] \tag{3.90}$$

Figure 3.41 General creep-compliance curve. (Sherman 1970)

In Eq. (3.85), J_m is the mean compliance of all the bonds and τ_m is the mean retardation time; τ_m equals $J_m\eta_m$ where η_m is the mean viscosity associated with elasticity. One can replace the mean quantities with a spectrum of retarded elastic moduli (G_i) and the viscosities (η_i), where $J_i = 1/G_i$. Typically, one or two Kelvin–Voigt elements can be used to describe the retarded elastic region.

Region C–D is a linear region of Newtonian compliance in which the units as a result of rupture of the bonds flow past one another. The compliance J_N and the viscosity are related by

$$J_N = t/\eta_N \tag{3.91}$$

The viscosity obtained from the above equation in the linear region of a creep experiment can be used to extend the low shear rate region of apparent viscosity versus shear rate data obtained in a flow experiment by about two decades (Giboreau et al. 1994; Rayment et al. 1998). The low shear rate region of about 10^{-3}–10^{0} is often used for the characterization and differentiation of structures in polysaccharide systems through the use of stress controlled creep and nondestructive oscillatory tests. The values of strain (γ) from the creep experiment can be converted to shear rate from the expression: $\dot{\gamma}(t) = \gamma(t)/t$.

When the stress is suddenly removed at D, the pattern consists of an elastic recovery (D–E) followed by a retarded elastic recovery (E–F). Because bonds between structural units are broken in region C–D, a part of the structure is not recovered. For values of time after the applied stress is removed, the recovered strain approaches a maximum value, S_r, called recoverable shear (Dealy 1982)

$$S_r = \sigma_c J_e^0 \tag{3.92}$$

where σ_c is the applied stress and J_e^0 is the value of the straight line portion of the creep-compliance curve extrapolated to zero time. A shear modulus, G_r, can be defined as the reciprocal of the steady state compliance, J_e^0

$$\sigma_c = G_r S_r \tag{3.93}$$

This equation is called "Hooke's law of shear."

The creep test is a simple and inexpensive test for viscoelastic foods which provides valuable information on the rheological parameters. Davis (1973) pointed out that indeed too much information can be obtained from the creep test. For example, an eight-parameter rheological model defined by eleven parameters was required for shortening and lard. One drawback with creep studies using concentric cylinder systems is that the material's structure is disturbed when the sample is being loaded.

The creep-compliance technique has been used extensively by Sherman and coworkers for the study of ice cream, model emulsions, margarine, and butter (Sherman 1966; Shama and Sherman 1969; Vernon Carter and Sherman 1980; Sherman and Benton1980). In these studies, the methodology employed was similar to that for ice cream, that is, the creep-compliance data on a sample were described in terms of mechanical models, usually containing four or six elements. Attempts were made to relate the parameters of the models to the structure of the samples studied. However, with increased emphasis on dynamic rheological tests and interpretation of results in terms of composition and structure, the use of mechanical models to interpret results of rheological tests has declined steadily.

Sherman (1966) studied the viscoelastic properties of ice cream mix and melted ice cream. The creep-compliance curve for the ice cream mix was described by the relation:

$$J(t) = J_0 + J_1(1 - e^{-t/\tau_1}) + (t / \eta_N) \tag{3.94}$$

The terms: J_0, J_1, and J_2, are the instantaneous compliance, and the compliances associated with retarded elastic behavior, respectively; $\tau_1 = \eta_1/G_1$ and $\tau_2 = \eta_2/G_2$ are the retardation times associated with retarded elasticity; η_1, η_2, and η_N are the viscosities associated with Newtonian flow.

For melted ice cream, the equation for creep-compliance data was

$$J(t) = J_0 + J_1\left(1 - e^{-t/\tau_1}\right) + J_2(1 - e^{-t/\tau_2}) + t / \eta_N \tag{3.95}$$

The mechanical analogs corresponding to Eqs. 3.94 and 3.95 are shown in Fig. 3.42. It is important to note that an association was established between the components of the ice cream and the components of the rheological model. This in turn permitted the justification of a six-element model for the mix and a four-element model for the melted ice cream.

The creep test can also be used for solid foods and as in the dynamic tests, a compressive or elongational stress can be employed (Rao et al. 1995). However, when

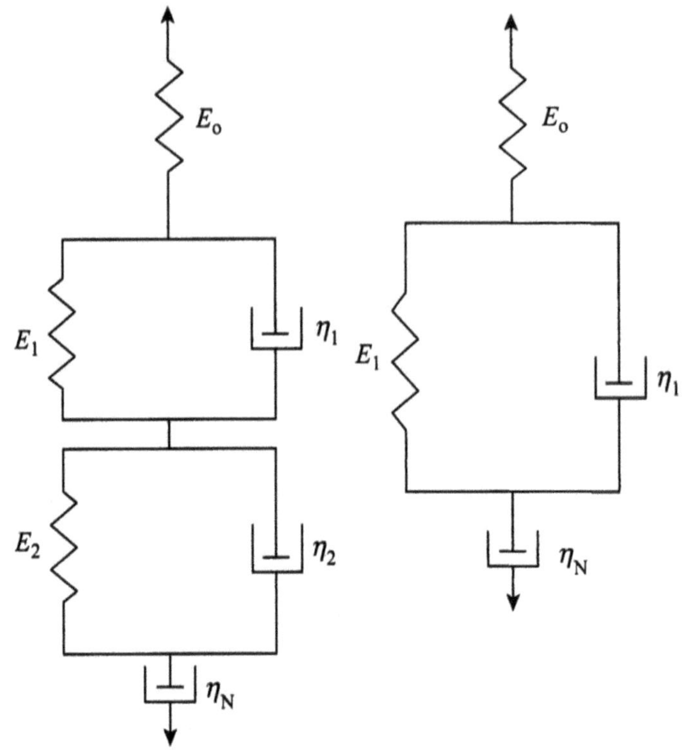

Fig. 3.42 Mechanical models for creep-compliance data of ice cream mix and melt. (Sherman 1966)

a compressive stress is imposed, the sample–metal platen interface should be either lubricated or bonded to minimize frictional effects.

Transient Viscoelastic Flow

One of the characteristics of viscoelastic foods is that when a shear rate is suddenly imposed on them, the shear stress displays an overshoot and eventually reaches a steady state value. Figure 3.43 illustrates stress overshoot data as a function of shear rate (Kokini and Dickie 1981; Dickie and Kokini 1982). The data can be modeled by means of equations which contain rheological parameters related to the stresses (normal and shear) and shear rate. One such equation is that of Leider and Bird (1974)

$$\sigma = K(\dot{\gamma})^n \left[1 + (b\dot{\gamma}t - 1)\exp(-t/an\lambda)\right] \tag{3.96}$$

where σ is the shear stress, K and n are power law parameters relating to the shear stress and the shear rate, $\dot{\gamma}$ is the imposed sudden shear rate, t is the time, λ is a time constant, and a and b are adjustable parameters. The time constant λ is related to

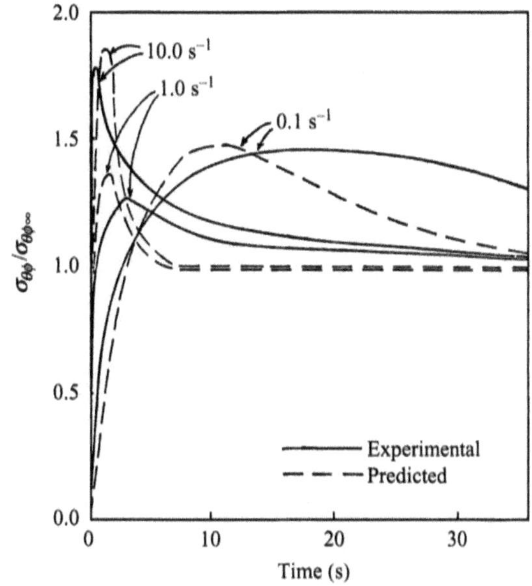

Fig. 3.43 Stress overshoot data of several foods as a function of shear rate and predicted values by Leider and Bird Model. (Dickie and Kokini 1982)

the power law parameters relating shear stress and shear rate and those relating the primary normal stress difference and shear rate

$$\lambda = (m'/2K)^{1/n'-n} \tag{3.97}$$

The Bird–Leider model predicted well the maximum and the steady state stresses as well as the times at which they occurred, but its prediction of the shear stress decay was not satisfactory (Dickie and Kokini 1982).

A number of other models have been developed to describe transient viscoelastic behavior. However, in order to test the applicability of the models, one must have at hand carefully obtained rheological data (e.g., obtained using low-inertia measurement systems). Another example of the applicability of models to viscoelastic data is the study of Leppard and Christiansen (1975) in which the models proposed by Bogue and Chen, Carreau, and Spriggs were evaluated. In the case of foods, Mason et al. (1982) developed an empirical model to describe the transient data on stick butter, tub margarine, and canned frosting (Kokini and Dickie 1981; Dickie and Kokini 1982). For mayonnaises, a model proposed by Larson (1985) was tested by Campanella and Peleg (1987) as described in Chapter 5 in the section on rheological behavior of mayonnaises and salad dressings.

Elliott and Ganz (1977) proposed a phenomenological analysis of the recorded stress overshoot data (Fig. 3.44): σ_A is the shear stress at the end of the initial linear portion of the plot, σ_p is the peak shear stress, σ_∞ is the equilibrium shear stress, and σ_R is the residual shear stress after shearing has stopped. The total strain that has occurred at different shearing times is the product of the shear rate and time, so that γt_A = strain at end of initial linear portion of the plot, γt_p = strain when peak shear

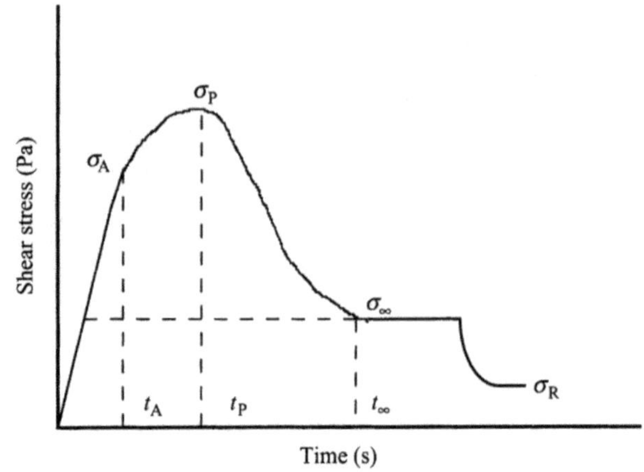

Fig. 3.44 Schematic diagram for phenomenological analysis of the recorded stress overshoot data at a constant shear rate. (Elliott and Ganz 1977)

stress is reached, and γt_∞ = strain when equilibrium shear stress is first reached. The following functions can then be calculated from the above measurements: (1) shear modulus $G = \sigma_A/\gamma t_A$, which is a measure of the elastic nature of the fluid, (2) equilibrium viscosity $\eta_\infty = \sigma_\infty/\dot{\gamma}$, which is a measure of the viscous nature of the fluid, and (3) work of structure breakdown W = area between the extrapolated equilibrium shear stress line and the curve above it. Stress growth and decay at a constant shear rate was analyzed as the excess work (W) of structure breakdown

$$W = \dot{\gamma} \int_A^B (\sigma - \sigma_\infty) \mathrm{d}t \tag{3.98}$$

where $\dot{\gamma}$ is the constant shear rate, σ_∞ is the equilibrium shear stress value, and A and B are the times at which the stress curve overshot the equilibrium stress value and the time at which the equilibrium value was reached, respectively. The analysis of Elliott and Ganz (1975) is simple and can provide useful information.

Stress-Relaxation

In a stress-relaxation experiment, the decay of stress as a function of time is recorded at a fixed magnitude of deformation. The experiments can be performed with an instrument such as the Instron UTM. However, in order to minimize errors due to friction between the test sample and the surfaces of the platens that are in contact with the sample, the sample must be either bonded to the platens or the contact surfaces must be well lubricated. In many studies, lubrication of the contact surfaces was employed as it is easier than bonding the sample to the platens. To further explain salient features of the stress-relaxation technique and the type of results that can be obtained, we consider a study of the rheological behavior of pectin sugar gels made from a commercial slow set pectin as a function of con-

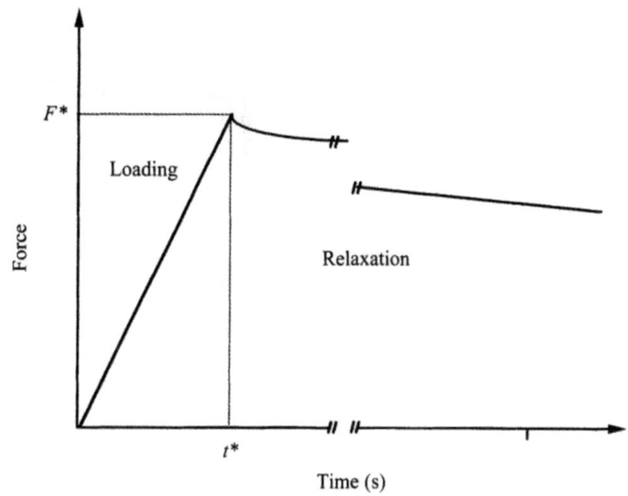

Fig. 3.45 Typical stress versus time curve obtained from a stress-relaxation experiment. (Comby et al. 1986)

centration, pH, and sugar content, and aged for 48 h, by means of stress-relaxation technique (Comby et al. 1986). A typical stress versus time curve is presented in Fig. 3.45. During the loading period ($t<t^*$), a linear relation between force and displacement was observed, and a compression modulus (E_0) was calculated from the relation

$$E_0 = \sigma^* / \varepsilon^* = (L_0 / A_0)(F^* / \Delta l) \tag{3.99}$$

where F^* is the force corresponding to the deformation ε^* and A_0 is the initial cross-section of the gel; $\varepsilon^* = (\Delta l / L_0)$, where Δl is the displacement and L_0 is the initial height of the gel sample; $F^* / \Delta l$ is the slope of the loading curve. From the relaxation curves, the Young's modulus $E(t)$ can be obtained after taking into consideration the relaxation phenomenon during the loading period. One such procedure is the Zappas approximation reported by Lin and Aklonis (1980)

$$E(t - t^* / 2) = \sigma(t) / \varepsilon^* \tag{3.100}$$

This method can be used to estimate $E(t)$ accurately using data from $t = 2t^*$; but from $t = 1.3t^*$ the approximation is good. The relaxation curves were characterized by an inflection point between 20 and 60 s. Assuming that the inflection point corresponds to the classical plateau zone, the value of the modulus at this point (E_p) may be viewed as a pseudoelastic modulus and interpreted according to theory of gel elasticity that was proposed for gels with low concentrations ($c<0.4\,\%$; Oakenfull 1984). An important determination was that the molecular weight of the junction

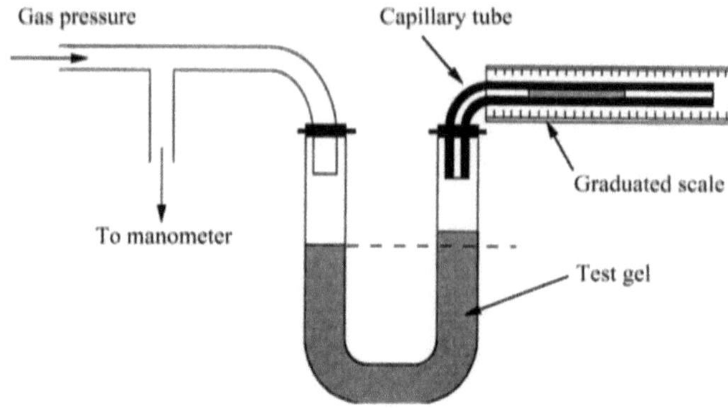

Fig. 3.46 U-tube apparatus for determination of gel modulus. (Redrawn from Saunders and Ward 1954)

zones was about (1/4) of the number average molecular weight, that is, each molecule can be involved in only two or three junction zones.

In addition, Comby et al. (1986) calculated the parameters of a two-element Maxwell model

$$E(t) = \sum E_i \exp(-t/\tau_i) \tag{3.101}$$

where τ_i is the relaxation time. Magnitudes of E_0, E_p, and E_1 increased with increase in pectin concentration. However, E_2 and τ_2 were not much affected up to pectin concentration (c) of 1.7 % suggesting that in turn the relaxation processes of chain segments between cross-linking zones were not affected; however, when the polymer concentration c was increased from 1.7 to 1.8 %, there was a sharp increase in τ_1 reflecting a decrease in chain mobility. The effect of pH on stress–relaxation curves was examined for 1.7 % pectin gels: E_p and E_0 increased steadily when pH decreased from 3.2 to 2.6, but remained constant below a pH of 2.6. Increasing sucrose content of 1.7 % pectin gels from 60.3 to 72.3 % resulted in increase in the elastic modulus from 6,750 to 28,300 Pa.

The stress-relaxation test can be used for characterizing solid foods after taking precautions to minimize friction at the platen–food interface (Rao et al. 1995). For deformations in the nonlinear range, the empirical data analysis technique of Peleg (1980) has been used in several studies (e.g., Nussinovitch et al. 1990)

$$\frac{F_0 t}{F_0 - F(t)} = k_1 + k_2 t \tag{3.102}$$

where F_0 and $F(t)$ are the initial force and that at any instant, and k_1 and k_2 are constants.

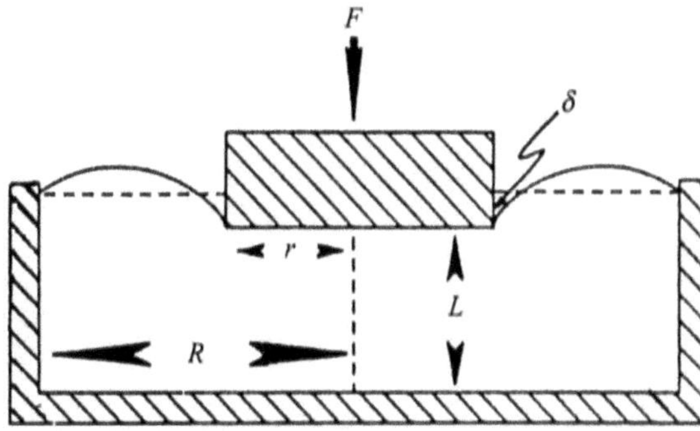

Fig. 3.47 Plunger-cup geometry for determination of modulus of weak gels. (from Oakenfull et al. 1989)

Static Measurement of Modulus

Saunders and Ward (1954) showed that a sample of gel placed in a U-tube together with a capillary containing an indexing fluid (Fig. 3.46) can be used to determine the rigidity modulus of weak gels. Assuming that the gel is entirely elastic, the shear modulus, E, can be calculated from the equation

$$E = Pr_0^4 / 8a^2hL \tag{3.103}$$

where P is the effective pressure (i.e., the applied pressure corrected for the back pressure of the indexing liquid), r_0 and a are the radii of the wide and narrow tubes, respectively, h is the displacement of the index liquid, and L is the length of the gel. Because the strain varies from a maximum value of $4a^2h/r_0^3$ at the tube wall to zero at the tube center, the volume average strain is obtained as $8a^2h/3r_0^3$. The stress is usually applied for 30 s and the displacement of the indexing liquid is recorded as a function of time. However, many food gels show a time-dependent strain (Stainsby et al. 1984) and, for gels that exhibit syneresis, the test sample as a whole slips along the tube when pressure is applied. Attempts to minimize slip of the sample include introducing dimples into the wall of the wide tube (Komatsu and Sherman 1974) and coating the tube with a single layer of antibumping granules about 1 mm in diameter (Stainsby et al. 1984). The latter technique, using tubes 6.25 and 0.478 mm in radii, was found to be suitable for determining the rigidity moduli of starch, amylose, κ-carrageenan, and egg white gels. Of further interest here is that with the modified apparatus, time-dependent strain was not observed for the studied gels so that it may not be possible to obtain creep data with this technique.

Oakenfull et al. (1989) presented a method for determining the absolute shear modulus (E) of gels from compression tests in which the force, F, the strain or relative deformation (δ/L) are measured with a cylindrical plunger with radius r, on samples in cylindrical containers of radius R, as illustrated in Fig. 3.47. Assuming that the gel is an incompressible elastic solid, the following relationships were derived:

Table 3.5 Factors (f) for converting apparent Young's modulus into absolute shear modulus for different sizes of dish and probe

r/R	L/R=0.1	L/R=0.2	L/R=0.5	L/R=1.0
0.05	0.157	0.0752	0.0381	0.0186
0.10	0.208	0.104	0.0634	0.0331
0.20	0.185	0.102	0.0912	0.0522
0.40	0.0549	0.0632	0.0930	0.0574
0.50	0.0264	0.0490	0.0803	0.0485
0.60	0.0138	0.0384	0.0622	0.0359

$$E = fY_e \tag{3.104}$$

where f is a parameter whose magnitude depends on the geometrical ratios r/R and R/L, as shown in Table 3.5, and

$$Y_e = (F/\pi r^2)/(\delta/L) \tag{3.105}$$

A correction must be applied to the measured force by subtracting the buoyancy effect, F_h

$$F_h = \pi q g \delta r^2 R^2/(R^2 - r^2) \tag{3.106}$$

Although the technique was illustrated for gelatin gels, it should be applicable to other gels also. We note here that because magnitudes of the force, F, were very small, they were measured with a sensitive electronic balance.

Relationship among Rheological Parameters

Cox–Merz Rule

The superimposition of the shear rate dependence of steady shear viscosity, that is, $\eta_a(\omega)$, and of the frequency dependence of the complex viscosity, that is, $\eta^*(\omega)$, at equal values of frequency and shear rate was first reported by Cox and Merz (1958) for polystyrene samples, and is known as the Cox–Merz rule.

$$\eta^*(\omega) = \eta_a(\omega)\big|_{\omega=\dot{\gamma}} \tag{3.107}$$

The Cox–Merz rule is an empirical correlation that has been confirmed experimentally for several synthetic polymers (Ferry 1980), and for solutions of several random-coil polysaccharides (Morris 1981). However, many exceptions to the Cox–Merz rule have also been found for synthetic polymers (Matsumoto et al. 1975; Kulicke and Porter 1980) as well as for biopolymer systems (Morris et al. 1981;

Mills and Kokini 1984; Yang and Rao 1998) and foods (Bistany and Kokini 1983; Rao and Cooley 1992). Relationships between the parameters were determined for equal magnitudes of the shear rate, $\dot{\gamma}$, and the frequency, ω (Bistany and Kokini 1983). Based on data on ten fluid and semisolid foods, power type relationships were found to be satisfactory between the parameters.

$$\eta^*(\omega) = C[\eta_a(\dot{\gamma})]^{\alpha} |_{\dot{\gamma}=\omega} \tag{3.108}$$

$$\eta'(\omega) = C'[\eta_a(\dot{\gamma})]^{\alpha'} |_{\dot{\gamma}=\omega} \tag{3.109}$$

Similarly, a power-type empirical equation was found for $G'/\omega^2(\omega)$ and the first normal stress coefficient, $\psi_1(\dot{\gamma})$

$$G'/\omega^2(\omega) = C^*[\psi_1(\dot{\gamma})]^{\alpha'} \tag{3.110}$$

Deviation from Cox–Merz rule appears to be an indication of structural heterogeneity in a food. For example, significant deviations from Cox–Merz rule were found in dispersed systems, such as tomato concentrates (Rao and Cooley 1992) and cross-linked WM (CWM) SDs (da Silva et al. 1997; Tattiyakul and Rao 2000). In contrast to the observation on several foods of dispersed nature, the Cox–Merz rule was found to be applicable to fluids with homogeneous structure, such as dispersions of guar gum (Mills and Kokini 1984) and LB gum (Lopes da Silva et al. 1993).

When departures from the Cox–Merz rule are attributed to structure decay in the case of steady shear, the complex viscosity is usually larger than the steady viscosity (Mills and Kokini 1984). Notwithstanding this feature, the relation between magnitudes of η_a and η^* can be dependent on the strain amplitude used (Lopes da Silva et al. 1993). Doraiswamy et al. (1991) presented theoretical treatment for data on suspensions of synthetic polymers. They suggested that by using effective shear rates, the Cox–Merz rule can be applied to products exhibiting yield stresses. The shift factors discussed earlier can be used to calculate effective shear rates.

Relationship between Viscosity and First Normal Stress Coefficient

In addition to relationships between apparent viscosity and dynamic or complex viscosity, those between first normal stress coefficient versus dynamic viscosity or apparent viscosity are also of interest to predict one from another for food processing or product development applications. Such relationships were derived for the quasilinear corotational Goddard–Miller model (Abdel-Khalik et al. 1974; Bird et al. 1974, 1977). It should be noted that a first normal stress coefficient in a flow field, $\psi_1(\dot{\gamma})$, and another in an oscillatory field, $\psi_1(\omega)$, can be determined. Further more, as discussed later, $\psi_1(\dot{\gamma})$ can be estimated from steady shear and dynamic rheological data.

The quasilinear corotational Goddard–Miller model can be written as (Abdel-Khalik et al. 1974, Bird et al. 1974, 1977)

$$\sigma = -\int_{-\infty}^{t} G(t-t')\dot{\Gamma}'dt' \tag{3.111}$$

For irrotational and small deformation flows, Eq. 3.84 reduces to the general linear viscoelastic model

$$\sigma = -\int_{-\infty}^{t} G(t-t')\dot{\gamma}dt' \tag{3.112}$$

Bird et al. (1974) derived relationships to predict $\psi_1(\dot{\gamma})$ from dynamic rheological (Eq. 3.113) and apparent viscosity (Eq. 3.114) data

$$\psi_1(\dot{\gamma}) = \frac{2}{\dot{\gamma}}\left\{\eta''\left[\left(1+\frac{C}{2}\right)-\frac{C}{2}\frac{\partial \ln \eta''}{\partial \ln \omega}\right]\right\}_{\omega=\dot{\gamma}} \tag{3.113}$$

$$\psi_1(\dot{\gamma}) = K\frac{4}{\pi}\int_{0}^{\infty}\frac{\eta_a(\dot{\gamma})-\eta_a(\dot{\gamma}')}{\dot{\gamma}'^2-\dot{\gamma}^2}d\dot{\gamma}' \tag{3.114}$$

In Eq. 3.113, $\eta''=G'/\omega$ is the dynamic rigidity and G' the elastic modulus.

In Eq. 3.114, η_a is apparent viscosity, $\dot{\gamma}$ is an assumed shear rate (e.g., $\dot{\gamma}=10,20,50\ \text{s}^{-1}$ and other values) at which $\eta_a\ (\dot{\gamma})$ is calculated, $\dot{\gamma}'$ is the dummy variable for integration, and K is an empirical factor that was found to be 2 for polymer solutions and 3 for polymer melts (Bird et al. 1974, 1977). Normal stresses were detected in CWM SDs of all concentrations studied: 3.5, 3.75, 4.0, 4.25, 4.5, and 5.0 % ww^{-1} and in tapioca only at the three highest concentrations: 3.5, 3.8, and 4.0 % ww^{-1} (Genovese and Rao 2003b). Double logarithmic plots of first normal stress coefficients, ψ_1, versus shear rates from down curves resulted in quasilinear plots (Fig. 3.48), as was also observed earlier (Youn and Rao 2003). Furthermore, values increased with starch concentration. The shear rate dependence of ψ_1, in terms of the slope of each curve, was almost the same for all the six concentrations of CWM SD, 4.0 % tapioca SD, and the sample of CWM granules mixed with tapioca serum; the average value of the slopes was -1.73 ± 0.05. This suggests that, in that range of concentrations, the number of starch granules did not have a significant effect on the variation of ψ_1 with $\dot{\gamma}$. The 3.8 and 3.5 % tapioca SDs did not follow this general behavior and the data showed more scatter, probably because of the weak stresses that were generated.

At the same concentrations: 4.0, 3.75–3.8, and 3.5 % values of ψ_1 of CWM SDs were moderately higher than those of tapioca SDs at low shear rates (not shown here), but the differences were negligible at high shear rates. It means that normal stresses depend not only on either volume fraction of starch granules or starch concentration but also on the nature of the granules. The dispersed phase has a greater

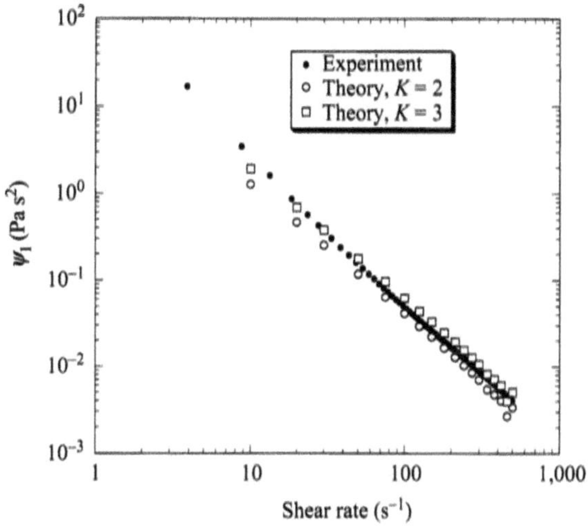

Fig. 3.48 Double logarithmic plot of first normal stress coefficient, ψ_1, versus shear rate from down ramp curve of 5 % cross-linked waxy maize starch dispersions exhibited quasilinear plots. Predictions of Bird et al. (1974, 1977) based on apparent viscosity data for values of the empirical factor $K=2$ and 3 are also shown

influence on the normal stresses of SDs than the continuous phase. Since normal forces are strongly related to the elastic component of viscoelastic materials, as previously mentioned, it is likely that in SDs the normal forces may be associated with the elastic modulus of the granules.

Also, based on the Goddard–Miller model, Bird et al. (1974) derived a relationship (Eq. 3.115) between the apparent viscosity η_a and the dynamic viscosity $\eta' = G''/\omega$, where G'' is the viscous modulus

$$\eta_a - B = \left\{ \eta' \left[1 - \frac{C}{2} \frac{\partial \ln \eta'}{\partial \ln \omega} \right] \right\} \Bigg|_{\omega=\dot{\gamma}} \qquad (3.115)$$

where C is a constant = 1. Experimental apparent viscosity up (η_{aup}) and down (η_{adw}) values of CWM SDs were compared with those of complex viscosity, η^*, and apparent viscosity predicted by Bird's Eq. 3.112 (η_{a-B}; Genovese and Rao 2003b). Not surprisingly, η_{aup} was in very good agreement with η_{a-B}, especially at low starch concentrations where both curves matched over the entire range of shear rates. This can be attributed to the fact that, unlike η_{adw}, when measuring η_{aup}, the sample has not been sheared extensively, as was also the case with the sample used to calculate η_{a-B} from dynamic data.

It should be noted that Eqs. 3.109–3.111 are unique in that they predict one rheological parameter from another. Because often it may be possible to obtain only one set of rheological data easily, such as shear rate versus shear stress or dynamic rheological data, those equations should be useful in estimating other rheological parameters. Specific relationships were also derived between the first normal stress coefficient on the one hand and dynamic rheological data on the other (Eq. 3.116).

$$\psi_1(\omega) = \psi_1^d + \psi_1' \cos(2\omega t) - \psi_1'' \sin(2\omega t) \qquad (3.116)$$

$$\psi_1^{\mathrm{d}} = \eta''/\omega \tag{3.117}$$

$$\psi_1'(\omega) = [-\eta''(\omega) + \eta''(2\omega)]/\omega \tag{3.118}$$

$$\psi_1'(\omega) = [-\eta'(\omega) - \eta'(2\omega)]/\omega \tag{3.119}$$

In Eq. 3.116, ψ_1 is rigorously defined as $[(\sigma_{11} - \sigma_{22})/(\dot{\gamma}_{21}^{0})^2]$, and is the sum of a constant term ψ_1^d and two oscillating terms, accounted by ψ_1' and ψ_1''; $\dot{\gamma}_{21}^{0}$ is the strain rate amplitude. Equations 3.6–3.8 suggest that oscillatory shear stress data are related to oscillatory primary normal stress difference data (Ferry 1980). Youn and Rao (2003) calculated values of $\psi_1(\omega)$ for SDs that is applicable to oscillatory shear fields.

Deborah Number

A unique dimensionless number in rheology is the Deborah (De) number (named after the song of Deborah, Book of Judges, the Old Testament) that compares the time scale of the deformation process (t_D) to the observation time (t_o)

$$\mathrm{De} = t_{\mathrm{D}}/t_{\mathrm{o}} \tag{3.120}$$

Typically, the time t_D for liquids is very small and is very large for solids. Small Deborah numbers signify liquid-like behavior and large numbers signify solid-like behavior. Furthermore, a material may exhibit solid-like character because either its characteristic time is large or the deformation process is very fast. For example, when a flat stone thrown at a suitable angle bounces several times on a calm body of water, because of the fast deformation process, the water is behaving like a solid. Characteristic deformation time scale can be estimated as the reciprocal of the shear rate or dynamic frequency, while characteristic observation times may be assumed to be the various relaxation and retardation times mentioned earlier. The Deborah number in terms of other variables has been used to better understand rheological behavior of dispersions (Chap. 4) and other phenomena.

Microrheology

Traditionally, viscoelastic properties of a fluid food have been obtained from small-amplitude oscillatory shear measurements in a mechanical rheometer. Most commercial rheometers require several milliliters of a material to probe frequencies up to tens of Hertz. The upper range is limited by the onset of inertial effects, when the oscillatory shear wave decays appreciably before propagating throughout the entire sample (Gardel et al. 2005). When it is difficult to obtain many biological materials in milliliter quantities, microrheology can probe a material's response at micrometer length scales with microliter sample volumes in a noninvasive manner (Mason and Weitz 1995). The techniques use embedded micron-sized probes to locally deform a sample (Rao 2013). There are two broad classes of microrheology techniques: (1) those involving the active manipulation of probes by the local application of stress and (2) those measuring the passive motions of particles due to Brownian fluctuations. In both cases, when the embedded particles are much larger than any structural size of the material, particle motions measure the macroscopic stress relaxation. Smaller particles measure the local mechanical response and also probe the effect of steric hindrances caused by local microstructure. The use of small colloidal particles theoretically extends the accessible frequency range by shifting the onset of inertial effects to the megahertz regime (Gardel et al. 2005).

Single-Particle Microrheology

One-particle microrheology assumes the local environment surrounding the bead reflects that of the bulk. In general, it is advisable to test for probe surface chemistry effects in a new system. It is noted that the bulk response is measured only if the probe size is larger than the length scale of heterogeneity in the sample. These length scales are often unknown prior to a microrheology experiment. Agarose is an example of a material containing many smaller voids, or pores, through which smaller probe particles may move.

The essential physics behind this technique was summarized by Dasgupta et al. (2001). The mean square displacement (MSD) of a spherical bead reflects the response of a fluid to the stress applied to it by the thermal motion of the bead. The MSD, $\langle \Delta r^2(\tau) \rangle$, of probe particles suspended in a sample as a function of time has been measured by dynamic light scattering. In a purely viscous fluid, the probe particles diffuse through it and the MSD increases linearly with time $\langle \Delta r^2(\tau) \rangle = 6D\tau$. After determining the diffusion coefficient D, one can calculate the viscosity of the material $\eta = k_B T / 6\pi D a$ where a is the radius of the beads.

In a viscoelastic fluid, the motion of the probe particles is constrained and the MSD reaches an average plateau value $\langle \Delta r_p^2 \rangle$ that is set by the elastic modulus of the material. By equating the thermal energy $k_B T$ of each bead with its elastic energy $\frac{1}{2}\kappa \langle \Delta r_p^2 \rangle$, where κ is the effective spring constant that characterizes the elasticity of the surrounding medium, an expression for the spring constant $\kappa \sim k_B T / \langle \Delta r_p^2 \rangle$ can be obtained. The elastic modulus $G'(\omega)$ is related to the spring constant by a factor of length, which is the bead radius, a, and one can obtain a relationship between the elastic modulus and the MSD, $G'(\omega) \sim k_B T / \langle \Delta r_p^2 \rangle a$ (Dasgupta et al. 2001).

In general, the full frequency dependence of the viscoelastic moduli is obtained from the MSD by using the generalized Stokes–Einstein relationship (GSER; Gardel et al. 2005)

$$\tilde{G}(s) = \frac{\kappa_B T}{\pi a s \left\langle \Delta r^2(s) \right\rangle} \tag{3.121}$$

Where $\langle \Delta r^2(s) \rangle$ is the Laplace transform of $\langle \Delta r^2(\tau) \rangle$ and $\tilde{G}(s)$ is the viscoelastic spectrum as a function of the Laplace frequency s.

For comparison with mechanical spectroscopy data, the Fourier space representation of the GSER is used:

$$G^*(\omega) = \frac{\kappa_B T}{\pi a i \omega \mathcal{F}\left\{\left\langle \Delta r^2(\tau) \right\rangle\right\}} \tag{3.122}$$

where $\mathcal{F}\{\langle \Delta r^2(\tau) \rangle\}$ is the Fourier transform of the MSD. Based on a local power law expansion for $\langle \Delta r^2(\tau) \rangle$ leads to expressions for the elastic $G'(\omega)$ and the loss moduli $G''(\omega)$ (Mason 2000).

Experimentally, in a dynamic light scattering experiment, a laser beam impinges on a sample and is scattered by the particles into a detector placed at an angle, θ, with respect to the incoming beam (Gardel et al. 2005). The intensity of light that reaches the detector fluctuates in time as the particles diffuse and rearrange in the sample. For example, Dasgupta et al. (2001) used three different bead sizes ranging between 0.46 and 0.97 μm at 0.0025 wt% concentration. The time-averaged intensity correlation functions were collected for 2 to 3 h at room temperature

$$g_1(\tau) = exp\left[\left\langle -q^2 \Delta r^2(\tau) \right\rangle / 6\right] \tag{3.123}$$

where $\langle \Delta r^2(\tau) \rangle$ is the ensemble-averaged three-dimensional MSD and q is the scattering wave vector given by

$$q = \frac{4\pi n}{\lambda} sin\left(\frac{\theta}{2}\right) \tag{3.124}$$

where n is the index of refraction of the sample and λ is the wavelength of the laser in vacuum, and θ is the scattering angle.

Once the MSD is obtained, as described earlier, the GSER can be applied to extract the frequency-dependent viscoelastic moduli. Single-light scattering techniques are typically sensitive to frequencies in the range of 0.01–10 Hz, similar to the frequency range available with a conventional mechanical rheometer. However, these techniques are practical only when a significant fraction of the scattered light that is collected has been scattered just once.

Diffusive Wave Spectroscopy (DWS)

In DWS, a laser beam impinges on an opaque sample and the light is scattered multiple times before exiting (Gardel et al. 2005). It is important to note that because DWS measures the properties of multiply scattered light in suspensions where all photons have been multiply scattered, all the scattering-vector information is lost. As a consequence, only two experimental modes: transmission and backscattering are used (Weitz and Pine 1992). DWS treats the photon path through the sample as a diffusive or random walk phenomenon. Also, it extends the conventional DLS to systems that are dominated by multiple scattering, to systems which are essentially opaque. Therefore, DWS requires a turbid or concentrated sample, so that the technique can be used to study practical phenomena, such as sol–gel transitions and glassy states (Alexander and Dalgleish 2006). In DWS experiments, the volume fraction of the embedded colloidal spheres is large, $\sim 10^{-2}$, in order to insure that the transport mean path, l^* is a small fraction of the length of the sample chamber, l. The ratio of l/l^* is typically greater than 5 (Gardel et al. 2005).

The MSD is obtained from the field autocorrelation function, $g_1(\tau)$ at a delay time (Gardel et al. 2005)

$$g_1(\tau) = \int_0^{\infty} P(s) exp\left[-k_o^2 \left\langle \Delta r^2(\tau)\right\rangle \frac{s}{3l^*}\right] ds \tag{3.125}$$

where $P(s)$ is the probability of light traveling a path of length s, $k_o = 2\pi/\lambda$ is the wave vector of the incident light, and λ is the wavelength of light in the medium. The transport mean free path of the light, l^*, is a characteristic of the sample itself and reflects the amount of scattering; it is the length light must travel before its direction is randomized. It is emphasized that the key to the solution of the above equation is the determination of $P(s)$ for the experimental geometry with the correct boundary conditions (Pine et al. 1988). The transport mean free paths for the samples used in these experiments are roughly four to ten times smaller than the cell thickness depending on the bead size used, ensuring strong multiple scattering. Because in DWS the number of scattering events is large, there is no appreciable angular dependence of the light emerging from the sample, and measurement at a precisely defined angle is not necessary.

Much of the research work on food systems with DWS has been done on milk proteins. DWS is capable of following the kinetics of destabilization of concentrated food colloids and detect differences at the microstructural level (Corredig and Alexander 2008). The onset of aggregation always appears earlier in DWS measurements than in oscillatory shear measurement, because DWS is a microscopic measurement and is more sensitive to particle changes than rheology (Alexander and Dalgleish 2006). Dhabi et al. (2010) determined that the jamming concentration of casein was 18.9 w/w%. Additionally, changes in l^* can be taken as indications of changing organization within the suspension (Alexander and Dalgleish 2006).

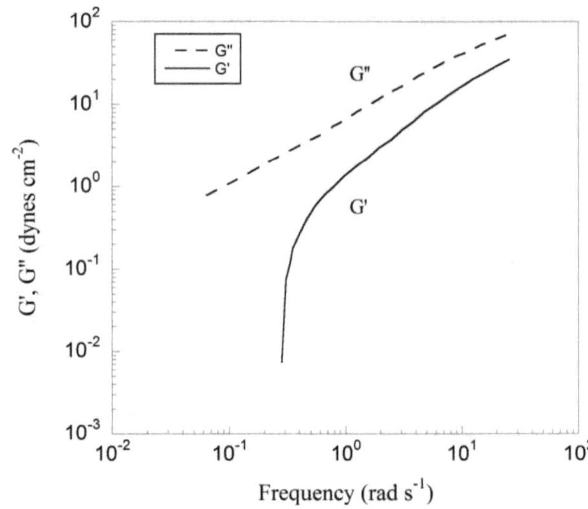

Fig. 3.49 The storage, G' (dynes cm^{-2}), and loss G'' (dynes cm^{-2}) moduli of a 400 mg mL^{-1} gliadin dispersion are shown; the units of the moduli are dynes cm^{-2} as given in the original reference. (Redrawn from Xu et al. 2002)

Based on microrheological experiments, Corrigan and Donald (2009) estimated the critical concentration for gelation of β-lactoglobulin was below 3 %, lower than the 5.2 % estimated by fitting the G'_{inf} versus concentration data to the cascade model (Gosal et al. 2004). They pointed out that oscillatory shear bulk rheological tests may disrupt either the fibril or gel formation. Moschakis et al. (2010) also noted that the particle tracking method has higher sensitivity and can detect changes in the structuring of the system before these are registered by the bulk rheological measurement.

Multiparticle Video Microscopy

In a homogeneous, isotropic material, it is sufficient to examine the MSD of a single particle. However, in heterogeneous materials, it would be useful to obtain the MSD of two or more particles. In addition, video microscopy is gaining increasing use. The main advantage in using video microscopy for microrheology is its potential for following the motions of as many as 100 colloidal particles simultaneously and the ability to obtain the ensemble-averaged MSD while still retaining each of the individual particle trajectories (Gardel et al. 2005). Algorithms have been developed to automate the process of particle tracking (finding particle centers and accurately finding particles; Gardel et al. 2005; Apgar et al. 2000; Crocker and Weeks, http://www.physics.emory.edu/~weeks/idl/tracking.html). Two-dimensional trajectories obtained using video microscopy have been used by Xu et al. (2007) and Moschakis et al. (2010).

Microstructure from Microrheology

Examples of studies using such direct particle tracking to determine the microheterogeneity include dispersions of wheat gliadin (Xu et al. 2002) and β-glucan (Xu et al. 2007), and a phase separating emulsion (Moschakis et al. 2006). Xu et al. (2002) evaluated the contributions of the 10, 25, and 50% highest MSD values to the ensemble-averaged MSD in 0.25, 0.50, 1.0, and 2% wheat gliadin dispersions. For β-glucan solutions of ≤1%, those contributions were about 11, 26, and 52%, respectively, very close to those of glycerol, a homogeneous liquid, In contrast, for a 2% β-glucan solution those contributions were 20, 39, and 65%, respectively, indicating that the 2% β-glucan solution exhibited a greater degree of heterogeneity than a homogeneous liquid. Examples of studies in which viscoelastic moduli were determined using direct particle tracking include dispersions of wheat gliadin (Xu et al. 2002) and β-glucan (Xu et al. 2007), and the gel points and kinetics of acid-induced sodium caseinate gelation (Moschakis et al. 2010). In Fig. 3.49, the storage, G' (dynes cm^{-2}), and loss G'', (dynes cm^{-2}) moduli of a 400 mg mL^{-1} gliadin dispersion (Xu et al. 2002) are shown. The necessary particle trajectories and their MSDs were obtained using a well-dispersed suspension of 0.97 μm d fluorescent polystyrene microspheres (0.1 volume%); the values of the viscoelastic moduli were obtained from the MSDs as described earlier.

Mechanical rheological tests require milliliter-volume samples. Furthermore, they can provide flow rheological parameters (apparent viscosity and yield stress) and oscillatory rheological parameters (G', G'') over a wide range of temperatures on samples with a wide range of concentrations; they are well suited for obtaining data on high-solids materials, such as mayonnaise. They can also be used to follow phase transitions, particularly gelation.

Single- and multiparticle light scattering techniques and DWS probe particle movements noninvasively at the molecular level. Together with the single-particle technique, the frequencies covered by DWS are about 10^{-1}–10^{5} rad s^{-1} (Dasgupta et al. 2001). DWS allows for the collection of very low-noise MSD spectra, from which rheological data can be generated over a wide frequency range. However, in contrast to video microscopy, much information is lost because DWS measurements are based on ensemble averages of the Brownian motion of thousands of particles and cannot be used to measure the MSD distribution (Apgar et al. 2000; Xu et al. 2002). Moschakis et al. (2010) obtained G'-time data on a 10% sodium caseinate dispersion. However, DWS and other light scattering experiments seem to have been limited to dispersions and polymer solutions having concentrations of ~4%. Invariably, they are restricted to fixed ambient temperatures, so that temperature sweep data (e.g., G'-temperature) have not been obtained. In addition, they also do not seem to provide data on flow rheology so that two important properties: apparent viscosity and, when applicable, yield stress cannot be obtained.

Microstructure from Micro Rheology. Examples of studies using such direct particle tracking to determine the microheterogeneity include on dispersions of wheat gliadin (Xu et al. 2002) and β-glucan (Xu et al. 2007), and a phase separating emulsion (Moschakis et al. 2006). Xu et al. (2002) evaluated the contributions of the 10%, 25%, and 50% highest MSD values to the ensemble-averaged MSD in 0.25%, 0.50%, 1.0%, and 2% wheat gliadin dispersions. For β-glucan solutions of ≤1%, those contributions were about 11%, 26%, and 52%, respectively, very close to those of glycerol, a homogeneous liquid. In contrast, for a 2% β-glucan solution those contributions were 20%, 39%, and 65%, respectively, indicating that the 2% β-glucan solution exhibited a greater degree of heterogeneity than a homogeneous liquid.

Commercial Rheometers

Many of the earlier commercial viscometers have been of the constant shear rate design where the speed of rotation was the controlled variable and the resulting shear stress was measured. However, viscometers and rheometers in which the shear stress is controlled and resulting shear rate or strain can be measured are available now. Because they provide an opportunity to conduct studies related to yield stress, creep-compliance, stress relaxation, and rate of breakdown of weak structures, it seems likely that their use will become more widespread in the future. In addition, because with most fluid foods the stresses to be measured are of relatively low magnitudes, a few low shear rheometers that are capable of measuring them have been introduced. Furthermore, to better understand how food structure deforms under elongational and other flows at the microscale, equipment is being developed to visualize food structure behavior under deformation combining confocal microscopy and shear/compression apparatus (e.g., Nicolas and Paques 2003). One can anticipate that more rheo-optical studies will be conducted which should provide both visual and rheological data on foods at the micro scale.

All the major manufacturers of viscometers and rheometers have Internet sites that illustrate and describe their products. In addition, many of the manufacturers are offering seminars on rheometers and rheology. Earlier lists of available models of rheometers and their manufacturers were given by Whorlow (1980), Mitchell (1984), and Ma and Barbosa-Cánovas (1995). It is very important to focus on the proper design of a measurement geometry (e.g., cone-plate, concentric cylinder), precision in measurement of strain and/or shear rate, inertia of a measuring system and correction for it, as well as to verify that the assumptions made in deriving the applicable equations of shear rate have been satisfied and to ensure that the results provided by the manufacturer are indeed correct.

Some Precautions with Foods

In addition to using a well-designed viscometer or a rheometer, many precautions should be taken to insure that the rheological data obtained on a food sample are reliable. The first step is to choose a measurement geometry that is amenable to analysis of the flow in the geometry so that a mathematical expression for the shear rate can be derived. A specific food and a measuring geometry can pose their unique set of problems during rheological experiments. Often, these problems can be identified by careful examination of the steps involved in the measurements and careful observation during the experiment; they should be analyzed and steps taken to ensure that they are overcome. Some of the factors to be considered in using the capillary, cone-plate, parallel plate, and concentric cylinder geometries were mentioned under each geometry in this chapter.

One serious problem in working with foods is dehydration of the test sample, such as from the edge of a parallel plate or a cone-plate geometry or the exposed annular ring of a concentric cylinder geometry. Moisture loss from a test sample can be a significant factor during either: (1) a long-duration low-temperature experiment, (2) a short-duration high-temperature rheological experiment, or (3) when the humidity of the surrounding atmosphere is low. In particular, the exposed sample surface area is large in the cone-plate and the parallel plate geometries. In dynamic rheological tests, silicone or mineral oil can be placed at the sample edge to prevent moisture loss but care must be taken that the oil does not penetrate into the food or dissolve a food component (e.g., lipids by a mineral oil). In the concentric cylinder geometry, only the annular ring-shaped food surface at the top is exposed.

Attempts to minimize moisture loss from a food sample during a shear rate versus shear stress include: placing a solvent trap above the cone-plate and parallel plate geometries, placing an annular ring/disc above the exposed surface of a concentric cylinder geometry, and saturating the surrounding atmosphere with moisture. It is emphasized that sometimes with rotating viscometer geometries: cone-plate, parallel plate, and concentric cylinder, one may at best minimize moisture loss, but may not be able to eliminate it. For example, Rayment et al. (1998) found moisture loss from guar gum solutions containing rice starch was very severe at concentrations greater than 33 % of the latter and could not obtain reliable data. The converse problem, that is, moisture gain due to condensation at low measurement temperatures is of concern because the moisture can dilute the test sample significantly. With a tube/capillary viscometer, a closed system can be designed to prevent both moisture loss and gain.

Another common problem with two-phase foods (e.g., mustard) and melted cheese is the formation of a thin water or oil layer next to the walls of the measuring geometry. Because this layer of water or oil has a finite velocity during the experiment (slip), the zero-velocity boundary condition (no slip) at the interphase between the food and the measuring geometry that is used in deriving the equations for shear rate is not satisfied.

It is tempting to study time-dependent rheological behavior of a food and apply different flow models to the data collected. However, during loading of the sample into a viscometer geometry, the structure of the food is altered and the degree to which the structure is altered depends on the shearing during sample loading. Therefore, the results obtained will depend on the extent of shearing the sample undergoes prior to data collection. Often, the remedy chosen is to allow the sample to rest for a period of time during which the structure recovers. Obviously, another option, when possible, is to create the test sample in situ in the test geometry and

Thixotropic/Antithixotropic Behavior Expected? Yes ⟶ CCG may be suitable if sample is prepared in situ or time is allowed for sample to recover and measuring system inertia is low

↓

No, and steady shear data are desired; perhaps a CCG can be used if end effects are minimized

↓

Is the fluid a suspension in which the solids tend to settle (e.g., ungelatinized starch granules in water) at any time in the CCG?

↓

Yes ⟶ Do not use CCG

No

↓

Is the gap between cylinders very small? Yes ⟶ Perhaps can use Newtonian $\dot{\gamma}$ as the true shear rate

↓

No. Correct Newtonian $\dot{\gamma}$ using Krieger or power law-based method

Fig. 3A.1 Decision tree for using a concentric cylinder geometry

conduct the rheological tests without going through the steps of loading the sample into the measuring geometry.

Appendix 3–A

Analysis of Flow in a Concentric Cylinder Geometry

One can derive applicable equations for shear rate starting from either the general equations of conservation of mass (continuity) and conservation of momentum (motion) or by conducting balances of mass or momentum on a differential shell. Numerous examples of using the conservation equations or shell balances can be found in Bird et al. (1960) and other texts on transport phenomena. For the concentric cylinder geometry shown in Fig. 3A.1, it will be assumed that the inner cylinder is rotating and the outer cylinder is stationary. Interested readers can derive the applicable equations for the case of outer cylinder rotating using appropriate boundary conditions.

First, from a simple force balance, the total torque (M) on the inner cylinder is

$$\sigma_r\theta \times 2\pi r_i h \times r_i = -M \tag{3A.1}$$

In terms of torque per unit height, $T = \frac{M}{h}$ Eq. 3A.1 becomes

$$\sigma_r\theta = -\frac{T}{2\pi r_i^2} \tag{3A.2}$$

To describe the flow between the inner and outer cylinders, we use cylindrical coordinates (r, θ, z) and note that the fluid moves in a circular motion; the velocities in the radial and the axial directions are zero: $v_\theta = r\Omega$, $v_r = 0$, $v_z = 0$, and due to symmetry $\partial/\partial\theta = 0$.

Also, at steady state $\partial\rho/\partial t = 0$. The equation of continuity in cylindrical coordinates is

$$\frac{\partial\rho}{\partial t} + \frac{1}{r}\frac{\partial}{\partial r}(\rho r v_r) + \frac{1}{r}\frac{\partial}{\partial\theta}(\rho v_\theta) + \frac{\partial}{\partial z}(\rho v_z) = 0$$

All terms in the above equation are zero.

The r-component of the equation of motion is

$$\rho\left(\frac{\partial v_r}{\partial t} + v_r\frac{\partial v_r}{\partial r} + \frac{v_\theta}{r}\frac{\partial v_r}{\partial\theta} - \frac{v_\theta^2}{r} + v_z\frac{\partial v_r}{\partial z}\right)$$
$$= -\frac{\partial p}{\partial r} + \left(\frac{1}{r}\frac{\partial}{\partial r}(r\sigma_{rr}) + \frac{1}{r}\frac{\partial\sigma_{r\theta}}{\partial\theta} - \frac{\sigma_{\theta\theta}}{r} + \frac{\partial\sigma_{rz}}{\partial z}\right) + \rho g r$$

The r-component of the equation of motion reduces to

$$-\rho\frac{v_\theta^2}{r} = \frac{1}{r}\frac{\partial}{\partial r}(r\sigma_{rr}) - \frac{\sigma_{\theta\theta}}{r} \tag{3A.3}$$

Equation 3A.3 deals with the normal stresses σ_{rr} and $\sigma_{\theta\theta}$. For the purpose of illustration, we consider the θ-component of the equation of motion in detail and note in particular that $v_\theta = f(r)$

θ-component

$$\rho\left(\frac{\partial v_\theta}{\partial t}+v_r\frac{\partial v_\theta}{\partial r}+\frac{v_\theta}{r}\frac{\partial v_\theta}{\partial_\theta}+\frac{v_r v_\theta}{r}+v_z\frac{\partial v_\theta}{\partial z}\right)$$
$$=-\frac{1}{r}\frac{\partial p}{\partial \theta}+\left(\frac{1}{r^2}\frac{\partial}{\partial r}\left(r^2\sigma_{r\theta}\right)+\frac{1}{r}\frac{\partial \sigma_{\theta\theta}}{\partial \theta}+\frac{\partial \sigma_{\theta z}}{\partial z}\right)+\rho g\theta \tag{3A.4}$$

Under steady flow condition $\partial v_\theta/\partial t=0$, $v_r=0$, $\partial v_\theta/\partial\theta=0$, and $\partial v_\theta/\partial z=0$, the LHS of Eq. 3A.3 reduces to zero. On the right-hand side, $\partial p/\partial\theta=0$ because there is no pressure gradient in the θ-direction and $\rho g_\theta=0$ because there is no θ-component of gravity. The $\partial\sigma_{\theta\theta}/\partial\theta$ and $\partial\sigma_{\theta z}/\partial z=0$ because there are no shear gradients in the θ and z directions. There is no pressure gradient in the θ direction. We, thus, have for the θ-component

$$\frac{\partial\left(r^2\sigma_{r\theta}\right)}{\partial r}=0 \tag{3A.5}$$

The z-component reduces to

$$0=-\frac{\partial P}{\partial z}+\rho g_z \tag{3A.6}$$

Equation 3A.6 simply describes the hydrostatic pressure in the gap between the two cylinders. Differentiation of Eq. 3A.5, which contains the shear stress of interest, $\sigma_{r\theta}$, results in

$$r^2 \mathrm{d}\sigma_{r\theta}+2r\sigma_{r\theta}\mathrm{d}r=0 \tag{3A.7}$$

Equation 3A.7 can be rearranged to:

$$\mathrm{d}r=-\frac{r}{2}\frac{\mathrm{d}\sigma_{r\theta}}{\sigma_{r\theta}} \tag{3A.8}$$

The boundary condition at the inner cylinder rotating at an angular velocity Ω is $v_\theta=r_i\Omega$ at $r=r_i$ and at the stationary outer cylinder is $v_\theta=0$ at $r=r_o$; both expressions are based again on the assumption of no slip condition at a solid–fluid interface. The velocity distribution is obtained from

$$\frac{v_\theta}{r}=\int_{r_i}^{r}\frac{\mathrm{d}(v_\theta/r)}{\mathrm{d}r}\mathrm{d}r \tag{3A.9}$$

Substituting from Eq. 3.8 for dr in Eq. 3A.9 and using the appropriate integration limits

$$\frac{v_\theta}{r}=\int_{\sigma_i}^{\sigma_{r\theta}}\left(\frac{r\mathrm{d}(v_\theta/r}{\mathrm{d}r}\right)\frac{\mathrm{d}\sigma_{r\theta}}{2\sigma_{r\theta}}=\int_{\sigma_{r\theta}}^{\sigma_i}\left(\frac{r\mathrm{d}(v_\theta/r)}{\mathrm{d}r}\right)\frac{\mathrm{d}\sigma_{r\theta}}{2\sigma_{r\theta}} \tag{3A.10}$$

By using the Leibnitz rule and noting to differentiate with respect to $\sigma_{r\theta}$, and with the boundary conditions: $\sigma_{r\theta}=\sigma_i$ at $r=r_i$ and $\sigma_{r\theta}=\sigma_o$ at $r=r_o$

$$\frac{d(v_\theta/r)}{d\sigma_i}=\frac{1}{2\sigma_i}\left[\left(r\frac{d(v_\theta/r}{dr}\right)_i-\left(r\frac{d(v_\theta/r)}{dr}\right)_o\right] \tag{3A.11}$$

Noting that $\Omega=(v_\theta/r)$ and $\dot{\gamma}=(r(d(v_\theta/r)/dr))$, Eq. 3A.9 can be written as

$$2\sigma_i\frac{d\Omega_i}{d\sigma_i}=\dot{\gamma}_i-\dot{\gamma}_o \tag{3A.12}$$

Therefore, while it is relatively easy to calculate the shear stress at the surface of the rotating cylinder from Eq. 3A.2, one can only derive an expression for the difference in shear rates at the surfaces of the inner and outer cylinders from the basic equations of flow. Additional work is required to calculate the corresponding shear rate $\dot{\gamma}_1$ and there have been several approaches to determine it. One approach has been to apply infinite series solution to the differential equation in 3A.12.

A popular method of calculating the shear rate at the surface of the rotating cylinder is to assume that the test fluid follows the simple power law model

$$\sigma=-K\left[r\frac{d(v_\theta/r)}{dr}\right]^n \tag{3A.13}$$

Equation 3A.12 can be solved after recalling Eq. 3A.2 for shear stress

$$\frac{v_\theta}{r}=-\left(\frac{T}{2nK}\right)^{1/n}\frac{n}{2}r^{-2/n}+C \tag{3A..14}$$

Using the boundary conditions $v_\theta=0$ at $r=r_o$ and $v_\theta=r_i\Omega$ at $r=r_i$, one can obtain the equation given earlier, Eq. 3.7

$$\dot{\gamma}=\frac{2\Omega r_i^2}{\left[1-\left(\frac{r_i}{r_o}\right)^2\right]}\left\{\frac{1-\left(\frac{r_i}{r_o}\right)^2}{n\left[1-\left(\frac{r_i}{r_o}\right)\right]^{2/n}}\right\} \tag{3A.15}$$

When $n=1$ (Newtonian fluid), the Margules equation for shear rate is obtained

$$\dot{\gamma}_i=\frac{2\Omega r_i^2}{\left[1-(r_i/r_o)^2\right]}$$

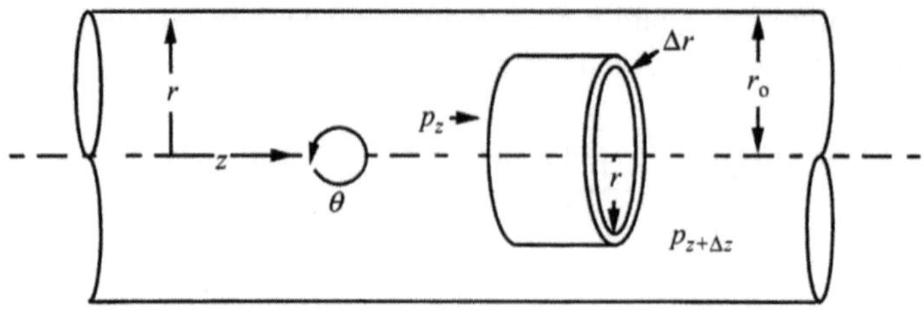

Fig. 3B.1 Schematic diagram for force balance in laminar capillary flow

Examining the limiting form of the second part of Eq. 3.7 as $r_i \rightarrow r_o$

$$\text{Limit}\left\{\frac{1-(r_i/r_o)^2}{n\left[1-(r_i/r_o)^{2/n}\right]}\right\} \underset{r_i \to r_o}{=^{\lim}} \left(\frac{2r_i/r_o}{-n(2/n)(r_i r_o)^{(2/n)-1}}\right) = 1$$

Therefore, when $r_i \rightarrow r_o$, the shear rate of a non-Newtonian fluid tends to that of a Newtonian fluid. In Table 3.1, values of the correction factor in parenthesis in Eq. 3.7 are given for several values of the flow behavior index, *n*, and the ratio (r_i/r_o) of the concentric cylinders, and they confirm that: (1) they are small when $r_i \rightarrow r_o$, and (2) they may be large when the fluids are substantially non-Newtonian and when $(r_i/r_o)<0.95$.

Although the concentric cylinder geometry is relatively easy to use in rheological studies, some of its limitations should be recognized as shown in Fig. 3.44.

Appendix 3–B

Analysis of Steady Laminar Fully Developed Flow in a Pipe

The shell balance method will be used to examine steady laminar flow of a fluid in a pipe. For the geometrical system illustrated in Fig. 3B.1 and for steady laminar fully developed flow of a fluid, a shell momentum balance can be conducted (Bird et al. 1960; Geankoplis 1983) using the cylindrical coordinates, *r, θ,* and *z*. The momentum balance is conducted on a control volume shell at a radius r with dimensions Δr and Δ*z*.

1. The balance of two opposing forces is

$$pA|_z - pA|_{z+\Delta z} = P(2\pi r\Delta r)|_z - p(2\pi r\Delta r)|_{z+\Delta z} \tag{3B.1}$$

2. The shear or drag force at radius $r = \sigma_{rz}\ (2\pi r\Delta z)$ which can be considered to be the momentum flow into the cylindrical surface of the shell. The net efflux of

momentum is the difference between the magnitudes the momentum out and momentum in

$$\sigma_{rz}(2\pi r\Delta z)|_{r+\Delta r}\ (\text{OUT}) - \sigma_{rz}(2\pi r\Delta z)|_{r}\ (\text{IN}) \tag{3B.2}$$

3. The momentum flux across the annular surface at z and $z+\Delta z$ is zero because the axial velocity v_z is of the same magnitude at z and $z+\Delta z$.
4. Recognizing that the sum of forces acting on control volume = rate of momentum out of control volume – rate of momentum into control volume

$$p(2\pi r\Delta r)|_{z} - p(2\pi r\Delta r)|_{z+\Delta z} = \sigma_{rz}(2\pi r\Delta z)|_{r+\Delta r}\ (\text{OUT}) - \sigma_{rz}(2\pi r\Delta r)|_{r}\ (\text{IN}) \tag{3B.3}$$

The above equation can be simplified to

$$r\Delta r(p|_{z} - p|_{z+\Delta z}) = \Delta z\left[(r\sigma_{rz})|_{r+\Delta r} - (r\sigma_{rz})|_{r}\right] \tag{3B.4}$$

Dividing both sides by $(\Delta r\Delta z)$, results in

$$\frac{r\left(p|_{z} - p|_{z+\Delta z}\right)}{\Delta z} = \frac{(r\sigma_{rz})|_{r+\Delta r} - (r\sigma_{rz})|_{r}}{\Delta r} \tag{3B.5}$$

As Δr and $\Delta z \rightarrow 0$, the above equation becomes

$$r\frac{\mathrm{d}p}{\mathrm{d}z} = -\frac{\mathrm{d}}{\mathrm{d}r}(r\sigma_{rz}) \tag{3B.6}$$

After integration, one gets

$$r\sigma_{rz} = -\frac{r^2}{2}\frac{\mathrm{d}p}{\mathrm{d}z} + c_1 \tag{3B.7}$$

which can be rearranged to

$$\sigma_{rz} = -\frac{r}{2}\frac{\mathrm{d}p}{\mathrm{d}z} + \frac{c_1}{r} \tag{3B.8}$$

By using the condition $\sigma_{rz} = 0$ at $r = 0$ (center line), the integration constant $c_1 = 0$, so that the stress distribution across the radius of the pipe is

$$\sigma_{rz} = -\frac{r}{2}\frac{\mathrm{d}p}{\mathrm{d}z} \tag{3B.9}$$

that for the wall of the pipe becomes: $\sigma_w = -[(r_o/2; dp/dz)]$, because at $r = r_o$, $\sigma_{rz} = \sigma_w$. When the pressure drop, Δp, is measured for fully developed flow over a length, L, the equation for the shear stress at the wall is

$$\sigma_w = \frac{D\Delta p}{4L} \tag{3B.10}$$

We will derive an expression for the volumetric flow rate, Q, and then another for the shear rate. Because the axial velocity, v_z, is dependent on the radial position

$$Q = \int_0^{r_o} 2\pi r v_z \mathrm{d}r = \pi \int_0^{r_o} v_z(r)\mathrm{d}(r^2) = \pi \int_0^{r_o} v_z 2r\,\mathrm{d}r + \pi \int_{v_z \mathrm{at} r=0}^{v_z \mathrm{at} r=r_o} r^2\,\mathrm{d}v_z \tag{3B.11}$$

Assuming a no slip boundary condition, that is, $v_z = 0$ at $r = r_o$, the right-hand side of the above equation can be written as

$$\pi = \int_0^{r_o} v_z 2r\,\mathrm{d}r - \pi \int_0^{v_z} r^2\,\mathrm{d}v_z \tag{3B.12}$$

The above equation can be integrated and noting that $v_z = 0$ at $r = r_o$, and $r = (r_o/\sigma_w)\sigma_{rz}$, one can obtain the expression

$$Q = \left(\frac{\pi r_o^3}{\sigma_w^3}\right) \int_0^{\sigma_w} \sigma_{rz}^2 \left(-\frac{\mathrm{d}v_z}{\mathrm{d}r}\right) \mathrm{d}\sigma_{rz} \tag{3B.13}$$

Multiplying both sides of the above equation by 4 and rearranging, we get

$$\frac{4Q}{\pi r_o^3} \left(\frac{4}{\sigma_w^3}\right) \int_0^{\sigma_w} \sigma_{rz}^2 \left(-\frac{\mathrm{d}v_z}{\mathrm{d}r}\right) \mathrm{d}\sigma_{rz} \tag{3B.14}$$

The above equation can be used for deriving a general solution for tube flow and specific expressions for the volumetric flow rates of fluids exhibiting different rheological behaviors. For a Newtonian fluid, noting that the shear rate, $-(\mathrm{d}v_z/\mathrm{d}r) = \frac{\sigma_{rz}}{\eta}$, it can be shown that

$$\frac{4Q}{\pi r_o^3} = \frac{\sigma_w}{\eta} = \frac{1}{\eta}\frac{D\Delta p}{4L} \tag{3B.15}$$

or that the viscosity of a Newtonian fluid

$$\eta = \frac{(D\Delta p/4L)}{(8\bar{v}_z/D)}, \text{that is, } \eta \sim \frac{1}{(\bar{v}_z)} \sim \text{flow time} \tag{3B.16}$$

Equation 3B.16 is the basis for calculation of viscosity of a Newtonian fluid using glass capillary viscometer. It should also be recognized that $(4Q/\pi r_0^3) = (32Q/\pi D^3)$ gives the shear rate for Newtonian fluids but not for non-Newtonian fluids and it is called pseudoshear rate. Additional steps are required to obtain an expression for the true shear rate.

Differentiating both sides of Eq. 3B.14 and setting the limits

$$\frac{d\left[\sigma_w^3 (4Q/\pi r_o^3)\right]}{d\sigma_w} = 4\sigma_w^2 \left(-\frac{dv_z}{dr}\right) \tag{3B.17}$$

Further differentiating the left-hand side and rearranging, one obtains the general solution for the true shear rate in tube flow

$$\left(\frac{dv_z}{dr}\right) = \left(\frac{3}{4}\right) \frac{4Q}{\pi r_0^3} + \frac{\sigma_W}{4} \frac{d(4Q/\pi r_0^3)}{d\sigma_W} \tag{3B.18}$$

Equation 3B.18 is known as the Weissenberg–Rabinowitsch–Mooney (WRM) equation in honor of the three rheologists who have worked on this problem. An alternate equation can be derived for fluids obeying the power law model between shear stress and the pseudoshear rate

$$\sigma = K' (4Q/\pi r_0^3)^{n'} \tag{3B.19}$$

$$\left(-\frac{dv_z}{dr}\right) = \frac{3n' + 1}{4n'} \left(\frac{4Q}{\pi r_0^3}\right) \tag{3B.20}$$

To use the WRM equation, the steps involved are: (1) using a tube/pipe/capillary of known diameter (D) and length (L), several sets of volumetric flow rate (Q) versus pressure drop (ΔP) data are obtained under isothermal fully developed and no slip at the pipe wall conditions.

(2) The quantities $(4Q/\pi r_o^3)$ and $(D\Delta p/4L)$ are calculated, plotted, and a smooth curve fitted.

(3) At each value of $(4Q/\pi r_o^3)$ the value of: $d(4Q/\pi r_o^3)/d\sigma_w$ is determined and the true shear rate calculated using Eq. 3B.18. Alternatively, for many foods, plots of the quantities: log $(4Q/\pi r_o^3)$, and log $(D\Delta p/4L)$ are straight lines with slopes n' that can be used in Eq. 3B.20 to obtain the true shear rate.

(4) Generally, for a given fluid, values of n' and of n of the power law model: $\sigma = K\dot{\gamma}^n$ will be the same. However, the value of K' in $\sigma = K'(4Q/\pi r_0^3)^{n'}$ is related to the consistency index K by the equation

$$K' = K\left(\frac{3n + 1}{4n}\right)^n \tag{3B.21}$$

As an example, for $n = 0.3$, $K' = 1.15\ K$.

The test fluid is low viscosity Newtonian? Yes → Use glass capillary

↓

No, Fluid is non-Newtonian; perhaps a capillary/tube viscometer can be used

↓

Is the fluid a suspension in which the solids tend to settle (e.g., ungelatinized starch granules in water) in the reservoir and tube?

↓

Yes → Do not use tube viscometer.

No

↓

Can Δp be measured over fully developed flow length? Yes → A single tube can be used

↓

No → Use several tubes with different L/D ratios and Bagley's end correction method

Fig. 3B.2 Decision tree for using a capillary/tube viscometer

From Eq. 3B.19, an expression can be derived for a pipe-flow apparent viscosity (η_{ap}) based on the diameter D and the average axial velocity in the tube ($\bar{v}_z$)

$$\eta_{\text{ap}} = K'\left(\frac{8\bar{v}_z}{D}\right)^{n'-1} \tag{3B.22}$$

When $n=n'$, together with Eqs. 3B.21 and 3B.22, one can obtain the generalized Reynolds number (GRe)

$$\mathrm{GRe} = \frac{D^n v_z^{2-n} \rho}{8^{(n-1)} K}\left(\frac{4n}{3n+1}\right)^n \tag{3B.23}$$

Some of the considerations in selecting a capillary/tube viscometer for viscosity measurement are shown in Fig. 3B.2.

Appendix 3–C

Analysis of Flow in a Cone-Plate Geometry

We assume that the cone of radius r_o and cone angle θ_o is on top over the plate rotating with an angular velocity Ω while the bottom plate is stationary as shown in Fig. 3.7. We consider the equation of motion in spherical coordinates *r*, *θ*, *ϕ*. For steady fully developed flow, the velocity components $v_\theta=0$ and $v_r=0$, and v_ϕ is a function of r and *θ*. In addition, we consider pressure variations (body forces) to be negligible and that the value of θ_o is very small, <0.1 rad. The *r*-component reduces

$$\rho \frac{v_\phi^2}{r}\phi = \frac{1}{r^2}\frac{\partial}{\partial r}(r^2\sigma_{rr}) - \frac{\sigma_{\theta\theta}+\sigma_{\phi\phi}}{r} \tag{3C.1}$$

The *θ*-component reduces to

$$\frac{1}{r\sin\theta}\frac{\partial}{\partial\theta}(\sigma_{\theta\theta}\sin\theta) - \frac{\cot\theta}{r}\sigma_{\phi\phi} = 0 \tag{3C.2}$$

The *ϕ*-component of the equation of motion reduces to

$$\frac{1}{r}\frac{\partial\sigma_{\theta\phi}}{\partial\theta} + \frac{2\cot\theta}{r}\sigma_{\theta\phi} = 0 \tag{3C.3}$$

We note that the *ϕ*-component contains the shear stress $\sigma_{\theta\phi}$. The boundary conditions are: $v_\phi=0$ at $\theta=\pi/2$ because the plate is stationary and $v_\phi=r\Omega\cos\theta_o$ at $\theta=(\pi/2)-\theta_o$. First, we will derive an expression for the shear stress by conducting a simple torque balance on the plate

$$M = \int_0^{2\pi}\int_0^{r_o} r^2\sigma_{\theta\phi}\,\mathrm{d}r\,\mathrm{d}\phi = \frac{2\pi r_o^3}{3}\sigma_{\theta\phi}\big|_{\pi/2} \tag{3C.4}$$

Integration of Eq. 3C.3 results in the expression: $\sigma_{\theta\phi}|_{(\pi/2)}=\sigma_{\theta\phi}(\theta)\sin^2\theta$. Therefore,

$$\sigma_{\theta\phi}(\theta)=\frac{3M}{2\pi r_o^3\sin^2\theta} \tag{3C.5}$$

When $\theta_o < 0.1\ \text{rad}, \sin^2\left(\frac{\pi}{2}-\theta_o\right)\sim 1$, so that

$$\sigma_{\theta\phi}=\frac{3M}{2\pi r_o^3} \tag{3C.6}$$

Defining T_{cn} as the torque per unit area, Eq. 3C.4 can be written in terms of the cone diameter D

$$\sigma_{\theta\phi}=\frac{3T_{cn}}{D} \tag{3C.7}$$

From Eqs. 3C.5 and 3C.6, it follows that the shear stress in a cone-plate geometry is essentially uniform.

To obtain an expression for shear rate, we can simply say that for low values of the cone angle θ_o we need not distinguish between $\sin\theta_o$ and θ_o (Whorlow 1980) when the shear rate at radius r will be

$$\dot{\gamma}_r=\frac{r\Omega}{r\theta_o}=\frac{\Omega}{\theta_o} \tag{3C.8}$$

Alternatively, we can consider the velocity distribution (Brodkey 1967) for v_ϕ

$$\frac{v_\phi}{r\sin\theta}=\int_{\theta=\pi/2}^{\theta}\frac{d(v_\phi/r\sin\theta)}{d\theta}d\theta \tag{3C.9}$$

We can also substitute for from Eq. 3C.3, $d\theta=\frac{1}{2}\frac{\sin\theta}{\cos\theta}\frac{1}{\sigma_{\theta\phi}}$ and also switch limits: $\sigma_{\theta\phi}$ at $\theta=0$ and σ_{plate} at $\theta=\pi/2$. For a Newtonian fluid, we can substitute in 3C.9 the relationship between shear rate and shear stress and get 3C.11

$$\sigma_{\theta\phi}=\eta\left[\sin\theta\frac{d(v_\phi/r\sin\theta)}{d\theta}\right] \tag{3C.10}$$

$$\frac{v_\phi}{r\sin\theta}=\int_{\theta}^{\pi/2}\frac{1}{\eta}\frac{\sigma_{\theta\phi}}{\sin\theta}d\theta \tag{3C.11}$$

After substituting for $\sigma_{\theta\phi}$ in 3C.11 and performing the integration results in

$$\frac{v_\phi}{r} = \frac{3T}{4\eta r_o}\left[\cot\theta_o - \ln\left(\tan\frac{\theta}{2}\right)\sin\theta\right] \tag{3C.12}$$

Equation 3C.12 can be written for the boundary condition at $\theta = \frac{\pi}{2} - \theta_o$:

$$\Omega\cos\theta_o = \frac{3T}{4\eta r_o}\left[\tan\theta_o - \cos\theta_o \ln\left(\frac{1-\tan\theta_o}{1+\tan\theta_o}\right)\right] \tag{3C.13}$$

For low values of the cone angle θ_o, the above equation becomes

$$\Omega = \frac{3T}{4\eta r_o}(\theta_o + \theta_o) = \frac{3T\theta_o}{\eta D}$$

Therefore, $\sigma_{\theta\phi} \cong \eta\frac{\Omega}{\theta_o}$.

References

Abdel-Khalik, S.1, Hasseger, O., and Bird, R. B. 1974. Prediction of melt viscosity from viscosity data. *Polym. Eng. Sei.* 14: 859–867.

Abdelrahim, K. A., Ramaswamy, H. S., and Van de Voort, F. R. 1995. Rheological properties of starch solutions under aseptic processing temperatures. *Food Res. Int.* 28: 473–480.

Alexander, M. and Dalgleish, D. G. 2006. Dynamic light scattering techniques and their applications in food science. *Food Biophysics*. 1: 2–33.

Apgar, J., Tseng, Y., Fedorov, E., Herwig, M. B., Almo, S. C. and Wirtz, D. 2000. Multiple-particle tracking measurements of heterogeneities in solutions of actin filaments and actin bundles. *Biophysical Journal*. 79: 1095–1106.

Arola, D. F., Powell, R. L., Barrall, G. A. and McCarthy, M. J. 1999. Pointwise observations for rheological characterization using nuclear magnetic resonance imaging. *J. Rheol.* 43: 9–30.

Barnes, H. A. and Carnali, J. O. 1990. The vane-in-cup as a novel rheometer geometry for shear thinning and thixotropic materials. *J. Rheol.* 34: 841–865.

Barnes, H. A., Hutton, J. F., and Walters, K. 1989. An Introduction to Rheology, Elsevier Science Publishers B.V., Amsterdam, The Netherlands.

Bird R. B., Armstrong, R. C., and Hasseger, O. 1977. Dynamics of Polymeric Liquids. John Wiley, New York.

Bird, R. B., Hasseger, O., and Abdel-Khalik S. I. 1974. Co-rotational rheological models and the Goddard expansion. *AIChE J* 20: 1041–1066.

Bistany, K. L. and Kokini, J. L. 1983. Dynamic viscoelastic properties of foods in texture control. *J. Rheol* 27: 605–620.

Bongenaar, J. J. T., Kossen, N. W. F., Metz, B., and Meijboom, F. W. 1973. A method for characterizing the rheological properties of viscous fermentation broths. *Biotechnol. Bioeng.* 15: 201–206.

Briggs, J. L. and Steffe, J. F. 1996. Mixer viscometer constant (k′) for the Brookfield small sample adapter and flag impeller. *J. Texture Stud.* 27: 671–677.

Brodkey, R. S. 1967. The Phenomena of Fluid Motions, Addison-Wesley, Reading, MA.

Campanella, O. H. and Peleg, M. 1987. Analysis of the transient flow of mayonnaise in a coaxial cylinder viscometer. *J. Rheol.* 31: 439–452.

Campanella, O. H., Popplewell, L. M., Rosenau, J. R., and Peleg, M. 1987. Elongational viscosity measurements of melting American process cheese. *J. Food Sei.* 52: 1249–1251.

Cannon Instrument Co. 1982. Instructions for the Use of the Cannon-Ubbelohde Dilution Viscometer, State College, PA.

Casiraghi, E. M., Bagley, E. B., and Christiansen, D. D. 1985. Behavior of mozzarella, cheddar and processed cheese spread in lubricated and bonded uniaxial compression. *J. Texture Stud.* 16: 281–301.

Castell-Perez, M. E., Steffe, J. F., and Morgan, R. G. 1987. Adaptation of a Brookfield (HBTD) viscometer for mixer viscometer studies. *J. Texture Stud.* 18: 359–365.

Chamberlain, E. K. 1999. Rheological properties of acid converted waxy maize starches: effect of solvent, concentration and dissolution time. Ph.D. thesis, Cornell University, Ithaca, NY.

Champenois, Y. C., Rao, M. A., and Walker, L. P. 1998. Influence of-amylase on the viscoelastic properties of starch-gluten pastes and gels. *J. Sei. Food Agric.* 127–133.

Chatraei, S. H., Macosko, C. W., and Winter, H. H. 1981. Anew biaxial extensional rheometer. *J. Rheol.* 25: 433–443.

Choi, Y. J., McCarthy, K. L., and McCarthy, M. J. 2002. Tomographic techniques for measuring fluid flow properties. *J. Food Sei.* 67(7): 2718–2724.

Clark, A. H. and Ross-Murphy, S. B. 1987. Structural and mechanical properties of biopoly. *gels. Adv. Polym. Sei.* 83: 57–192

Clark, R. 1997. Evaluating syrups using extensional viscosity. *Food Technol.* 511: 49–52.

Cogswell, F. N. 1972. Converging flow of polym. melts in extrusion dies. *Polym. Eng. Sei.* 12: 64–73.

Cogswell, F. N. 1978. Converging flow and stretching flow: a compilation. *J. Non-Newtonian FluidMech.* 4: 23–38.

Comby, S., Doublier, J. L., and Lefebvre, J. 1986. Stress-relaxation study of high-methoxyl pectin gels, in Gums and Stabilisers for the Food Industry 3, eds., G. O. Phillips, D. J. Wedlock, and P. A. Williams, pp. 203–212. Elsevier Science Publishers, New York.

Corredig, M. and Alexander, M. 2008, Food emulsions studied by DWS: recent advances. *Trends in Food Science & Technology.* 19:67–75

Corrigan, A.M. and Donald, A.M. 2009. Particle tracking microrheology of gel-forming amyloid fibril networks. *European Physical Journal E* 28 (4):457–462.

Cox, W. P. and Merz, E. H. 1958. Correlation of dynamic and steady flow viscosities. *J. Polym. Sei.* 28: 619–622.

Crocker, J. C.; Weeks, E. R. Software package for particle tracking, available from: http://www.physics.emory.edu/~weeks/idl/tracking.html, (cited in Moschakis et al. 2006).

Da Silva, P. M. S., Oliveira, J. C., and Rao, M. A. 1997. The effect of granule size distribution on the rheological behavior of heated modified and unmodified maize starch dispersions. *J. Texture Stud.* 28: 123–138.

Dail, R. V. and Steffe, J. F. 1990. Rheological characterization of crosslinked waxy maize starch solutions under low acid aseptic processing conditions using tube viscometry techniques. *J. Food Sei.* 55: 1660–1665.

Dasgupta, B. R., Tee, S-Y., Crocker, J. C., Frisken, B. J. and Weitz, D. A. 2001. Microrheology of polyethylene oxide using diffusing wave spectroscopy and single scattering. *Physical Review E* 65: 051505.

Dealy, J. M. 1982. Rheometers for Molten Polymers. A Practical Guide to Testing and Property Measurement, Van Nostrand Reinhold, New York.

Dickie, A. M. and Kokini, J. L. 1982. Use of the Bird-Leider equation in food rheology. *J. Food Process Eng.* 5: 157–174.

Diehl, K. C., Hamann, D. D., and Whitfield, J. K. 1979. Structural failure in selected raw fruits and vegetables. *J. Text. Stud.* 10:371–400.

Dogan, N., McCarthy, M. J., and Powell, R. L. 2002. In-line measurement of rheological parameters and modeling of apparent wall slip in diced tomato suspensions using ultrasonics. *J. Food Sei.* 67(6): 2235–2240.

Dogan, N., McCarthy, M. J., and Powell, R. L. 2003. Comparison of in-line consistency measurement of tomato concentrates using ultrasonics and capillary methods. *J. Food Process Eng.* 25(6): 571–587.

Doraiswamy, D., Mujumdar, A. N., Tsao, I., Beris, A. N., Danforth, S. C., and Metzner, A. B. 1991. The Cox-Merz rule extended: a rheological model for concentrated suspensions and other materials with a yield stress. *J. Rheol.* 35: 647–685.

Dzuy, N. Q. and Boger, D. V. 1983. Yield stress measurement for concentrated suspensions. *J. Rheol.* 27: 321–349.

Dzuy, N. Q. and Boger, D. V. 1985. Direct yield stress measurement with the vane method. *J. Rheol.* 29: 335–347.

Elliott, J. H. and Ganz, A. J. 1971. Modification of food characteristics with cellulose hydrocolloids, I. Rheological characterization of an organoleptic property. *J. Texture Stud.* 2: 220–229.

Elliott, J. H. and Ganz, A. J. 1977. Salad dressings-preliminary rheological characterization. *J. Texture Stud.* 8: 359–371.

Ferry, J. D. 1980. Viscoelastic Properties of Polymers, John Wiley, New York.

Gardel, M. L., Valentine, M.T. and Weitz, D. A. 2005. Microrheology, in Microscale Diagnostic Techniques, ed. K. Breuer., p. p. 1–54, New York: Springer-Verlag.

Genovese, D. B. and Rao, M. A. 2003a. Vane yield stress of starch dispersions. *J. Food Sei.* 68(7): 2295–2301.

Genovese, D. B. and Rao, M. A. 2003b. Apparent viscosity and first normal stress of starch dispersions: role of continuous and dispersed phases, and prediction with the Goddard-Miller model. *Appl. Rheol.* 13(4): 183–190.

Genovese, D. B., Acquarone, V. M., Youn, K.-S., and Rao, M. A. 2004. Influence of fructose and sucrose on small and large deformation rheological behavior of heated Amioca starch dispersions. *Food Sei. Technol. Int.* 10(1): 51–57.

Giboreau, A., Cuvelier, G., and Launay, B. 1994. Rheological behavior of three biopolymer/water systems with emphasis on yield stress and viscoelastic properties. *J. Texture Stud.*, 25: 119–137.

Gosal WJ, Clark AH, Ross-Murphy SB. 2004. Fibrillar -lactoglobulin gels: Part 2. Dynamic mechanical characterization of heat-set systems. Biomacromolecules 5(6):2420–9.

Grikshtas, R. and Rao, M. A. 1993. Determination of slip velocities in a concentric cylinder viscometer with Mooney and Kiljanski methods. J. Texture Stud. 24: 173–184.

Grosso, C. R. F. and Rao, M. A. 1998. Dynamic rheology of structure development in low-methoxyl pectin+Ca2++sugar gels. *Food Hydrocolloids* 12: 357–363.

Hamann, D. D. 1983. Structural failure in solid foods, in Physical Properties of Foods, eds. M. Peleg, and E. B. Bagley, pp. 351–383 AVI Publ., Westport, CT.

Hamann, D. D. 1987. Methods for measurement of rheological changes during thermally induced gelation of proteins. *Food Technol.* 41(3): 100, 102–108.

Hansen, L. M., Hoseney, R. C., and Faubion, J. M. 1990. Oscillatory probe rheometry as a tool for determining the rheological properties of starch-water systems. *J. Texture Stud.* 21: 213–224.

James, A. E., Williams, D. J. A., and Williams, P. R. 1987. Direct measurement of static yield properties of cohesive suspensions. *Rheol. Acta* 26: 437–446.

Jao, Y. C., Chen, A. H., Lewandowski, D., and Irwin, W. E. 1978. Engineering analysis of soy dough rheology in extrusion. *J. Food Process Eng.* 2: 97–112.

Keentok, M. 1982. The measurement of the yield stress of liquids. *Rheol. Acta* 21: 325–332.

Khagram, M., Gupta, R. K., and Sridhar, T. 1985. Extensional viscosity of xanthan gum solutions. *J. Rheol.* 29: 191–207.

Kokini, J. L. and Dickie, A. 1981. An attempt to identify and model transient viscoelastic flow in foods. *J. Texture Stud.* 12: 539–557.

Komatsu, H. and Sherman, P. 1974. A modified rigidity modulus technique for studying the rheological properties of w/o emulsions containing microcrystalline wax. *J. Texture Stud.* 5: 97–104.

Kulicke, W.-M. and Porter, R. S. 1980. Relation between steady shear flow and dynamic rheology. *Rheol. Acta* 19: 601–605.

Lai, K. P., Steffe, J. F, and Ng, P. K. W. 2000. Average shear rates in the Rapid Visco Analyser (RVA) mixing system. *Cereal Chem.* 77(6): 714–716.

Larson, R. G. 1985. Constitutive relationships for polymeric materials with power-law distributions of relaxation times. *Rheol. Acta* 24: 327–334.

Leider, P. J. and Bird, R. B. 1974. Squeezing flow between parallel disks-I. Theoretical analysis. *Ind. Eng. Chem. Fundam.* 13: 336–341.

Leppard, W. R. and Christiansen, E. B. 1975. Transient viscoelastic flow of polymer solutions. *Am. Inst. Chem. Engrs. J.* 21: 999–1006.

Liao, H.-J. 1998. Simulation of continuous sterilization offluid food products: theroleofthermorheological behavior of starch dispersion and process, Ph.D. thesis, Cornell University, Ithaca, NY.

Lopes da Silva, J. A. L., Gonpalves, M. P., and Rao, M. A. 1993. Viscoelastic behavior of mixtures of locust bean gum and pectin dispersions. *J. Food Eng.* 18: 211–228.

Lopes da Silva, J. A. L., Gonsalves, M. P., and Rao, M. A. 1994. Influence of temperature on dynamic and steady shear rheology of pectin dispersions. *Carbohydr. Polym.* 23: 77–87.

Lopes da Silva, J. A., Rao, M. A., and Fu, J.-T. 1998. Rheology of structure development and loss during gelation and melting, in Phase/State Transitions in Foods: Chemical, Rheological and Structural Changes, eds. M. A. Rao and R. W. Hartel, pp. 111–156, Marcel Dekker, Inc., NY.

Macosko, C. W. 1994. Rheology: Principles, Measurements and Applications, VCH Publishers, New York.

Maranzano, B. J. and Wagner, N. J. 2002. Flow-small angle neutron scattering measurements of colloidal dispersion microstructure evolution through the shear-thickening transition. *J. Chem. Phys.* 117: 10291–10302.

Mason, P. L., Bistany, K. L., Puoti, M. G., and Kokini, J. L. 1982. A new empirical model to simulate transient shear stress growth in semi-solid foods. *J. Food Process Eng.* 6: 219–233.

Mason, T. G. 2000. Estimating the viscoelastic moduli of complex fluids using the generalized Stokes-Einstein equation. *Rheol. Acta* 39: 371–378.

Mason, T. G. and Weitz, D. A. 1995. Optical measurements of frequency-dependent linear viscoelastic moduli of complex fluids. *Physical Review Letters.* 74(7): 1250–1253.

Mason, T. G., Ganesan, K., van Zanten, J. H., Wirtz, D. and Kuo, S. C. 1997. Particle Tracking Microrheology of Complex Fluids. *Physical Review Letters.* 79 (17): 3282–3285.

Matsumoto, T., Hitomi, C., and Onogi, S. 1975. Rheological properties of disperse systems of spherical particles in polystyrene solution at long time-scales. *Trans. Soc. Rheol.* 194: 541.

McCarthy, K. L. and Seymour, J. D. 1993. A fundamental approach for the relationship between the Bostwick measurement and Newtonian fluid viscosity. *J. Texture Stud.* 24(1): 1–10.

McCarthy, K. L. and Seymour, J. D. 1994. Gravity current analysis of the Bostwick consistometer for power law foods. *J. Texture Stud.* 25(2): 207–220.

McKelvey, J. N. 1962. Polymer Processing, John Wiley and Sons, New York.

Metz, B., Kossen, N. W. F., and van Suijdam, J. C. 1979. The rheology of mould suspensions in Advances in Biochemical Engineering, eds. Ghose, T. K. A. Fiechter, and N. Blakebrough, Vol. 2, pp. 103–156, New York: Springer Verlag.

Metzner, A. B. and Otto, R. E. 1957. Agitation of non-Newtonian fluids. *Am. Inst. Chem. Eng.* J. 3: 3–10.

Michaels, A. S. and Bolger, J. C. 1962. The plastic flow behavior of flocculated kaolin suspensions. Ind. Eng. Chem. Fund. 1: 153–62.

Mills, R and Kokini, J. L. 1984. Comparison of steady shear and dynamic viscoelastic properties of guar and karaya gums. J. Food Sei. 49: 1–4 and 9.

Mitchell, J. R. 1984. Rheological techniques, in Food Analysis: Principles and Techniques, eds. D. W. Gruenwedel and J. R. Whitaker, pp. 151–220, Marcel Dekker, New York.

Mooney, M. 1931. Explicit formulas for slip and fluidity. *J. Rheol.* 2: 210–222.

Morris, E. R. 1981. Rheology of hydrocolloids, in Gums and Stabilisers for the Food Industry 2, eds. G. O. Philips, D. J. Wedlock, and P. A. Williams, pp. pp. 57–78, Pergamon Press Ltd., Oxford, Great Britain.

Morris, E. R., Cutler, A. N., Ross-Murphy, S. B. and Rees, D. A. 1981. Concentration and shear rate dependence of viscosity in random coil polysaccharide solutions. *Carbohydr. Polym.* 1: 5–21.

Moschakis, T., Murray B. S. & Dickinson, E. 2006. Particle tracking using confocal microscopy to probe the microrheology in a phase-separating emulsion containing nonadsorbing polysaccharide. *Langmuir* 22:4710–4719.

Moschakis, T., Murray B. S. & Dickinson, E. 2010. On the kinetics of acid sodium caseinate gelation using particle tracking to probe the microrheology. *J. Colloid and Interface Science* 345(2: 278–285.

Nicolas, Y. and Paques, M. 2003. Microrheology: an experimental technique to visualize food structure behavior under compression-extension deformation conditions. *J Food Sei.* 68(6): 1990–1994.

Nussinovitch, A., Kaletunc, G., Normand, M. D., and Peleg, M. 1990. Recoverable work versus asymptotic relaxation modulus in agar, carrageenan and gellan gels. *J. Texture Stud.* 21: 427–438.

Oakenfull, D. 1984. A method for using measurements of shear modulus to estimate the size and thermodynamic stability of junction zones in non-covalently cross-linked gels. *J. Food Sei.* 49: 1103–1104, 1110.

Oakenfull, D. G., Parker, N. S., and Tanner, R. I. 1989. Method for determining absolute shear modulus of gels from compression tests. *J. Texture Stud.* 19: 407–417.

Okechukwu, P. E., Rao, M. A., Ngoddy, P. O., and McWatters, K. H. 1991. Rheology of sol-gel thermal transition in cowpea flour and starch slurry. *J. Food Sei.* 56: 1744–1748.

Owen, S. R., Tung, M. A., and Paulson, A. T. 1992. Thermorheological studies of food polymer dispersions. *J. Food Eng.* 16: 39–53.

Padmanabhan, M. 1995. Measurement of extensional viscosity of viscoelastic liquid foods. *J. Food Eng.* 25:311–327.

Padmanabhan, M. and Bhattacharya, M. 1993. Planar extensional viscosity of corn meal dough. *J. Food Eng.* 18: 389–411.

Peleg, M. 1980. Linearization of relaxation and creep curves of solid biological materials. *J. Rheol.* 24:451–463.

Perkins, T. T., Smith, D. E., and Chu, S. 1997. Single polymer dynamics in an elongational flow. *Science* 276:2016–2021.

Pine, D. J., Weitz, D. A., Chaikin, P. M. and Herbolzheimer, E. 1988. Diffusing-wave spectroscopy. *Physical Review Letters* 60(2): 1134–1137.

Plazek, D. J. 1996. 1995 Bingham medal address: Oh, thermorheological simplicity, wherefore art thou? *J. Rheol.* 40: 987–1014.

Qiu, C.-G. and Rao, M. A. 1988. Role of pulp content and particle size in yield stress of apple sauce. *J. Food Sei.* 53: 1165–1170.

Qiu, C.-G. and Rao, M. A. 1989. Effect of dispersed phase on the slip coefficient of apple sauce in a concentric cylinder viscometer. *J. Texture Stud.* 20: 57–70.

Rao, M. A. 1975. Measurement of flow properties of food suspensions with a mixer. *J. Texture Stud.* 6: 533–539.

Rao, M. A. 1977a. Rheology of liquid foods-a review. *J. Texture Stud.* 8: 135–168.

Rao, M. A. 1977b. Measurement of flow properties of fluid foods-developments, limitations, and interpretation of phenomena. *J. Texture Stud.* 8: 257–282.

Rao, M. A. 1992. Measurement of viscoelastic properties of fluid and semisolid foods, in Viscoelastic Properties of Food, eds. M. A. Rao and J. F. Steffe, pp. 207–232, Elsevier Applied Science Publishers, London.

Rao, M. A. 2005. Rheological properties of fluid foods, in Engineering Properties of Foods, eds. M. A. Rao and S. S. H. Rizvi, and A. K. Datta, 3rd ed., 41–99, CRC Press, Boca Raton, FL.

Rao, M. A. 2013. Food microstructure and rheology, Chapter 13, in Food Microstructures: Microscopy, Measurement and Modelling, eds: Vic Morris and Kathy Groves, Woodhead Publishing Ltd., Cambridge, UK.

Rao, M. A. and Cooley, H. J. 1984. Determination of effective shear rates of complex geometries. *J. Texture Stud.* 15: 327–335.

Rao, M. A. and Cooley, H. J. 1992. Rheology of tomato pastes in steady and dynamic shear. *J. Texture Stud.* 23: 415–425.

Rao, M. A. and Cooley, H. J. 1993. Dynamic rheological measurement of structure development in high-methoxyl pectin/fructose gels. *J. Food Sei.* 58: 876–879.

Rao, M. A., Cooley, H. J., and Liao, H.-J. 1999. High temperature rheology of tomato puree and starch dispersion with a direct-drive viscometer. *J. Food Process Eng.* 22: 29–40.

Rao, V. N. M., Delaney, R. A. M., and Skinner, G. E. 1995. Rheological properties of solid foods, in Engineering Properties of Foods, eds. M. A. Rao and S. S. H. Rizvi, 2nd ed., pp. 55–97, Marcel Dekker, Inc., New York.

Rayment, P., Ross-Murphy, S. B., and Ellis, P. R. 1998. Rheological properties of guar galactomannan and rice starch mixtures. II. Creep measurements. *Carbohydr. Polym.* 35: 55–63.

Rieger, F. and Novak, V. 1973. Power consumption of agitators in highly viscous non-Newtonian liquids. *Trans. Inst. Chemi. Eng.* 51: 105–111.

Roberts, I. 2003. In-line and on-line rheology measurement of food, in "Texture in Food, Volume 1: SemiSolid Foods," pp. 161–182, edited by Brian M. McKenna, Woodhead Publishing Ltd., Cambridge, UK.

Saunders, P. R. and Ward, A. G. 1954. An absolute method for the rigidity modulus of gelatine gel, in Proceedings of the Second International Congress on Rheology, ed. V. G. W. Harrison, pp. 284–290. Academic Press, New York.

Schlichting, H. 1960. Boundary Layer Theory, McGraw-Hill, New York.

Senouci, A. and Smith, A. C. 1988. An experimental study of food melt rheology. I. Shear viscosity using a slit die viscometer and a capillary rheometer. *Rheol. Acta* 27: 546–554.

Sestak, J., Zitny, R., and Houska, M. 1983. Simple rheological models of food liquids for process design and quality assessment. *J. Food Eng.* 2: 35–49.

Shama, F. and P. Sherman. 1969. The influence of work softening on the viscoelastic properties of butter and margarine. *J. Texture Stud.* 1: 196–205.

Shama, F. and Sherman, P. 1973. Identification of stimuli controlling the sensory evaluation of viscosity. II. Oral methods. *J. Texture Stud.* 4: 111–118.

Sharma, S. K., Hill, A. R., and Mittal, G. S. 1992. Evaluation of methods to measure coagulation time of ultrafiltered milk. *Milchwissenschaft* 47(11): 701–704.

Sharma, S. K., Hill, A. R., Goff, H. D., and Yada, R. 1989. Measurement of coagulation time and curd firmness of renneted milk using a Nametre viscometer. *Milchwissenschaft* 44(11): 682–685

Sherman, P. 1966. The texture of ice cream 3. Rheological properties of mix and melted ice cream. *J. Food Sei.* 31: 707–716.

Sherman, P. 1970. Industrial Rheology, Academic Press, New York.

Sherman, P. and Benton, M. 1980. Influence of skim milk powder/recodan R S ratio on the viscoelasticity of groundnut oil-in-water imitation milks. *J. Texture Stud.* 11: 1–13.

Shomer, I., Rao, M. A., Bourne, M. C., and Levy, D. 1993. Rheological behavior of potato tuber cell suspensions during temperature fluctuations and cellulase treatments. *J. Sei. Food. Agric.* 63: 245–250.

Smith, T. L., Ferry, J. D., and Schremp, F. W. 1949. Measurement of the mechanical properties of polymer solutions by electromagnetic transducers. *J. App. Phys.* 20: 144–153.

Sridhar, T., Tirtaatmadja, V., Nguyen, D. A., and Gupta, R. K. 1991. Measurement of extensional viscosity of polymer solutions. *J. Non-Newtonian Fluid Mech.* 40: 271–280.

Stainsby, G., Ring, S. G., and Chilvers, G. R. 1984. A static method for determining the absolute shear modulus of a syneresing gel. *J. Texture Stud.* 15: 23–32.

Steffe, J. F. 1996. Rheological Methods in Food Process Engineering, Freeman Press, East Lansing, Michigan.

Steiner, E. H. 1958. A new rheological relationship to express the flow properties of melted chocolate. *Revue Internationale de la Chocolaterie* 13: 290–295.

Tamura, M. S., Henderson, J. M., Powell, R. L., and Shoemaker, C. F. 1989. Evaluation of the helical screw rheometer as an on-line viscometer. *J. Food Sei.* 54: 483–484.

Tanner, R. I. 1988. Recoverable elastic strain and swelling ratio, in Rheological Measurements, eds. A. A. Collyer and D. W. Clegg, pp. 93–118, Elsevier Applied Science, New York.

Tattiyakul, J. 1997. Studies on granule growth kinetics and characteristics of tapioca starch dispersion during gelatinization using particle size analysis and rheological methods. M. S. thesis, Cornell University, Ithaca, NY.

Tattiyakul, J. and Rao, M. A. 2000. Rheological behavior of cross-linked waxy maize starch dispersions during and after heating. *Carbohydrate Polymers* 43: 215–222.

Truong, V. D. and Daubert, C. R. 2000. Comparative study of large strain methods for assessing failure characteristics of selected food gels. *J. Texture Stud.* 31: 335–353.

Truong, V. D. and Daubert, C. R. 2001. Textural characterization of cheeses using vane rheometry and torsion analysis. *J. Food Sei.* 66: 716–721.

Van Wazer, J. R., Lyons, J. W., Kim, K. Y., and Colwell, R. E. 1963. Viscosity and Flow Measurement, Interscience Publishers, New York.

Vernon Carter, E. J. and Sherman, P. 1980. Rheological properties and applications of mesquite tree Prosopis juliflora gum 2. Rheological properties and stability of o/w emulsions containing mesquite gum. *J. Texture Stud.* 11:351–365.

Vitali, A. A. and Rao, M. A. 1982. Flow behavior of guava puree as a function of temperature and concentration. *J. Texture Stud.* 13:275–289.

Weitz, D. A. and Pine, D. J. 1992. Dynamic Light Scattering, edited by W. Brown, Oxford University Press, Oxford.

Whorlow, R. W. 1980. Rheological Techniques, Halsted Press, New York.

Wood, F. W. and Goff, T. C. 1973. The determination of the effective shear rate in the Brabender Viscograph and in other systems of complex geometry. *Die Starke* 25: 89–91.

Wu, M. C., Lanier, T. C. and Hamman, D. D. 1985b. Thermal transitions of admixed starch/fish protein systems during heating. *J. Food Sei.* 50: 20–25.

Wu, M. C., Lanier, T. C., and Hamann, D. D. 1985a. Rigidity and viscosity changes of croacker actomyosin during thermal gelation. *J. Food Sei.* 50: 14–19.

Xu, J., Tseng, Y., Carriere, C. J. and Wirtz, D. 2002. Microheterogeneity and micro-rheology of wheat gliadin suspensions studied by multiple particle tracking. *Biomacromolecules*, 3(1): 92–99.

Xu, J., Chang, T., Inglett, G. E., Kim, S., Tseng, Y. and Wirtz, D. 2007. Micro-heterogeneity and micro-rheological properties of high-viscosity oat -glucan solutions. *Food Chemistry* 103: 1192–1198.

Yang, W. H. and Rao, M. A. 1998. Complex viscosity-temperature master curve of cornstarch dispersion during gelatinization. *J. Food Proc. Eng.* 21: 191–207.

Yoo, B. and Rao, M. A. 1995. Yield stress and relative viscosity of tomato concentrates: effect of total solids and finisher screen size. *J. Food Sei.* 60: 777-779, 785.

Yoo, B. and Rao, M. A. 1996. Creep and dynamic rheological behavior of tomato concentrates: effect of concentration and finisher screen size. *J. Texture Studies* 27: 451–459.

Yoo, B., Rao, M. A., and Steffe, J. F. 1995. Yield stress of food suspensions with the vane method at controlled shear rate and shear stress. *J. Texture Stud.* 26: 1–10.

Yoshimura, A. and Prud'homme, R. K. 1988a. Wall slip corrections for Couette and parallel disk viscometers. *J. Rheol.* 32: 53–67.

Yoshimura, A. and Prud'homme, R. K. 1988b. Wall slip effects on dynamic oscillatory measurements. *J. Rheol.* 32: 575–584.

Youn, K.-S. and Rao, M. A. 2003. Rheology and relationship among rheological parameters of cross-linked waxy maize starch dispersions heated in fructose solutions. *J. Food Sei.* 68: 187–194.

Zhou, Z., Solomon, M. J., Scales, P. J., and Boger, D. V. 1999. The yield stress of concentrated flocculated suspensions of size distributed particles. *J. Rheol.* 43: 651–71.

Chapter 4
Rheology of Food Gum and Starch Dispersions

Gums and starches are used extensively as thickening and gelling agents in foods. Therefore, understanding their rheological characteristics is of considerable interest. Because many food gums in dispersions have random-coil configuration and starch dispersions (SDs) have granules, it would be better to study their rheological behavior separately.

Rheology of Food Gum Dispersions

Several water-soluble gums, also called hydrocolloids, are being used in foods and their chemical composition and applications have been well documented (Glicksman 1969; Davidson 1980), and some of their characteristics are summarized in Table 4.1. Imparting viscosity, gelation, and film formation are the main functionalities of food gums. The predominant interest in the food industry is in dispersions of relatively low gum concentrations because of their use in practice. It is emphasized that considerable care should be exercised in preparation of the gum/SDs and the references mentioned in this chapter should be consulted for suitable preparation techniques. In particular, a reasonable period of hydration should be allowed for all polymer dispersions and for charged polymers, to control their conformation, the pH, and the ionic strength of the dispersion should be controlled.

Rheological Models for Apparent Viscosity–Shear Rate Data

Apparent viscosity–shear rate data of food polymer dispersions have been reviewed by Launay et al. (1986); Lopes da Silva and Rao (1992), and others. The general log η_a versus log $\dot{\gamma}$ curve, discussed in Chap. 2, has been used to characterize food polymer dispersions. For example, Lopes da Silva et al. (1992) found that both the modified Carreau and the Cross models, wherein the infinite shear viscosity was considered to be negligible, described the apparent viscosity–shear rate data of

M. A. Rao, *Rheology of Fluid, Semisolid, and Solid Foods*, Food Engineering Series,
DOI 10.1007/978-1-4614-9230-6_4,

Table 4.1 Principal biopolymers used in foods and their classification based on origin (Adapted from Glicksman 1979)

Natural	Semisynthetic	Synthetic
Plant exudates	*Cellulose derivatives*	Polyvinylpyrolidone (PVP)
Gum Arabic	Sodium carboxymethyl cellulose	Polymers from polyethylene
Adraganta gum	Methylcellulose	oxide (Polyox)
Karaya gum	Hydroxyethylcellulose	
Gatti gum	Hydroxypropylmethyl cellulose	
Algal extracts	*Other derivatives*	
Agar	Modified starches	
Alginates	Low-methoxyl pectins	
Carrageenans	Propylene glycoalginates	
Furcellaran		
Gums from seeds		
Guar		
Locust bean gum		
Mesquite		
Gums from cereals		
Starches		
Plant extracts		
Pectins		
Gums from fermentation		
Xanthan		
Dextran		
Gums of animal origin		
Gelatin		
Cassei nates		

locust bean (LB) gum (Fig. 4.1) and pectin dispersions (Fig. 4.2) well, especially when there were experimental data in the plateau region at low shear rates

$$\eta_a = \frac{\eta_0}{\left[1+(\alpha_c\dot{\gamma})^m\right]} \tag{4.1}$$

$$\eta_a = \frac{\eta_0}{\left[1+(\lambda_c\dot{\gamma})^2\right]^N} \tag{4.2}$$

where α_c and λ_c are time constants related to the relaxation times of the polymer in solution and m and N are dimensionless exponents. However, when experimental data were scarce in the plateau region, as was the case with low-concentration pectin solutions of pH 7.0, ionic strength 0.1 M, the Carreau model was adept at predicting the zero shear rate viscosity. In general, the deviations between predicted values by the Carreau model and the experimental data were at high shear rates. It is

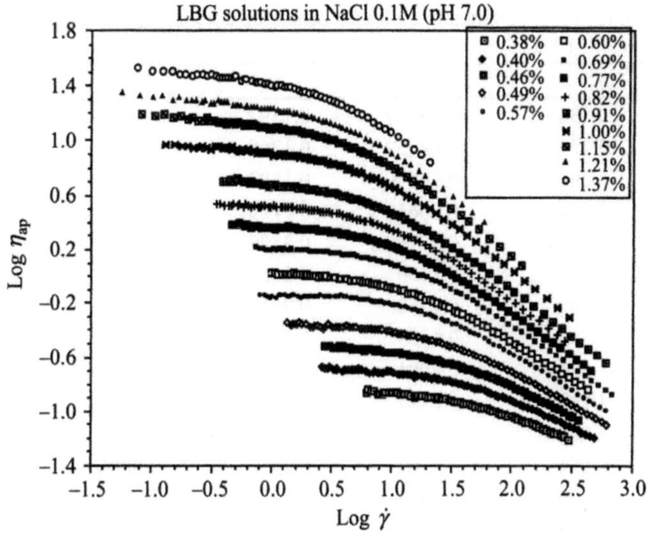

Fig. 4.1 Apparent viscosity (η_a) versus shear rate ($\dot{\gamma}$) of locust bean gum dispersions (0.1 N, pH 7) of different concentrations (Adapted from Lopes da Silva et al. 1992; Lopes da Silva 1994)

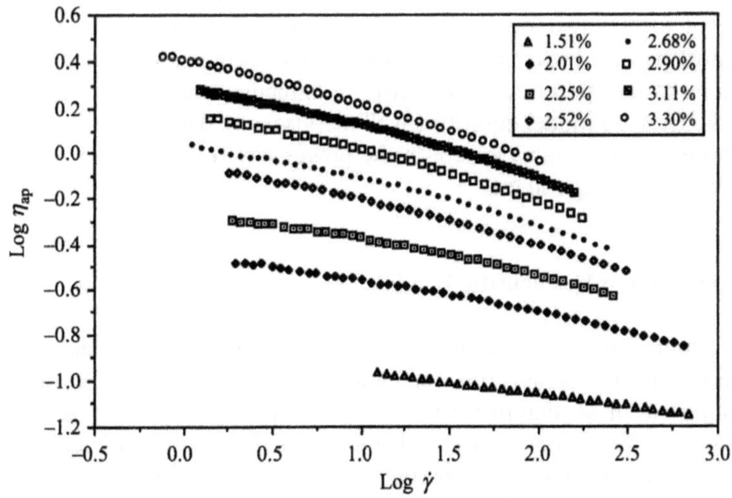

Fig. 4.2 Apparent viscosity (η_a) versus shear rate ($\dot{\gamma}$) of high-methoxyl pectin (0.1 N, pH 7) dispersions with different concentrations (Adapted from Lopes da Silva et al. 1992; Lopes da Silva 1994)

Table 4.2 Carreau model parameters for locust bean gum solutions (ionic strength 0.1 M, pH 7.0; Lopes da Silva et al. 1992)

Sample (%)	η_0	λ	N	R^2	Residual
0.38	0.1386	0.03799	0.1593	1.000	1.844×10^{-4}
0.40	0.1992	0.04552	0.1769	0.9999	1.147×10^{-3}
0.46	0.2995	0.06076	0.1882	0.9999	1.645×10^{-3}
0.49	0.4241	0.06618	0.2125	0.9998	4.851×10^{-3}
0.57	0.7134	0.07798	0.2399	0.9998	7.762×10^{-3}
0.60	1.020	0.1200	0.2265	0.9995	2.886×10^{-2}
0.69	1.589	0.1250	0.2554	0.9995	4.819×10^{-2}
0.77	2.345	0.1755	0.2504	0.9985	2.094×10^{-1}
0.82	3.321	0.2091	0.2515	0.9986	2.988×10^{-1}
0.91	4.917	0.2727	0.2590	0.9954	1.193
0.98	7.086	0.3584	0.2459	0.9896	4.592
1.00	8.681	0.4064	0.2456	0.9886	5.769
1.15	14.00	0.5973	0.2341	0.9738	20.20
1.21	19.24	0.8044	0.2152	0.9606	39.72
1.37	30.63	1.0739	0.2006	0.9638	52.24

emphasized that, as illustrated by the high magnitudes of R^2 in Table 4.2 for Carreau model and in Table 4.3 for Cross model with data on LB gum dispersions, pH 7.0, ionic strength 0.1 M, both models followed the experimental data well.

Yoo et al. (1994) also found that the Cross and Carreau models were satisfactory for the most part in describing the data that covered the zero shear and the power law regions of mesquite seed gum. Also, in agreement with Lopes da Silva et al. (1992), the Carreau model described well the viscosity data shown in Fig. 4.3 over a wide range of shear rates except in the power law region at high shear rates in concentrations > 1.4 g 100 mL^{-1}.

Time Constants of Viscoelastic Foods

Time constants are related to the relaxation times and can be found in equations based on mechanical models (phenomenological approaches), in constitutive equations (empirical or semiempirical) for viscoelastic fluids that are based on either molecular theories or continuum mechanics. Equations based on mechanical models are covered in later sections, particularly in the treatment of creep-compliance studies while the Bird–Leider relationship is an example of an empirical relationship for viscoelastic fluids.

Time constants based on molecular theories have been derived for rod-like and bead-spring models (Bird et al. 1977b; Ferry 1980). For example, Whitcomb and Macosko (1978) showed that the conformation of xanthan gum in solution is rod-like with some flexibility. The bead-spring model has found extensive use in the literature on polymers. The development of molecular theory for dilute solutions of linear polymers credited to Rouse relates the time constant to molecular weight

Table 4.3 Cross model for locust bean gum solutions (ionic strength 0.1 M, pH 7.0; Lopes da Silva et al. 1992)

Sample (%)	η_0	α	m	R^2	Residual	MRD ($\times 10^2$)
0.38	0.1513	0.01939	0.7640	1.000	1.180×10^{-4}	0.1269
0.40	0.2155	0.02396	0.7606	1.000	5.184×10^{-4}	0.1747
0.46	0.3284	0.03167	0.7726	1.000	4.279×10^{-4}	0.1276
0.49	0.4562	0.03118	0.8088	0.9999	2.998×10^{-3}	0.3061
0.57	0.7516	0.02768	0.8965	0.9996	1.733×10^{-2}	0.5405
0.60	1.113	0.05566	0.8133	0.9999	5.443×10^{-3}	0.2515
0.69	1.708	0.05134	0.8773	0.9997	2.889×10^{-2}	0.6125
0.77	2.516	0.06845	0.8712	0.9995	6.941×10^{-2}	0.4222
0.82	3.526	0.07147	0.9043	0.9995	9.541×10^{-2}	0.3555
0.91	5.361	0.1132	0.8564	0.9992	0.2124	0.4375
1.00	9.340	0.1459	0.8526	0.9983	0.8833	0.1653
1.15	15.23	0.2098	0.8217	0.9944	4.362	0.2999
1.21	21.00	0.2552	0.7965	0.9904	9.725	0.3782
1.37	33.25	0.2894	0.8158	0.9921	11.46	0.2539

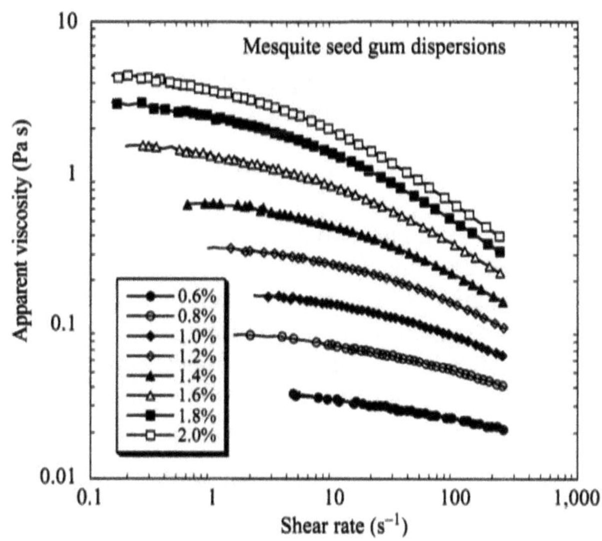

Fig. 4.3 Apparent viscosity (η_a) versus shear rate ($\dot{\gamma}$) of mesquite seed gum dispersions with different concentrations (Adapted from Yoo et al. 1994)

and other basic parameters. The theory assumed that a polymer molecule is made up of beads joined together by flexible spring segments. The equation for the first spring in terms of the zero shear rate viscosity of the polymer solution (η_0) and the viscosity of the solvent (η_S):

$$\tau_R = \frac{6(\eta_0 - \eta_s)M}{\pi^2{}_c RT} \tag{4.3}$$

where τ_R is the Rouse time constant, M is the molecular weight, c is the concentration, R is the gas constant, and T is the absolute temperature. Equations for the time constant in terms of intrinsic viscosity of a polymer solution (Ross-Murphy 1984) and zero shear viscosity of a polymer melt (Elbirli and Shaw 1978) can be found; in the case of polydisperse polymers, it was suggested that a weight average molecular weight can be used in the equation. A time constant due to Zimm derived a few years after Rouse can also be used.

Because shear rate–shear stress data can be obtained relatively easily, a number of rheologists explored the possibility of determining the time constants from these data. Most shear-thinning (pseudoplastic) fluids behave as Newtonian fluids at low shear rates and at a particular shear rate they begin to show their pseudoplastic behavior; the reciprocal of the shear rate at which the transition from Newtonian to pseudoplastic behavior occurs is the characteristic time or the time constant.

Data must be obtained at very low shear rates in order to observe the aforementioned transition and because obtaining such data may not be an easy task, a number of models have been proposed and 14 of these were evaluated by Elbirli and Shaw (1978) for experimental data on polymer melts. Some of the models containing two parameters were found to fit the data and also predict characteristic times as well as some of the models containing three and four parameters. The predicted time constants were higher than those based on Rouse's equation (Eq. 4.3), but the correlation between the two was quite good for the data examined. Among the two-parameter models, the Eyring equation was found to have a slight edge over others in terms of correlation with the Rouse constant

$$(\eta/\eta_0) = \left[\sinh_{-1} \lambda\dot{\gamma}\right]/\lambda\dot{\gamma} \tag{4.4}$$

where η_0 and λ are the two parameters of the model.

The Carreau model not only described well the flow data of LB gum solutions, but the magnitudes the time constant (λ_c) were in good agreement with those of Rouse time (τ_R) constant derived from solution viscosity data while the Cross (α_c) time constants were lower in magnitudes; both the Carreau and the Cross time constants followed well power relationships with respect to the concentration (c) of the solutions (Lopes da Silva et al. 1992)

$$\lambda_c = 0.672_c{}^{2.55} \quad R^2 = 0.981 \tag{4.5}$$

$$\alpha_c = 0.425_c{}^{2.12} \quad R^2 = 0.958 \tag{4.6}$$

The lower magnitudes of the empirical constants in Eq. 4.6 reflect the lower magnitudes of the Cross time constant discussed earlier.

Rheological Properties of Binary Mixtures of Equal Concentration

Equations 4.7 (Kaletunc-Gencer and Peleg 1986) and 4.8 (Miller and Mann 1944) have been used to calculate the "expected" viscosity of the mixtures of equal concentration and to find a blend equation to predict the viscosity properties of LB gum and pectin mixtures (Lopes da Silva et al. 1992):

$$\eta_{\text{mix}} = X_A \eta_A + X_B \eta_B \tag{4.7}$$

$$\eta_{\text{mix}} = \eta_A^{X_A} . \eta_B^{XB} \tag{4.8}$$

where X_A and X_B are the weight fractions of gum A and gum B, respectively, and η_A and η_B are the apparent viscosities (fitted values) of the pure dispersions of gum A and B at the same concentration, respectively. Equation 4.7 does best when the difference between molecular weights of the components is not too great (Bird et al. 1977a). This can be one of the possible reasons for the divergence of the predicted values relative to the experimental values.

Equation 4.7 does indicate the absence of synergistic effect, but it does not predict the viscosity of the pectin-LB gum polymer blends with concentrations of 1.5%. Equation 4.8 was first used by Miller and Mann (1944) in calculations of power requirements for agitation of mixtures of immiscible liquids. Harris (1970) has used the logarithmic form of this equation to relate the steady zero shear viscosity of two-component blends of polystyrene solutions with the steady zero shear viscosity of the separate components and for CMC/guar gum blends with satisfactory results (Plutchok and Kokini 1986). Equation 4.8 does best when the difference between molecular weights of the components is not too great (Bird et al. 1977a) and predicted a geometric mean viscosity that provided better approximation to the experimental values for the viscosity of pectin-LB gum solutions (Lopes da Silva et al. 1992).

A possible and most likely explanation for the low value of viscosity for the mixtures based on Eq. 4.7 is the very different viscosity dependence on concentration for the pectin and LB gum solutions, and not a true antagonism. To work with these gum solutions at equal concentration and to detect some possible synergism or antagonism between them, we must take into account the viscosity–concentration dependence of dilute gum solutions over the range of concentrations of each gum in the mixture.

Morris (1981) suggested that when the apparent viscosities of two gum dispersions at equal concentrations are substantially different, in order to account for the different viscosity–concentration relationships for the two gums, the apparent viscosity of mixtures of solutions of two gums be studied by mixing solutions of the two biopolymers with concentrations adjusted to give similar viscosities. The steps in this procedure for pectin and LB gum dispersions were (Lopes da Silva et al. 1992): (1) the shear rate—apparent viscosity data of solutions of pectin and LB gum with different concentrations were obtained and (2) the concentrations of pectin

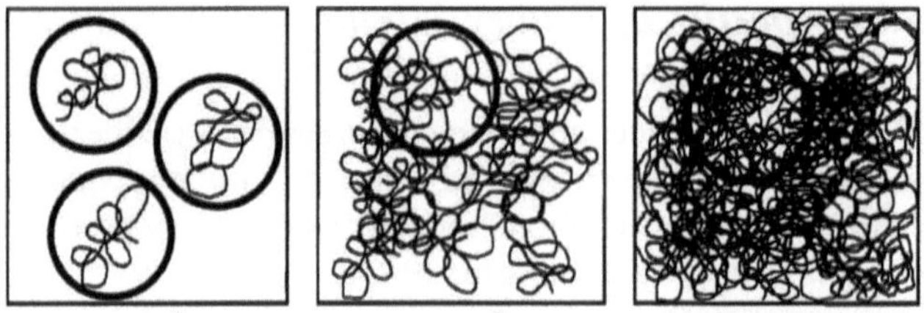

Fig. 4.4 Polymer chain entanglement in dispersions At low concentrations, the polymer chains are not in contact with each other. At high concentrations, the polymer chains are in contact with each other contributing to a large increase in viscosity

and LB gum solutions that had nearly equal zero shear viscosities and that permitted gathering of apparent viscosity–shear rate data over a wide range of shear rates were selected.

Effect of Biopolymer Concentration on Zero-Shear Viscosity

In earlier studies on solutions of synthetic polymers (Ferry 1980), the zero shear viscosity was found to be related to the molecular weight of the polymers. Plots of log η versus log M often resulted in two straight lines with the lower M section having a slope of about 1 and the upper M section having a slope of about 3.4. Because the apparent viscosity also increases with concentration of a specific polymer, the roles of both molecular size and concentration of polymer need to be understood. In polymer dispersions of moderate concentration, the viscosity is controlled primarily by the extent to which the polymer chains interpenetrate that is characterized by the coil overlap parameter $c[\eta]$ (Graessley 1980). Determination of intrinsic viscosity $[\eta]$ and its relation to molecular weight were discussed in Chap. 1. The product $c[\eta]$ is dimensionless and indicates the volume occupied by the polymer molecule in the solution.

The various concentration regimes of importance following de Gennes and coworkers (Tirrell 1994) are: dilute ($c < c^*$), semidilute ($c^* < c < c^‡$), and concentrated ($c > c^‡$). At low concentrations ($c < c^*$), the polymer chains are not in contact with each other, the polymer coils have infinite dilution radii, and the viscosity is relatively low (Fig. 4.4). A good rule of thumb for determining that concentration effects will become important is when the magnitude of $c[\eta]$ is about unity. At the overlap threshold concentration ($c = c^*$), the coils begin to overlap and there is no contraction. It can be shown that the functional dependence of c^* on M is: $c^* \sim M^{-4/5}$. In the semidilute region, the coils contract (Fig. 4.4), but the shrinking does not continue indefinitely and the polymer chain reaches a minimum (θ) dimensions at a concentration $c^‡$ that is independent of molecular weight.

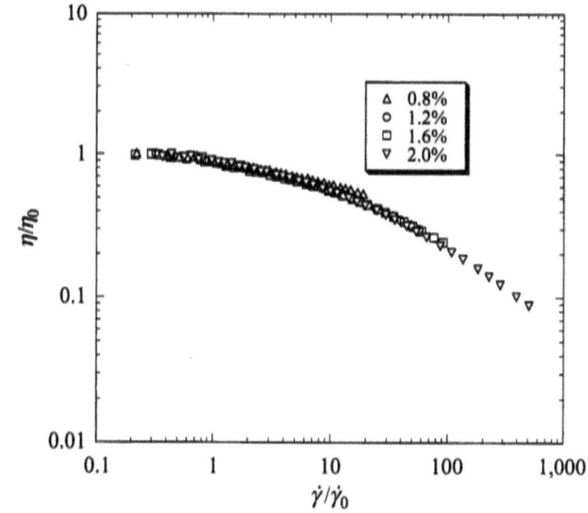

Fig. 4.5 5 Plot of the reduced variables: η/η_0 versus $\dot{\gamma}/\dot{\gamma}_0$ ($\dot{\gamma}_0$ is the shear rate when $\eta_a=0.9\eta_0$) mesquite seed gum dispersions (data of Yoo et al. 1994)

In contrast to $c[\eta]$, the quantity $M[\eta]$ is termed the hydrodynamic volume and is directly proportional to the volume occupied by each molecule as can be deduced from the equation to rotate a polymer molecule in a complex spherical shape:

$$[\eta] = \Phi \frac{(\bar{r^2})^{3/2}}{M} \tag{4.9}$$

where, Φ is a universal constant (Rodriguez 1989).

Graessley (1967) introduced the use of reduced variables, η/η_0 and $\dot{\gamma}/\dot{\gamma}_0$, to represent viscosity data at several concentrations and temperatures. Tam and Tiu (1989) determined $\dot{\gamma}_0$ as the shear rate where $\eta_a = 0.9\ \eta_0$ while Morris et al. (1981) determined $\dot{\gamma}_0$ when $\eta_a = 0.1\ \eta_0$. Figure 4.5 illustrates applicability of the reduced variables technique to data on mesquite seed gum dispersions, where data shown in Fig. 4.3 were converted using the following values of η_0 and $\dot{\gamma}_0$: 0.088 Pa s and 7.63 s^{-1} for the 0.80 % dispersion, 0.33 Pa s and 4.29 s^{-1} for the 1.2 % dispersion, 1.56 Pa s and 0.89 s^{-1} for the 1.6 % dispersion, and 4.50 Pa s and 0.45 s^{-1} for the 2 % dispersion, respectively.

A versatile reduced parameter approach of Graessley (1974) accounted for not only the shear rate but also polymer concentration and temperature

$$\frac{\eta - \eta_s}{\eta_0 - \eta_s} = \left[\frac{(\eta_0 - \eta)M}{cRT}\right](\dot{\gamma}) \tag{4.10}$$

where η_s is the solvent viscosity, M is the molecular weight, c is the polymer concentration, R is the gas constant, and T is the temperature. The molecular relaxation time, τ_M, is defined by the terms in the bracket. When $\eta_0 >> \eta_s$, the above equation can be simplified to

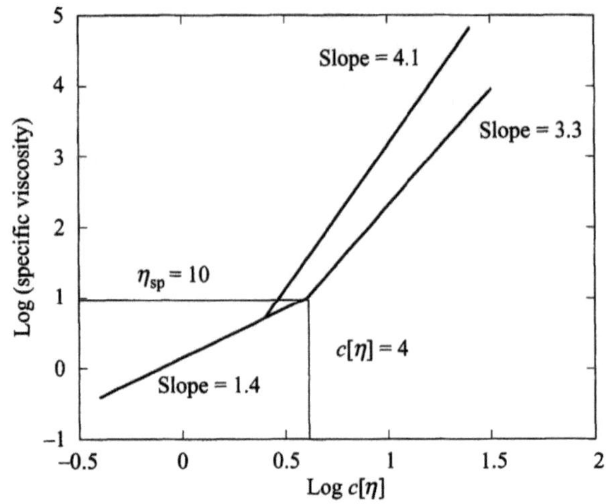

Fig. 4.6 Illustration of dilute and concentrated regimes in terms of log $c[\eta]$ (coil overlap parameter) against log $\eta_{Sp} = [(\eta_0 - \eta_s)/\eta_s]$ (η_{sp} = specific viscosity); slope of 3.3 for entangled polysaccharide chains dissolved in good solvents and 4.1 for polymers with specific intermolecular associations

$$\frac{\eta}{\eta_0} = \left[\frac{(\eta_0 - \eta)M}{cRT}\right](\dot{\gamma}) \tag{4.11}$$

For polyelectrolytes (charged polymers), Tam and Tiu (1993) utilized the expression for specific viscosity in the equation of Fuoss–Strauss

$$\frac{\eta_{sp}}{c} = \frac{[\eta]}{(1 + Bc^{1/2})} \tag{4.12}$$

Tam and Tiu (1993) derived the expression

$$\frac{\eta}{\eta_s} = (1 + Bc^{1/2})\left[\frac{(\eta_0 - \eta_s)M}{cRT}\right](\dot{\gamma}) \tag{4.13}$$

To determine the value of B, (c/η_{sp}) is plotted against $c^{1/2}$ resulting in an intercept of $1/[\eta]$ and a slope of $B/[\eta]$.

A plot of log $c[\eta]$ against log $[(\eta_0 - \eta_s)/\eta s]$ of several food gum solutions at 25°C adapted from Morris et al. (1981) is shown schematically in Fig. 4.6. In the figure, there are two concentration regions: (1) a region of dilute dispersions where the viscosity dependence on concentration follows a 1.4 power and (2) a region of concentrated dispersions where the viscosity dependence on concentration follows a 3.3 power. The latter value is typically found for entangled polysaccharide chains dissolved in good solvents. The transition from the dilute to the concentrated region occurred at a value of $c[\eta] = 4$. The concentration at which the transition from dilute region is denoted as c^*. However, data on a few biopolymers such as guar gum and LB gum were found to deviate from the above observations. First, the region of

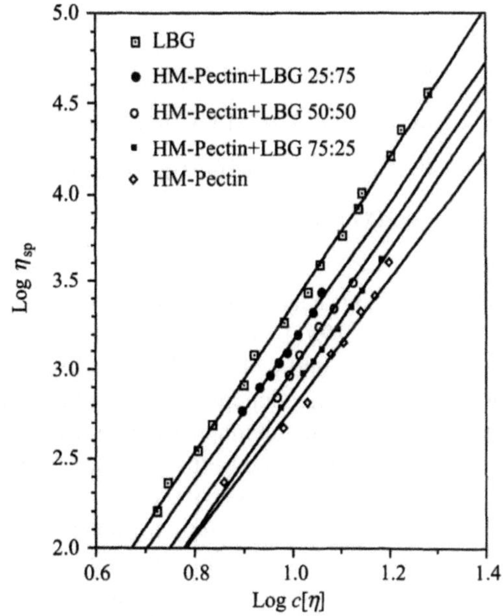

Fig. 4.7 Plot of $c[\eta]$ against $\log \eta_{sp} = [(\eta_0 - \eta_s)/\eta_s]$ for high-methoxyl pectin and locust bean gum, and blends of solutions with similar viscosity, at pH 7.0 and ionic strength 0.1 M (Lopes da Silva et al. 1992)

concentrated solution behavior began at lower values of the coil overlap parameter $c[\eta]=2.5$. Second, the viscosity showed a higher dependence on concentration with a slope of 4.1 instead of about 3.3. These deviations were attributed to specific intermolecular associations (hyperentanglements) between regular and rigid chain sequences in addition to the simple process of interpretation.

Launay et al. (1986) suggested that there could be two transitions, instead of one transition shown in Fig. 4.6, before the onset of high-concentration viscosity behavior. The critical concentration at the boundary between the semidilute and concentrated regimes is denoted as c^{**}. Such behavior was also found for citrus pectin samples with different values of DE at pH 7 and 0.1 M NaCl (Axelos et al. 1989; Lopes da Silva and Rao 2006). Because the intrinsic viscosity of a biopolymer can be determined with relative ease, Fig. 4.6 can be used to estimate the zero shear viscosity of that biopolymer at a specific polymer concentration at 25 °C.

Concentration Dependence of the Zero-Shear Viscosity of Gum Mixtures

Figure 4.7 shows viscosity data of HM pectin and LB gum dispersions, and blends of dispersions with similar viscosity, at pH 7.0 and ionic strength 0.1 M. All the solutions studied were above the critical concentration (c^*), corresponding to $c[\eta]\sim4.5$ for random-coil polysaccharides (Morris and Ross-Murphy 1981). For this concentration regime, relationships with different exponents for the dependence of the specific viscosity on the coil overlap parameter ($c[\eta]$) were obtained:

for LB gum solutions: $\eta_{sp} \propto c[\eta]^{4.2}$ (4.14)

for LB gum 38:HM Pectin 62 blend solutions: $\eta_{sp} \propto c[\eta]^{3.9}$ (4.15)

for LB gum 18:HM Pectin 82 blend solutions: $\eta_{sp} \propto c[\eta]^{4.0}$ (4.16)

for LB gum 6:HM Pectin 94 blend solutions: $\eta_{sp} \propto c[\eta]^{4.0}$ (4.17)

for HM pectin solutions: $\eta_{sp} \propto c[\eta]^{3.6}$ (4.18)

The slope 3.6 is near the expected value for disordered random-coil polysaccharides (Morris et al. 1981), and consistent with the value of 3.3 obtained for citrus pectin samples (Chou and Kokini 1987; Axelos et al. 1989). For the LB gum solutions, a higher coil overlap dependence was obtained (slope≈4.2) in good agreement with the data for galactomannan samples (Morris et al. 1981; Robinson et al. 1982), which can be attributed to the presence of more specific polymer–polymer interactions. Intermediate values of exponents were obtained for the mixtures suggesting that the hydrodynamic volume of each polymer was not significantly affected by the presence of the other, even in concentrated solutions. However, the slopes obtained for the mixtures were slightly greater than expected by simple additivity, which can be due to the errors in the η_0 or to the very important presence of galactomannan entanglements.

Concentration Dependence of the Zero-Shear Viscosity of Amylopectin Solutions

Figure 4.8 is a plot of log $c[\eta]$ against log $[(\eta_0 - \eta)/\eta_s]$ of hydrolyzed and unhydrolyzed amylopectin samples dissolved in 90% DMSO and 10% water (Chamberlain 1999; Chamberlain and Rao 2000). The Newtonian viscosities used to determine η_{sp} in the dilute region were those determined in intrinsic viscosity measurements, while the zero shear viscosities determined from steady shear flow tests were used in the concentrated region. The slope in the dilute region again was about 1.4 as found for random-coil polymer dispersions, while in the concentrated region the slope was 3.6. Based on the intersection of the lines fitted to the dilute and concentrated regions, the c^* region for acid-hydrolyzed Amioca starches in 90% DMSO was found at $c[\eta] = 2$ and $\eta_{sp} = 3$, these values are lower than in Morris et al. (1981; $c[\eta] = 4$ and $\eta_{sp} = 10$). For guar and LB gum, the region of concentrated solution behavior began at lower values of the coil overlap parameter $c[\eta] = 2.5$. However, in the concentrated region, the viscosity of the acid-hydrolyzed Amioca starches showed only a slightly higher dependence on concentration with a slope of 3.6.

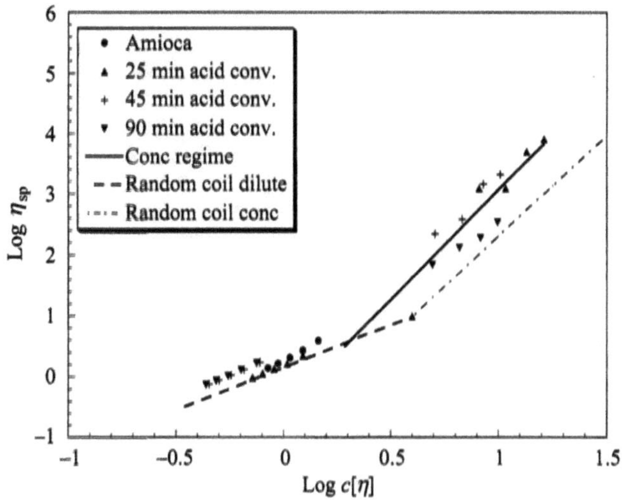

Fig. 4.8 Plot of $c[\eta]$ against log $\eta_{sp} = [(\eta_0 - \eta_s)/\eta_s]$ for hydrolyzed and unhydrolyzed amylopectin dissolved in 90 % DMSO/10 % water (Chamberlain 1999)

Viscoelastic Behavior of Food Gum Dispersions

The magnitudes of G' and G'' obtained for LB gum solutions of 0.75, 1.02, and 1.22 % in distilled water are shown in Fig. 4.9; similar behavior was observed with 2.19 and 3.97 % HM pectin solutions (Lopes da Silva et al. 1993). For both LB gum and HM pectin, the magnitudes of G' were lower than those of G'' over the range of gum concentrations and over a large range of the oscillatory frequencies studied. Log G' and log G'' versus log ω plots followed either a single or two straight lines. The frequency at which G' became equal to G'' will be called crossover frequency, while, in the case of data that followed two straight lines, the frequency at which the break occurred will be denoted ω^*.

The crossover frequency for LB gum and LM pectin decreased with increase in concentration; for LB gum from 27.4 to 12.0 rad s^{-1} as the concentration was increased from 0.75 to 1.22 % and for LM pectin from 64.4 to 31.9 rad s^{-1} as the concentration was increased from 2.78 to 3.98 % (Table 4.4). In contrast, the crossover frequency for HM pectin solutions increased from 49.2 to 231 rad s^{-1} when the concentration was increased from 2.19 to 3.93 % (Table 4.5).

In the study of mixtures of LB gum and HM pectin solutions, the single-gum solutions were selected so that the apparent viscosities were equal at a shear rate of about 10 s^{-1}. The LB gum and HM pectin concentrations of the solutions with pH 7.0, ionic strength 0.1 M, were 0.72 and 3.38 %, while those made with dissolved in distilled water were 0.96 and 3.51 %, respectively; the proportion of LB gum in the mixtures was 25, 50, and 75 %. The crossover frequency for the mixtures decreased

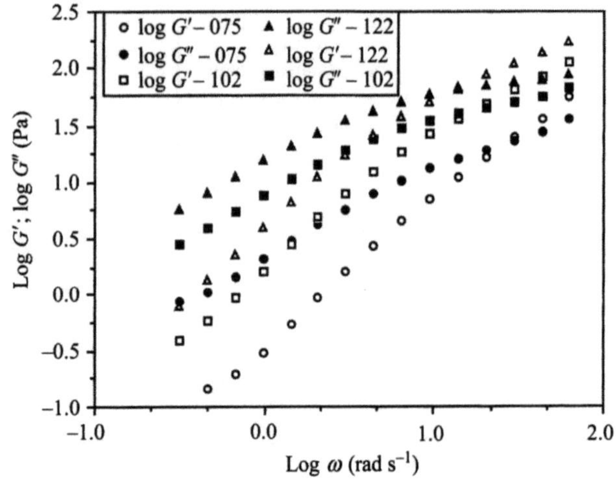

Fig. 4.9 Magnitudes of G' and G'' for locust bean gum solutions of 0.75, 1.02, and 1.22 % in distilled water (Lopes da Silva et al. 1993)

as the proportion of LB gum in the mixture was increased (Lopes da Silva et al. 1993).

Cox–Merz Rule for Biopolymer Dispersions

The Cox–Merz rule (Cox and Merz 1958) is useful in predicting steady shear viscosity from complex viscosity and vice versa

$$\eta^*(\omega) = \eta_a(\omega)\big|_{\omega=\dot{\gamma}} \tag{4.19}$$

Lopes da Silva et al. (1993) obtained the flow data by increasing the shear stress from zero to a maximum value, that depended on the concentration of the sample, and reducing it over the same range in a continuous manner. The time taken for the up and down curves was standardized. For examining the correlation between dynamic and steady shear properties, the down flow curves (decreasing shear stress) and the dynamic data obtained also by decreasing the oscillatory frequency (from 10 to 0.01 Hz) were used.

A common feature observed was the departure between the steady shear viscosity (η_a) and the real component of the dynamic viscosity (η') at large values of shear rate and frequency, with the expected more rapidly decrease of η' with frequency than η_a does with shear rate (Bird et al. 1977a), which can be attributed to the very different molecular motions involved in the dynamic and steady shear at high ω and (Ferry 1980). Because of the relatively high value of strain amplitude used in our tests (36 %) and the two-phase nature of our HM dispersions, the observations with respect to η^* and η_a are in agreement with those of Matsumoto et al. (1975).

Table 4.4 Power law correlation between G' and G'', frequency (rad s^{-1}), and the crossover frequency for solutions of LBG (natural pH and ionic strength)

Sample	A′	B′	R^2	A″	B″	R^2	Crossover frequency	ω^* G'	ω^* G''
LBG 0.75%	0.316	1.46	1.00	2.15	0.907	1.00			
($\omega<\omega^*$)								6.6	5.2
LBG 0.75%	0.668	1.07 0	999	4.05	0.522	0.995	27.4		
($\omega>\omega^*$)									
LBG 1.02%	1.72	1.42	1.00	7.85	0.867	1.00			
($\omega<\omega^*$)								5.0	4.4
LBG 1.02%	5.05	0.75	10.999	18.3	0.298	0.997	17.3		
($\omega>\omega^*$)									
LBG 1.22%	4.05	1.42	1.00	15.4	0.852	1.00			
($\omega<\omega^*$)								4.5	4.0
LBG 1.22%	13.9	0.605	0.999	38.0	0.199	0.989	12.0		
($\omega>\omega^*$)									

Table 4.5 Power law correlation between G' and G'' and frequency (rad s^{-1}) solutions of HM pectin (natural pH and isoionic dilutions)

Sample	A′	B′	R^2	A″	B″	R^2	Crossover frequency
HMP 2.19%	0.0349	1.61	0.998	0.758	0.824	0.997	49.2
HMP 2.62%	0.199	1.20	0.995	1.33	0.844	1.00	197
HMP 2.76%	0.186	1.28	0.996	1.66	0.821	0.999	116
HMP 3.07%	0.386	1.17	0.997	2.55	0.813	1.00	211
HMP 3.48%	0.878	1.07	0.999	4.66	0.773	1.00	261
HMP 3.79%	1.91	0.934	0.997	6.53	0.741	1.00	584
HMP 3.93%	2.04	0.960	0.998	3.12	0.554	0.998	231
				1.66	0.912	0.999	($\omega^*=7.5$ rad s^{-1})

High-Methoxyl Pectin Dispersions

For high-methoxylpectin dispersions, magnitudes of $\eta^*(\omega)$ were almost always lower than those of $\eta_a(\dot{\gamma})$, only converging at higher values of frequency and shear rate (Fig. 4.10). Such behavior is not very common in biopolymeric systems and was reported for semidilute solutions of xanthan gum in 0.5% NaCl (Rochefort and Middleman 1987) and aqueous solutions of hydroxy ethyl guar gum (Lapasin et al. 1991). In addition, Kulicke and Porter (1980) reported that for clustered synthetic ionomers also that magnitudes of $\eta^*(\omega)$ were < $\eta_a(\dot{\gamma})$. However, most of the deviations from the Cox–Merz rule are attributed to structure decay due to the effect of strain deformation applied to a system being low in oscillatory shear and sufficiently high in steady shear to break down intermolecular associations. In these

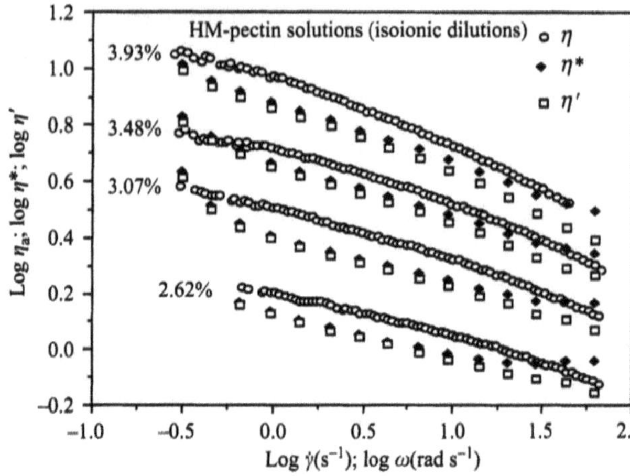

Fig. 4.10 Cox-Merz plot for rheological data on high-methoxyl pectin dispersions (Lopes da Silva et al. 1993)

cases, the magnitudes of $\eta^*(\omega)$ are higher than those of $\eta_a(\dot{\gamma})$ (Lopes da Silva et al. 1993). The observed departure from the Cox–Merz rule may be attributed to aggregation phenomena of the HM pectin in solution (Davis et al. 1980; Paoletti et al. 1986; Sawayamaetal. 1988). Consequently, the HM pectin dispersions were not true solutions, but were two-phase systems with pectin microaggregates dispersed into the solvent.

In addition, for dispersed polymer systems at small strain amplitudes, Matsumoto et al. (1975) observed a pronounced departure from the Cox–Merz rule, with η_a much smaller than η^* at the same rate of shear and angular frequency. They also observed that η^* decreased with increase in the strain amplitude such that for high strain amplitudes η^* was smaller than η_a.

Low-Methoxyl Pectin Dispersions

The LM pectin solutions followed quite satisfactorily the correlation of Cox and Merz (Lopes da Silva et al. 1993; Fig. 4.11). The different behavior in comparison to the HM pectin solutions can be attributed to the less extension of intermolecular association, as a result of electrostatic repulsions due to the presence of ionized free carboxyl groups (higher charge density).

The viscosity values in oscillatory shear at low frequency, especially for the LM pectin solutions for which the oscillatory tests were extended to lower frequency magnitude, may be suggestive of either the existence of a yield stress, which can be substantiated by the results obtained in our earlier study in steady shear flow. On the other hand, the effects of friction in our controlled stress rheometer (Krieger 1990) also cannot be ruled out.

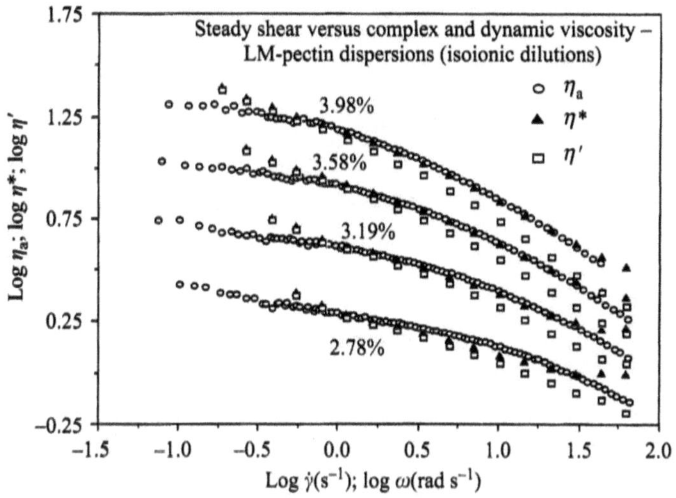

Fig. 4.11 Cox-Merz plot for rheological data on low-methoxyl pectin dispersions (Lopes da Silva et al. 1993)

Locust Bean Gum Dispersions

For LB gum aqueous solutions, at low shear rate and frequency, η^*, η' and η_a have similar magnitudes but η^* and η' *are* slightly smaller than η_a; at higher shear rate and frequency, η^* and η' diverged while η^* and η_a converged, with η^* becoming higher than η_a (Lopes da Silva et al. 1993; Fig. 4.12). Similar behavior at high shear rate and frequency was previously reported for guar gum solutions (Mills and Kokini 1984). Departures from the Cox–Merz rule were observed for galactomannan samples at low shear rates and frequencies (Morris et al. 1981), which were attributed to the presence of high density of entanglements "hyperentanglements" resultant from very specific polymer–polymer interactions. In the LB gum solutions, the departure seems to increase with decreasing concentration. Mesquite seed gum dispersions (Yoo et al. 1994) showed reasonable agreement with the Cox–Merz rule unlike the departures found for galactomannan samples at low shear rates and frequencies.

Constitutive Equations Based on Molecular Structure

Constitutive equations relate stress tensor to various kinematic tensors and can be used to solve flow and other engineering problems. The development of constitutive equations to describe the state of stress continues to be an active area of research.

They have been developed based on either molecular structure or continuum mechanics where the molecular structure is not considered explicitly and the response

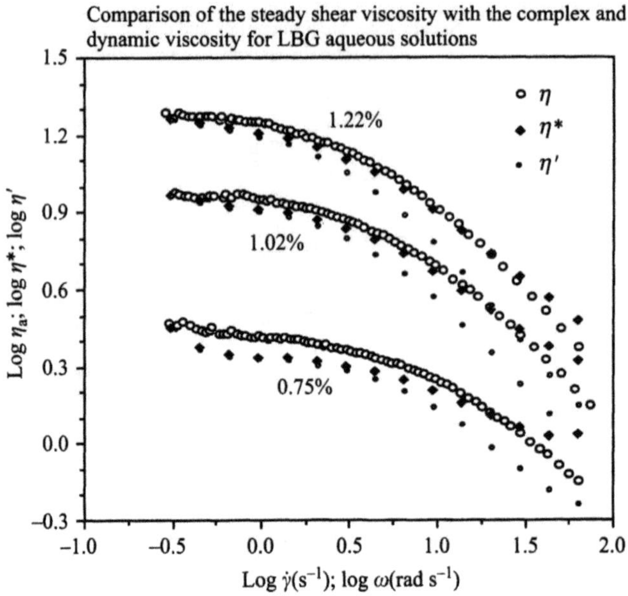

Fig. 4.12 Cox-Merz plot for rheological data on locust bean gum dispersions (Lopes da Silva et al. 1993)

of a material is independent of the coordinate system (principle of material indifference). In the former, the polymer molecules are represented by mechanical models and a probability distribution of the molecules, and relationships between macroscopic quantities of interest are derived. Three models have found extensive use in rheology: the bead-spring model for dilute polymer solutions, the transient network, and the reptation models for concentrated polymer solutions and polymer melts.

Dilute Polymer Solution Molecular Theory

In a dilute polymer solution, there are no entanglements of the individual molecules and the molecules can be considered to be essentially isolated from each other, so that treatment of an individual molecule is sufficient. Much work was done to understand the rheological behavior of dilute polymer solutions using the concept that a polymer molecule can be idealized as dumbbells, that is, two beads connected by a spring. The hydrodynamic, Brownian, and intramolecular forces acting on the beads are considered when the dumbbells are suspended in a Newtonian fluid. Because experimental data cannot be taken at concentrations considered to be sufficiently low for the absence of intermolecular interactions, data obtained at low polymer concentrations (e.g., $c < 10^{-2}$ gm L^{-1}) must be extrapolated to infinite dilution (Ferry 1980).

The appropriate viscoelastic functions are the dynamic rheological properties (storage modulus G' and the loss modulus G'', and the dynamic viscosities η' and η'') extrapolated to infinite dilution and are called the intrinsic dynamic rheological properties

$$[G'] = \lim_{c\to 0} (G'/c) \tag{4.20}$$

$$[G'] = \lim_{c\to 0} [(G'' - \omega\eta_s)/c] \tag{4.21}$$

$$[\eta'] = \lim_{c\to 0} [(\eta' - \eta_s)/\eta_s c] \tag{4.22}$$

$$[\eta''] = \lim_{c\to 0} [\eta''/\eta_s c] \tag{4.23}$$

The results of theories of several molecular models (e.g., rigid and flexible molecules) are predicted to be proportional to *RT/M,* where *M* is the molecular weight that is often assumed to be uniform. The theoretical predictions are conveniently expressed in terms of the reduced dimensionless intrinsic moduli

$$[G'']_R = [G']Mpt/RT \tag{4.24}$$

$$[G'']_R = [G'']M/RT \tag{4.25}$$

Two theories based on the bead-spring model have found extensive use in the polymer literature. In Rouse's theory, the hydrodynamic interaction between the motions of the submolecule junctions is ignored and is based on the change in free energy associated with the entropy decrease for nonrandom configurations and the tendency of a system to diffuse toward a random state (Ferry 1980). In Zimm's theory, inertial forces are neglected and the force on each bead is calculated as the sum of: (1) hydrodynamic drag force by the solvent, (2) the force associated with Brownian motion, and (3) the Hookean springlike forces exerted by two neighboring submolecules (Ferry 1980). Rouse's theory predicts

$$[G']_R = [G'']_R = \frac{\pi}{2\sqrt{2}}(\omega\tau_1)^{1/2} \tag{4.26}$$

The predictions of Zimm's theory are

$$[G']_R = 1.05(\omega\tau_1)^{2/3} \tag{4.27}$$

$$[G'']_R = 1.82(\omega\tau_1)^{2/3} \tag{4.28}$$

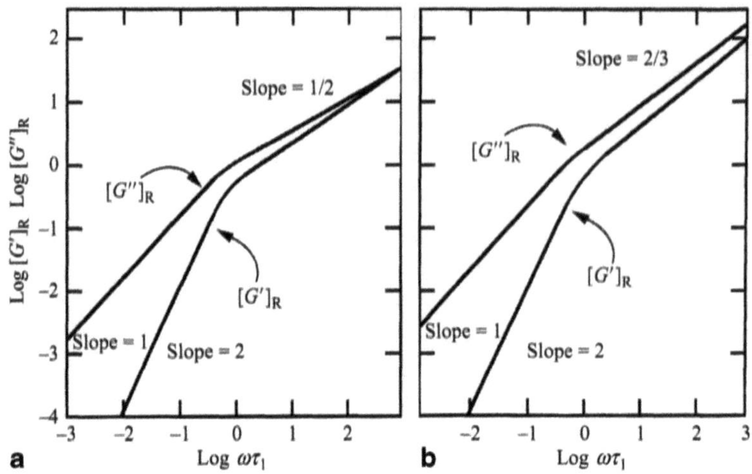

Fig. 4.13 Predictions of reduced storage and loss moduli by models of Rouse (*left*) and Zimm (*right*)

Figure 4.13 contains the predictions of the Rouse theory on the left and of the Zimm theory on the right. As is to be expected, the prediction of the Zimm theory that takes the hydrodynamic interactions into account predicts experimental data well.

Concentrated Polymer Dispersions

For a long time, based on experimental rheological data, some kind of intermolecular coupling was suspected. Pioneering studies by Tobolsky and others led to the concept of an entangled network structure with two sets of relaxation or retardation times. The transient network model is an adaptation of the network theories of rubber elasticity. In concentrated polymer solutions and polymer melts, the network junctions are temporary and not permanent as in chemically cross-linked rubber, so that existing junctions can be destroyed to form new junctions. Earlier, it was mentioned that a critical concentration, c^*, exists below which a polymer dispersion exhibits dilute solution behavior and above which it exhibits concentrated solution behavior. In terms of molecular weight dependence, a low-molecular weight region where η_0 increases almost in direct proportion to molecular weight M and a high molecular weight region where η_0 increases with the 3.4 power of the molecular weight. Further studies led to the conclusion that polymer chain entanglements are topological and not dissociating junctions (Ferry 1980).

The transient network model is an adaptation of the network theory of rubber elasticity. In concentrated polymer solutions and polymer melts, the network junctions are temporary and not permanent as in chemically cross-linked rubber, so that existing junctions can be destroyed to form new junctions. It can predict many of

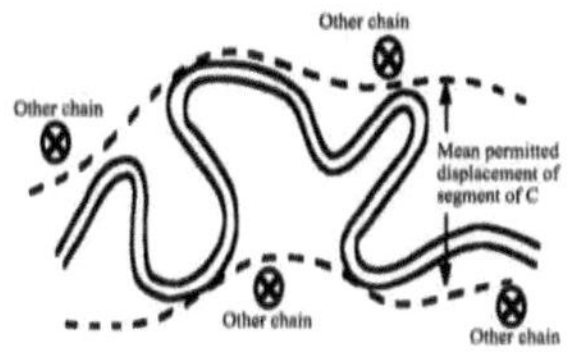

Fig. 4.14 Illustration of reptation concept, credited to de Gennes and Doi and Edwards, of a polymer chain in a concentrated polymer solution. The cross-sections of the constraining chains and the tube for reptation of the polymer chain are shown

the linear viscoelastic phenomena and to predict shear-thinning behavior, the rates of creation and loss of segments can be considered to be functions of shear rate.

The concept of polymer entanglements represents intermolecular interaction different from that of coil overlap type interaction. However, it is difficult to define the exact topological character of entanglements. The entanglements concept was aimed at understanding the important nonlinear rheological properties, such as the shear rate dependence of viscosity. However, viscoelastic properties could not be defined quantitatively as is possible with the reptation model. Because an entanglement should be able to slide along the contours of the participating molecules and change its position with thermal motions, for better understanding of rheological behavior entanglement points or loci are not very useful. Instead, one can consider a model with topological restraints such as a tube placed to surround a polymer molecule and that conforms to the molecule's contortions (Fig. 4.14). Such a model was introduced by Edwards and mathematically treated by de Gennes who pointed out that the easiest escape for sufficiently long chains is achieved by sliding along the contorted contour, that is, by reptation.

The tube model avoids dealing with specification of an average molecular weight between entanglement points (M_e), but the diameter of the tube (d) is an equivalent parameter. Furthermore, the ratio of the contour length to diameter (L/d) is taken to be approximately the magnitude of the entanglements per molecule (M/M_e). Some of the important relationships predicted from the reptation model include (Tirrell 1994)

$$\lambda_{\text{rep}} \sim M^3 \tag{4.29}$$

where λ_{rep} is the longest relaxation time that is time for the chain to diffuse one chain length or renew its configuration. The self-diffusion coefficient

$$D_s \sim M^{-2} \tag{4.30}$$

and viscosity

$$\eta \sim M^3 \tag{4.31}$$

The prediction of the above equation is the closest of any molecular theory to experimentally observed 3.4 power. A constitutive equation based on the reptation model can be written as

$$\bar{\bar{\tau}} = \int_{\infty}^{t} \left\{ M_1[t-t', I_1(t,t'), I_2(t,t')] \bar{\bar{C}}_t^{-1} + M_2[t-t', I_1(t,t'), I_2(t,t')] (\bar{\bar{C}}_t^{-1})^2 \right\} \mathrm{d}t \qquad (4.32)$$

where $\bar{\bar{\tau}}$ is the extra stress tensor, and I_1 and I_2 are the invariants of the Finger deformation tensor $\bar{\bar{C}}_t^{-1}$ that incorporates both stretch and rotation in the deformation gradient. The memory functions M_1 and M_2 can be written as products of functions of the past time $(t - t')$ and of deformation only. In continuum mechanics, the above equation is known as the Kaye–BKZ equation (De Kee and Wissburn 1998). More discussion on constitutive equations can be found in the texts of Bird et al. (1977b); Dealy and Wissburn (1990); and Carreau et al. (1997).

Rheology of Heated Starch Dispersions

Starch Granules and Composition

Starch is the major energy reserve of seeds, roots, and tubers, and is produced in small individual packets called granules (Whistler and Daniel 1985). The composition and characteristics of starches have been well covered elsewhere (Whistler and Daniel 1985; Hoseney 1998) and they will be reviewed only briefly here. The characteristics of starch granules vary in an individual plant depending on the tissue environment. For example, granules in corn's central floury endosperm are round, yet those in the outer, proteinacious, horny endosperm are angular (Whistler and Daniel 1985). As discussed under the rheological behavior of dispersions in Chap. 2, the rheological behavior of SDs is also influenced very strongly by the volume fraction and state of the granules in the dispersion.

Starch granules have a nucleation point, the hilum, around which the granule is formed. A Maltese cross-pattern, characteristic of birefringency, with its center at the hilum can be seen when the granule is viewed with polarized light. Granules contain molecules of amylose and/or amylopectin that seem to be uniformly arranged (Whistler and Daniel 1985). All starch molecules are natural polymers of anhydroglucose units and are present in either a linear or branched configuration. Linkages between the glucose molecules occur through enzymatic condensation in the plant at carbons 1 and 4 (forming a linear polymer) or carbons 1 and 6 (forming a branched polymer; Langan 1986).

Amylose, the linear starch polymer, typically has molecular weights ranging from 10^5 to 10^6 and 500 to 5,000 glucose residues per molecule. It has been determined that some α-(1 Æ6)-branching exists in the amylose molecule on the order of 2–8 branch points per molecule, with side chains ranging in length from 4 to >100 glucose units (Galliard and Bowler 1986). Based on published data on $[\eta]$ in DMSO, Norisuye (1996) suggested that amylose is a random coil as a whole, but that it has certain irregular helical conformation. In addition, a relationship between intrinsic viscosity and molecular weight (M) was suggested for amylose samples with $M > 5 \times 10^4$

$$[\eta] = 1.70 \times 10^{-6}(M)^{0.70}$$

Native starches contain approximately 25 % amylose, yet there are some mutant varieties of corn that contain 85 % amylose (high-amylose corn) or no amylose (waxy maize; Whistler and Daniel 1985). Amylose leaches from starch granules during the gelatinization stage and affects both the viscosity of the continuous phase and the rate of retrogradation on cooling. During gelatinization, a starch is transformed from an ordered, crystalline, state to a disordered, amorphous state. Starch retrogradation is a time- and temperature-dependent process in which, during cooling, the linear regions return to a crystalline and reduced solubility state causing bread staling and syneresis in starch gels.

Amylopectin characteristically has a molecular weight greater than 10^8 with 5 % of its structure containing α-(1 → 6)-branch points (Galliard and Bowler 1986). It is generally agreed that the amylopectin chains are assembled in a cluster structure yet a consensus has not been found for the fine structure, nature of branching, and proportions of side-chains in amylopectin (Galliard and Bowler 1986).

In food products, starch can be either a mere ingredient used in low concentrations to thicken the product texture or the main component responsible for the texture. The former kind of products usually contain water in excess of gelatinization requirements while, in the latter, such as baked products, water is available in limited amount. In rheological studies on dispersions of low starch concentrations, because of settling of the ungelatinized granules, considerable caution should be exercised to ensure that different samples taken from a dispersion contain the same concentration of starch granules. Similarly, granule settling will occur in viscometer geometries that are vertically oriented, for example, concentric cylinder and vertical capillary/tube.

Starch Gelatinization

Starch granules show birefringence in the form of a "Maltese cross" when viewed under polarized light. The amylopectin molecules in the starch granule are partially crystalline. At room temperature, the polymers in the amorphous phase of the starch granule are glassy under dry conditions. At about 22 % moisture and at 25°C, the polymer will go through a glass transition (Hoseny 1998). Irreversible changes, of-

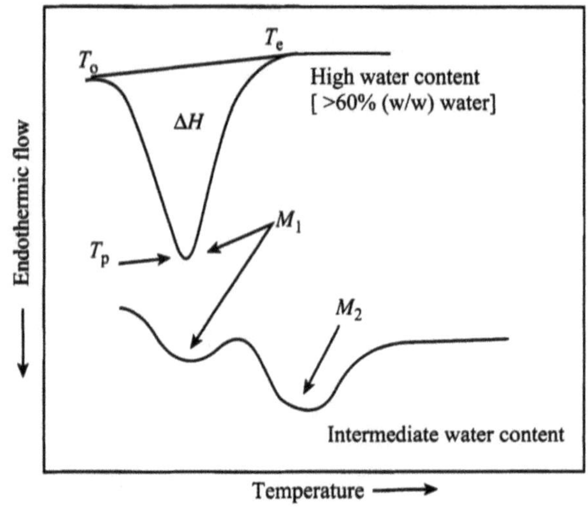

Fig. 4.15 Schematic of DSC thermograms of starch heated in excess and limited water content. At low water content, the endotherm shows two peaks: M_1 and M_2 (Hoseney 1998)

ten referred to as gelatinization, occur when a starch–water mixture is heated above a critical temperature that depends on the source of starch. These changes include loss of birefringence and, more importantly, crystallinity that allows the granules to swell under excess water conditions. For illustration, the DSC output for gelatinization of starch in a Russet Burbank potato sample under excess water conditions (Bu-Contreras 2001), and an idealized output curve of starch with intermediate water content are shown in Fig. 4.15. For the excess moisture curve, drawing a tangent at the start and end points of transition, one can determine well-defined gelatinization initiation, T_0, peak, T_p, and end, T_e, temperatures, respectively. The area under the curve is a measure of the energy required for the transition, ΔH. For gelatinization at intermediate water condition, two endothermic peaks are observed. Often, the temperature at the end of gelatinization is called the starch melting temperature.

Sucrose and other sugars increase the gelatinization temperature of starches; the increase depends on the water activity of the sugar solution and the sugar: the gelatinization temperature of starch increased with decrease in water activity of the sugar solution, and at the same water activity, sugars with longer chain lengths (sucrose, maltose, and maltotriose) increased the gelatinization temperature more than sugars with shorter chain lengths (fructose and glucose). At very high sugar content, the starch does not gelatinize at all and the product collapses during baking, as with cookies and fudge brownies (Hoseney 1998). In limited water ($<\approx 30\,\%$), the gelatinization temperature increases as the moisture content decreases. In a sugar solution, the sugar displaces part of the water and, in effect, decreases the moisture content; therefore, the gelatinization temperature increases in sugar solutions. Hoseney (1998) pointed out that the increase in the onset gelatinization temperature is also due to lowering of the water activity.

Chaotropic salts, such as calcium chloride, decrease the starch gelatinization temperature to the extent that gelatinization may take place at room temperature. In contrast, nonchaotropic salts raise the gelatinization temperature. Although fats and emulsifiers have minimal affect on the starch gelatinization temperature, they affect the rate of starch granule swelling after gelatinization (Hoseney 1998).

Gelatinization Temperature and Extent of Gelatinization

Temperature versus enthalpy profiles similar to those shown in Fig. 4.15 are obtained when starches are heated either in excess water, that is, when water to starch ratios are equal to 2 or greater, or under intermediate water content. The start of the endothermic peak at T_O corresponds to loss of birefringence, in the form of the typical "Maltese cross," when the starch granule is viewed under polarized light. A single endotherm, such as that obtained under excess water conditions, is referred to as the Ml endotherm. Based on data obtained with DSC (Roos 1995), it can be said that many starches heated in excess water conditions exhibit gelatinization temperature ranges of about 7–10°C and the gelatinization initiation temperatures (T_O) range from about 50 to 68°C.

As the amount of water is reduced, the DSC peak for starch gelatinization widens. For example, for wheat starch, the range of temperatures for gelatinization is about 7°C in excess water, and at low water content (water to starch ratio, 0.35:1) it is greater than 30°C and the endotherm shows two peaks (Hoseney 1998), M_1 and M_2, respectively. In addition, the starch does not gelatinize completely even at 100°C. Therefore, when foods containing starch, such as rice and spaghetti, are boiled, the starch granules at or near the surface gelatinize easily, but those in the interior of the food may not gelatinize. The extent of gelatinization (EG) of starch inside a food will depend on the temperature and water content that in turn depends on diffusion of water. For wheat starch and water, based on enthalpies of gelatinization, Fukuoka et al. (2002) proposed an empirical equation for EG as a function of moisture content (m, g water/g starch) and temperature (T,°C) which is valid for $0.54 \leq m \leq 1.5$

$$\mathrm{EG}(m,T) = \frac{3.15m/(1+m) - 0.946}{1+\exp[-0.1792(T-69.1)]} \tag{4.33}$$

After gelatinization, a starch dispersion (SD) may be regarded as a composite material consisting of swollen granules and granular fragments dispersed in a continuous biopolymer matrix (Evans and Haisman 1979; Eliasson 1986; Morris 1990; Noel et al. 1993). Therefore, the properties of the dispersed phase (granules), the continuous phase, and interactions between them are three important considerations for insights into the rheology of gelatinized starch suspensions. Often, these properties are affected by the experimental variables during gelatinization.

Viscous and Viscoelastic Properties During Heating of Starch Dispersions

Rheological changes in heated SDs have been studied under isothermal and non-isothermal heating conditions. Under isothermal conditions, the apparent first-order reaction rate and the Arrhenius models were used to describe the rate of starch gelatinization and the influence of temperature on the rate, respectively (Kubota et al. 1979; Lund 1984; Okechukwu et al. 1991). However, because of the important role of temperature history, changes in η_a during gelatinization cannot be described accurately by the models obtained under isothermal conditions. Temperature and heating rates are important variables during the sol–gel transition period; in addition, the shear rate ($\dot{\gamma}$) or dynamic frequency (ω) has a significant effect on η_a or complex viscosity (η^*), respectively. A good appreciation of the transition phase is vital to a full understanding of the rheological properties of gelatinized SDs.

Because foods containing starches are heated continuously over a range of temperatures, viscosity versus temperature or viscosity versus time profiles are of interest. Results on an 8% (w/w) unmodified corn SD obtained with a DSC are compared with dynamic rheological data in Fig. 4.16. The initial temperature of gelatinization from both methods was about the same, ~67°C, but the differences in peak and conclusion temperatures from the two methods were considerable. The DSC peak temperature, due to the melting of starch crystallites cooperatively in excess water, coincided with that of first rise in viscosity where the rate of development of viscosity was the greatest. Because of the continued swelling of the starch granules, the viscosity increased until 90°C. After the peak viscosity, further increase in temperature resulted in a decrease in viscosity (Yang and Rao 1998).

Model of Dolan et al.

Dolan et al. (1989) developed a model (Eqs. 4.34 and 4.35) to describe apparent viscosity as a function of time during starch gelatinization under nonisothermal conditions and additional discussion on its application to SDs can be found in Dolan and Steffe (1990). The model contains an exponential function of the temperature-time history and the Arrhenius equation to describe the gelatinization reaction. The special form of the model for constant shear rate and starch concentration is

$$\eta_{\mathrm{dim}} = \frac{\eta_{\mathrm{a}} - \eta_{\mathrm{ug}}}{\eta_{\infty} - \eta_{\mathrm{ug}}} = [1 - \exp(-k\Psi)]^{\alpha} \tag{4.34}$$

$$\Psi = \int T(t) \exp\left(\frac{-\Delta E_g}{RT(t)}\right) \mathrm{d}t \tag{4.35}$$

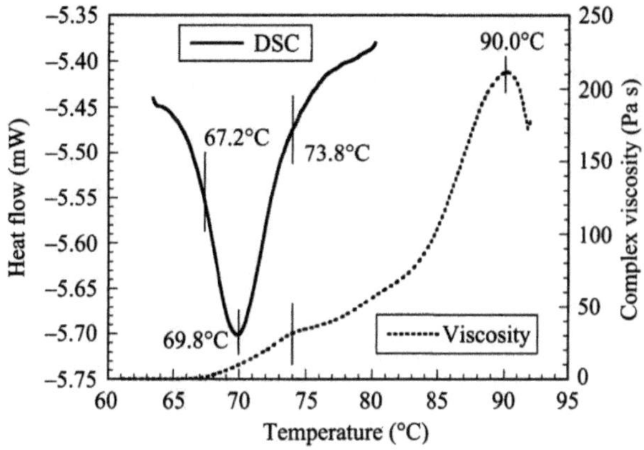

Fig. 4.16 Results on an 8% (w/w) starch dispersion obtained by differential scanning calorimetry are compared with dynamic rheological data (Yang et al. 1997)

where η_a is apparent viscosity (Pa s), η_{dim} dimensionless apparent viscosity, η_{ug} ungelatinized apparent viscosity (Pa s), η_∞ highest magnitude of η_a during gelatinization, Ψ is integral of time-temperature history, α is a constant, E_g *is* activation energy of gelatinization (J mol^{-1}), R is gas constant (8.314 J mol^{-1}K^{-1}), and T temperature of fluid (Kelvin).

Figure 4.17 illustrates applicability of Eq. 4.34 to the η^* -time data obtained at 1.26 rad s^{-1} on an 8% com SD, magnitudes of E_g, k and α in Eqs. 4.34 and 4.35, over the range of heating rates studied were: 240 kJ mol^{-1}, 1.95 E+30, and 0.668, respectively (Yang and Rao 1998). These values are in reasonable agreement with those of Dolan et al. (1989): 210 kJ mol^{-1}, 0.846 E+26, and 0.494, respectively, considering that η^* instead of η_a and starch from a different source were used. The value of R^2 was relatively high (R^2=0.96) but the experimental data showed deviations from the predicted curve for time-temperature history values 5.0×10^{-31} to 1.25×10^{-30} K s. Equations 4.33 and 4.34 with the calculated constants can be used to obtain apparent viscosity from any temperature-time history during gelatinization, provided that the initial and maximum viscosities are known; however, because of granule sedimentation, the former is difficult to determine experimentally. Better fit may be obtained by expressing E_g as a function of temperature and the Arrhenius relationship to be temperature-time history-dependent (Dolan et al. 1989; Dolan and Steffe 1990).

Model for η-Temperature Data (Yang and Rao 1998)*

A η^* versus temperature master curve was derived for com SDs (Yang and Rao 1998) based on separate characteristics: (1) The η_a versus temperature profiles of dispersions of a specific starch (e.g., com, bean) are usually similar over narrow ranges of starch concentrations (Launay et al. 1986; Champenois et al. 1998), so that it may be possible to develop simple relationships for estimating viscosity over specific ranges of concentration. (2) Because η^* data are obtained at low strains

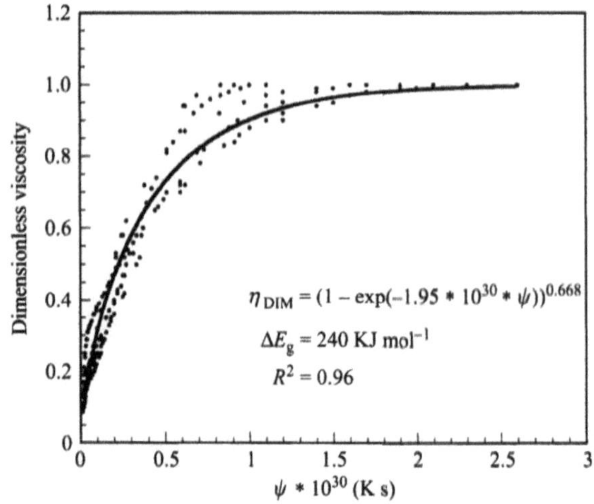

Fig. 4.17 Applicability of the model of Dolan et al. (1989) to an 8 % corn starch dispersion η^* data (Yang and Rao 1998)

with minimal alteration of the SD structure, they provide unique opportunities for studying applicable models. (3) Empirically obtained frequency shift factor (Ferry 1980) has been used successfully in time-temperature superposition studies on food polymer dispersions (Lopes da Silva et al. 1994), and the applicability of similar, if not identical, scaling of frequency was explored for SDs.

Two critical factors for obtaining reliable η^* data were (Yang and Rao 1998): (1) control of the sample volume that was loaded carefully on the bottom rheometer plate and removal of excess dispersion from the edge of the top plate and (2) minimizing water evaporation from the sample when the temperature was high and avoiding displacement of paraffin oil by tiny air bubbles from the edge of the top plate. Dynamic rheological data were obtained using 4-cm-diameter parallel plate (gap 500 μm) geometry of a Carri-Med CSL-100 rheometer (TA Instruments, New Castle, DE). A plate-cone geometry was also tried, but the narrow gap (52 μm at the center) contributed to capillary suction of the few drops of paraffin oil that were placed on the exposed edge of the plate to minimize loss of water vapor during heating into the corn SD. In contrast, with the parallel plate geometry, the drops of paraffin oil placed soon after the storage modulus (G') reached values that were measurable did not penetrate into the corn SD. Because of the narrow gap (500 μm) between the parallel plates, settling of starch granules was not a problem as might have been in a concentric cylinder geometry and the structure formation was not disrupted because of the low magnitude of oscillatory strain.

Influence of Heating Rates

The η^* data obtained at a fixed frequency (1.26 rad s^{-1}) and at different heating rates plotted against temperature (Fig. 4.18) over the range of heating rates: 1.6–6.0°C min^{-1} collapsed to a single curve. The maximum difference in the magnitudes of

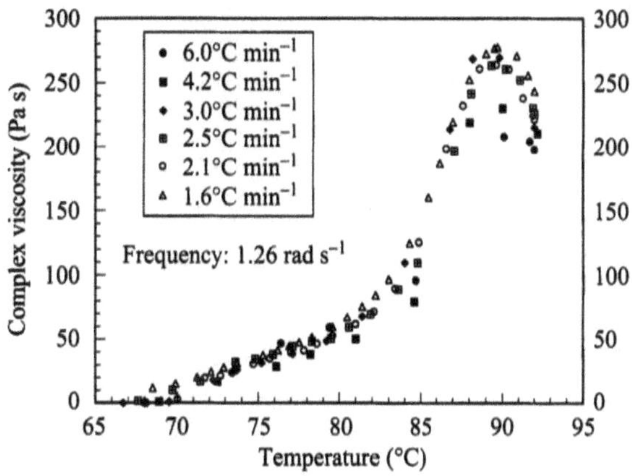

Fig. 4.18 The η^* data of an 8 % corn starch dispersion at a fixed frequency (1.26 rad s⁻1) and at different heating rates plotted against temperature (Yang and Rao 1998)

complex viscosity between the highest (6.0°C min^{-1}) and the lowest (1.6°C min^{-1}) heating rates was about 25 %.

The slight decrease in η^* that occurred after the first increase (Fig. 4.18), was found to be concomitant with end of the DSC endothermic peak due to melting of crystallites (Eliasson 1986). Because of continued swelling of the starch granules, the viscosity increased until about 90° C leading to the formation of a transient network of granules touching each other and sometimes resulting in granule disruption (Blanshard 1987) and amylose leaching (Eliasson 1986). Similar viscosity–temperature behavior has been noted for other cereal starches (Launay et al. 1986; Doublier 1987). However, a complete shear- and temperature-dependent curve, such as in Fig. 4.18, was not presented earlier.

Heating at a Fixed or Increasing Temperature

After the peak viscosity was reached, increasing the temperature further resulted in decrease in viscosity (Eliasson 1986; Doublier 1987) that can be attributed to rupture of the granules. Two distinct types of curves were obtained after the peak (Fig. 4.19) viscosity was reached depending on whether: (1) the temperature was maintained at 92°C for about 2.5 min or at 95°C for about 5.5 min or (2) the temperature was increased up to 95°C continuously after the peak viscosity temperature. The former was characterized by a slower rate of decrease in η^* after the temperature was held constant and the latter by a rapid rate of decrease in η^* up to the final temperature (95°C) suggesting that the network structure formed during gelatinization was drastically weakened by increase in the temperature.

It is theoretically possible that at very high heating rates, starch granule swelling would be limited by moisture diffusion. However, a more practical limitation would be to obtain reliable rheological data at a high heating rate because the temperature range over which viscosity of the corn SD increases would be covered in a short time period.

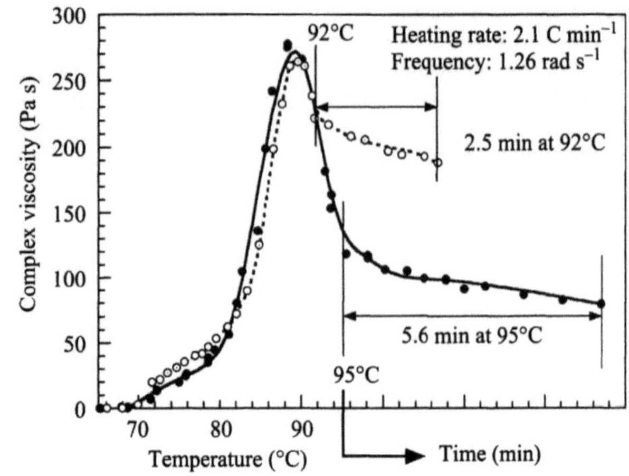

Fig. 4.19 Two distinct curves were obtained after the peak viscosity was reached depending on whether: (1) The temperature was maintained at 92°C for about 2.5 min or at 95°C for about 5.5 min or (2) the temperature was increased up to 95°C continuously after the peak viscosity temperature (an 8% corn starch dispersion)

Influence of Frequency

At a fixed heating rate of 2.1°C min^{-1} and over the range of frequencies employed from 1.26 to 47.12 rad s^{-1}, the profiles of η^* at a specific frequency versus temperature were similar in shape, so that by choosing an arbitrary reference frequency ω_r, all the η^*–temperature curves at the different frequencies could be reduced to a single curve. A master curve of reduced complex viscosity $\eta^*_R = (\omega / \omega_r)$ versus temperature was obtained using $\omega_r = 6.28$ rad s^{-1} (1.0 Hz) at the heating rate of 2.1°C min^{-1}. When data from a limited number of experiments conducted at high ω values: 62.83 rad s^{-1} (10 Hz) and 78.54 rad s^{-1} (12.5 Hz) at the heating rate of 2.16°C min^{-1} were considered, scaling of frequency could be achieved by a more general relationship

$$\eta^*_R = \left(\frac{\omega}{\omega_r}\right)^\beta \tag{4.36}$$

where $(\omega/\omega_r)^\beta$ is the frequency shift factor and the magnitude of the exponent, β, needs to be determined from experimental data at ω and ω_r. For complex viscosity data at 62.83 rad s^{-1} (10 Hz) and 78.54 rad s^{-1} (12.5 Hz), with ω_r=6.28 rad s^{-1}, values of β were 0.913 and 0.922, respectively. It is noteworthy that both increasing and decreasing segments of the viscosity–temperature data at all the frequencies were reduced to a single curve. Complex viscosity data on 8% corn SDs from experiments conducted at different heating rates and frequencies were also superposed using 6.28 rad s^{-1} (1.0 Hz) as ω_r (Fig. 4.20). Figure 4.20 is the master curve of reduced complex viscosity η^*_R data obtained on the 8% corn SD at different heating rates and frequencies and it represents data at several shear rates as a function of temperature. The above modeling procedure was found to be applicable to the η^* data obtained on 6% corn SDs at a heating rate of 3.0°C min^{-1} and at dynamic frequencies: 3.14, 12.57, and 47.12 rad s^{-1}. The superposition of the η^* versus tem-

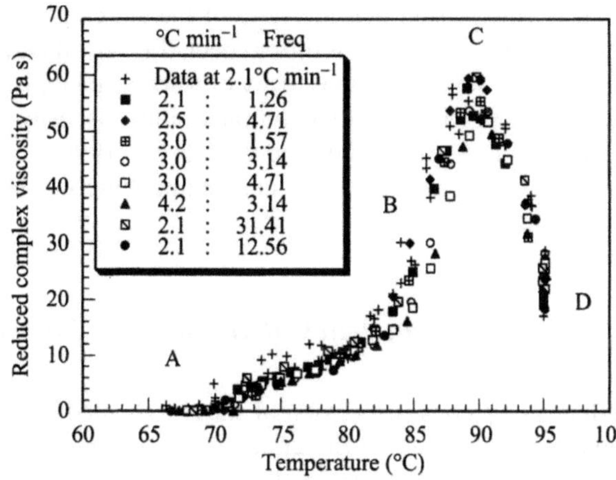

Fig. 4.20 Master curve of reduced complex viscosity $\eta_R^* = \eta^*(\omega/\omega_r)$ data obtained on the 8 % corn starch dispersion at several heating rates and frequencies (Yang and Rao 1998). The letters A–D refer to stages in granule growth

perature data suggests that the molecular mechanisms involved were similar at the different heating rates and oscillatory frequencies used, and depended mainly on the temperature history of the corn SD. It also suggests that the studied corn SDs exhibited thermorheologically simple behavior (Plazek 1996).

The same modeling procedure was applied to 3.5 % corn SDs and the results used to numerically simulate heat transfer to SDs in a can (Yang 1997). From recent studies (Yang and Rao 1998; Liao et al. 1999), the general shape of viscosity versus temperature for native, as opposed to cross-linked, SDs is as shown in Fig. 4.20. It is important to note that the values of viscosity of the gelatinized dispersions at the higher temperatures are substantially higher than those of the ungelatinized dispersions.

State of Granules during Heating

The important role of suspended solids in the viscosity of a dispersion was discussed in Chap. 2. The shape of the curve in Fig. 4.20 reflects the changes in the volume fraction of starch granules, that in turn is due to the changes in the size of the granules and release of amylose, as a dispersion is heated. Initially, at low temperatures, the granules are in the raw state and the volume fraction is low. As the granules swell due to water absorption as they are heated (Fig. 4.20, segment A–B–C), the volume fraction increases and reaches a maximum value (Fig. 4.20, C). With further heating, the granules rupture and disintegrate resulting in a gradual decrease in the volume fraction of the granules in the dispersion (Fig. 4.20, segment C–D). The rupture of granules also results in the release of additional amylose that contributes to the viscosity of the continuous phase of the SD. Thus, in Fig. 4.20, the segment C–D is not a mirror image of the segment A–B–C. The leached amylose and the granule remnants contribute to a viscosity at point D that is higher than that at point A.

It is emphasized that the state of the starch granules has a major influence on the viscosity of an SD. In this respect, Tattiyakul and Rao (2000) reported that the decrease in the apparent viscosity of a cross-linked waxy maize (CWM) SD did not decrease significantly after the maximum viscosity was attained. However, both Tattiyakul and Rao (2000) and Liao et al. (1999) confirmed that master curves of reduced apparent viscosity versus temperature can be generated for application in heat transfer studies.

Rheological Properties of Gelatinized Starch Dispersions

When an aqueous suspension of starch is heated isothermally above the gelatinization onset temperature, the granules absorb large amounts of water and swell considerably to impart substantial increase to the viscosity of the suspension. As granule swelling progresses, the suspension is transformed into a paste which exhibits rheological properties very different from the initial suspension. The swelling of granules is also accompanied by leaching or solubilization of granular constituents, mainly amylose, into the continuous suspending matrix. Granules swell to a maximum value at elevated temperatures, followed by granular disruption and exudation of granule contents into the suspension matrix.

Rheology of Starch Dispersions with Intact Granules

Starch gelatinization up to peak viscosity studied by various techniques such as viscometry and particle size measurement by laser diffraction, is essentially a first-order kinetic process whose extent is determined principally by pasting temperature and time (Kubota et. al. 1979; Lund 1984. p. 59; Okechukwu et al. 1991; Okechukwu and Rao 1995).

Mild thixotropy has been observed in gelatinzed wheat SDs characterized at low temperatures such as 25°C but not at 60°C and above (Doublier 1981; Harrod 1989; Bagley and Christianson 1982). The time-dependent rheological behavior shown by starch is attributed to structure formation by amylose leached into the suspending matrix on cooling to low gelling temperatures (Ellis et al. 1989).

At low starch concentrations, most authors have used the simple power law model without a yield stress for describing shear responses over a wide range of shear. The flow behavior index n and the consistency coefficient K are useful parameters for describing the flow behavior of starch suspensions. Gelatinized starch pastes preheated to temperatures of about 90° C are generally reported to be shear-thinning (pseudoplastic) fluids with values of n considerably less than 1.0 (Evans and Haisman 1979; Doublier 1981; Colas 1986; Ellis et al. 1989; Noel et al. 1993). This value was also reported to decrease linearly with increasing starch concentration in cross-linked corn starches (Colas 1986). This observation appears consistent with those of Evans and Haisman (1979) who showed that in suspensions of corn, potato, modified corn, and tapioca starches as well as wheat flour, the index decreased rapidly from unity and approached a constant value at high concentration. Yield stress of SDs is covered after discussion on their flow behavior.

Shear Thickening in Starch Dispersions

Several studies on starch gelatinization have indicated that some starches, notably wheat, corn, waxy maize, and CWM may show shear-thickening (dilatant) behavior with values of the power law flow behavior index, *n*, greater than 1.0 during the early stages of gelatinization (Bagley and Christianson 1982; Dail and Steffe 1990a, b; Okechukwu and Rao 1995). Shear-thickening flow behavior was reported in wheat SDs heated at 60°C for 15 min (Bagley and Christianson 1982) using a 2.61-mm-gap concentric cylinder viscometer measured at 23°C. Dintzis and Bagley (1995) reported shear-thickening behavior in Amioca (98% amylopectin; 7.7%) dispersions in cans heated at 140°C and the rheological behavior examined at 50°C in a 2.61-mm-gap concentric cylinder viscometer. It was also reported in CWM starch (National 465; 1.82 and 2.72%) at 121–143°C (Dail and Steffe 1990a, b) using a tube (1.27 cm diameter) flow viscometer.

Typical granule size distribution (GSID) data of heated corn SDs (Okechukwu and Rao 1996a) are shown in Fig. 4.21. Similar data were presented for other SDs: cowpea (Okechukwu and Rao 1996b), CWM (Chamberlain 1996), and tapioca (Tattiyakul 1997). These data seem to suggest a relationship between *n* and the change in polydispersity of granules during the sol–gel transition process. Specifically, the flow behavior index appears to decrease linearly with increase in the standard deviation (SD) of heated corn starch granules (Okechukwu and Rao 1995). With an R^2 of 0.98, Eq. 4.37, which is independent of pasting temperature, provides the basis for resolving the changes in the value of *n* observed at various stages of starch gelatinization

$$n = 2.17 - 0.0968\,\mathrm{SD} \qquad (4.37)$$

The above equation suggests a maximum *n* value of about 2.2 for corn starch. The minimum values of *n*, usually within the pseudoplastic range, would depend, in part, on the particle size distribution of the gelatinizing granules which increases as gelatinization progresses. By arbitrarily choosing gelatinization to be complete at 95°C, it was shown that a range of values of *n* can be obtained depending on the extent of granule swelling of corn starch (Okechukwu and Rao 1995).

Apparently, the different values of *n* found in the literature for suspensions of the same starch pasted under various temperature-time profiles reflect various extents of starch granule swelling. Early investigations on the rheology of raw starch suspensions reveal that dispersions of corn starch in water and aqueous mixtures of ethylene glycol and glycerin exhibited dilatant behavior with values of *n* in the range 1.2–2.5 (Griskey and Green 1971). It would, therefore, appear that gelatinization

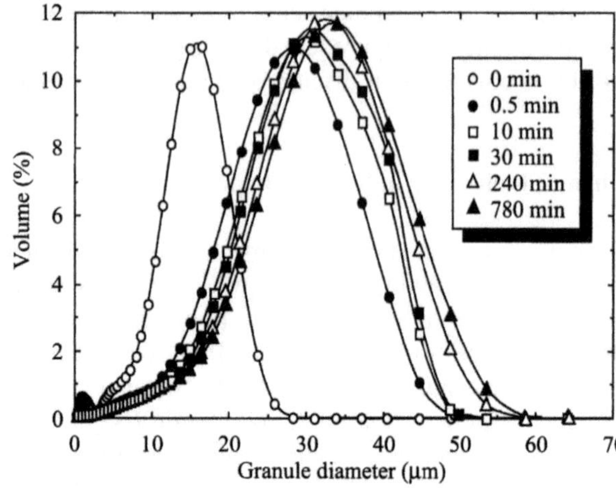

Fig. 4.21 Typical granule size distribution data of heated corn starch dispersions are shown; significant increase in granule size resulted due to heating (Okechukwu and Rao 1996a)

may be described as a transition process that transforms the rheological behavior of starch suspensions from dilatant to pseudoplastic.

Hoffman's study (1972) on polyvinyl chloride spheres in dioctyl phthalate dispersions pointed out that shear-thickening flow behavior is a reproducible phenomenon. Hoffman suggested that: "The discontinuous viscosity behavior which occurs at solids concentrations above 0.50 is caused by a flow instability in which surfaces of spheres, packed in two dimensional hexagonal packing at low shear rates, breakup into less ordered arrays of spheres. In any given viscometric flow, the ordered surfaces of spheres are parallel to surfaces of constant shear and one axis of the packing always points in the direction of flow." However, computer simulation studies on dispersions of rigid particles suggest formation of clusters of particles that leads to initiation of shear-thickening behavior (Bossis and Brady 1989). As shear rate is increased, the loss of long-range orientational order that results in discontinuous viscosity behavior is also referred to as "shear melting" (Chow and Zukoski 1995b). Particle size distribution, volume fraction, shape, particle–particle interaction, continuous-phase viscosity, and type, rate, and time of deformation are important factors in dilatancy (Barnes 1989; Boersma et al. 1991, 1992; Laun et al. 1991).

Role of Particles in Shear-Thickening Nonfood Dispersions

In shear-thickening behavior of nonfood colloidal dispersions, mostly TiO_2 particles, the role of particle size distribution has been examined by D'Haene et al. (1993). These studies on rigid particles with diameter 1–600 nm and relatively high viscosity suspending fluids (e.g., dioctyl phthalate) reveal several important features of shear thickening (Barnes 1989; Boersma et al. 1991, 1992; Laun et al. 1991; Chow and Zukoski 1995a, b). The smaller-diameter particles and higher-viscosity suspending fluids, such as dioctyl phthalate, result in very low particle Reynolds numbers, defined as (Russel et al. 1989)

$$\mathrm{Re_p} = \frac{D^2 \dot{\gamma} \rho}{\eta} \tag{4.38}$$

To illustrate the order of magnitudes, at a fixed $\dot{\gamma} = 100\ \mathrm{s^{-1}}$, the particle Reynolds number for typical 500 nm particles in dioctyl phthalate ($\rho = 980\ \mathrm{kg\ m^{-3}}$, $\eta = 0.054$ Pa s) is 4.54×10^{-7}, while that for a typical 30 μm starch granule in water ($\rho = 1{,}000\ \mathrm{kg\ m^{-3}}$, $\eta = 0.001$ Pa s) at the same $\dot{\gamma}$ is 9.0×10^{-2}, so that the flow regimes are substantially different.

From electrostatic interaction between two particles and the hydrodynamic interaction between two particles approaching each other along their line of centers, Boersma et al. (1991) developed an equation for $\dot{\gamma}_c$ of high volume fraction dispersions, that is, dispersions with small interparticle distances, h

$$\dot{\gamma}_c = \frac{2\pi \varepsilon_0 \varepsilon_1 \Psi_0}{6\pi \eta_s a^2} \frac{kh}{2} \tag{4.39}$$

where ε_0 is permitivity of vacuum, ε_1 is relative dielectric constant, ψ_0 is surface potential of sphere, here equal to ζ-potential of the particles in the medium, η_s is the continuous medium's viscosity, a is particle radius, κ is the reciprocal Debye double layer thickness, and h is distance between two particles. This equation restricted to small values of h, is valid ($h/a < 0.05$, where a is the particle radius) within ±25 % error. A simpler estimate of the shear rate at which shear thickening appears was derived using a dimensionless number (N_d) from the characteristic magnitudes of the interparticle and shear forces. Shear thickening will occur at values of $N_d > \sim 1.0$

$$\dot{\gamma}_c = \frac{2\pi \varepsilon_0 \varepsilon_1 \Psi_0}{6\pi \eta_s a^2} \tag{4.40}$$

Boersma et al. (1991) suggest that experimental and calculated values of $\dot{\gamma}_c$ be compared at a volume fraction ϕ_c given by

$$\phi_c = \left(\frac{a}{a + 1/K}\right)^3 (0.605) \tag{4.41}$$

However, the model overpredicted the shear rate for thickening and also failed to capture the volume fraction dependence of the critical shear rate (Chow and Zukoski 1995a, b).

Chow and Zukoski (1995a, b) observed that although all suspensions that were ordered at rest melted with increase in shear rate, thickening was observed only at a volume fraction greater than a critical value ϕ_c, which occurred near 0.5 for particles with diameters 250 nm. Shear melting occurred for all samples with $\phi > \phi_{melt}$ (=0.38) while shear thickening occurred only for $\phi > \phi_c$ (~0.5). Near ϕ_c, the thickening was weakened and reversible and as ϕ approached 0.6, the thickening became "nearly violent." For shear-thickening samples, starting at a critical stress the shear rate remained at a nearly constant value over a range of applied stresses that was limited by fracture of the sample (D'Haene et al. 1993). This suggests that controlled stress rheometers are better suited to study the transition to shear thickening (D'Haene et al. 1993).

For oscillatory strains less than approximately 0.03–0.04, the suspensions showed linear viscoelastic response with the plateau modulus, G_0, being independent of frequency (Chow and Zukoski 1995b). Because G_0 increases exponentially with volume fraction, by investigating properties over a wide volume fraction range, a correlation can be established indicating that stresses marking the onset of microstructural or rheological transitions were independent of ϕ if scaled on the suspensions elastic modulus G_0. By scaling stress on G_0, the constitutive response of the latex suspension was reduced to a single, volume fraction-independent curve when plotted as a function of the Deborah number

$$\mathrm{De} = (\eta_s \dot{\gamma} / G_0) \tag{4.42}$$

where η_s is the continuous-phase viscosity and $\dot{\gamma}$ is the shear rate. Chow and Zukoski (1995b) suggested that this correlation results from η_s/G_0 represented the natural relaxation time of the suspension.

Size Distribution of Intact Starch Granules

Granule swelling during starch gelatinization progresses in a manner that increases the GSID when measured by the SD of granules in suspension (Okechukwu and Rao 1995). This change in the polydispersity of the starch granules introduces important rheological consequences on the suspension behavior. Over a wide range of swelling of cowpea starch granules heated at 67–80°C, n decreased from 1.3 to 0.56 with increase in ($\mathrm{SD}/\mathrm{SD}_0$) of unruptured granules in the SD, where SD_0 is the SD of the unheated granules. Over a granule SD of 15–32 μm, a power law equation correlated the two dimensionless variables n and ($\mathrm{SD}/\mathrm{SD}_0$)

$$n = 2.36\left(\frac{\mathrm{SD}}{\mathrm{SD}_0}\right)^{-1.4} \quad R^2 = 0.95 \tag{4.43}$$

Therefore, in addition to the concentration of starch that influences the magnitude of n (Colas 1986; Evans and Haisman 1979), increase in polydispersity of the gelatinized granules appears to be an important factor in decreasing the severity of shear thickening and increasing pseudoplasticity of gelatinized SDs.

Antithixotropic Behavior of Cross-Linked Starch Dispersions

The antithixotropic nature of the thermally processed for 8, 15, and 30 min at 120°C CWM SDs was observed clearly in shear cycles: 5 min ascending and 5 min descending. The response curves of each sample (Fig. 4.22) indicate that each sample

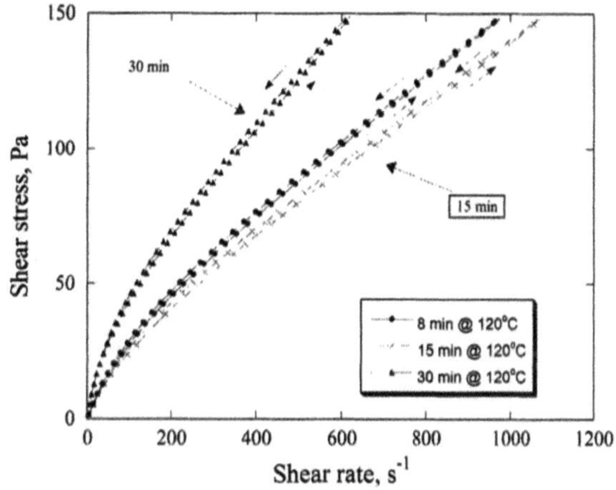

Fig. 4.22 The flow curves of cross-linked waxy maize samples heated at 120°C for 5, 15, and 30 min were more viscous after shearing, that is, they exhibited time-dependent shear-thickening (antithixotropic) behavior

became more viscous with shearing, that is, it exhibited time-dependent shear-thickening (antithixotropic) behavior. Furthermore, the antithixotropic behavior was irreversible in that after the shearing was stopped the structure did not return to the structure prior to shearing.

Although it is possible that the hysteresis loops in Fig. 4.22 were affected by sample preparation procedure, for example, the stirring prior to rheological testing, there were distinct differences between the sheared and unsheared samples, thus, suggesting shear-induced structure. CWM SDs have visible, possibly shear-induced, structure in the form of clusters. The CWM starch granule heated at 120°C for 30 min (Fig. 4.23) had dents and crevices throughout the surface and appeared to have pores on the surface. It is possible that the granule crevices served as linkage sites to aid in attachment to other granules. The membranes also have many microscopic finger-like protrusions on the surface. After shearing for three cycles, the thermally processed samples did not appear to have sustained any surface damage reflecting the strength and resilience of the granules due to cross-linking. However, the morphology of the gelatinized granule population has been modified by shearing with concomitant time-dependent shear thickening.

The aqueous dispersions of CWM heated at 120°C for 8, 15, and 30 min all showed a shear-thinning behavior (Chamberlain 1996; Chamberlain et al. 1998). In addition, as described earlier, an antithixotropic loop was found in the shear rate versus shear stress curves for the three samples with the stresses of the descending shear curve being higher than those of the descending shear curve for each sample. The loops for 8 and 15 min were similar, but the stress values of 30 min loop were significantly higher than those of the 8 and 15 min loops indicating that the latter was more viscous than the other two dispersions. In Fig. 1.5, the time-dependent shear-thickening (antithixotropic) behavior during three successive shear cycles of a CWM SD heated at 120°C for 30 min was shown.

Time-dependent shear-thickening behavior was also seen with autoclaved 2.0% dispersions of unmodified waxy maize starch in 0.2 *N* KOH (Dintzis and Bagley 1995). Two hysteresis loops were seen in the first shear cycle: one loop, at shear

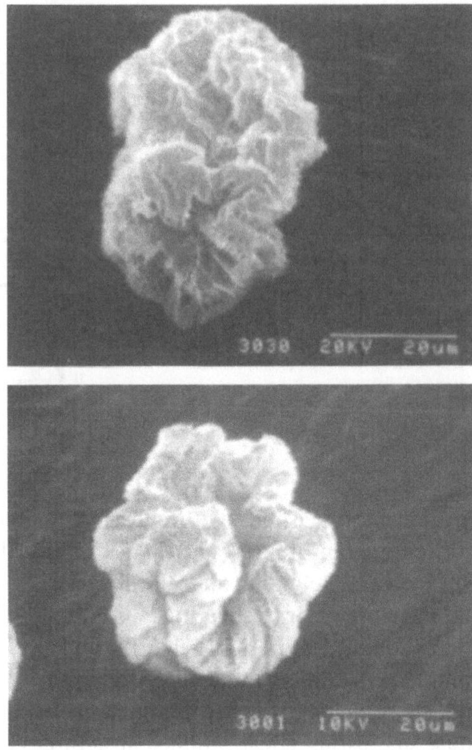

Fig. 4.23 A Sheared cross-linked waxy maize granule heated at 120°C for 30 min had dents and crevices throughout the surface and appeared to have pores on the surface

rates < 200 s^{-1}, was anticlockwise indicating time-dependent shear thickening, and the other loop, at shear rates > 250 s^{-1}, was clockwise indicating thixotropic behavior. A second shear cycle resulted in only one clockwise loop. The unmodified waxy maize starch dispersed in DMSO showed the same behavior as that in 0.2 N KOH. Dintzis and Bagley (1995) showed that a 2% gently dispersed waxy maize SD exhibited both thixotropy (at shear rates > 60 s^{-1}) and antithixotropy (at shear rates < 40 s^{-1}). With additional stirring, the 2% waxy maize SD had an anticlockwise loop indicative of antithixotropic behavior. Phase microscopy revealed that the sheared waxy maize starch preparations had a mottled appearance indicative of refractive index inhomogeneities.

Tattiyakul and Rao (2000) carefully examined the shearing conditions in studies on starches that reported antithixotropic behavior (Table 4.6). From the data in the table, she suggested that antithixotropic behavior resulted from shear-induced structure formation that occurred predominantly when the level of shear stress imposed on the sample was less than 120–150 Pa.

Starch Granule Mass Fraction and Viscosity

The mass fraction of granules plays an important role in the magnitudes of rheological properties of SDs (Evans and Haisman 1979). Because of their deformable

Table 4.6 Relationships between $(D_t/D_0)^3$ and volume fraction of heated 2.6% starch dispersions

Starch	Intercept	Slope	R^2	Reference
Unmodified corn	0.16	0.011	0.86	Okechukwu and Rao (1995)
Cowpea	0.09	0.013	0.84	Okechukwu and Rao (1996)
Cross-linked waxy maize	0.23	0.0015	0.98	Chamberlain et al. (1998)
Tapioca $(15 < (D_t/D_0)^3 < 40)$	~0*	0.012	NA	Rao and Tattiyakul (1998)
Tapioca $(40 < (D_t/D_0)^3 < 50)$	~0*	0.044	NA	Rao and Tattiyakul (1998)

D_t diameter of the heated starch granule, D_0 diameter of the unhealed granule
Physically meaningless values −0.07 and −1.4 were obtained

nature, it is difficult to determine volume fraction of SDs accurately and it is preferable to work with starch granule mass (SGM) fraction, (cQ), where c is dry starch concentration, w/w, and Q is the mass of hydrated starch granules per unit weight of dry starch. To determine the mass of the hydrated granules, a 20 ml sample of evenly dispersed starch suspension is centrifuged for 15 min at 1,000 × g (Eliasson 1986; Leach et al. 1959) and the top clear supernatant is removed leaving behind the swollen sediment. In addition, the liquid in the interstices should be drained carefully; otherwise, high values of cQ will result. SGM fraction takes into consideration both solids concentration and volume of starch granules after water absorption, and plays an important role in consistency and yield stress (Bagley and Christianson 1982; Okechukwu and Rao 1995) as well as the viscoelastic properties (G', G'') of SDs, and has been referred to as "notional volume fraction" (Evans and Lips 1992).

The mean diameter of the granules and the size distribution of the granules influence the magnitude of consistency index and the flow behavior (shear-thinning or shear-thickening), respectively (Okechukwu and Rao 1995). Increase in the volume fraction of solids in an SD can be achieved by either increasing the concentration of the solids or through volume expansion of the solids. Linear relationships were found between cQ and $(D_t/D_0)^3$ (D_t is the diameter of the heated starch granule and D_0 is that of the unheated granule) over a wide range of the latter for heated SDs of corn

(Okechukwu and Rao 1996b), cowpea (Okechukwu and Rao 1996), and CWM starches (Chamberlain 1996; Table 4.7). For the tapioca SDs (Rao and Tattiyakul 1999), two linear relationships provided a better fit of the data: one for values of $(D_t/D_0)^3$ from about 15 to 40 and the other from 40 to about 50 (Fig. 4.24). In determination of Q, centrifuging and siphoning off the top layer of supernatant fluid are not precise procedures that contributed to the spread of the cQ data points in Fig. 4.24. For tapioca SDs, a large increase in cQ at high values of $(D_t/D_0)^3$ is noteworthy. Because tapioca starch had a higher value of slope, especially at high $(D_t/D_0)^3$ values, the starch granule diameter ratio contributed to a higher increase of cQ of tapioca SDs.

Table 4.7 Shear stress range at which antithixotropic behavior of starch dispersions was reported (Tattiyakul 2000)

Material	Shear stress range (Pa)	Shear rate range (s^{-1})	Reference
1.5–2.5 % Gelatinized waxy maize starch in 90 % DMSO, 25°C	$0<\sigma<\sim 100$	$0.1<\dot{\gamma}<200\ s^{-1}$ (2 & 2.5 %) and $0.1<\dot{\gamma}<500\ s^{-1}$ (1.5 %)	Carriere (1998)
Autoclaved 2.6 % cross-linked waxy maize starch, 20°C	$0<\sigma<\sim 150$	$0.1<\dot{\gamma}<1000\ s^{-1}$	Chamberlain (1996)
3 % waxy rice starch in 0.2 N NaOH, 25°C	$0<\sigma<\sim 40$	$0<\dot{\gamma}<225\ s^{-1}$	Dintzis **et al.** (1996)
2 % waxy maize starch in 0.2 N KOH, 30° C	$0<\sigma<30$	$0<\dot{\gamma}<750\ s^{-1}$	Dintzis and Bagley (1995)
7.7 % autoclaved waxy maize starch, 50° C	$0<\sigma<150$	$0<\dot{\gamma}<750\ s^{-1}$	Dintzis and Bagley (1995)
3, 5, and 8 % modified potato starch, 10–90°C	$10\pm 5<\sigma<150\pm 50$	$6<\dot{\gamma}<540\ s^{-1}$	Härröd (1989)
5 % modified waxy maize starch, 20°C	~90	$\dot{\gamma}=200\ s^{-1}$	Rao et al. (1997)
5 % cross-linked waxy maize starch (Purity W), 20°C	$0<\sigma<\sim 150$	$0.1<\dot{\gamma}<125.7\ s^{-1}$	This work
6 % Purity W, 20°C	$0<\sigma<\sim 150$	$0.1<\dot{\gamma}<40\ s^{-1}$	This work
5 % Purity W, 20–80°C	$0<\sigma<\sim 150$	$0.1<\dot{\gamma}<500\ s^{-1}$	This work

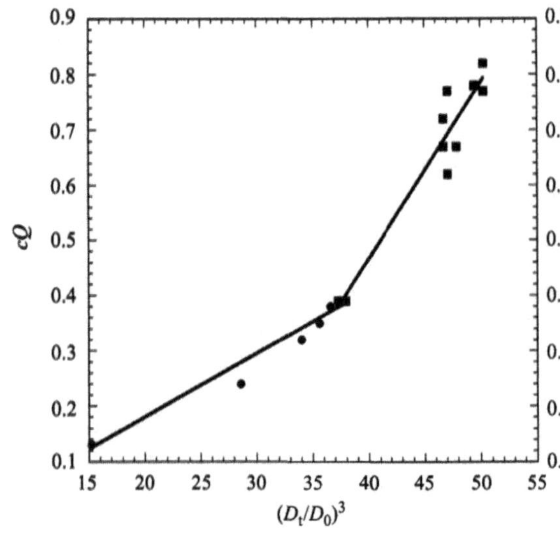

Fig. 4.24 For tapioca starch dispersions, two linear equations or a single exponential equation (not shown) provided a better fit of the data between cQ and $(D_t/D_t)^3$ (D_t is the diameter of the heated starch granule and D_0 is that of the unheated granule; Rao and Tattiyakul 1999)

The ratio of mean granule diameters (D_t/D_0) can be used to correlate consistency coefficient data of SDs. At low values of cQ, the granules were sufficiently apart and did not affect dispersion viscosity substantially. As cQ increased above a threshold value, granule–granule interaction increased accompanied by large increase in dispersion viscosity. Figure 4.25 shows the relationship between the granule diameter ratio (D_t/D_0) and the consistency coefficient (K) of gelatinized 2.6% tapioca SDs that can be expressed by the equation

$$K = K_0 \exp\{\varepsilon(D_t/D_0)\} \tag{4.44}$$

where K_0 and ε are empirical constants. K increased gradually at the beginning and rapidly as the granule diameter ratio approached its maximum value. At $(D_t/D_0)=3.4$, values of K started to increase rapidly. The estimated values of K_0, ε, and R^2 for various 2.6% SDs are given in Table 4.8. The higher value of ε for tapioca starch indicates that, among these four starches, the starch granule ratio had the highest influence on the consistency coefficient (K) of the tapioca SDs. During the later stages of gelatinization involving the bursting of granules, a progressive reduction in suspension viscosity takes place and Eq. 4.43 does not describe the suspension viscosity data.

Mass Fraction versus Relative Viscosity

The relative viscosity of the gelatinized 2.6% tapioca SDs was calculated using Eq. 4.45

$$\eta_{r\infty} = \frac{\eta_\infty}{\eta_s} \tag{4.45}$$

where $\eta_{r\infty}$ is the relative viscosity, η_∞ is the Casson viscosity (mPa s; Table 4.9), and η_s is the viscosity (mPa s) of the supernatant determined at 20°C; Table 4.9 also contains yield stress values calculated using different exponents in the Casson model. The η_r values of tapioca SDs strongly depended on their volume fraction and there was considerable scatter in the values as a result of uncertainties in estimating η_∞ and cQ values (Fig. 2.8). The line in Fig. 2.8 represents values predicted by the model (Eq. 4.46) of Quemada et al. (1985) with the structural parameter $k = 1.80$

$$\eta_r = \left(1 - \frac{1}{2}k(cQ)\right)^{-2} \tag{4.46}$$

Quemada et al. (1985) reported values of k in the range 2.50–3.82 for dispersions of rigid solids, and 1.70–1.85 for red blood suspensions. Because heated starch granules are deformable, the value $k=1.80$ of red blood cells was selected for the data on tapioca SDs. Given that the Krieger–Dougherty model predictions were substantially different than from experimental η_r values of SDs (Ellis et al. 1989; Noel et al. 1993; Tattiyakul 1997), the reasonable applicability of Eq. 4.45 (Quemada et al.

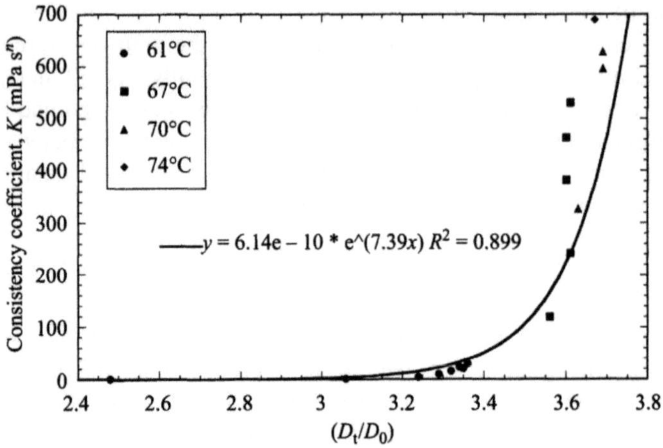

Fig. 4.25 The relationship between the granule diameter ratio (D_t/D_0) and the consistency coefficient (K) of gelatinized 2.6% tapioca starch dispersions (Rao and Tattiyakul 1999)

Table 4.8 Relationships between (D_t/D_0) and the power law consistency index of heated starch dispersions: $K = K_0 \exp\{\varepsilon\,(D_t/D_0)\}$

Starch	K_0 (mPas^n)	ε	R^2	Reference
Unmodified corn	2.07×10^{-5}	3.69	0.99	Okechukwu and Rao (1995)
Cowpea	2.40×10^{-5}	4.84	0.96	Okechukwu and Rao (1996)
Cross-linked waxy maize	2.13×10^{-7}	3.36	0.92	Chamberlain **et al.** (1998)
Tapioca	6.14×10^{-10}	7.40	0.90	Rao and Tattiyakul (1998)

D_t diameter of the heated starch granule, D_0 diameter of the unheated granule

Table 4.9 Volume fraction (cQ), Casson yield stress (σ_{0c}, Pa, $m=0.25$ and $m=0.50$) and infinite shear viscosity (η_∞, mPa s), viscosity of supernatant (η_s, mPa s), and relative viscosity (η_r) of tapioca starch dispersions

Dispersion[a]	***cQ***	σ_{0c} $m=0.25$	η_∞, $m=0.25$	σ_{0c}, $m=0.50$	η_s	η_r
61°C, 240 **min**	0.39	0.002	7.2	0.11	1.53	4.7
67°C, 5 **min**	0.62	0.120	12.1	1.13	2.02	6.0
67°C, 15 **min**	0.67	0.071	27.0	1.48	2.48	10.9
67°C, 30 **min**	0.72	0.181	13.9	1.64	2.60	5.3
67°C, 60 **min**	0.77	0.188	14.3	1.78	2.76	5.2
70°C, 5 **min**	0.77	0.078	30.1	1.96	1.97	15.3
70°C, 15 **min**	0.82	0.098	32.1	2.10	2.82	11.4
74°C, 1 **min**	0.78	2.77	139.45	2.25	2.85	48.9

[a] Heating temperature and time

1985) to tapioca SDs is noteworthy. However, the model of Quemada et al. (1985) is a phenomenological model, while that of Krieger–Dougherty is based on intrinsic viscosity of a single sphere.

Effect of Starch Concentration

Power law relations of the form shown in Eq. 4.46 have been used to describe the influence of starch concentration on the viscosity of gelatinized wheat starch suspensions over the shear rate range 20 and 200 s^{-1}.

$$K = K_c c^b \tag{4.47}$$

where K is in Pa s^n and starch concentration was defined in mg ml^{-1} (Doublier 1981). For gelatinized starch characterized at 70°C, $K_c = 8.74 \times 10^{-7}$ and $b = 3.88$ for starch concentration, c, in the range 2.5–8%. Evans and Haisman (1979) reported that the apparent viscosity at any shear rate $\eta_{a,\dot{\gamma}}$ raised to power 0.4 was linearly correlated with concentration for potato, tapioca, and corn starches within the shear rates 10^1–10^3 s^{-1}

$$(\eta_{a,\dot{\gamma}})^{0.4} = K^{11}(c - c_0) \tag{4.48}$$

where c_0 is the concentration at which the viscosity extrapolated to zero and K^{11} is a proportionality constant. Thus, at various shear rates, it is possible to construct linear plots relating apparent viscosity with starch concentration. Plots of apparent viscosity as a function of starch concentration were later shown by Bagley and Christiansen (1982) to be greatly affected by the pasting temperature and time. However, they obtained a temperature-independent master plot which correlated the viscosity over cQ ratio (η/cQ) with the mass fraction of swollen starch granules in gelatinized wheat starch suspension, cQ. A similar master plot was also obtained by plotting the ratio K/cQ against volume fraction of swollen corn starch granules since at low starch concentrations volume fraction approximates mass fraction (Okechukwu and Rao 1995) for a 2.6% corn starch suspension the plot could be represented by Eq. 4.48

$$\frac{K}{cQ} = 0.48\exp(16.94cQ)R^2 = 0.85 \tag{4.49}$$

Yield Stress and Structural Characteristics of Dispersions

It is known that concentrated gelatinized SDs exhibit yield stresses at low shear rates (Evans and Haisman 1979; Kubota et al. 1979; Doublier 1981; Christianson and Bagley 1984; Colas 1986; Harrod 1989; Giboreau et al. 1994). In these and other earlier studies, values of yield stresses were determined by extrapolation of

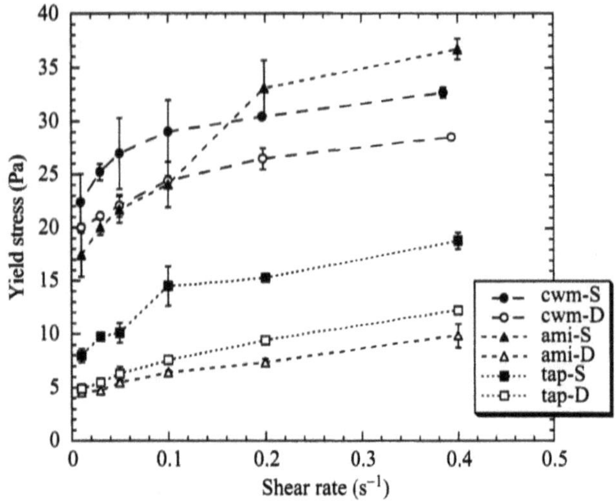

Fig. 4.26 Static (-S) and dynamic (-D) yield stress values of cross-linked waxy maize (CWM), tapioca, and Amioca 5 % (w/w) starch dispersions at different shear rates Filled symbols are values of static yield stress (σ_{0s}) and open symbols are values of dynamic yield stress (σ_{0d})

shear rate versus shear stress data using a flow model such as the Hershel Bulkey. Yield stresses of gelatinized corn starch suspensions at granule volume fraction (cQ) of about 0.6 were shown to increase with granule swelling (Christianson and Bagley 1984). Yield stresses so determined are influenced by the flow model used and the number of data points used.

In contrast to yield stresses calculated using flow models, as discussed in Chap. 3, vane yield stresses are more reliable and additional information on the structure of the dispersions can be obtained from them (Genovese and Rao 2003c). Values of the static and dynamic vane yield stresses of heated 5 % (w/w) CWM, tapioca, and Amioca SDs determined at different vane rotational speeds (shear rates) are shown in Fig. 4.26 (Genovese and Rao 2003c). Those values reflect the structural differences between the SDs as well as their responses to structure breakdown. Additional valuable information on the behavior of the dispersions before and after structure breakdown was obtained by plotting values of the static and dynamic yield stresses against the corresponding values of strain at which values of the yield stress were recorded. Such a plot, which may be called a texture plot, for the three SDs is shown in Fig. 4.27.

Effect of Temperature on Flow Behavior

The viscosity of fully gelatinized starch suspensions generally decreases with increase in temperature. This rheological response of heated SDs has been shown to follow an Arrhenius dependency of the consistency coefficient on temperature as shown in Eq. 4.49

$$K = K_{TO} \exp(E_a/RT) \tag{4.50}$$

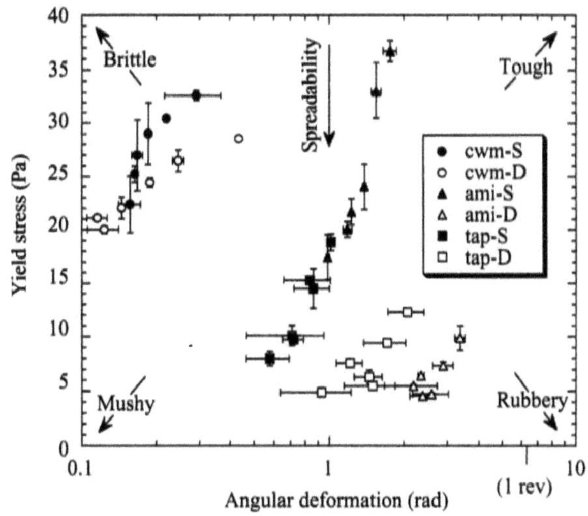

Fig. 4.27 Texture map based on static (-S) and dynamic (-D) values of yield stress of cross-linked waxy maize (CWM), tapioca, and Amioca 5 % (w/w) starch dispersions at 20°C

where E_a is the activation energy and K_{TO} an empirical constant. For a wheat SD heated to 96°C, values of $E_a = 5.13$ kcal mol^{-1} and $K_{TO} = 1.51 \times 10^{-4}$ Pa were reported (Doublier 1981).

Dynamic Rheological Behavior of Starch Dispersions

Viscoelastic properties of SDs are of interest to better understand their structure (Biliaderis 1992). In addition, studies on synthetic particle dispersions (Boersma et al. 1992; Chow and Zukoski 1995b) showed that they may play a role in understanding their flow behavior. As a reminder, one can anticipate that the viscoelastic behavior of SDs is also influenced very strongly by the volume fraction and state of the granules in the dispersion.

Effect of Particle Characteristics on Modulus

Evans and Lips (1992) extended the work of Buscall et al. (1982a, b) to relate the shear modulus, *G*, of a dispersion of spherical particles to various characteristics of the particles and the continuous medium

$$G = \frac{\phi_{cp} n G_p}{5\pi(1-\mu_p)} \left[\phi_{red}^{1/3} \left(1-\phi_{red}^{-1/3}\right)^{1/2} - \frac{8}{3}\phi_{red}^{2/3}\left(1-\phi_{red}^{-1/3}\right)^{3/2} \right] \tag{4.51}$$

where G_p is shear modulus of material of spherical particle, μ_p is Poisson's ratio, ϕ_{cp} is the volume fraction for close-packing and ϕ_{red} is the reduced phase volume fraction (ϕ/ϕ_{cp}) that is given as $\phi_{red}=(a_0/h)^3$; a_0 is undeformed particle radius, h is distance to nearest particle. Evans and Lips suggested that for fully swollen starch granules, phase volume fraction $\phi=cQ$. However, agreement between experimental data on SDs and theory was poor but the theoretical trends were followed by Sephadex spheres used in chromatography columns (Evans and Lips 1992, 1993). The difference between theory and experiment for heated SDs was because the granules were not truly spherical and rigid "macroscopic elastic spheres" as well as they were polydisperse in nature. Also, it was not possible to determine the representative viscoelastic behavior of all the granules from the stress–strain data on a few granules using the cell poker technique (McConnaughey et al. 1980) that still provides mainly qualitative results (Zahalak et al. 1990). Plots of SGM fraction versus normalized, based on value of G' at $cQ=1$, G' data on starches resulted in three groups: chemically modified, root and tuber, and cereal (Evans and Lips 1992) that were attributed to "structural origins" of the starches.

Role of Continuous and Dispersed Phases on Viscoelastic Properties of Starch Dispersions

To examine the role of the starch granule and of the continuous phase on the dynamic rheological properties, two starches were selected (Genovese and Rao 2003a) whose characteristics were substantially different from each other: a CWM and tapioca, a tuber starch, with 19.3% amylose. The volume fraction occupied by the starch granules (ϕ) was determined using the Blue Dextran exclusion method (Evans and Haisman 1979; Tester and Morrison 1990).

Experiments were carried out in which the continuous phase was interchanged. Two SDs with similar volume fractions of granules, $\phi\sim0.64$, were prepared as described in detail in Genovese and Rao (2003a). One dispersion was prepared by mixing the continuous phase of the CWM sample (mainly water) to the tapioca granules, in the same weight proportion as in the original sample of the tapioca SD. In a similar manner, the supernatant of the tapioca SD, essentially a solution of amylose in water, was mixed with the CWM sediment. Rheological experiments were performed at 20°C on these samples under the same conditions used for the original tapioca 3% and CWM 4.25% SDs, respectively.

The dynamic rheological characteristics of the SD made up of tapioca granules in CWM continuous-phase are compared with those of the original tapioca 3% *w/w* in Fig. 4.28. The absence of amylose from the continuous phase did not have a noticeable effect on the storage (G') and the loss (G'') moduli. In Fig. 4.29, the dynamic rheological behaviors of SDs made up of CWM granules in tapioca continuous-phase and the original CWM 4.25% are compared. The addition of amylose to the continuous phase produced a moderate increase in G', but G'' remained almost the same. Thus, it can be stated that the presence of amylose in the continuous phase. The loss modulus did not change noticeably due to the presence of amylose,

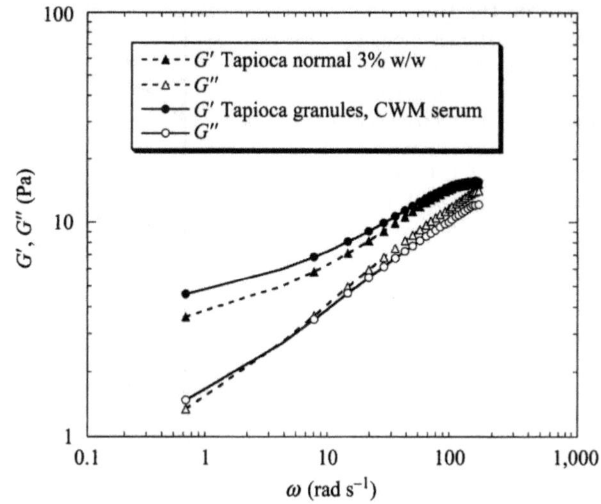

Fig. 4.28 Storage (G') and loss (G'') moduli versus angular frequency (ω) of tapioca granules mixed with the cross-linked waxy maize (CWM) continuous phase, compared with the original tapioca 3 % w/w

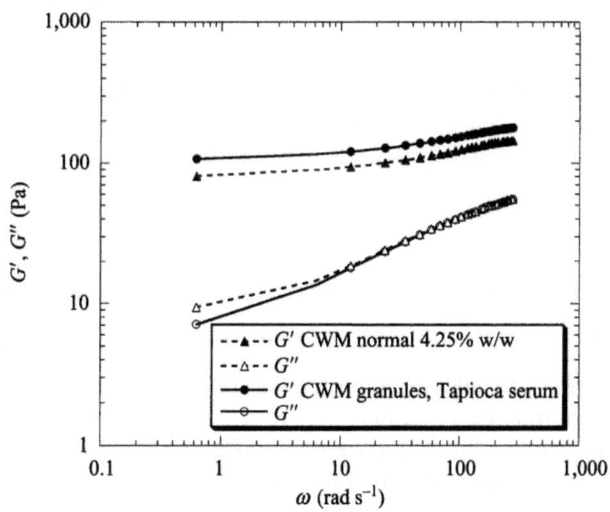

Fig. 4.29 Storage (G') and loss (G'') moduli versus angular frequency (ω) of cross-linked waxy maize (CWM) granules mixed with the tapioca continuous-phase, compared with the original CWM 4.25 % w/w

and the effect on the storage modulus, if any, is a moderate increase. Therefore, the higher elasticity of the CWM SD, with lower amylose content than tapioca, may be attributed exclusively to the structural properties of the granules.

Cross-Linked Waxy Maize Starch Dispersions

Figure 4.30 shows the dynamic properties, storage modulus (G', Pa) and loss modulus (G'', Pa) versus frequency ω (rad s^{-1}) of gelatinized CWM SDs of 3, 4, and 5% solids. In all dispersions, G' was much higher than G'', that is, the dispersions

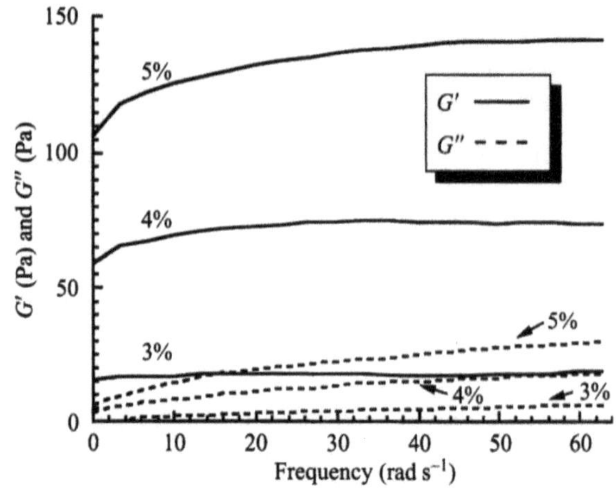

Fig. 4.30 The dynamic properties, storage modulus (G', Pa) and loss modulus (G'', Pa), versus frequency (rad s^{-1}) of gelatinized cross-linked waxy maize starch dispersions of 3, 4, and 5% Solids (da Silva et al. 1997)

exhibited gel-like behavior over the studied range of frequencies. Both G' and G'' increased with concentration and showed a weak dependence on frequency at low values. This dependency was more pronounced in dispersions of higher concentration.

The Cox–Merz rule (Eq. 4.51) was not applicable to gelatinized CWM SDs in the range of concentrations studied in that the complex viscosity (η^*) was higher than the flow viscosity over the range of frequencies and shear rates studied (Fig. 4.31).

$$\eta^*(\omega) = \eta_a(\omega)|_{\omega=\dot{\gamma}} \tag{4.52}$$

where $\eta=(G^*/\omega)=\eta' - i(G'/\omega)$, ω is the frequency of oscillation (rad s^{-1}) and $|G^*| = \sqrt{(G')^2 + (G'')^2}$ Deviation from Cox–Merz rule also reflects the gel-like nature of the gelatinized dispersions (Ross-Murphy 1984). The complex viscosity can be further expressed in terms of real and imaginary parts, $\eta^* = \eta' - i\eta''$; it is interesting to note that the dynamic viscosity (η') is in better agreement with the apparent viscosity than with η^* (Fig. 4.31).

4% Tapioca Starch Dispersions

Double logarithmic plots of η^* versus dynamic frequency (ω, rad s^{-1}) and η_a versus shear rate ($\dot{\gamma}$) of a 4% tapioca SD heated at 70°C for 30 min, were described by Eqs. 4.52 and 4.53, respectively

$$\eta^* = 18.3\omega^{-0.74} \quad R^2 = 1.00 \tag{4.53}$$

$$\eta_a = 8.36\dot{\gamma}^{0.73} \quad R^2 = 1.00 \tag{4.54}$$

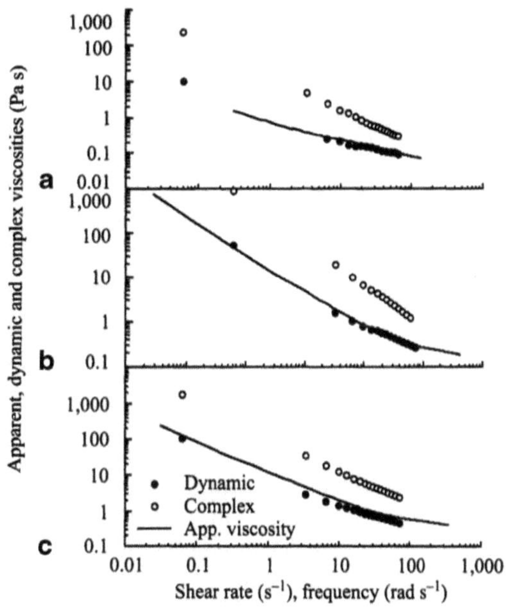

Fig. 4.31 The Cox–Merz rule was not applicable to gelatinized cross-linked waxy maize starch dispersions in the range of concentrations studied in that the complex viscosity (η^*) was higher than the flow viscosity over the range of frequencies and shear rates studied

A modified Cox–Merz relationship was found to be applicable between η^* and η_a

$$\eta^*(\omega) = 2.24[\eta_a(\dot{\gamma})]^{1.04}|_{\omega=\dot{\gamma}} \quad R^2 = 1.00 \tag{4.55}$$

This relationship is useful for calculation of apparent viscosity data from complex viscosity data or vice versa.

For the 4% tapioca SD heated at 70°C for 5 min, values of η_a decreased with $\dot{\gamma}$, but those of η^* increased with ω after an initial decrease (Fig. 4.32). For the data shown in Fig. 4.32, the critical frequency (ω_c) at which the transition occurred was about 100 rad s^{-1}. In comparison to $\dot{\gamma}_c$, values for shear-thinning to shear-thickening flow behavior of 2.6% tapioca and cowpea starch (Okechukwu and Rao 1996) dispersions, the value of $\omega_c = 100$ rad s^{-1} is lower. It is possible that the η_a versus $\dot{\gamma}$ data of the 4% dispersion heated at 70°C for 5 min had a $\dot{\gamma}_c$ at a very high shear rate. However, because the objective was to examine applicability of the Cox–Merz relationship, very high values of $\dot{\gamma}$ were not used. Given that heated SDs were found to exhibit shear-thinning and shear-thickening (Bagley and Christianson 1982; Okechukwu and Rao 1995, 1996a, b), as well as time-dependent shear-thickening (Chamberlain 1996; Dintzis and Bagley 1995; da Silva et al. 1997) flow behavior, perhaps it should not be surprising that shear thickening should be seen in η^* versus ω data.

Spherulites were created by first cooking high-amylose corn starch in an excess-steam jet cooker in the presence of 5% oleic or palmitic acid, based on amylose, and the cooked product was rapidly cooled in an ice bath and then freeze-dried or drum-dried; after the amylose was removed from solution by forming helical in-

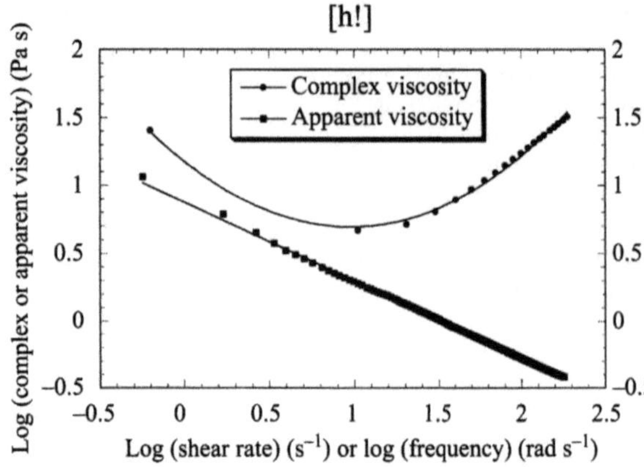

Fig. 4.32 For the 4 % tapioca starch dispersion heated at 70°C for 5 min, values of η_a decreased with, $\dot{\gamma}$, but those of η^* increased with η^* after an initial decrease

clusion complexes with the fatty acid, the inclusion complexes formed submicron spherical particles upon rapid cooling. The spherulites formed aggregates, which in turn formed a network throughout the sample. The dried material was reconstituted to form a paste that properties of these pastes were measured to determine the effects on the flow exhibited gel-like properties upon standing, but that flowed readily when shear was applied; at the highest concentrations (~ 15 %), their flow properties were similar to a commercial shortening (Byars et al 2009). It is noted that the pastes can be prepared from either isolated and undried spherulites, or from dried and reconstituted material containing uncomplexed amylopectin. A wide range of material properties were obtained, depending on the concentration of material, the drying and reconstitution method, and the fatty acid used to form the spherulites. The flow properties of the pastes at the highest concentrations (~ 15 %) were similar to those of a commercial shortening (Byars et al. 2009).

Effect of Sugar on Rheology of Starch Dispersion

Addition of a sugar was shown to affect the rheological behavior of dispersions of several starches, including: wheat (Cheer and Lelievre 1983), potato (Evans and Haisman 1982), corn (Evageliou et al. 2000), and sago (Ahmad and Williams 1999). Addition of either sucrose or fructose resulted in increase in the storage (G') and loss (G'') moduli of heated 5 % *w*/*w* Amioca SDs; in Fig. 4.33 data, on dispersions containing sucrose are shown (Genovese et al. 2004). The static vane yield stress (σ_{0s}) also increased with addition of either sucrose or fructose (Table 4.10), suggesting that sugars increased the strength of the internal bonds or cohesiveness of the SDs; in general, fructose was slightly more effective than sucrose (Genovese et al 2004). Ahmad and Williams (1999) explained the influence of the sugars on the structure of native SDs in terms of inhibition of chain organization: The sugar molecules with equatorial hydroxyl groups (e.g., ribose) prevent chain reordering,

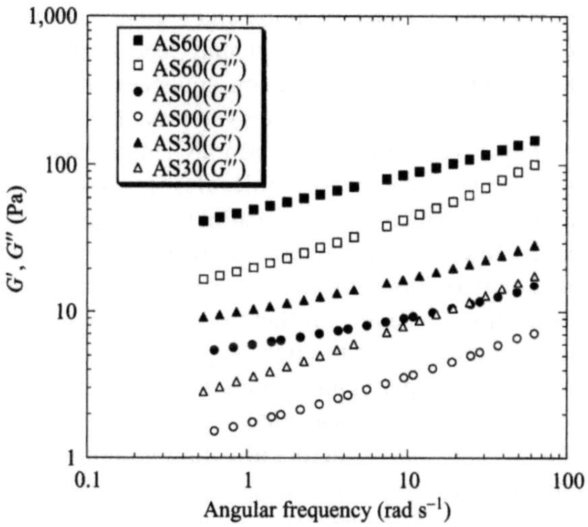

Fig. 4.33 Magnitudes of G' and G'' increased with addition sucrose. AS60 is Amioca-60 % sucrose, AS30 is Amioca-30 % sucrose, and AS30 is Amioca-0 % sucrose (Based on data in Genovese et al. 2004)

Table 4.10 Magnitudes of static (σ_{0s}) and dynamic (σ_{0d}) yield stress and angular deformation (θ) of 5 % Amioca—30 % sugar SDs, pasted in rotavapor (RV) or retort (RT)

Sugar (30 % w/w)	Pasting Procedure	Undisrupted Sample		Disrupted Sample	
		θ (rad)	σ_{0s}(Pa)	θ (rad)	σ_{0d} (Pa)
None	RV	0.99±0.01	17.4±2.1	2.39±0.27	4.5±0.0
Sucrose	RV	0.95±0.00	37.5±1.6	2.52±0.35	6.1±0.1
Sucrose	RT	1.85±0.28	35.8±2.7	2.51±0.35	5.4±0.4
Fructose	RV	1.14±0.04	34.1±1.5	2.72±0.11	5.7±0.4
Fructose	RT	1.53±0.13	38.1±2.5	3.14±0.37	5.1±0.0
Fructose	RT (t=0′)	1.44	39.9	3.41	5.0
Fructose	RT(t=135′)	2.31	49.5	3.00	5.2
Fructose	RT (t=265′)	3.66	52.5	3.28	5.1

making the whole structure weaker, while the sugars with axial hydroxyl groups (e.g., fructose) have the opposite effect.

Rheological Behavior of Starch–Protein Dispersions

In many foods, both starch and protein can be encountered so that understanding interactions between them would be useful. The selectivity in interaction between proteins and starches is best seen in results of dynamic rheological studies. The results depend upon the molecular structure of protein, the starch: state of the granules and the amylose/amylopectin ratio, the composition of protein and starch, as well as the phase transition temperatures are important factors influencing protein–starch interaction. Because proteins and starches are thermodynamically different polymers, their presence together may lead to phase separation, inversion, or mutual interaction with significant consequences on texture (Morris 1990).

In an aqueous medium, a given protein and polysaccharide can be compatible or incompatible. If they are thermodynamically incompatible, a two-phase system is obtained in which the two molecules are mostly in separate phases. However, if they are thermodynamically compatible, two types of mixtures can be formed: (1) a homogeneous and stable system in which the two molecules coexist in a single phase or (2) a two-phase system where the molecules interact and both are in essentially concentrated phases. Furthermore, there is a coacervation due to an electrostatic association between the two polymers. Anionic polysaccharide and protein are incompatible at pH values above the protein's isoelectric point (point of minimum solubility, pH≈5.1) and completely compatible below it due to the net opposite charges they carry. Factors affecting protein–polysaccharide compatibility and the characteristics of their complexes include the molecular characteristics of the two molecules (e.g., molecular weight, net charge, and chain flexibility), the pH, ionic strength, temperature, the protein/polysaccharide ratio, rate of acidification, and rate of shear during acidification (Tolstoguzov 1985, 1991). The rheological behavior of specific starch/protein systems are discussed next.

Gluten/Starch

In food products, starch can be either a mere ingredient used in low concentrations to thicken the product texture or the main component responsible for the texture. The former kind of products usually contain water in excess of gelatinization requirements while, in the latter, such as baked products, water is available in limited amount. In these products, made from wheat flour, gluten is considered to be the main component responsible for viscoelastic properties. Nevertheless, experiments on dough made with only starch and gluten revealed that starch does not act only as an inert filler but interacts with gluten and is active in determining the viscoelastic behavior of both the dough and the cooked product (Faubion and Hoseney 1990; Petrofsky and Hoseney 1995).

Addition of gluten to 6.5–6.6 % (w/w) dispersions of different starches at a 1 % level calculated on dry starch basis increased the G' of wheat and rye starches, lowered that of maize starch but did not affect the modulus of barley, triticale, and potato starches (Lindahl and Eliasson 1986). The G' was also found to increase with increasing amounts of gluten (1–4 %) added to the wheat starch. Studies on starches with different amylose/amylopectin ratios suggest that gelatinization of amylopectin and not amylose favors a synergistic interaction with proteins leading to an increase in dispersion viscosity (Chedid and Kokini 1992; Madeka and Kokini 1992).

Addition of gluten strongly changed the viscoelastic properties of starch pastes and gels. In pure SDs, the rheological unit was a starch granule that determined the viscoelastic behavior of the whole network due to its ability to be deformed or packed (Champenois et al. 1998). When gluten was present, the rheological units were no longer the starch granules but the "cells" formed by gluten fibrils and filled by starch granules. In the case of flour, the internal lipids obviously play an important role so that it cannot be considered theologically to be a simple blend of starch and gluten. The range of gluten/starch ratios studied (0–0.20) included the natural ratio in flour of 12/88 (Chedid and Kokini 1992). When gluten was added to starch, the water content was calculated based on the weight of both starch and gluten. Therefore, compared with a pure SD of the same concentration, there was less starch in a gluten/SD, but the water/starch ratio was almost constant. The gluten delayed or partly reduced the contact between granules.

The probable effect of adding gluten was to hinder granule–granule interactions because, at the temperatures corresponding to the early stage of swelling, the amylose had not yet leached out of the granules. It is hypothesized that the gluten fibrils became embedded between the starch granules and consequently hindered the formation of the granule network (Fig. 4.34). This structure may be described as a phase-separated network of starch granules and gluten (Morris 1986). The gluten fibrils hinder the formation of the starch network that normally occurs either by granule–granule interactions (early stage) or by amylose chain entanglements (later stage; Fig. 4.34). The presence of these fibers weakened the starch network, so that gluten probably plays the role of fault zones in the network structure. It is, thus, possible to imagine that the presence of both gluten and an emulsifier or a lipid could result in a synergistic effect among them. Indeed, values of G' and G'' for flour were much lower than those for gluten/starch mixtures even though gluten to starch ratio was equal to 0.20 (Table 4.11).

Soy Protein/Corn Starch

The shape of the G' versus ω curves for all corn starch-SP isolate dispersions was similar, where G' exhibited a small dependence on frequency (Fig. 4.35). The 10 % corn SD had the highest and the 10 % SP isolate dispersion the lowest value of G'. The values of the other dispersions were between these two limits. Gel compositions in order of decreasing G' values were: 10 % CS > 1SI:9CS≈3S1:7CS≈5SI:5C S>7SI:3CS>9SI:1CS> 10 % SI, where CS is corn starch and SI is SP isolate (Liao et al. 1996).

The corn starch-SP isolate dispersions displayed true gel-like behavior. Table 4.12 contains magnitudes of the slopes (n' and n'') and intercepts (K' and K'') in the equations

$$G' = K'(\omega)^{n'} \tag{4.56}$$

$$G'' = K''(\omega)^{n''} \tag{4.57}$$

In the starch-SP isolate dispersions, the starch continuous network was formed first as a result of its lower transition temperature. Because the denaturation temperature

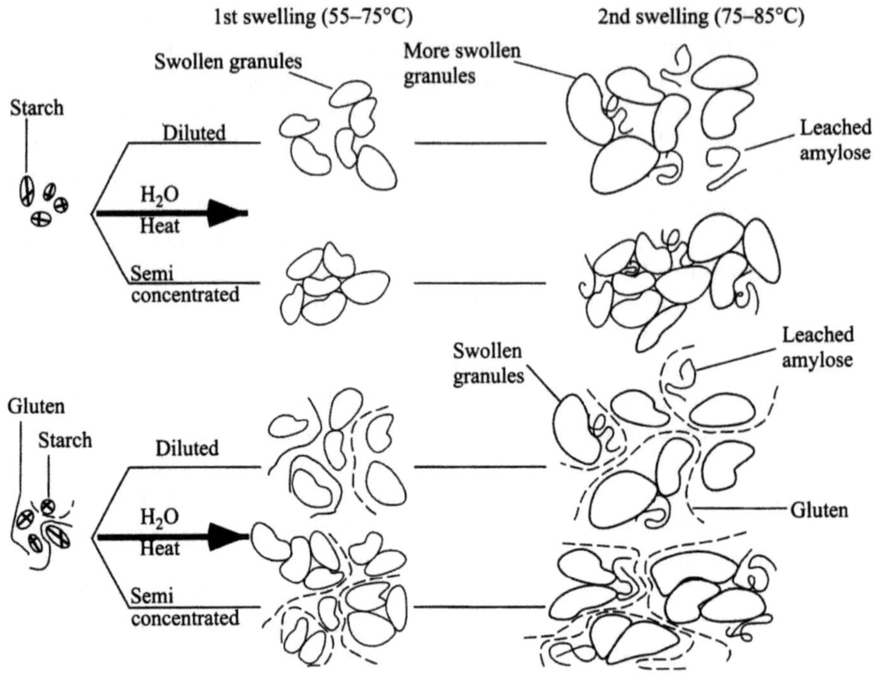

Fig. 4.34 Schematic of a network of wheat starch granules and gluten (Champenois et al. 1998)

Table 4.11 Storage modulus G' (Pa) and tan δ at the end of heating (EH) of dispersions of wheat starch, gluten/starch blends, and flour (Champenois et al. 1998)

Sample	Starch/Flour Concentration, g kg^{-1}					
	60	80	100	150	200	300
Storage modulus (*G'*, *Pa*)						
Starch	140	342	779	2,531	7,565	31,898
G/S=0.12	LV	295	616	1,422	3,713	19,150
G/S=0.20	LV	LV	533	986	1,101	6,606
Flour	LV	68	102	320	1,567	5,894
tanδ						
Starch	0.36	0.19	0.21	0.19	0.09	0.04
G/S=0.12	>1	0.23	0.20	0.15	0.10	0.06
G/S=0.20	>1	>1	0.30	0.18	0.22	0.13
Flour	>1	0.44	0.46	0.40	0.22	0.21

G/S gluten to starch ratio, *LV* low value (< 10 Pa)

of 7S globulin was close to the gelatinization temperature of corn starch, the diffusion and aggregation of amylose molecules and the swelling of starch granules was a little bit hindered and reduced. This could be the reason for the value of G' of starch-SP isolate dispersions being lower than that of 10 % SD. However, com-

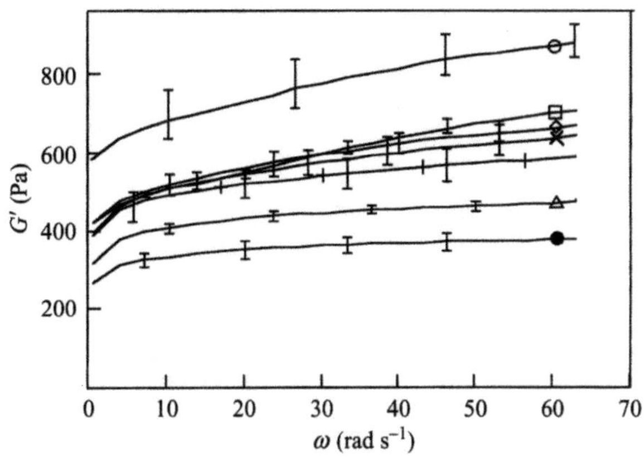

Fig. 4.35 The shape of the G' versus ω curves for all corn starch-soyprotein isolate dispersions was similar, where G' exhibited a small dependence on frequency. In order of decreasing G' values were: 10% CS > ISI:9CS ≈ 3 SI:7CS ≈ 5SI;5CS > 7SI:3CS > 9SI:1CS > 10% SI, where CS is corn starch and SI is SP isolate (Liao et al. 1999)

Table 4.12 Magnitudes of intercepts (K', K'') and Slopes *(n', n'')* from linear regression of log ω versus log G' and log G'' of different ratios of corn starch (CS) and soybean protein isolate (SPI) mixtures

Mixture	G' (Pa)		G'' (Pa)	
	K'	n'	K''	n''
10% Corn starch **(CS)**	562.3	0.10	29.5	0.24
10% **SPI**	281.8	0.07	39.8	0.002
9% **CS** + 1% **SPI**	398.1	0.12	29.5	0.23
7% **CS** + 3% **SPI**	426.6	0.11	44.7	0.19
5% **CS** + 5% **SPI**	416.9	0.10	51.3	0.14
3% **CS** + 7% **SPI**	398.1	0.09	53.7	0.08
1% **CS** + 9% **SPI**	338.8	0.08	47.9	0.013

parison with G' values of 6–10% SDs reveals the substantial contribution of the SP isolate. The unfolding protein molecules continued to penetrate the amylose aggregates, especially in the early stages of gel formation when the starch paste was still weak and two continuous networks were formed. The two structures supplement each other and the properties of proteins add to those of the starch network already present.

Figure 4.36 is a plot of the experimental values of G', at a frequency of 30 rad s^{-1}, of dispersions of various SP isolate/corn starch ratios and of corn starch with 6–10% solids. It also has a straight line that interpolates G' values between those of corn starch and SP isolate. When the proportion of starch was less than 5%, the measured values were higher than the interpolated values. Although above 5% starch content, the interpolated values were higher than the measured.

The asymmetry in the G' versus corn starch content (Fig. 4.36) is typical of several protein–polysaccharide systems in which phase separation and inversion have occurred (Ross-Murphy 1984; Tolstoguzov 1985). However, the magnitudes of G' of the individual components with 10% solids differ only by a factor of 2 and are lower than in other such systems (Tolstoguzov 1985). More severe asymmetry was observed in dispersions containing corn starch on the one hand, and 11S and 7S

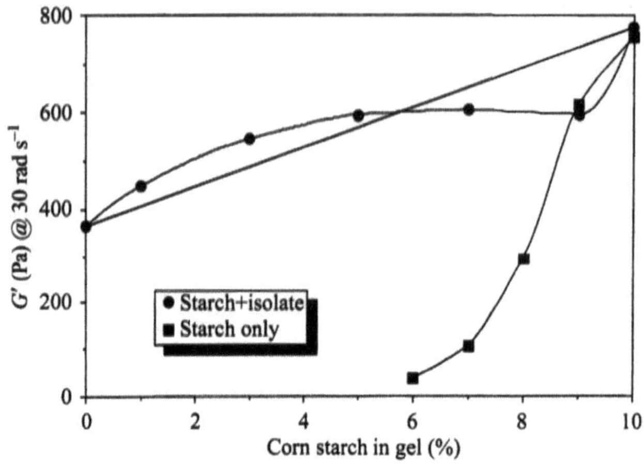

Fig. 4.36 Plot of the experimental values of G', at a frequency of 30 rad s^{-1}, of dispersions of various soyprotein isolate/corn starch ratios and of corn starch with 6–10% solids. It also has a straight line that interpolates G' values between those of corn starch and soyprotein isolate

globulins on the other (Chen et al. 1996). At starch concentrations less than 5%, the continuous phase made up predominantly by SP isolate is weaker, and the effect of adding the stronger (higher modulus) starch granules is to increase G', the so-called isostress (Ross-Murphy 1984; Morris 1986) condition. As the starch concentration was increased further, phase inversion occurred in that the stronger SD was the continuous phase and the SP isolate dispersion became the dispersed phase (Fig. 4.36). Additional insight into the interaction of corn starch-SP isolate was found in creep-compliance data (Liao et al. 1996).

Cowpea Protein/Cowpea Starch

The development of G' in 10% solids gels from blends of cowpea protein and cowpea starch over a 10 h aging period at 20° C is shown in Fig. 4.37 (Okechukwu and Rao 1997); G' increased from an initial value, G_0', to a maximum plateau value, G'_{max}, that increased with the proportion of starch in the mixture. A slight decrease in G' was observed during the later part of aging, probably due to weakening of starch granules. The rise in G' up to the maximum value followed an apparent first-order rate process, Eq. 4.58.

$$\frac{\mathrm{d}X}{\mathrm{d}t} = -kX \tag{5.58}$$

where $X = \dfrac{G'_{max} - G(t)}{G'_{max} - G_0'}$, k is a rate constant, and t is aging time. Estimated values of the apparent rate constants at 20°C were: 0.016, 0.022, and 0.023 min^{-1} for the

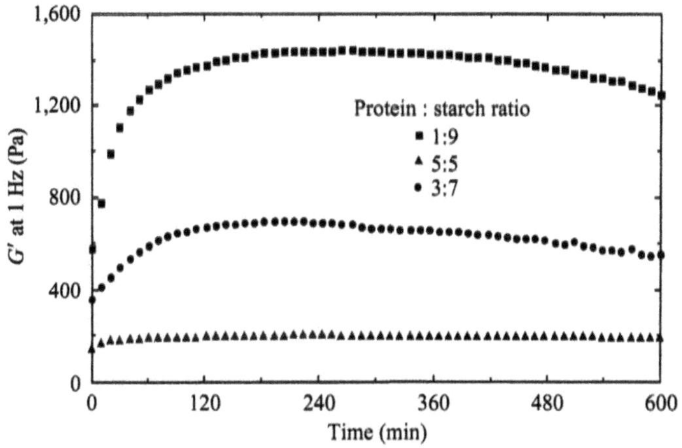

Fig. 4.37 Development of G' in 10 % solids gels from blends of cowpea protein and cowpea starch over a 10 h aging period at 20°C (Okechukwu and Rao 1997)

blends at protein/starch ratios (R) of 5/5, 3/7, and 1/9, with R^2 values 0.96, 1.00, 1.00, respectively. The variation of G'_{max} (Pa) with mass fraction (x_s) of starch in the blend was described by the equation

$$G'_{\max} = 7.69(x_s)^{3.35}; \quad R^2 = 0.998 \tag{4.59}$$

The nearly equal values of the exponents for the variation of G' with starch concentration in cowpea starch/protein mixtures (within the range of $R < 1/9$) and in cowpea starch gels emphasizes the important contribution of starch to the structure of the former. Development of G' in the protein–starch gels followed an essentially apparent first-order process and did not show either the dominant two-stage growth pattern seen in starch gels or the almost linear pattern observed in the cowpea protein gels.

At pH values close to 7.0 the cowpea protein solubilized and formed the continuous phase in which raw starch granules were dispersed. On heating, the granules absorbed water and as they swelled they exuded some amylose into the continuous matrix prior to the protein denaturation. The higher transition temperature of the protein compared with that of starch created the environment for the leaching of amylose into the solution of the protein (Muhrbeck and Eliasson 1991). The coexistence of the protein and the amylose in the continuous phase may be responsible for the distinct kinetic pattern in the observed development of G' in the cowpea protein–starch blends. For such a system, the protein/starch ratio (R) would reflect the protein/amylose ratio in the continuous matrix and may be considered to be an important parameter for assessing G' of the mixture gels.

Figure 4.38 shows the variation of G' at 1 Hz of protein/starch gels (10 % solids) from frequency sweeps on gels after 1 h of aging with starch fraction (x_s). Up to

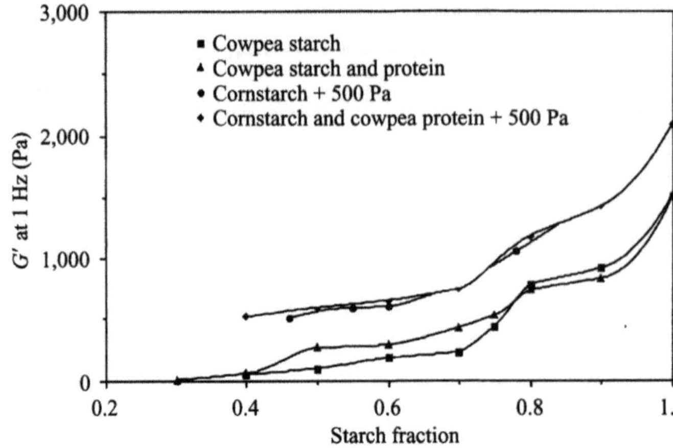

Fig. 4.38 Variation of G' at 1 Hz of protein/starch gels (10% solids) from frequency sweeps on gels after 1 h of aging with starch fraction (x_s)

$x_s < 0.8 (R < 2/8)$, G' of the mixed gels was higher than that of cowpea starch gel at the same concentration. For $x_s > 0.8$, G' of the starch gel was higher than G' of the mixed gel. Figure 4.38 also contains G' at 1 Hz from frequency sweeps on gels of cowpea protein and corn starch after 1 h of aging at 25°C. Addition of the cowpea protein to corn starch at high levels did not affect the magnitude of G' of gel and at low levels resulted in a slight decrease. Similar decrease in G' was observed after the addition of 1% gluten to corn starch (Lindahl and Eliasson 1986). Recognizing that G' of 10% solids cowpea protein gels was too low to be measured, the shape of the modulus–starch fraction curves of the mixed gels containing either corn starch or cowpea starch suggests phase separation. The higher G' values of the protein/cowpea starch gels at low levels of starch appear to be due to favorable kinetics during the early stages of G' development (Fig. 4.39). For aging periods less than 250 min in 5% solids gels, G' of the mixed gel (5% solids and 5 parts protein and 5 parts starch) was more than that of the starch gel of equivalent solids content. At a higher starch and solids level (9%), G' of the mixed gel (9% solids and 1 part protein and 9 parts starch) was lower throughout the aging period.

Whey Protein Isolate/Cross-Linked Waxy Maize Starch Dispersions

Whey protein isolate (WPI) is a by-product from cheese and casein manufacture, and has excellent gelation properties, a high nutritional value, and functional properties that can be utilized to meet the demands of value-added food applications. Thermal gelation of whey protein involves an initial denaturation unfolding step

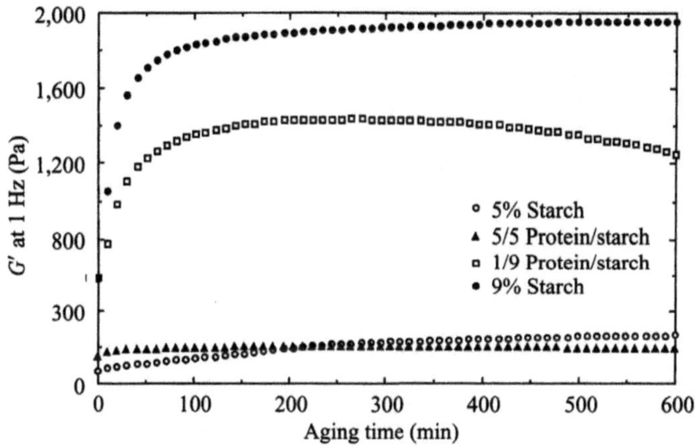

Fig. 4.39 The higher G' values of the protein/cowpea starch gels at low levels of starch appear to be due to favorable kinetics during the early stages of G' development. For aging periods less than 250 min in 5 % solids gels, G' of the mixed gel (5 % solids and 5 parts protein and 5 parts starch) was more than that of the starch gel of equivalent solids content. At a higher starch and solids level (9 %), G' of the mixed gel (9 % solids and 1 part protein and 9 parts starch) was lower throughout the aging period

followed by aggregation into a protein network (Aguilera and Rojas 1996; Zasypkin et al. 1997). Aggregation of unfolded protein molecules occurs by hydrophobic and sulfhydryl–disulfide interactions (McSwiney et al. 1994; Mleko and Foegeding 1999). The β-lactoglobulin is the main protein in whey that dominates the overall gelling behavior (McSwiney et al. 1994).

Thermal gelation of a protein is influenced by many factors, such as the protein concentration, pH, and presence of salts. Protein–protein interactions are generally favored at conditions that reduce the net charge on the molecules, that is, pH values near the isoelectric point (Boye et al. 1995). Monovalent and divalent salt ions screen electrostatic interactions between charged protein molecules (Bryant and McClements 2000). Specifically, calcium is reported to promote heat-induced aggregation and gelation of whey proteins (Van Camp et al. 1997).

Dynamic and flow rheological characteristics of heated mixed WPI and CWM starch dispersions: 5 % solids, pH = 7.0, 75 mM NaCl, were examined at starch mass fractions (x_s) from 0 (pure WPI) to 1 (pure CWM; Ravindra et al. 2004). The mixed dispersions had lower values of G' than the pure WPI dispersion, primarily due to the disruptive effect of CWM granules on the WPI network. The point of phase inversion (minimum G' value) was at about $x_s = 0.65$ (Fig. 4.40).

With respect to flow behavior (not shown here), pure WPI dispersions ($x_s = 0$) up and down shear stress ramp curves showed thixotropic behavior, as also reported by Mleko and Foegeding (2000). Dispersions with $x_s = 0.2$ were almost superimposed over a wide range of shear rates higher than 50 s^{-1} and showed a slight anti-thixotropic possibly due to an order–disorder transition, as reported by Mleko and

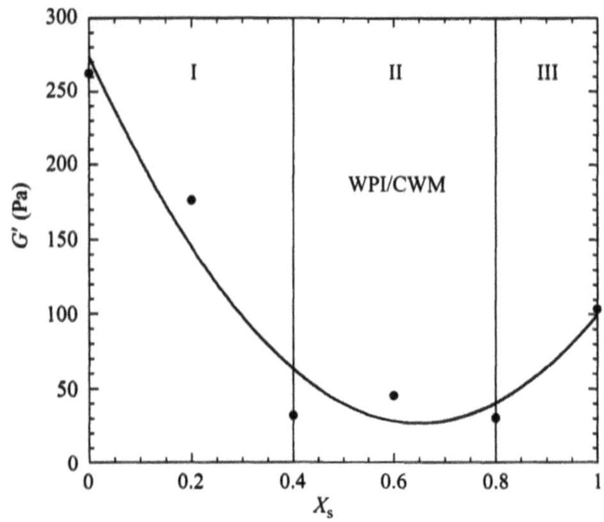

Fig. 4.40 Influence of starch mass fraction (x_s) on the storage modulus (G') at 1 Hz of heated mixed WPI/CWM dispersions, 5 % total solids, after aging for 4 h at 20°C

Foegeding (1999) in WPI dispersions ≥3 % heated at 80°C in a two-step process, first at pH 8.0 and second at pH 7.0. Dispersions with $x_s = 0.4$ and 0.6 exhibited shear-thinning thixotropic behavior. Dispersions with $x_s = 0.8$ and 1 exhibited shear-thinning antithixotropic behavior typical of pure CWM dispersions suggesting that at this high x_s the heated mixed WPI/CWM dispersions behaved more like a CWM dispersion.

After shearing of the heated mixed WPI/CWM dispersions during the up ramp in the rheometer, the apparent viscosity at 300 s^{-1} increased with starch mass fraction. That increase reflected the dominant effect of the still intact CWM granules in the structure disrupted during the shearing of the blends (Fig. 4.41).

Rheology of Starch–Gum Dispersions

In order to understand the rheological behavior of a starch–gum system well, the molecular characteristics of the gum and the structural characteristics of the starch should be determined. In addition, the starch–gum mixture should be heated under controlled conditions. Christianson et al. (1981) reported synergistic interaction between wheat starch and the gums: xanthan, guar, and methyl cellulose. Synergistic interactions were also observed between corn and wheat starches on the one hand and guar and LB gum on the other (Alloncle et al. 1989). They suggested that because SDs are suspensions of swollen granules dispersed in a macromolecular medium, it was suggested that the galactomannans were located within the continuous phase. Therefore, the volume of this phase, accessible to the galactomannan was reduced.

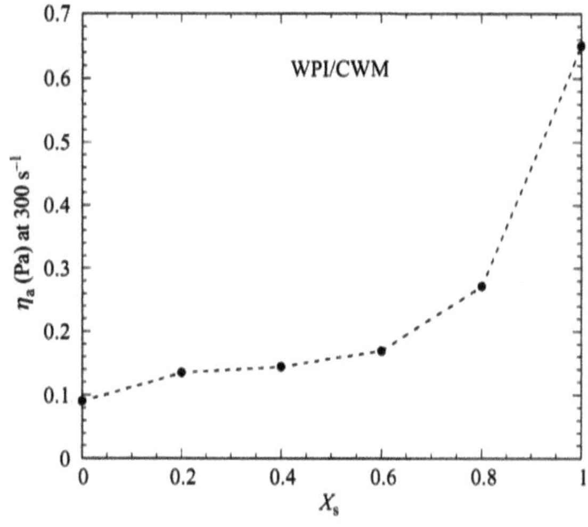

Fig. 4.41 Influence of starch mass fraction (x_s) on the apparent viscosity (η_a) at the end of the up curve (300 s^{-1}) of heated mixed WPI/CWM dispersions, 5 % total solids, 20°c

Two-phase liquid separation occurred in mixtures of potato maltodextrin with LB gum, gum arabic, and carboxymethyl cellulose (Annable et al. 1994). Rheological properties of potato maltodextrin were greatly influenced by the addition of gum arabic. At low additions, the storage modulus (G') was greatly enhanced, while at higher additions, corresponding to compositions in the two phase region of the phase diagram, lower values of G' were obtained. The rate of PMD gelation followed first-order kinetics and the rate constant increased with increasing gum arabic concentration.

Achayuthakan et al. (2006) studied vane yield stress of xanthan gum SDs. The intrinsic viscosity of xanthan gum was determined to be 112.3 dL/g in distilled water at 25°C. In addition, the size of the granules in the dispersions of the studied starches: waxy maize (WXM), CWM, and cold water swelling (CWS) were determined. The values of yield stress of the starch–xanthan dispersions relative to those of the starch–water dispersions (YS/YS0) and relative mean granule diameters (D/D0) plotted against values of $c[\eta]$ of xanthan gum are shown in Fig. 4.42. With the values of YS/YS0 being less than 1.0, there was no synergism between CWM starch and xanthan gum.

Table 4.13 contains the static yield stresses of the xanthan dispersions in water (column A), of the three starches in water (column B) and their sum (column C), and those of the mixed starch–xanthan dispersions (column D). One can say that there is synergism between xanthan and starch if values in column D are higher than those in column C; if they are lower, there is antagonism. From the table, it seems that WXM (except with 1 % xanthan concentration) and CWS starches exhibited synergistic interaction with xanthan, while CWM starch–xanthan dispersions exhibited antagonistic behavior.

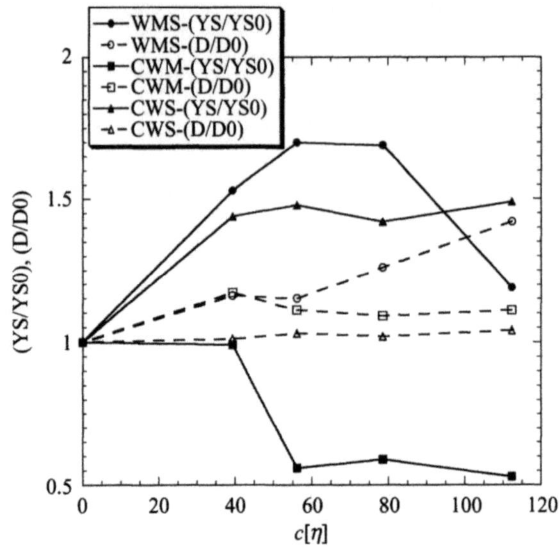

Fig. 4.42 Values of yield stress of starch–xanthan dispersions relative to those of the starch–water dispersions (YS/YS0) and relative mean granule diameters (D/D0) plotted against values of $c[\eta]$ of xanthan gum; waxy maize (WXM), cross-linked waxy maize (CWM), and cold water swelling (CWS)

Table 4.13 Comparison of static yield stress (YSS), Pa, of xanthan gum, starch–water and starch–xanthan gum mixtures, and of interactions

Xanthan (%)	YSS xanthan (A)	YSS starch–water (B)	C=A+B	YSS (D)	D–C	Interaction
Waxy maize						
0.00	–	23.9	23.9	23.9	0.0	–
0.35	1.9	23.9	25.7	36.6	10.9	S
0.50	3.5	23.9	27.4	40.5	13.1	S
0.70	5.1	23.9	28.9	40.2	11.3	S
1.00	8.6	23.9	32.5	28.4	−4.1	A
Cross-linked waxy maize						
0.00	–	59.3	59.3	59.3	0.0	–
0.35	1.9	59.3	61.2	59.0	−2.2	A
0.50	3.5	59.3	62.9	33.4	−29.5	A
0.70	5.1	59.3	64.4	35.2	−29.2	A
1.00	8.6	59.3	68.0	31.7	−36.3	A
Cold water-swelling						
0.00	–	39.2	39.2	39.2	0.0	–
0.35	1.9	39.2	41.1	56.4	15.3	S
0.50	3.5	39.2	42.7	58.1	15.4	S
0.70	5.1	39.2	44.3	55.6	11.3	S
1.00	8.6	39.2	47.8	58.6	10.7	S

A=static yield stress of xanthan gum dispersion, *B*=static yield stress of each starch dispersion, *C*=A+B, *D* static yield stress of starch-xanthan dispersion Interaction S (synergistic), no interaction A (antagonistic)

References

Achayuthakan, P., Suphantharika, M., and Rao, M. A. 2006. Yield stress components of waxy corn starch-xanthan mixtures: effect of xanthan concentration and different starches. *Carbohydr. Polym.* 65: 469–478.

Aguilera, J. M. and Rojas, E. 1996. Rheological, thermal and microstructural properties of whey protein-cassava starch gels. *J. Food Sci.* 61: 962–966.

Ahmad, F. B. and Williams, P. A. 1999. Effect of sugars on the thermal and rheological properties of sago starch. *Biopolymers* 50: 401–412.

Alloncle, M., Lefebvre, J., Llamas, G., and Doublier, J. L. 1989. A rheological characterization of cereal starch-galactomannan mixtures. *Cereal-Chem.* 66(2): 90–93.

Annable, P., Fitton, M. G., Harris, B., Phillips, G. O., and Williams, P. A. 1994. Phase behaviour and rheology of mixed polymer systems containing starch. *Food-Hydrocolloids* 8(3/4): 351–359.

Axelos, M. A. V., Thibault, J. F., and Lefebvre, J. 1989. Structure of citrus pectins and viscometric study of their solution properties. *Int. J. Biol. Macromol.* 11:186–191.

Bagley, E. B. and Christianson, D. D. 1982. Swelling capacity of starch and its relationship to suspension viscosity: effect of cooking time, temperature and concentration. *J. Texture Stud.* 13:115–126.

Barnes, H. A. 1989. Shear thickening “Dilatancy” in suspensions of non aggregating solid particles dispersed in Newtonian liquids. *J. Rheol.* 33: 329–366.

Biliaderis, C. G. 1992. Characterization of starch networks by small strain dynamic rheometry, in *Developments in Carbohydrate Chemistry*, eds. R. J. Alexander and H. F. Zobel, American Association of Cereal Chemists, St. Paul, MN.

Bird, R. B., Armstrong, R. C., and Hassager, O. 1977a. *Dynamics of Polymeric Liquids-Fluid Mechanics*, John Wiley and Sons, New York.

Bird, R. B., Hassager, O., Armstrong, R. C., and Curtiss, C. F. 1977b. *Dynamics of Polymeric Liquids-Kinetic Theory*, John Wiley and Sons, New York.

Blanshard, J. M. V. 1987. Starch granule structure and function: a physicochemical approach, in *Starch: Properties and Potential*, ed. T. Galliard pp. 16–54, John Wiley & Sons, New York.

Boersma, W. H., Baets, P. J. M., Laven, J., and Stein, H. N. 1991. Time-dependent behavior and wall slip in concentrated shear thickening dispersions. *J. Rheol.* 35: 1093–1120.

Boersma, W. H., Laven, J., and Stein, H. N. 1992. Viscoelastic properties of concentrated shear-thickening dispersions. *J. Colloid and Interface Sci.* 149: 10–22.

Bossis, G. and Brady, J. F. 1989. The rheology of Brownian suspensions. *J. Chemical Phys.* 91: 1866–1879.

Boye, J. I., Alii, I., Ismail, A. A., Gibbs, B. F., and Konishi, Y. 1995. Factors affecting molecular characteristics of whey protein gelation. *Int. Dairy J.* 5: 337–353.

Bryant, C. M. and McClements, D. J. 2000. Influence of NaCI and $CaCl_2$ on cold-set gelation of heat-denatured whey protein. *J. Food Sci.* 65: 801–804.

Bu-Contreras, R. 2001. Influence of physico-chemical factors on the firmness of potatoes and apples. Ph.D. thesis, Cornell University, Ithaca, New York, USA.

Buscall, R., Goodwin, J. W., Hawkins, M. W., and Ottewell, R. H. 1982a. Viscoelastic properties of concentrated lattices I. Methods of examination. *J. Chem. Soc. Fraday Trans.* 78: 2873–2887.

Buscall, R., Goodwin, J. W., Hawkins, M. W., and Ottewell, R. H. 1982b. Viscoelastic properties of concentrated lattices II. *Theor. Anal.* 78: 2889–2899.

Byars, J. A., Fanta, G. F. and Felker, F. C. 2009. Rheological properties of dispersions of spherulites from jet-cooked high-amylose corn starch and fatty acids. *Cereal Chem.* 86(1):76–81.

Carreau, P. J., De Kee, D., and Chhabra, R. P. 1997. *Rheology of Polymeric Systems: Principles and Applications*, Hanser, New York.

Chamberlain, E. K. 1996. Characterization of heated and thermally processed cross-linked waxy maize starch utilizing particle size analysis, microscopy and rheology. M.S. thesis, Cornell University, Ithaca, NY.

Chamberlain, E. K. 1999. Rheological properties of acid converted waxy maize starches: effect of solvent, concentration and dissolution time. Ph.D. thesis, Cornell University, Ithaca, NY.

Chamberlain, E. K. and Rao, M. A. 2000. Concentration dependence of viscosity of acid-hydrolyzed amylopectin solutions. *Food Hydrocolloids* 14: 163–171.

Chamberlain, E. K., Rao, M. A., and Cohen, C. 1998. Shear thinning and antithixotropic behavior of a heated cross-linked waxy maize starch dispersion. *Int.. J. Food Properties* 2: 63–77; errata, 2: 195–196.

Champenois, Y. C., Rao, M. A., and Walker, L. P. 1998. Influence of gluten on the viscoelastic properties of starch pastes and gels. *J. Sci. Food Agric.* 78: 119–126.

Chedid, L. L. and Kokini, J. L. 1992. Influence of protein addition on rheological properties of amylose-and amylopectin-based starches in excess water. *Cereal Chem.* 69: 551–555.

Cheer, R. L. and Lelievre, J. 1983. Effects of sucrose on the rheological behavior of wheat-starch pastes. *J. Appl. Polym. Sci.* 28(6): 1829–1836.

Chen, C.-J., Okechukwu, P. E., Damodaran, S., and Rao, M. A. 1996. Rheological properties of heated corn starch + soybean 7S and 11S globulin dispersions. *J. Texture Stud.* 27: 419–432.

Chou, T. D. and Kokini, J. L. 1987. Rheological properties and conformation of tomato paste pectins, citrus and apple pectins. *J. Food Sci.* 52: 1658–1664.

Chow, M. K. and Zukoski, C. F. 1995a. Gap size and shear history dependencies in shear thickening of a suspension ordered at rest. *J. Rheol.* 39: 15–32.

Chow, M. K. and Zukoski, C. F. 1995b. Nonequlibrium behavior of dense suspensions of uniform particles: volume fraction and size dependence of rheology and microstructure. *J. Rheol.* 39: 33–59.

Christianson, D. D. and Bagley, E. B. 1984. Yield stresses in dispersions of swollen deformable cornstarch granules. *Cereal Chem.* 61: 500–503.

Christianson, D. D., Hodge, J. E., Osborne, D., and Detroy, R. W. 1981. Gelatinization of wheat starch as modified by xanthan gum, guar gum, and cellulose gum. *Cereal Chem.* 58(6): 513–517.

Colas, B. 1986. Flow behavior of crosslinked cornstarches. Lebensmittel Wissenschaft u. *Technol.* 19: 308–311.

Cox, W. P. and Merz, E. H. 1958. Correlation of dynamic and steady flow viscosities. *J. Polymer Sci.* 28(118): 619.

Da Silva, P. M. S., Oliveira, J. C., and Rao, M. A. 1997. The effect of granule size distribution on the rheological behavior of heated modified and unmodified maize starch dispersions. *J. Texture Stud.* 28: 123–138.

Dail, R. V. and Steffe, J. F. 1990a. Dilatancy in starch solutions under low acid aseptic processing conditions. *J. Food Sci.* 55: 1764–1765.

Dail, R. V. and Steffe, J. F. 1990b. Rheological characterization of crosslinked waxy maize starch solutions under low acid aseptic processing conditions using tube viscometry techniques. *J. Food Sci.* 55: 1660–1665.

Davidson, R. L. 1980. *Handbook of Water-Soluble Gums and Resins*, McGraw-Hill Book Co., New York.

Davis, M. A. F., Gidley, M. J., Morris, E. R., Powell, D. A., and Rees, D. A. 1980. Intermolecular association in pectin solutions. *Int. J. Biol. Macromol.* 2: 330.

Dealy, J. M. and Wissburn, K. F. 1990. *Melt Rheology and Its Role in Plastics Processing: Theory and Applications*, Van Nostrand Reinhold, New York.

De Kee, D. and Wissburn, K. F. 1998. Polymer rheology. *Physics Today* 51, no. 6: 24–29.

D'Haene, P., Mewis, J., and Fuller, G. G. 1993. Scattering dichroism measurements of flow-induced structure of a shear thickening suspension. *J. Colloid Interface Sci.* 156: 350–358.

Dintzis, F. R. and Bagley, E. B. 1995. Shear-thickening and transient flow effects in starch solutions. *J. Appl. Polymer Sci.* 56: 637–640.

Dolan, K. D. and Steffe, J. F. 1990. Modeling rheological behavior of gelatinizing starch solutions using mixer viscometry data. *J. Texture Stud.* 21: 265–294.

Dolan, K. D., Steffe, J. F., and Morgan, R. G. 1989. Back extrusion and simulation of viscosity development during starch gelatinization. *J. Food Process Eng.* 11: 79–101.

Doublier, J. L. 1981. Rheological studies on starch. Flow behavior of wheat starch pastes. *Starch/Stärke* 33:415–420

Doublier, J. L. 1987. A rheological comparison of wheat, maize, faba bean and smooth pea starches. *J. Cereal Sci.* 5:247–262.

Elbirli, B. and M. T. Shaw. 1978. Time constants from shear viscosity data. *J. Rheol.* 22: 561–570.

Eliasson, A. C. 1986. Viscoelastic behavior during the gelatinization of starch: 1. Comparison of wheat, maize, potato and waxy barley starches. *J. Texture Stud.* 17: 253–265.

Ellis, H. S., Ring, S. G., and Whittam, M. A. 1989. A comparison of the viscous behavior of wheat and maize starch pastes. *J. Cereal Sci.* 10: 33–44.

Evageliou, V., Richardson, R. K., and Morris, E. R. 2000. Effect of sucrose, glucose and fructose on gelation of oxidized starch. *Carbohydr. Polym.* 42: 261–272.

Evans, I. D. and Haisman, D. R. 1979. Rheology of gelatinized starch suspensions. *J. Texture Stud.* 10: 347–370.

Evans, I. D. and Haisman, D. R. 1982. The effect of solutes on the gelatinization temperature range of potato starch. *Starch/Stäerke* 34(7): 224–231.

Evans, I. D. and Lips, A. 1992. Viscoelasticity of gelatinized starch dispersions. *J. Texture Stud.* 23: 69–86.

Evans, I. D. and Lips, A. 1993. Influence of soluble polymers on the elasticity of concentrated dispersions of deformable food microgel particles, in *Food Colloids and Polymers: Stability and Mechanical Properties*, eds. E. Dickinson and P. Walstra, The Royal Society of Chemistry, Cambridge, England.

Faubion, J. M. and Hoseney, R. C. 1990. The viscoelastic properties of wheat flour doughs, in *Dough Rheology and Baked Product Texture*, eds. H. Faridi and J. M. Faubion, Van Nostrand Reinhold, New York, USA, pp. 29–66.

Ferry, J. D. 1980. *Viscoelastic Properties of Polymers*, John Wiley, New York.

Fukuoka, M., Ohta, K., and Watanabe, H. 2002. Determination of the terminal extent of starch gelatinization in a limited water system. *J. Food Eng.* 53: 39–42.

Galliard, T. and Bowler, P. 1987. Morphology and composition of starch, in *Starch: Properties and Potential, Critical Reports on Applied Chemistry*, ed. T. Galliard, Vol. 13, pp. 54–78, John Wiley and Sons, New York.

Genovese, D. B. and Rao, M. A. 2003a. Role of starch granule characteristics (volume fraction, rigidity, and fractal dimension) on rheology of starch dispersions with and without amylose. Cereal Chem. 80: 350–355.

Genovese, D. B. and Rao, M. A. 2003b. Apparent viscosity and first normal stress of starch dispersions: role of continuous and dispersed phases, and prediction with the Goddard-Miller model. *Appl. Rheol.* 13(4): 183–190.

Genovese, D. B. and Rao, M. A. 2003c. Vane yield stress of starch dispersions. *J. Food Sci.* 68(7): 2295–2301.

Genovese, D. B., Acquarone, V. M., Youn, K.-S., and Rao, M. A. 2004. Influence of fructose and sucrose on small and large deformation rheological behavior of heated Amioca starch dispersions. *Food Science and Technology International* 10(1): 51–57.

Giboreau, A., Cuvelier, G., and Launay, B. 1994. Rheological behavior of three biopolymer/water systems with emphasis on yield stress and viscoelastic properties. *J. Texture Stud.* 25: 119–137.

Glicksman, M. 1969. *Gum Technology in the Food Industry*, Academic Press, New York.

Graessley, W. W. 1967. Viscosity of entangling polydisperse polymers. *J. Chem. Phys.* 47: 1942–1953.

Graessley, W. W. 1974. The entanglement concept in polymer rheology. *Adv. Polymer Sci.* 16: 1–179, Springer-Verlag, Berlin.

Graessley, W. W. 1980. Polymer chain dimensions and the dependence of viscoelastic properties on concentration, molecular weight and solvent power. *Polymer* 21: 258–262.

Griskey, R. G. and Green, R. G. 1971. Flow of dilatant shear-thickening fluids. *Am. Inst. Chem. Engrs. J.* 17: 725–728.

Harris, E. K. Jr. 1970. Viscometric properties of polymer solutions and blends as functions of concentration and molecular weight. Ph.D thesis, University of Wisconsin, Madison.

Harrod, M. 1989. Modelling of flow properties of starch pastes prepared by different procedures. *J. Food Process Eng.* 11: 257–275.

Hoffman, R. L. 1972. Discontinuous and dilatant viscosity behavior in concentrated suspensions. 1. Observation of a flow instability. *Trans. Soc. Rheol.* 16: 155–173.

Hoseney, R. C. 1998. Gelatinization phenomena of starch, in *Phase/State Transitions in Foods: Chemical, Structural, and Rheological Changes*, eds. M. A. Rao and R. W. Hartel, pp. 95–110, Marcel Dekker, Inc., New York.

Kaletunc-Gencer, G. and Peleg, M. 1986. Rheological characteristics of selected food gum mixtures in solution. *J. Text. Stud.* 17: 61–70.

Krieger, I. J. 1985. Rheology of polymer colloids, in *Polymer Colloids*, eds. R. Buscall, T. Corner, and J. F. Stageman, pp. 219–246, Elsevier Applied Science, New York.

Kubota, K., Hosakawa, Y., Suziki, K., and Hosaka, H. 1979. Studies on the gelatinization rate of rice and potato starches. *J. Food Sci.* 44: 1394–1397.

Kulicke, W.M. and Porter, R.S. 1980. Relation between steady shear flow and dynamic rheology. *Rheologica Acta* 19: 601–605.

Langan, R. E. 1986. Food industry, in *Modified Starches: Properties and Uses*, pp. 199–212, CRC Press, Boca Raton, FL.

Lapasin, R., Priel, S., and Tracanelli, P. 1991. Rheology of hydroxyethyl guar gum derivatives. *Carbohydr. polym.* 14: 411–427.

Laun, H. M., Bung, R., and Schmidt, F. 1991. Rheology of extremely shear thickening polymer dispersions passively viscosity switching fluids. *J. Rheol.* 35: 999–1034.

Launay, B., Doublier, J. L. and Cuvelier, G. 1986. Flow properties of aqueous solutions and dispersions of polysaccharides, in *Functional Properties of Food Macromolecules*, eds. J. R. Mitchell and D. A. Ledward, Chapter 1, pp. 1–78, Elsevier Applied Science Publishers, London.

Leach, H. W., McGowen, L. D., and Schoch, T. J. 1959. Structure of starch granule. I. Swelling and solubility patterns of various starches. *Cereal Chem.* 36: 534–544.

Liao, H.-J., Okechukwu, P. E., Damodaran, S., and Rao, M. A. 1996. Rheological and calorimetric properties of heated corn starch-soybean protein isolate dispersions. *J. Texture Stud.* 27: 403–418.

Liao, H.-J., Tattiyakul, J., and Rao, M. A. 1999. Superposition of complex viscosity curves during gelatinization of starch dispersion and dough. *J. Food Proc. Eng.* 22: 215–234.

Lindahl, L. and Eliasson, A. C. 1986. Effects of wheat proteins on the viscoelastic properties of starch gels. *J. Sci. Food Agric.* 37: 1125–1132.

Lopes da Silva, J. A. L. 1994. Rheological characterization of pectin and pectin- galactomannan dispersions and gel. Ph.D thesis, Escola Superior de Biotecnologia, Porto, Portugal.

Lopes da Silva, J. A. L., Gonçalves, M. P., and Rao, M. A. 1992. Rheological properties of high-methoxyl pectin and locust bean gum solutions in steady shear. *J. Food Sci.* 57: 443–448.

Lopes da Silva, J. A. L. and Rao, M. A. 1992. Viscoelastic properties of food gum dispersions, in *Viscoelastic Properties of Foods*, eds. M. A. Rao and J. F. Steffe, pp. 285–316, Elsevier Applied Science Publishers, London.

Lopes da Silva, J. A. L., Gonçalves, M. P., and Rao, M. A. 1993. Viscoelastic behavior of mixtures of locust bean gum and pectin dispersions. *J. Food Eng.* 18:211–228.

Lopes da Silva, J. A. L., Gonsalves, M. P., and Rao, M. A. 1994. Influence of temperature on dynamic and steady shear rheology of pectin dispersions. *Carbohydr. Polym.* 23: 77–87.

Lopes da Silva, J. A. L. and Rao, M. A. 2006. Pectins: Structure, functionality, and uses, in *Food Polysaccharides and Their Applications: Second Edition, Revised and Expanded*, eds. A. M. Stephen, G. O. Phillips, and P. A. Williams, pp. 353–411, CRC Press, Inc., Boca Raton, New York.

Lund, D. 1984. Influence of time, temperature, moisture, ingredients and processing conditions on starch gelatinization. *Crit. Rev. Food Sci. and Nutr.* 20: 249–273.

Madeka, H. and Kokini, J. L. 1992. Effect of addition of zien and gliadin on the rheological properties of amylopectin starch with low-to-intermediate moisture. *Cereal Chem.* 69: 489–494.

Matsumoto, T., Hitomi, C., and Onogi, S. 1975. Rheological properties of disperse systems of spherical particles in polystyrene solution at long time scales. *Trans. Soc. Rheol.* 19: 541–545.

McConnaughey, W. B. and Petersen, N. O. 1980. Cell poker: an apparatus for stress-strain measurements on living cells. *Rev. Sci. Instrum.* 51: 575–580.

McSwiney, M., Singh, H., and Campanella, O. H. 1994. Thermal aggregation and gelation of bovine β-lactoglobulin. *Food Hydrocolloids* 8: 441–453.

Miller, S. A. and Mann, C. A. 1944. Agitation of two-phase systems of immiscible liquids. *Trans. Am. Inst. Chem. Engrs.* 40: 709.

Mills, P. L. and Kokini, J. L. 1984. Comparison of steady shear and dynamic viscoelastic properties of guar and karaya gums. *J. Food Sci.* 49: 1–4 and 9.

Mleko, S. and Foegeding, E. A. 1999. Formation of whey protein polymers: effects of a two-step heating process on rheological properties. *J. Texture Stud.* 30: 137–149.

Mleko, S. and Foegeding, E. A. 2000. pH induced aggregation and weak gel formation of whey protein polymers. *J. Food Sci.* 65: 139–143.

Morris, E. R. 1981. Rheology of hydrocolloids, in *Gums and Stabilisers for the Food Industry 2*, eds. G. O. Philips, D. J. Wedlock, and P. A. Williams, p. 57, Pergamon Press Ltd., Oxford, Great Britain.

Morris, V. J. 1986. Multicomponent gels, in *Gums and Stabilisers for the Food Industry 3*, eds. G. O. Philips, D. J. Wedlock, and P. A. Williams, pp. 87–99, Elsevier Applied Science Publishers, London.

Morris, V. J. 1990. Starch gelation and retrogradation. *Trends Food Sci. Technol.* July, 1: 2–6.

Morris, E. R. and Ross-Murphy, B. 1981. Chain flexibility of polysaccharides and glicoproteins from viscosity measurements, in *Techniques in Carbohydrate Metabolism*, ed. D. H. Northcote, B310, pp. 1–46, Elsevier, Amsterdam.

Morris, E. R., Cutler, A. N., Ross-Murphy, S. B., and Rees, D. A. 1981. Concentration and shear rate dependence of viscosity in random coil polysaccharide solutions. *Carbohydr. Polym.* 1: 5–21.

Muhrbeck, P. and Eliasson, A. C. 1991. Rheological properties of protein/starch mixed gels. *J. Texture Stud. 22*: 317–332.

Noel, T. R., Ring, S. G., and Whittam, M. A. 1993. Physical properties of starch products: structure and function, in *Food Colloids and Polymers: Stability and Mechanical Properties*, eds. E. Dickinson and P. Wolstra, pp. 126–137, Royal Society of Chemistry, Cambridge, UK.

Norisuye, T. 1996. Conformation and properties of amylose in dilute solution. *Food-Hydrocolloids* 10(1): 109–115.

Okechukwu, P. E. and Rao, M. A. 1995. Influence of granule size on viscosity of cornstarch suspension. *J. Texture Stud.* 26: 501–516.

Okechukwu, P. E. and Rao, M. A. 1996a. Kinetics of cornstarch granule swelling in excess water, in *Gums & Stabilisers for the Food Industry 8*, eds. G. O. Phillips, P. A. Williams, and D. J. Wedlock), pp. 49–57, The Oxford University Press, Oxford, U.K.

Okechukwu, P. E. and Rao, M. A. 1996b. Role of granule size and size distribution in the viscosity of cowpea starch dispersions heated in excess water. *J. Texture Stud.* 27: 159–173.

Okechukwu, P. E. and Rao, M. A. 1997. Calorimetric and rheological behavior of cowpea protein plus starch cowpea and corn gels. *Food Hydrocolloids* 11: 339–345.

Okechukwu, P. E., Rao, M. A., Ngoddy, P. O., and McWatters, K. H. 1991. Flow behavior and gelatinization of cowpea flour and starch dispersions. *J. Food Sci.* 56: 1311–1315.

Paoletti, S., Cesaro, A., Delben, F., and Ciana, A. 1986. Ionic effects on the conformation, equilibrium, properties, and rheology of pectate in aqueous solution and gels, in *Chemistry and Function of pectins*, eds. M. L. Fishman and J. J. Jen, pp. 73–87, ACS Symposium Series, American Chemical Society, Washington, DC.

Petrofsky, K. E. and Hoseney, R. C. 1995. Rheological properties of dough made with starch and gluten from several cereal sources. *Cereal Chem.* 72(1): 53–58.

Plazek, D. J. 1996. 1995 Bingham medal address: Oh, thermorheological simplicity, wherefore art thou? *J. Rheology* 40: 987–1014.

Plutchok, G. J. and Kokini, J. L. 1986. Predicting steady and oscillatory shear rheological properties of CMC and guar gum blends from concentration and molecular weight data. *J. Food Sci.* 515: 1284–1288.

Quemada, D., Fland, P., and Jezequel, P. H. 1985. Rheological properties and flow of concentrated diperse media. *Chem. Eng. Comm.* 32: 61–83.

Rao, M. A. and Tattiyakul, J. 1999. Granule size and rheological behavior of heated tapioca starch dispersions. *Carbohydrate Polymers* 38: 123–132.

Ravindra, P., Genovese, D. B., Foegeding, E. A., and Rao, M. A. 2004. Rheology of mixed whey protein isolate/cross-linked waxy maize starch gelatinized dispersions. *Food Hydrocolloids* 18: 775–781.

Robinson, G., Ross-Murphy, S. B., and Morris, E. R. 1982. Viscosity-molecular weight relationships, intrinsic chain flexibility and dynamic solution properties of guar galactomannan. *Carbohydr. Res.* 107: 17–32.

Rochefort, W. E. and Middleman, S. 1987. Rheology of xanthan gum: salt, temperature and strain effects in oscillatory and steady shear experiments. *J. Rheol.* 31: 337–369.

Rodriguez, F. 1989. *Principles of Polymer Systems*, 3rd ed., Hemisphere Publishing Corp., New York.

Roos, Y. H. 1995. *Phase Transitions in Foods*, Academic Press, New York.

Ross-Murphy, S. B. 1984. Rheological methods, in *Biophysical Methods in Food Research*, ed. H. W.-S. Chan, pp. 138–199, Blackwell Scientific, London.

Russel, W. B., Saville, D. A., and Schowalter, W. R. 1989. *Colloidal Dispersions*, Cambridge University Press, Cambridge, U. K.

Sawayama, S., Kawabata, A., Nakahara, H., and Kamata, T. 1988. A light scattering study on the effects of pH on pectin aggregation in aqueous solution. *Food Hydrocolloids* 2: 31–37.

Tam, K.C. and Tiu, C. 1989. Steady and dynamic shear properties of aqueous polymer solutions. *Journal of Rheology* 33: 257–280.

Tam, K. C. and Tiu, C. 1993. Improved correlation for shear-dependent viscosity of polyelectrolyte solutions. *J Non-Newtonian Fluid Mech.* 46: 275–288.

Tattiyakul, J. 1997. Studies on granule growth kinetics and characteristics of tapioca starch dispersion during gelatinization using particle size analysis and rheological methods. M.S. thesis, Cornell University, Ithaca, NY.

Tattiyakul, J. and Rao, M. A. 2000. Rheological behavior of cross-linked waxy maize starch dispersions during and after heating. *Carbohydr. Polym.* 43: 215–222.

Tester, R. F. and Morrison, W. R. 1990. Swelling and gelatinization of cereal starches. I. Effects of amylopectin, amylose and lipids. *Cereal Chem.* 67(6): 551–557.

Tirrell, M. 1994. Rheology of polymeric liquids, in *Rheology: Principles, Measurements, and Applications*, ed. Macosko, C. W. 1994. VCH Publishers, New York.

Tolstoguzov, V. B. 1985. Functional properties of protein-polysaccharide mixtures, in *Functional Properties of Food Macromolecules*, eds. J. Mitchell and D. A. Ledward, pp. 385–415, Elsevier Applied Science Publishers, London.

Tolstoguzov, V. B. 1991. Functional properties of food proteins and role of protein-polysaccharide interaction—review. *Food Hydrocolloids* 4: 429–468.

Van Camp, J., Messens, W., Clément, J. and Huyghebaert, A. 1997. Influence of pH and calcium chloride on the high-pressure-induced aggregation of a whey protein concentrate. *J. Agric. Food Chem.* 45: 1600–1607.

Whistler, R. L. and Daniel, J. R. 1985. Carbohydrates, in *Food Chemistry*, ed. O. R. Fennema, pp. 69–138, New York, Marcel Dekker.

Whitcomb, P. J. and Macosko, C. W. 1978. Rheology of xanthan gum. *J. Rheol.* 22:493–505.

Yang, W. H. 1997. Rheological behavior and heat transfer to a canned starch dispersion: computer simulation and experiment. Ph.D thesis, Cornell University, Ithaca, NY.

Yang, W. H., Datta, A. K., and Rao, M. A. 1997. Rheological and calorimetric behavior of starch gelatinization in simulation of heat transfer, in *Engineering and Food at ICEF 7/Part 2*, ed., pp. K1–K5. Sheffield Academic Press, London.

Yang, W. H. and Rao, M. A. 1998. Complex viscosity-temperature master curve of cornstarch dispersion during gelatinization. *J. Food Proc. Eng.* 21: 191–207.

Yoo, B., Figueiredo, A. A., and Rao, M. A. 1994. Rheological properties of mesquite seed gum in steady and dynamic shear. Lebensmittel Wissenschaft und Technologie 27: 151–157.

Zahalak, G. I., McConnaughey, W. B., and Elson, E. L. 1990. Determination of cellular mechanical properties by cell poking, with an application to leukocytes. *J. Biomechanical Eng.* 112: 283–294.

Zasypkin, D. V., Braudo, E. E., and Tolstoguzov, V. B. 1997. Multicomponent biopolymer gels. *Food Hydrocolloids* 11: 159–170

Chapter 5
Rheological Behavior of Processed Fluid and Semisolid Foods

In this chapter, the rheological properties of processed fluid and semisolid foods will be discussed. Where data are available, the role of the composition of the foods on their rheological behavior will be emphasized. In addition, literature values of data on several foods, many of which are discussed here and some that are not discussed, are given at the end of this chapter.

Fruit Juices and Purees: Role of Soluble and Insoluble Solids

Fruit juices and plant food (PF) dispersions are important items of commerce. Single strength and concentrated clear fruit juices, that is, juices from which insoluble solids and dissolved polymers (e.g., pectins) have been removed, exhibit Newtonian flow behavior or close to it and the sugar content plays a major role in the magnitude of the viscosity and the effect of temperature on viscosity (Saravacos 1970; Rao et al. 1984; Ibarz et al. 1987, 1992a, b). Magnitudes of flow parameters of fruit juices and purees are summarized in Table A5.2, of tomato pastes in Table A5.3, and the effect of temperature on apparent viscosity and consistency index is summarized in Table A5.4. It should be noted that the viscosity (or apparent viscosity, consistency index) of concentrated fruit juices decreases with increase in temperature and the magnitude of the activation energy of clear fruit juices increases markedly with increase in sugar content.

Because depectinized and filtered apple and grape juices are Newtonian fluids, equations were derived by Bayindirli (1992, 1993) that can be used to estimate their viscosities as a function of concentration and temperature. The equation for viscosity of both apple and grape juices is of the form:

$$\frac{\eta}{\eta_{\mathrm{wa}}} = \exp\left\{\frac{A(^{\circ}\mathrm{Brix})}{100 - B(^{\circ}\mathrm{Brix})}\right\} \tag{5.1}$$

M. A. Rao, *Rheology of Fluid, Semisolid, and Solid Foods*, Food Engineering Series, DOI 10.1007/978-1-4614-9230-6_5, © Springer Science+Business Media New York 2014

where, η is the viscosity (mPa s) of depectinized and filtered apple juice, η_{wa} = viscosity of water at the same temperature (mPa s), and A and B are constants. For apple juice in the concentration range 14–39 °Brix and temperature 293–353 K (20–80 °C), the constants A and B are:

$$A = -0.24 + (917.92/T) \quad B = 2.03 - 0.00267T \tag{5.2}$$

For grape juice in the concentration range 19–35 °Brix and temperature 293–353 K (20–35 °C), the constants A and B are:

$$A = -3.79 + (1821.45/T) \quad B = 0.86 + 0.000441T \tag{5.3}$$

In both Eqs. 5.2 and 5.3, the temperature T is in K. For the purpose of illustration, from Eqs. 5.2 and 5.3, the viscosity of 30 °Brix apple juice at 20 °C was estimated to be 4.0 mPa s and that of grape juice 2.8 mPa s, respectively.

Rao et al. (1984) studied the role of concentration and temperature on the viscosity of concentrated depectinized and filtered apple and grape juices. At a constant temperature, the effect of concentration could be well described by an exponential relationship. For example, at 20 °C, the effect of concentration (~41–68 °Brix) on the viscosity of apple concentrates was described by the equation (R^2=0.947):

$$\eta = 1.725 \times 10^{-5} \exp[0.136(^{\circ}\text{Brix})] \tag{5.4}$$

For concentrated grape juice samples (~41–68 °Brix) at 20 °C, the effect of concentration was described by (R^2=0.966):

$$\eta = 2.840 \times 10^{-5} \exp[0.137(^{\circ}\text{Brix})] \tag{5.5}$$

The effect of temperature and concentration on the viscosity of concentrated apple juice can be combined to obtain a single approximate equation that can be used for estimating viscosity as a function of both temperature and concentration (°Brix):

$$\eta_a = A \exp(\bar{E}_a/RT) \exp[\beta(^{\circ}\text{Brix})] \tag{5.6}$$

For apple juice, the constants A, $\bar{E}_a$, and β were: 3.153×10^{-14} Pa s, 45.0 kJ mol^{-1}, and 0.167 °Brix^{-1}, respectively, and the R^2=0.940. For the viscosity of concentrated grape juice concentrates, the constants A, $\bar{E}_a$, and β were: 6.824×10^{-14} Pa s, 46.6 kJ mol^{-1}, and 0.151 °Brix^{-1}, respectively, and the R^2=0.971. The above equation should be considered to provide only approximate values of viscosity because $\bar{E}_a$ is an average activation energy for the entire range of concentrations studied. In fact, the $\bar{E}_a$ of concentrated apple and grape juices are exponentially dependent on the °Brix values (Fig. 5.1). Figure 5.1 illustrates the important role of sugar content on the activation energy of concentrated apple and grape juices. The respective equations describing the influence of °Brix on the $\bar{E}_a$ (kJ mol^{-1}) of the apple and grape juice are:

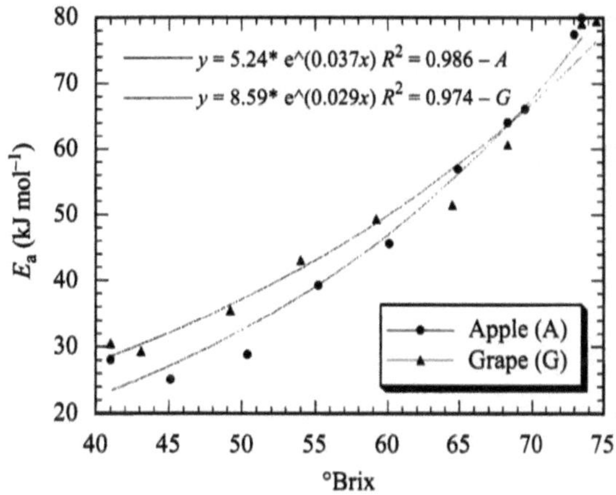

Fig. 5.1 Values of the flow activation energy (E_a) of concentrated apple and grape juices are exponentially dependent on their °Brix values

$$E_a = 5.236\exp[0.0366(^{\circ}\text{Brix})];\ R^2 = 0.94 \tag{5.7}$$

$$E_a = 8.587\exp[0.0294(^{\circ}\text{Brix})];\ R^2 = 0.97 \tag{5.8}$$

Thus a more accurate estimate of the viscosities of concentrated apple and grape juices can be obtained by incorporating the value of E_a calculated using either Eq. 5.7 or 5.8 at a specific concentration in Eq. 5.6. Although two separate equations were derived for the apple and grape juice concentrates, one could easily justify a single equation that would be reasonably applicable to both concentrates.

Role of PF Insoluble Solids

Several observations useful to concentrated purees can be found in studies on the rheological behavior of nonfood dispersions that have been reviewed earlier in Chap. 2 and elsewhere (Jinescu 1974; Jeffrey and Acrivos 1976; Metzner 1985); other useful studies include on the rheology of rigid fiber suspensions by Ganani and Powell (1985) and on slurries of irregular particles (Wildemuth and Williams 1984). Also, there has been what may be called a kinetic or structural approach to rheology of suspensions (Michaels and Bolger 1962; Hunter and Nicol 1968) where the basic flow units are assumed to be small clusters, or aggregates, that at low shear rates give the suspension a finite yield stress.

The solid particles in PF dispersions are not of simple shapes (e.g., spheres, rods) and they are deformable, and have multimodal size distributions (Tanglertpaibul and Rao 1987a). Also, the particles are hydrated and are in physical and chemical equilibrium with the continuous medium so that they differ significantly from artificial fibers such as of glass or of synthetic polymers. The continuous phases of PF dispersions also have features that are different than those of nonfood suspensions.

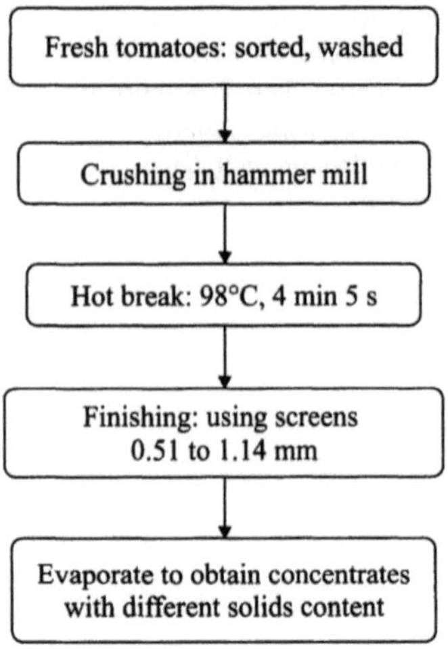

Fig. 5.2 Illustration of the processing steps in tomato concentrates; finisher screens with different diameter openings were used to vary pulp particle size and content (Tanglertpaibul and Rao 1987a)

The continuous medium of a typical food dispersion, usually called serum, is an aqueous solution of sugars, organic acids, salts, and pectic substances. The chemical composition of the continuous medium depends on the particular commodity, the cultivar, and factors such as the extent of ripening.

The differences in characteristics of PF dispersions can lead to significant interparticle forces compared to viscous forces, and Brownian motion can be important, resulting in different flow and rheological characteristics. For example, while negligible wall slip effects were reported for pipe flow of glass fiber suspensions in silicone oils (Maschmeyer and Hill 1977), they were of concern for PF dispersions in capillary flow (Kokini and Plutchok 1987) and in a concentric cylinder geometry (Higgs 1974; Qiu and Rao 1989). Food dispersions are complex materials whose characteristics with respect to the nature of the insoluble solids as well as those of the fluid media often are determined a priori to experimentation. Nonfood suspensions, such as glass fibers in mineral oils and in polymers, can be custom made with one's choice of continuous and dispersed media having the desired characteristics. In contrast, for preparing a set of engineered PF dispersion samples, the source for the continuous and dispersed phases is another PF dispersion.

Nevertheless, in order to understand the role of various components of PF dispersions, systematic studies on the role of the amount and size of insoluble solids can be conducted by careful preparation of the test samples. Figure 5.2 illustrates the processing steps used to vary pulp particle size distribution in tomato concentrates (Tanglertpaibul and Rao 1987a). Similar processing technique was used in the preparation of apple sauce samples by Rao et al. (1986) and Qiu and Rao (1988). The amount and the size distribution of the insoluble solids in the PF dispersions depend to some extent on the size of the screen employed in the finishing operation during their manufacture (Tanglertpaibul and Rao 1987a) and to a limited extent on the speed of the finisher (Rao et al. 1986).

In the rheological behavior of concentrated pulpy fruit juices and purees (e.g., orange and grapefruit juices), both the pulp and sugar contents were found to be the key components (Mizrahi and Berk 1972; Durán and Costell 1982; Vitali and Rao 1984). In tomato concentrates also, where sugar content is not as high as in concentrated fruit juices (e.g., orange juice), both pulp and sugars are key components (Harper and El Sahrigi 1965; Rao et al. 1981; Fito et al. 1983).

The viscosity (or apparent viscosity) of a suspension can be related to the viscosity of the continuous medium in terms of the relative viscosity, η_r:

$$\eta_r = \frac{\text{Viscosity of suspension}}{\text{Viscosity of continuous medium}} \tag{5.9}$$

and the volume fraction of solids, ϕ (Jinescu 1974; Metzner 1985). Because of the compressible nature of PF dispersions, the direct determination of the magnitude of ϕ is difficult as it depends on the centrifugal force employed in the separation of the phases. In samples of tomato concentrates, volume fraction of solids was determined from volumetric measurements to be about 0.15–0.45 (Tanglertpaibul and Rao 1987b). In the case of concentrated orange juice (COJ), for the purpose of quality control, the amount of dispersed solid matter (pulp content) in a 12 °Brix sample is determined at 360 g so that the pulp content in different samples is determined at a standard centrifugal force and at a standard concentration that sets, to a reasonable extent, the continuous medium's viscosity (Praschan 1981). The pulp content of COJ, as determined by the standard method, ranges from about 2–16 %. Because of the highly viscous nature of apple sauce and tomato concentrates as well as the large amount of pulp content, the technique used for COJ may not be satisfactory for these products. Therefore, ratios such as pulp: total sample weight (Takada and Nelson 1983) and pulp: serum (Tanglertpaibul and Rao 1987b) have been employed.

Size Distribution of Fruit Juice Solids

Size distribution data on solid particles of PF dispersions are sparse. Carter and Buslig (1977) determined size distribution of relatively small pulp particles in COJ with a Coulter counter. Wet sieving technique was employed on apricot puree, apple sauce, and tomato concentrates (Rao et al. 1986; Trifiro et al. 1987; Tanglertpaibul and Rao 1987a) where volume fractions of solids retained on standard sieves were used to calculate the approximate average particle sizes. It is clear that the solid particles of PF dispersions have multimodal size distributions. Also, changes in particle sizes induced by homogenization of apricot puree affected consistency of the puree (Durán and Costell 1985), while changes induced by using finisher screens of different sizes affected the rheological properties of a tomato puree (Tanglertpaibul and Rao 1987a). Because the volume fraction of some of the tomato concentrates was in the range 0.15–0.45 (Tanglertpaibul and Rao 1987a), these results are consis-

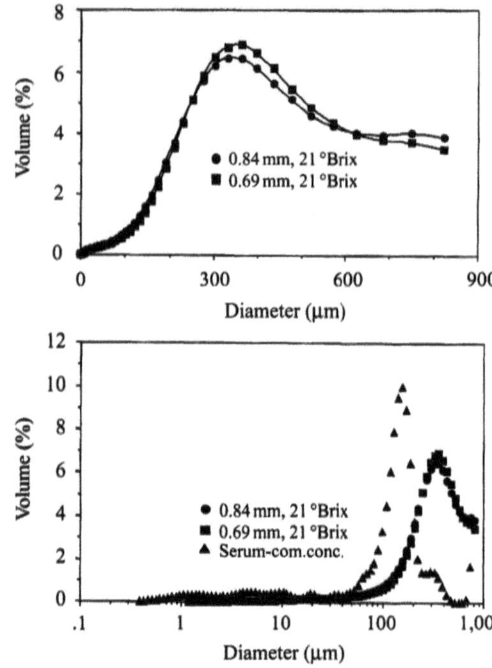

Fig. 5.3 Particle size distributions of 21 °Brix tomato concentrates (*top*) derived from using 0.69 and 0.84 mm finisher screens, and in serum obtained from a commercial concentrate after centrifugation at 100,000 × g for 1 h (*bottom*) (Yoo and Rao 1996)

tent with the general observation (Metzner 1985) that distribution of particle sizes affects dispersion viscosities when volumetric loading of solids is above 20 %.

Because relatively low-cost powerful desktop computers and instruments have become available, size distribution of tomato pulp particles can be examined based on laser diffraction (den Ouden and van Vliet 1993). Although most tomato pulp particles were smaller than 900 μm, the upper particle size limit of the LS130 instrument, a few particles were larger (den Ouden and van Vliet 1993; Yoo and Rao 1996). Particle size distributions of 21 °Brix concentrates (Fig. 5.3, top) illustrate the general shape of size distribution versus particle diameter profiles. The distributions reveal differences, although minor, between the concentrates from 0.69 and 0.84 mm finisher screens. Interestingly, serum obtained from a commercial concentrate after centrifugation at 100,000 × g for 1 h also contained relatively large particles, > 100–200 μm as measured by laser diffraction (Fig. 5.3, bottom). The presence of the few large particles could be due to the finishing operation in which the tomato pulp is forced through finisher screen holes resulting in elongated shreds of cell wall and skin of the fruit.

While the size distribution of reasonably well-defined smaller (max. dia ~ 100 μm) gelatinized corn starch granules could be determined reliably with the Coulter LS130 and its role in rheological behavior established (Okechukwu and Rao 1995), similar efforts for tomato concentrates requires further work. Also, what role these few large particles may play in the rheological behavior of the concentrates is not very obvious so that determination of their exact sizes (e.g., by using an instrument with a wider range of particle sizes) may provide only marginal understanding of the size distribution but not of the rheology of the concentrates. Because of the relatively wide range of size distribution of the tomato pulp particles (Fig. 5.3), it is unlikely that shear-thickening flow behavior will be encountered with tomato concentrates as in gelatinized starch dispersions that can exhibit relatively narrow particle size distributions (Okechukwu and Rao 1995).

Serum Viscosity of PF Dispersions

The viscosities of the continuous media of PF dispersions depend on their composition, viz., sugar and pectin content that in turn may be affected by cultivars, fruit maturity, and processing conditions. The continuous media of apple sauce with sugar concentration of about 16 °Brix and tomato concentrates with sugar concentrations in the range 6–25 °Brix were Newtonian fluids (Rao et al. 1986; Tanglertpaibul and Rao 1987b). At high concentrations, where the plant polymers are concentrated, they can be expected to be mildly shear-thinning fluids as was the case with the serum of COJ with concentration of about 65 °Brix (Vitali and Rao 1984). Caradec and Nelson (1985) found that heat treatment of tomatoes resulted in loss of serum viscosity. Correlations in terms of the relative viscosity, η_{rel}, for relatively low magnitudes and narrow ranges of sugar concentrations, such as the continuous media of apple sauce and tomato concentrates, may be possible with just the pectin content as the independent variable. Information regarding serum intrinsic viscosity, which is a measure of the hydrodynamic volume and is related to the molecular weight of the biopolymer in solution, will be very useful. Magnitudes of intrinsic viscosity can be determined either from extrapolation techniques, such as with the Huggins' equation or from the slopes of straight lines of quantities based on relative viscosity. Magnitudes of intrinsic viscosity varied slightly with the method of determination, the extent of heat treatment, and to a lesser extent on the shearing experienced by the food suspension (Tanglertpaibul and Rao 1987c).

Rheological Properties of Frozen Concentrated Orange Juice (FCOJ)

COJ (65 °Brix) is a major item of commerce with over a million tons being exported by Brazil and another million tons being produced in Florida. Figure 5.4 illustrates the steps in the manufacture of 65 °Brix COJ and Fig. 5.5 the preparation samples with different amounts of pulp (Vitali and Rao 1984a, b). At a fixed temperature, the flow behavior of concentrated 65 °Brix orange juice serum is nearly Newtonian while that of 65 °Brix orange juice is a shear-thinning fluid (Fig. 5.6). Also at a fixed temperature, the effect of concentration (c) of soluble solids (°Brix) and insoluble solids (pulp) on either apparent viscosity or the consistency index of the power law model of FCOJ can be described by exponential relationships (Vitali

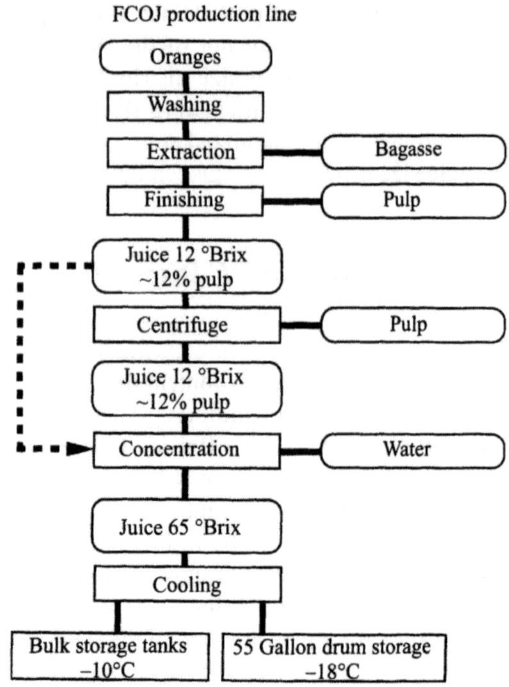

Fig. 5.4 Unit operations in the manufacture of 65 °Brix concentrated orange juice samples with different amounts of pulp (Vitali 1983)

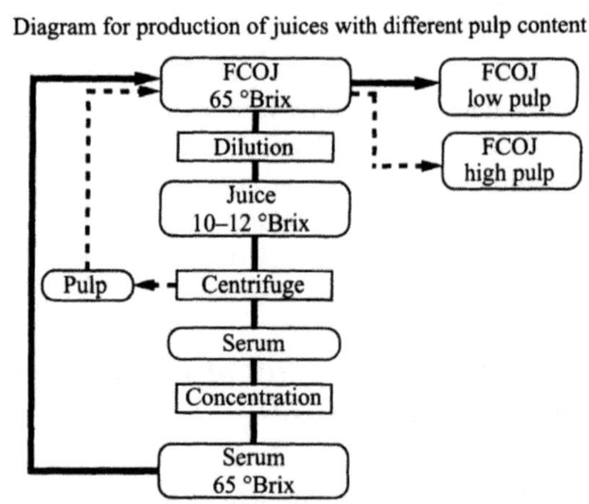

Fig. 5.5 Steps (top to bottom) in the preparation samples with different amounts of pulp from 65 °Brix sample (Vitali 1983)

and Rao 1984a, b). Earlier, the influence of soluble solids on the consistency index of FCOJ was shown in Fig. 2.10. The influence of pulp content on the consistency index is shown in Fig. 5.7, where ln K is plotted versus % pulp content. Equations 5.10 and 5.11 are applicable to the consistency index (K) of the power law model (in Chap. 2, Eqs. 2.35 and 2.36, respectively). In the case of FCOJ, insoluble solids are expressed in terms of pulp content determined on a 12 °Brix sample by centrifugation at for 10 min at 360 × g.

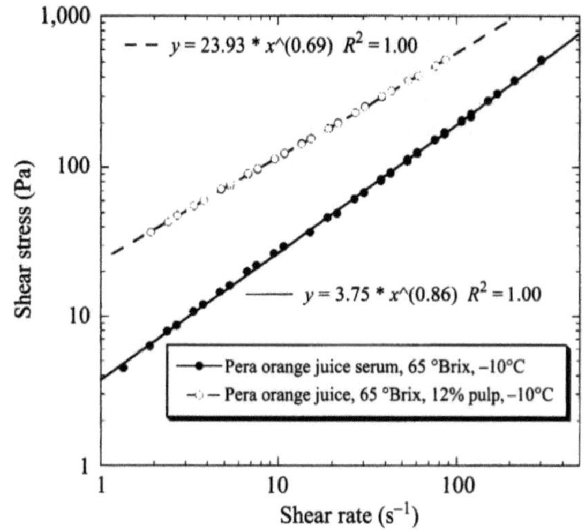

Fig. 5.6 The flow behavior of concentrated 65 °Brix orange juice serum is nearly Newtonian (slope = 0.86) while that of 65 °Brix orange juice is a mildly shear-thinning fluid (slope = 0.69) (based on data in Vitali 1983)

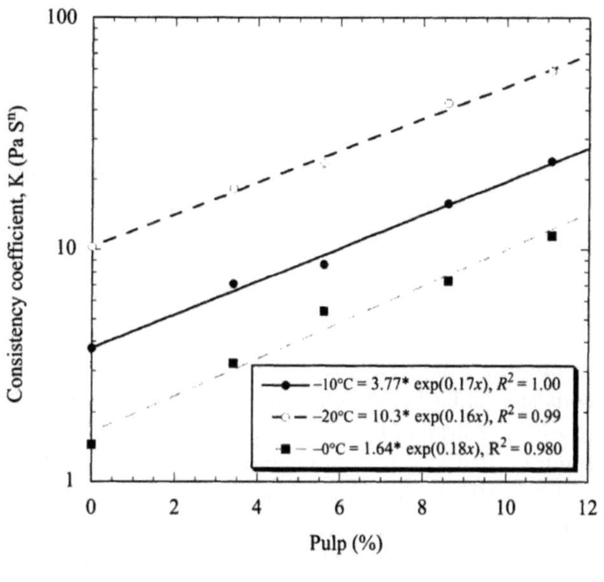

Fig. 5.7 The influence of pulp content on the consistency index (K) is shown, where in K is plotted versus % pulp content (Vitali and Rao 1984b)

$$K = K^c \exp\left[B_{\mathrm{K}}^c(^\circ\mathrm{Brix})\right] \tag{5.10}$$

$$K = K^P \exp\left[B_{\mathrm{K}}^P(\mathrm{Pulp})\right] \tag{5.11}$$

where, K^c, K^p, B_K^c and B_K^P are constants. As pointed out in Chap. 2, to study the effect of composition variables, one can also work with apparent viscosity instead of the power law consistency index.

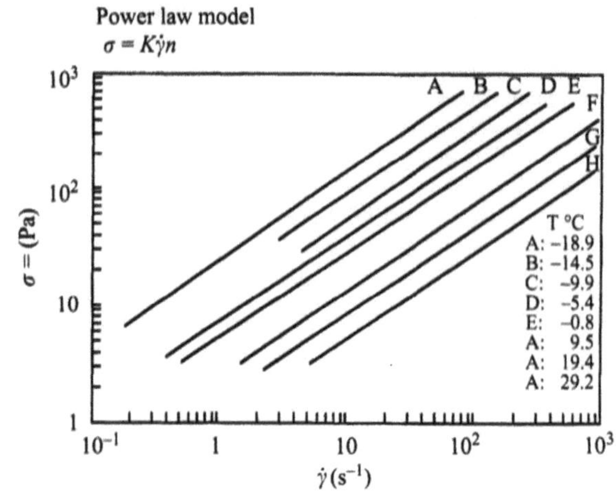

Fig. 5.8 The flow behavior of FCOJ is strongly influenced by temperature; lines representing data at several temperatures from Vitali and Rao (1984a) are shown

Effect of Temperature

The flow behavior of FCOJ is strongly influenced by temperature as seen in Fig. 5.8 where data at several temperatures from Vitali and Rao (1984a, b) are shown. A wide range of temperatures are encountered during FCOJ processing and storage so that the effect of temperature on rheological properties needs to be documented. The effect of temperature on either apparent viscosity or the consistency index, K, of the power law model (Eq. 5.12) of FCOJ can be described by the Arrhenius relationship. The equation for the consistency index is:

$$K = K_{\infty} \exp(E_{aK}/RT) \tag{5.12}$$

where, K_{∞} is the frequency factor, E_{aK} is the activation energy (J mol^{-1}), R is the gas constant, and T is temperature (K). A plot of In K (ordinate) versus (T^{-1}) (abscissa) results in a straight line, and E_{aK} = (slope × R), and K_{∞} is exponential of the intercept. Figure 5.9 illustrates applicability of the Arrhenius relationship for FCOJ (Vitali and Rao 1984b). The activation energy should be expressed in joules (J), but in the earlier literature it has been expressed in calories (1 calorie = 4.1868 J). Instead of the consistency index of the power law model, the apparent viscosity has been used in some studies.

Literature Values of Rheological Properties of FCOJ

It is clear from the above discussion that the rheological properties of FCOJ will depend on the °Brix, pulp content, size and shape of the pulp particles, and pectin content of FCOJ sample, and the temperature at which the data were obtained. Carter and Buslig (1977) studied particle size distribution in commercial frozen FCOJ samples. Mizrahi and coworkers conducted systematic studies on FCOJ that they described as a physicochemical approach (Mizrahi and Berk 1970; Mizrahi

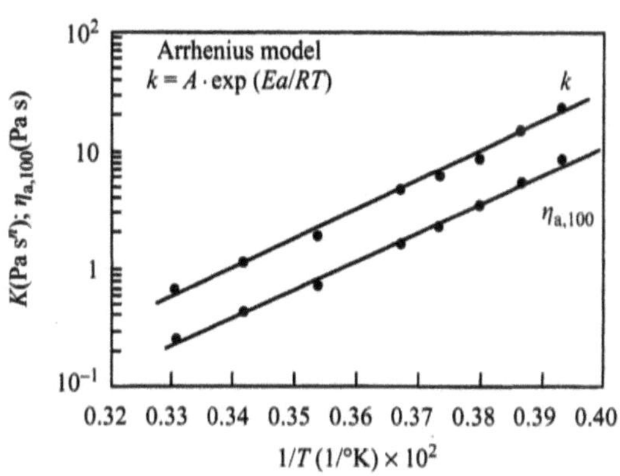

Fig. 5.9 Arrhenius plots for the effect of temperature on apparent viscosity at 100 s^{-1} (η_a,100) and consistency coefficient (K) of Pera FCOJ with 5.7 % pulp content (Vitali and Rao 1984b)

and Firstenberg 1975; Mizrahi 1979). They used a modified Casson equation to describe the flow behavior of FCOJ.

Because the power law model (Eq. 2.3) is used in determination of pumping and mixing power requirements, literature values of the power law parameters of several FCOJ samples reported by Crandall et al. (1982) and Vitali and Rao (1984a, b) are given in Table A5.2. The different values in Table A5.2 reflect the influence of fruit varieties, °Brix, pulp content, and temperature. In particular, the strong influence of temperature on the consistency index of the power law model should be noted. The constants in Eqs. 5.10–5.12 that describe the effect of temperature, °Brix, and pulp content on the consistency index of the power law model presented by Vitali and Rao (1984a, b) are summarized in Table A5.2. These data should be useful, albeit with caution, in estimation of ball park values of the power law parameters of FCOJ from temperature, °Brix, and pulp content data.

Rheological Properties of Tomato Concentrates

Numerous studies have been conducted on tomato concentrates using quality control instruments, such as the Bostwick consistometer. Unfortunately, for a specific set of test samples, values of Bostwick consistency decrease with increase in solids concentration making it impossible to obtain data at high solids loadings. Attempts at correlating Bostwick readings and basic viscosity data on Newtonian fluids in terms of theoretical analysis based on gravity wave flow were successful (McCarthy and Seymour 1994a); however, similar attempt with non-Newtonian fluids (McCarthy and Seymour 1994b) and attempt at empirical correlation met with moderate success (Rao and Bourne 1977). However, an approach based on dimensionless groups to correlate Bostwick readings with power law parameters seems to have had success (Perona 2005) that merits further evaluation.

The power law parameters (consistency index and flow behavior index) of tomato concentrates reported in the literature can be found in Table A5.2. Tanglertpaibul and Rao (1987b) examined the contribution of soluble solids to the rheological properties of tomato concentrates (Fig. 5.10) with concentrates made from a single

[t]

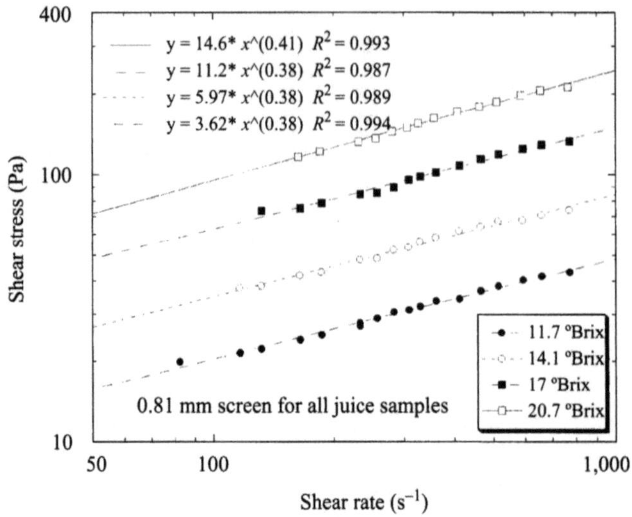

Fig. 5.10 Influence of soluble solids on shear rate versus shear stress data of tomato concentrates using the same finisher screen (based on data of Sornsrivichai 1986)

finisher screen. They also (Tanglertpaibul and Rao 1987b) examined the influence of different finisher screen sizes (0.51–1.14 mm) on flow properties. From the shear rate versus shear stress data obtained on 20 % solids concentrates made from juices that were produced using different finisher screen openings shown in Fig. 5.11, it can be seen that using a 0.69 mm screen resulted in concentrates with the highest viscosity; use of the 0.69 mm screen probably resulted in juice with highest pulp content. Because the composition of tomatoes and other fruits depends on the particular varieties being employed and their maturity, the effect of finisher screen size on the pulp content and size distribution should be further explored.

The role of pulp content on flow behavior was examined by adding tomato pulp to concentrates with different soluble solids: 5.6, 10.0, 15.0, and 20.0 °Brix. A plot of the pulp content versus the apparent viscosity at 100 s^{-1} (Fig. 5.12) is shown where the linear relationship is valid only over the narrow range of pulp contents. Because the influence of pulp content was approximated and the apparent viscosity at 100 s^{-1}, a meaningless negative intercept was obtained. More importantly, the data in Fig. 5.12 reflect the important role of solids loading on rheological behavior.

Rao et al. (1981) studied the influence of temperature and total solids content (c) on the apparent viscosity at 100 s^{-1} ($\eta_{a,100}$) of tomato concentrates prepared from the varieties grown at the New York State Agricultural Experiment Station: Nova, New Yorker, #475, #934 hot break process, and #934 cold break process. The results of Nova hot break process concentrates were expressed by a model slightly different than that used to examine concentrated apple and grape juices.

$$\eta_{a,100} = \alpha \exp(E_a/RT)(C^{\beta}) \tag{5.13}$$

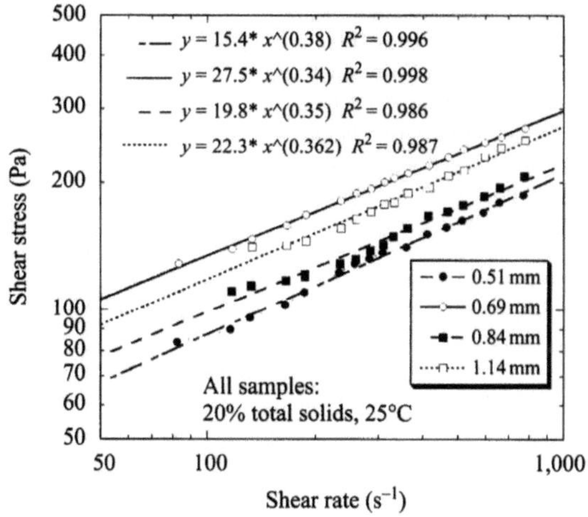

Fig. 5.11 Shear rate versus shear stress data at 25 °C obtained on 20 % solids concentrates made from juices that were produced using different finisher screen openings show that using a 0.69 mm screen resulted in concentrates with the highest viscosity (Tanglertpaibul and Rao 1987a)

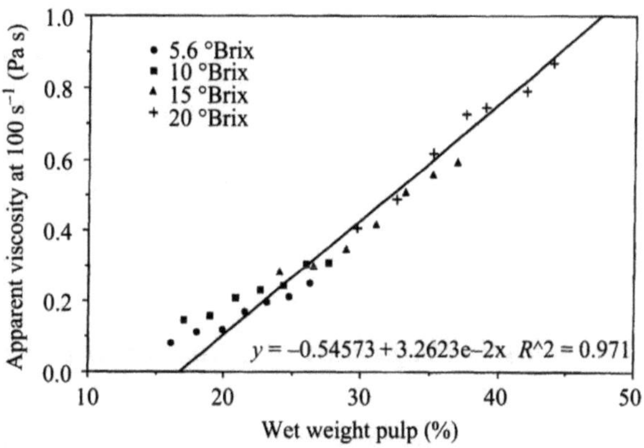

Fig. 5.12 Plot of the pulp content versus the apparent viscosity at 100 s^{-1} of tomato concentrates (Tanglertpaibul and Rao 1987b) illustrates the important role of pulp content. The meaningless negative intercept Indicates that the linear relationship cannot be extended to low pulp contents

The results for Nova concentrates were: $\alpha = 1.3 \times 10^{-5}$, Ea = 9.21 kJ mol^{-1}, and $\beta = 2.6$. It is very encouraging that these values were close to those reported by Harper and El Sahrigi (1965) for the tomatoes of variety VF6: $\alpha = 1.05 \times 10^{-5}$, Ea = 9.63 kJ mol^{-1}, and $\beta = 2.0$. Magnitudes of β for the concentrates made from New Yorker hot break process, #475 hot break process, #934 hot break process, and #934 cold break process were: 2.5, 2.4, 2.5, and 2.4, respectively, while that of Nova tomato concentrates was slightly higher, ~2.6 (Figs. 5.13 and 5.14).

Thus, a useful general result is that the viscosity of tomato concentrates can be scaled by the factor $(\text{totalsolids})^{2.5}$. For example, at a fixed temperature, one can predict the viscosity η_2 of a tomato concentrate having the concentration c_2 from the magnitude η_1 of a concentrate of the same variety having the concentration c_1 from the relationship:

$$\eta_2 = \eta_1 (c_2 / c_1) \tag{5.14}$$

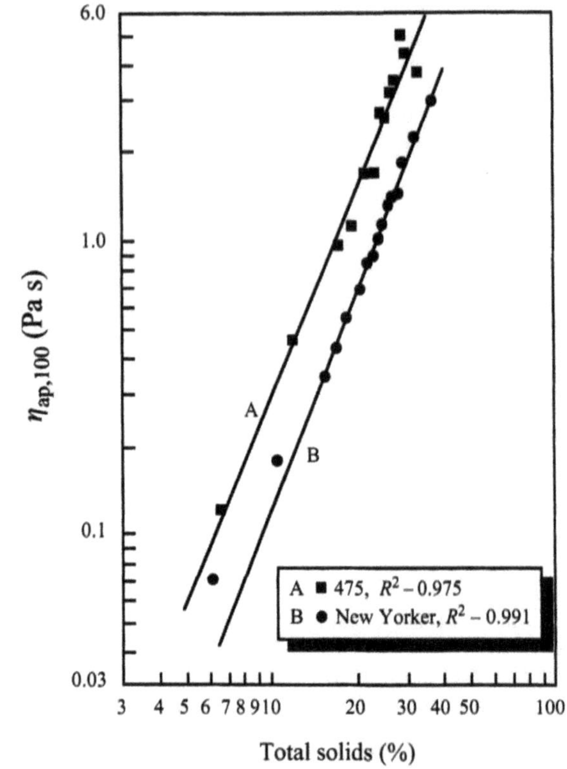

Fig. 5.13 Log-log plot of the apparent viscosity at 100 s^{-1} versus total solids of tomato concentrates from #475 and New Yorker tomatoes using hot-break juice; the slopes of the lines were 2.4 and 2.5, respectively

Viscoelastic Properties of Tomato Concentrates

Dynamic rheological and creep compliance, $J(t)$, data were obtained on tomato concentrates made from Pinto #696 hot-break juice prepared and two finisher screens having holes of diameter 0.69 mm (0.027 in) and 0.84 mm (0.033 in) were used (Yoo and Rao 1996); the finisher was operated at 1,000 rpm. The magnitudes of G' and G'' of three concentrates from juice using a 0.84 mm screen as a function of dynamic frequencies are shown in Fig. 5.15. Values of G' were higher than those of G'', but the values of the slopes were about the same (0.17–0.19). In general values of the power law consistency coefficients, K' and K'', between G' and G'' against the dynamic frequencies, shown in Table A5.5, were higher for the concentrates made from the juice using a 84 mm screen.

$$G' = K'(\omega)^{n'} \tag{5.15}$$

$$G'' = K''(\omega)^{n''} \tag{5.16}$$

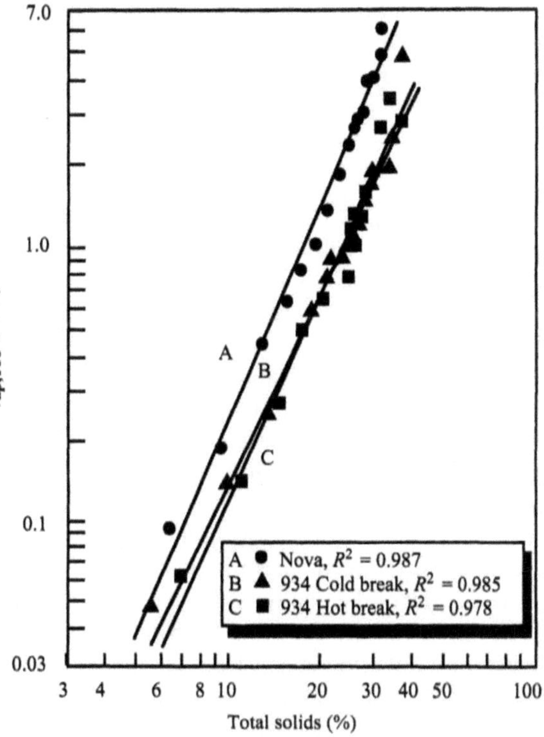

Fig. 5.14 Log-log plot of the apparent viscosity at 100 s^{-1} versus total solids of tomato concentrates from hot-break Nova tomato concentrates, and #934 hot-break and cold-break process; the slopes of the lines were 2.7, 2.4, and 2.5, respectively

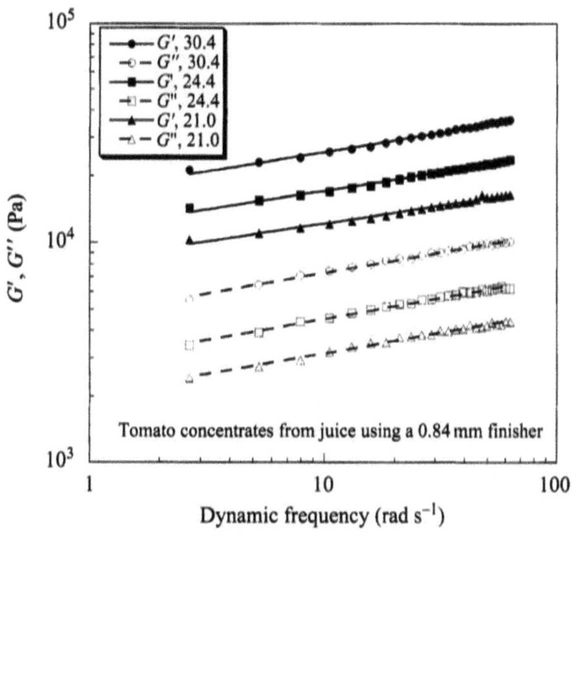

Fig. 5.15 Magnitudes of *G*′ and *G*″ of three tomato concentrates from juice using a 0.84 mm screen as a function of dynamic frequencies

The compliance $J(t)$ at any time was described by a six-element model made up of an instantaneous compliance, two Kelvin–Voigt bodies, and Newtonian compliance (Sherman 1970):

$$J(t) = J_0 + J_1\left(1 - e^{-t/\tau_1}\right) + J_2\left(1 - e^{-t/\tau_2}\right) + t/\eta \tag{5.17}$$

where, $J_0 = 1/G_0$ is the instantaneous elastic compliance, $J_1 = 1/G_1$ $J_1 = 1/G_1$ and $J_2 = 1/G_2$ are the retarded elastic compliances, τ_1 and τ_2 are retardation times, η is viscosity, and G_0, G_1, and G_2 are moduli (Sherman 1970; Rao 1992).

The auto mode of analysis of the experimental $J(t)$ versus t data provided magnitudes of model parameters that followed the experimental data better than those obtained in the manual mode of data analysis. The latter were very sensitive to estimates of instantaneous modulus and, in some instances, there were considerable differences between experimental data and model predictions (not shown here). Therefore, the results of auto mode data analysis are presented in Table A5.5; the standard deviations of the parameters are also in Table A5.6 in parentheses. The standard deviation of J_1 was higher than those of the other parameters reflecting greater uncertainty in its estimation.

The effect of the two finisher screen sizes on the $J(t)$ versus t data was examined at 21 °Brix and the concentrate from 0.84 mm screen was more viscoelastic than that from 0.69 mm screen. Because the particle size distributions of the two concentrates were similar, it appears that the higher pulp content, 38.6 % in the former as opposed to 32.5 % in the latter, is the primary reason for the difference in the magnitudes of the rheological parameters.

As expected, magnitudes of instantaneous elastic modulus, $G_0 = (1/J_0)$, and of storage modulus G' at 1 rad s^{-1} (Yoo and Rao 1996). Also, as noted by Giboreau et al. (1994) for a gel-like modified starch paste, the two parameters were of the same order of magnitude. Further, as can be inferred from data in Table A5.6, the other moduli from creep-compliance data ($G_1 = 1/J_1$, $G_2 = 1/J_2$) and the Newtonian viscosity increased with increase in °Brix.

Rheological Properties of Tomato Pastes

Rao and Cooley (1992) examined the flow properties of tomato pastes and their serum samples, as well as the dynamic rheological properties of the pastes. It should be noted that tomato pastes are of higher concentration than tomato concentrates. From the steady shear rheological data, obtained with a concentric cylinder viscometer, given in Table A53, it can be seen that the tomato paste samples had high magnitudes of yield stress. Further, they were highly shear-thinning in nature: the range of magnitudes of the flow behavior index (n) was 0.13–0.40; the average value was 0.28 with $\sigma = 0.062$. Magnitudes of Casson yield stresses were in the range: 78–212 Pa. The tomato paste serum samples were low viscosity Newtonian fluids; their viscosities ranged from 4.3 to 140 mPa s.

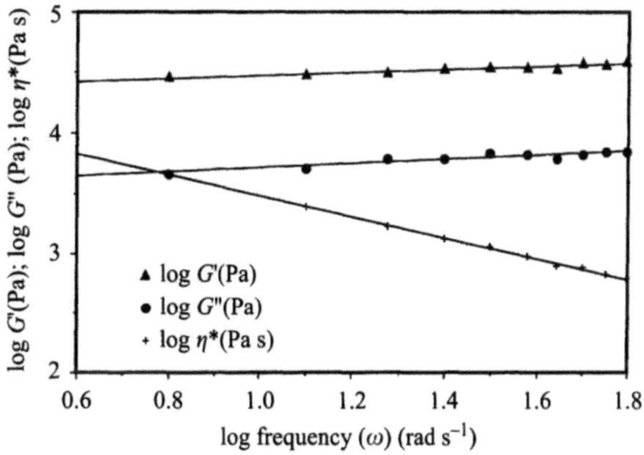

Fig. 5.16 Magnitudes of log ω versus log G', log G'', and log η^* of a tomato paste (Rao and Cooley 1992)

Plots of log ω versus log G' and log G'', such as that shown in Fig. 5.16, showed that G' was higher than G'' at all values of ω employed; in addition, the decrease in η^* confirmed the shear-thinning nature of the TP samples. Further, linear regression of log G' and log G'' versus log ω data showed that the resulting straight lines had low values of slopes (0.075–0.210 for G' and 0.110–0.218 for G'').

From a structural point of view, it is known that for true gels log ω versus log G' or log G'' plots have very low slopes (Ross-Murphy 1984), while for weak gels and highly concentrated solutions such plots have positive slopes and G' is higher than G'' over large ranges of ω. In addition, as discussed in detail below, magnitudes of η^* were greater than η_a for all magnitudes of shear rates and oscillatory frequencies. This means that TP samples did not behave as true gels, but exhibited weak gel properties.

Applicability of Cox–Merz Rule

With automated rheometers it is relatively easy to obtain dynamic shear data than steady shear data. For this reason, the interrelationship between ω and η^* on one hand and $\dot{\gamma}$ and η_a on the other is of interest. The Cox–Merz rule, that is, equal magnitudes of η_a and η^* at equal values of $\dot{\gamma}$ and ω, respectively (Eq. 5.17), was obeyed by several synthetic and biopolymer dispersions (Lopes da Silva and Rao 1992).

$$\eta^*(\omega) = \eta_a(\dot{\gamma})\,|_{\omega=\dot{\gamma}} \tag{5.18}$$

However, it is not obeyed by biopolymer dispersions with either hyperentanglements or aggregates (Lopes da Silva and Rao 1992). Because of the strong network structure with interspersed tomato particles, there was considerable deviation between the log $\dot{\gamma}$ versus log η_a and log ω versus log η^* data on the TP samples. However, the two lines were parallel to each other. Further, by multiplying ω by a shift factor (α), the two sets of data could be made to follow a single line (Fig. 5.17). Shift factors for correcting the frequency were calculated for the 25 tomato paste

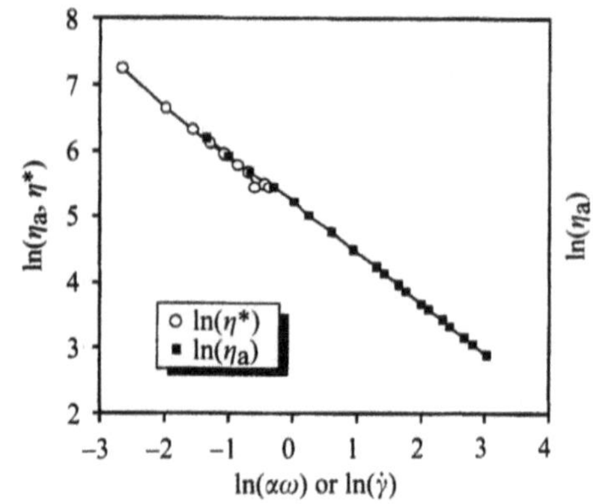

Fig. 5.17 Modified Cox–Merz plot for tomato concentrates (Rao and Cooley 1992)

samples to be in the range: 0.0029–0.029. It is of interest to note that the shift factors for 18 of the tomato paste samples were in the relatively narrow range: 0.0041–0.012; the average value of the shift factor for these samples was 0.0074 with a standard deviation of 0.0019. Therefore, from dynamic shear data and tabulated shift factors, it would be possible to estimate steady shear data from dynamic shear data, and vice versa.

Role of Composition of Tomato Pastes

Earlier studies have shown that solids loading plays an important role in the rheological behavior of tomato concentrates (Takada and Nelson 1983; Tanglertpaibul and Rao 1987b). However, it is emphasized that tomato pastes cannot be classified as distinct suspensions in which the solids are suspended in continuous fluid media. The difference in the structures of tomato concentrates and pastes is very well indicated by the magnitudes of centrifugal forces required to separate the solids (dispersed phase) from the serum (continuous phase). For the tomato concentrates, magnitudes of centrifugal force <30,000 × g were sufficient to separate the solids and the serum (Takada and Nelson 1983; Tanglertpaibul and Rao 1987b). In contrast, a centrifugal force of 100,000 × g was necessary to separate the two phases of TP samples in this study. Therefore, the classical relative viscosity-volume fraction relationships used for correlating data on suspensions would not be appropriate for TP samples. Indeed, attempts at correlating relative viscosity, based on apparent viscosity of tomato pastes at 10 s^{-1} and serum viscosity, with volume fraction of solids alone were not successful in that very low values of R^2 were obtained. It is likely that the shape of the particles should be also taken into consideration (Yoo and Rao 1994).

The rheological properties of the complex network structure of tomato pastes may be assumed to be made up of two contributions: (1) one network structure contributed by the solids phase, in proportion to ϕ_s, and (2) another network structure contributed by the liquid (continuous phase), in proportion to $\phi_1 = 1 - \phi_s$. However, the effective continuous phase is not the low-viscosity serum itself, but a highly viscous liquid that is an integral part of the tomato paste. Also implicit is that the solids fraction plays a major role in the structure of the TP samples, that is, it can be considered to be the structuring component. These assumptions are also in line with the weak gel behavior indicated by the dynamic shear data.

Model for Apparent Viscosity of PF Dispersions

Because no general theories exist even for concentrated nonfood suspensions of well-defined spherical particles (Jeffrey and Acrivos 1976; Metzner 1985), approaches to studying the influence of the viscosity of the continuous medium (serum) and the pulp content of PF dispersions, just as for nonfood suspensions, have been empirical. In PF dispersions, the two media can be separated by centrifugation and their characteristics studied separately (Mizrahi and Berk 1970). One model that was proposed for relating the apparent viscosity of food suspensions is (Rao 1987):

$$\eta_{\mathrm{a,PFD}} = \eta_{\mathrm{a,\,serum}} + A\,(\mathrm{pulp})^{B} \tag{5.19}$$

where, $\eta_{\mathrm{a,\,PFD}}$ is the apparent viscosity of suspension, $\eta_{\mathrm{a,\,serum}}$ is the apparent viscosity of serum, and pulp refers to the pulp content. The model is simple and predicts the viscosity of the serum when the pulp content is zero. The coefficient A indicates the contribution of a unit amount of pulp content and the coefficient B is the slope of a plot of log pulp versus log η_{specific} of the dispersion and it reflects the relative intensity of the effect of pulp content in different food suspensions or the effect of different processes and/or cultivars that could influence the pulp content. However, innovative and broadly applicable models that describe the role of pulp content are needed for PF dispersions.

Structure Based Models for Steady Shear Properties

For correlating the consistency coefficient (K) of the power law model and the Casson yield stress (σ_{0c}), again the lower bound form of Takayanagi model (Ross-Murphy 1984) was satisfactory:

$$(1/\mathrm{SSP}) = [(\phi_s/A)] + [(1-\phi_s)/B] \tag{5.20}$$

where, SSP is a steady shear rheological parameter, either the consistency coefficient (K) of the power law model or the Casson yield stress (σ_{0c}), and A and B are constants. For the consistency coefficient, magnitudes of A, B, and R^2 were: 900, 72, and 1.00, respectively, and for σ_{0c} they were: 135, 112, and 1.00, respectively. For the consistency index, 21 (out of 25) values predicted by Eq. (5.18) were within

±20% of the experimental results, while for the Casson yield stress, 19 values were within ±25%.

Structure Based Models for Dynamic Properties

Based on the above assumptions, correlations were developed first for the intercepts (K', K'') and slopes (n', n'') of log ω–log G' and log ω–log G'' data. The correlation form that was successful was again the lower bound form of Takayanagi model (Ross-Murphy 1984):

$$(1/DRP) = [(\phi_s/A)] + [(1-\phi_s)/B] \tag{5.21}$$

where, DRP is a dynamic rheological property, and A and B are constants. Magnitudes of the constants and R^2 were determined by nonlinear regression (GENSTAT, Numerical Algorithms Group, Oxford, UK). For K' values of A, B, and R^2 were: 9540, 4807, and 1.00, respectively, and for K'' they were: 1552, 6197, and 1.00, respectively. We note that the mean value of n' was 0.177 and that of n'' was 0.152. Agreement between predicted K' and K'' values was not as good as for the steady shear properties: 17 predicted K' and 19 K'' values were within ±30%.

Because of the wide variability in tomato pastes due to different cultivars and processing methods, the constants in the structure-based correlations would not be universally applicable. However, the correlations can be used to obtain order of magnitude estimates of the rheological properties of tomato pastes without extensive and expensive experimentation. The various constants are valid within the ranges of the variables from which they were derived. More importantly, it is our hope that studies will be conducted to make use of the concepts presented here with respect to: the weak gel nature of tomato pastes, shift factors for Cox–Merz rule, and structure based (ϕ_s and $\phi_l = 1-\phi_s$) relationships for correlating the rheological parameters.

Yield Stress of Structured Food Products

The vane method has become an accepted means of measuring yield stresses of foods since it was used by Qiu and Rao (1988). It was also observed that the magnitudes of the yield stress of apple sauce from the vane method (Table A57) were much higher than those from the flow models (not shown); in a few instances the former were more than twice the latter. In an attempt to understand the difference between the two types of yield stresses, we turn to the model presented by Michaels and Bolger (1962). Because the underlying assumptions of the model have been reviewed by others (Metz et al. 1979; Duran and Costell 1982; Genovese and Rao 2003), only the pertinent aspects of the model will be covered here. In the model, the stress necessary to produce deformation at a constant rate is divided into three parts:

$$\sigma_0 = \sigma_b + \sigma_n + \sigma_v \tag{5.22}$$

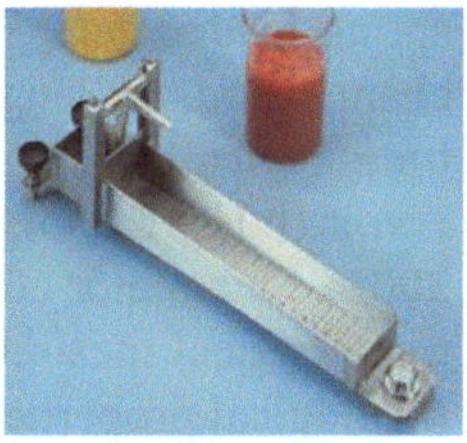

Fig. 5.18 The Bostwick consistometer is used extensively in quality control of pureed foods (Figure courtesy of CSC Scientific Company, Inc.)

where, σ_b is the stress required to break the structural bonds between the floes, σ_n is the stress required to break the aggregate network, and σ_v' is the stress dissipated due to purely viscous drag. The extension of this model to vane yield stress data on dispersions (Genovese and Rao 2003) will be discussed under structural analyses.

Another yield stress, the Bingham yield stress (σ_B) can be obtained by extrapolation of only the linear portion of the shear rate versus shear stress diagram to zero shear rate (Michaels and Bolger 1962). The Bingham yield stress is part of the total stress at high shear rates. One contribution of the model of Michaels and Bolger (1962) is that σ_0 is proportional to the volume fraction of solids raised to the power of three and that σ_B is proportional to the square of the volume fraction of solids. Also, it is known that the magnitude of σ_B is higher than that of σ_0. In addition to the two yield stresses, the viscosity at infinite shear rate (η_∞) also has been shown to be important.

Influence of Pulp Content on Yield Stress

One result of the work of Michaels and Bolger (1962) is the prediction that σ_0 is proportional to the volume fraction of solids raised to the power of three and that σ_B is proportional to the square of the volume fraction of solids. Also, they confirmed their predictions with data on dispersions that had magnitudes of σ_B less than 0.9 Pa. The apple sauce samples of Qiu and Rao (1988) were highly concentrated dispersions with some magnitudes of σ_B of about 80 Pa. In addition, the pulp content determined by centrifugation at 360 g had a relatively narrow range of magnitudes. For this reason, verification of the Michaels-Bolger model was shown with data on the samples in which the amount of pulp was adjusted.

Correlation of Bostwick Consistometer Data in Terms of Property-Based Dimensionless Groups

As noted earlier, the Bostwick consistometer, illustrated in Fig. 5.18, is used extensively to determine the consistency of fruit and vegetable purees, including many baby foods. It is used mainly as a quality control instrument. Essentially, it is a rectangular trough whose floor is graduated in millimeters; one end of the trough has a reservoir 100 ml in capacity in which the test sample is held in place by a spring-loaded gate. When the gate is released, the pureed food product flows in the channel and, the length of flow, the Bostwick consistency, is recorded in a specified time, typically 30 s. The ratio between width, b, and height, H, of the sample reservoir (b/H = 1.3 for a narrow consistometer and b/H = 6.6 for a wide one) (Perona 2005). One major concern with the Bostwick consistency is that it does not have a true physical meaning and does not allow for an unequivocal identification of the rheological properties of a test fluid.

McCarthy and Seymour (1994a) evaluated the flow of Newtonian fluids in a Bostwick consistometer as a balance of gravitational and viscous forces dependent on the height of the fluid, fluid density, and fluid viscosity. They derived an expression for the Bostwick measurement, L, as a function of trough geometry and fluid properties:

$$L = \xi_N \left(\frac{gq^3}{3\upsilon} \right)^{0.2} t^{0.2} \tag{5.23}$$

where, ξ_N is a similarity variable with a theoretical value of 1.411, g is the gravitational constant, q is fluid volume per unit width, υ is the kinematic viscosity, and t is time of flow. Experimental measurements with four Newtonian fluids (corn syrup and silicone oils of different viscosities) verified the theoretically predicted dependence of the extent of flow on kinematic viscosity.

Additional work (McCarthy and Seymour 1994) showed that the extent of flow at a given time was greater in a wider consistometer than in the standard consistometer. After the experimental measurements for Newtonian and power law fluids were compared to theoretical predictions, it was suggested that the theory could be used to evaluate values for power law parameters. The Bostwick measurement length, L, after 30 s was linearly related to: $(\eta_a/\rho)^{-02}$.

The Bostwick consistometer is designed to sit at a specified small angle, which is usually neglected. Before conducting a test, one ensures that the sloped flowing lane is at the right position by adjusting a series of screws until the leveling bubble on the front of the instrument is centered. The sample reservoir of the narrow instrument has a capacity $V = b^2H$ of approximately 100 ml (typically, $b \approx 0.05$ m and $H \approx 0.04$ m) and connects the flow lane through a sluice gate. While an internal structure is responsible for the yield stress or, more generally for an apparent viscosity, it cannot be quantified by Bostwick measurements.

Perona (2005) pointed out that it is important to consider the yield stress of the fluid when analyzing the flow in a Bostwick consistometer. It was pointed out that the Bostwick length L should be related to the variables: the mass of the sample M, gravity g, rheological parameters $(\sigma_{0H}, \mathrm{K}, \mathrm{n})$, initial height of the fluid at rest H, and time t. Therefore:

$$L = f(M,\, g,\, \sigma_0.K,\, t,\, H,\, n) \tag{5.24}$$

It was shown by applying the Buckingham Pi-theorem (Perona 2005) and using the three variables: M, g, H the above equation can be expressed in terms of the dimensionless groups: $\Pi_1 = \frac{L}{H}$, $\Pi_2 = \frac{H^2\sigma_0}{Mg}$, $\Pi_3 = t\sqrt{\frac{g}{H}}$, $\Pi_4 = \frac{K g^{\frac{n-2}{2}}}{MH^{\frac{n-4}{2}}}$ as:

$$\Pi_1 = \Phi(\Pi_2, \Pi_3, \Pi_4, n) \tag{5.25}$$

The apparent viscosity of a Herschel–Bulkley fluid may be expressed as the sum of the two components: one due to the yield value σ_{0H} and the other due to the viscous terms K and n, i.e.,

$$\eta = \eta_{\sigma_{0H}} + \eta_{K,n} \tag{5.26}$$

It was pointed out that when the pseudoplastic nature of the fluid is less the higher is the contribution given by the yield stress is when compared with that of the consistency index at low shear rates. In particular, such a condition means that: $\eta_{K,n} \ll \eta_{\sigma_{0H}}$

$$\Pi_1 = \Psi(\Pi_2, \Pi_3) \tag{5.27}$$

After sufficient time, the spreading mass will be arrested and, for geometrically similar devices, the final extent is only dependent on the fluid properties. This standstill condition is characterized by a free surface profile that is no longer horizontal. This characteristic behavior suggests that as the group Π_3 increases, it is less important in the dynamics, and the finite limit for Π_1 is a function of Π_2:

$$\lim_{\Pi_3 \to \infty} \Pi_1 = \Psi(\Pi_2) \tag{5.28}$$

The final form of the correlation was:

$$\Pi_1 = a_1\, \Pi_3^{a_2} \Pi_2^{b_1 \Pi_3^{b_2}} \tag{5.29}$$

Although the standard Bostwick test B30 requires that the position of the front is read at 30 s, Perona (2005) obtained values at different times: 5, 15, 30, and 60 s. The rheological characterization was done by using an experimental setup based on a temperature-controlled rectilinear pipe viscometer

The value of yield stress, σ_{0H}, was determined by using the direct U-tube technique, which furnished an accuracy that matched with the industrial needs of cost, rapidity, and sufficient reliability. This method was preferred to other indirect methods, such as those based on the extrapolation of rheological models to make the measure objective and, at the same time, independent of the choice of the adopted constitutive model and of errors due to the fitting procedure.

Results from experiments on 48 different samples of yield stress fluids were considered and included to check the adequacy of the proposed correlation. Both rheological data and Bostwick consistometer data were obtained on the fruit purees. It is noted that the rheological characterization was done by using an experimental setup based on a temperature-controlled rectilinear pipe viscometer (Perona 2005). For the studied fluid food samples, the flow curves were fitted to data by using the Herschel–Bulkley model with the yield parameter being set equal to that measured:

$$\sigma - \sigma_{0H} = K_K \dot{\gamma}^{n_H}$$

Values of the constants in the final correlation Eq. 5.29 were found to be: $a_1 = 1.30 \times 10^{-3}$, $a_2 = 2.46 \times 10^{-1}$; $b_1 = 6.71 \times 10^{-1}$, and $b_2 = 2.97 \times 10^{-2}$.

Rheological Properties of Chocolate

The basic ingredients of chocolate are crystalline sugar, cocoa solids, a liquid phase, and an emulsifier, lecithin (Hartel 1998). Most chocolate coatings contain less than 1 % water. In melted chocolate, the solid particles are dispersed throughout the lipid carrier and the rheological properties depend on the interactions between the solid particles and the continuous lipid phase. The lecithin coats the solid particles thus affecting the Casson yield stress and plastic viscosity. The polar group of lecithin interacts at the solid surfaces, particularly the sucrose crystals, where most of the water is located, and provides a bridge between the hydrophilic surface and the lipid phase. Thus, lecithin allows the movement of solids under shear (Hartel 1998). Increasing the amount of lecithin results in an initial decrease in yield stress, but a significant increase subsequently (Chevalley 1975). Addition of small amounts of cocoa butter reduce chocolate viscosity (Chevalley 1975) and different emulsifiers produce a wide range of plastic viscosities and yield stresses (Wayland 1994). As the average size of cocoa solids and sugar crystals was increased from 2 to 50 μm at a constant level of fat and emulsifier, the Casson plastic viscosity and yield stress decreased significantly (Musser 1973). Also of interest is that both plastic viscosity and yield stress increase as water content is increased in the range of 0.1–2 %. The Quemada et al. (1985) model (Eq. 2.11) was used to analyze data on cocoa dispersions (Fang et al. 1996) and the role of cocoa butter replacers (Fang et al. 1997). Selected values of rheological properties of chocolate are given in Table A58. In addition, the data of Fang et al. (1996) (Table A59) before and after degasification as a function of temperature are noteworthy.

An alternate method, instead of using the Casson model parameters, for measuring the yield stress and viscosity of chocolate and related cocoa products was proposed (Servais et al. 2004). Instead of obtaining shear rate versus shear stress data and conducting linear regression on the square roots of those values, the method is based on recording the value of the stress at a shear rate of 5 s^{-1} as representative of the yield stress of chocolate and the value of the viscosity at a shear rate of 40 s^{-1}

as representative of the infinite-shear (Casson plastic) viscosity. In addition, the difference between the viscosities measured at a shear rate of 40 s^{-1} during the ramp up and down in shear rate is a representative measure of thixotropy. The proposed method was validated with both research grade rheometers and factory grade viscometers. However, it should be noted that the value of the stress at a shear rate of 5 s^{-1} may not be in agreement with the vane yield stress.

Rheology of Milk and Milk Concentrates

The study of Fernandez-Martin (1972) on the viscosity of milk and its concentrates pointed out that, over a wide range of concentrations, the milk concentrates were Newtonian fluids. Data presented much later by others (Kokini 1992; Velez-Ruiz and Barbosa-Canovas 1998) support that finding (Table A510). The flow behavior of milk concentrates, and the effect of temperature and concentration on it were modeled in a manner similar to those of COJ (Vitali and Rao 1984a, b) and tomato concentrates (Rao et al. 1981; Tanglertpaibul and Rao 1987b). Velez-Ruiz and Barbosa-Canovas (1998) studied the flow behavior of milk concentrates with 17.2, 19.6, 22.3, 24.9, 30.5, 42.4, and 48.6% total solids at 5, 15, and 25 °C over a 4-week storage period. The samples with less than 19.6% solids content showed Newtonian behavior up to the third week of storage (Table A511). The samples above this level of concentration, especially above 24.9% solids content, exhibited mild shear-thinning non-Newtonian behavior. Only the two concentrates with 42.4 and 48.6% solids contents exhibited yield stress, albeit relatively low magnitudes. The power law model was used to describe the flow behavior of concentrates up to 30.5%, and the Herschel–Bulkley model to describe the behavior of the concentrates with 42.4 and 48.6% solids contents. Over the range of temperatures studied, the effect of temperature on the flow behavior index was minimal, which is in agreement with reports by other researchers (Rao et al. 1981), whereas the influence of temperature on the consistency coefficient was significant, in agreement with Rao et al. (1981). The effect of temperature on the consistency of milk concentrates followed the Arrhenius equation. Magnitudes of the energy of activation for flow (E_a) relating the consistency coefficient of the concentrates to temperature ranged from 2.42 to 11.8 kcal mol^{-1}. The magnitude of *E*a (kcal $mole^{-1}$) increased slightly with storage time: 1 week, 5.90 ± 1.02; 2 weeks, 5.94 ± 1.56; 3 weeks, 6.22 ± 2.99; 4 weeks, 7.91 ± 2.52. The combined effect of temperature and concentration was modeled using an equation similar to 5.13 and the results are given in Table A5.11.

Rheology of Mayonnaise, Salad Dressing, and Margarine

Mayonnaise, salad dressing, and margarine are o/w emulsions that are used as spreads in conjunction with salads, breads, and other foods. The discussion in Chap. 2 under emulsions should be consulted with respect to the basic concepts of stability, that is, separation of phases. Regular, as opposed to low-fat, mayonnaise has a dispersed phase (oil) concentration of about 70–80 %. Often the emulsifying agent in these foods is egg yolk. The role of oil, water, egg yolk, thickening agents (usually gums), salt, and sugar have been studied by food technologists and summarized in recent publications (Harrison and Cunningham 1985, 1986). The physical changes in the dispersed droplets occur through the processes of creaming, flocculation, and coalescence. Creaming denotes the movement of oil droplets under the action of gravity; flocculation denotes the clustering of droplets, and coalescence means the spontaneous joining of small droplets into larger ones. Unless phase separation occurs, creaming and flocculation in o/w emulsions may not lead to detectable changes in the appearance and rheological behavior. However, coalescence of droplets which is associated with the free fat from o/w emulsions is generally unacceptable at all levels (Dickinson and Stainsby 1982, 1987).

Small droplet size is created by high pressure homogenization and the premature coalescence of the droplets is minimized by adsorption of an emulsifier at the interface. Generally egg yolk is used as the emulsifier, that is, to lower the o/w interfacial tension. Gums such as CMC, propylene glycol, guar gum, and xanthan gum are used as emulsion stabilizers in mayonnaise and salad dressings. Therefore, the time required to hydrate gums is an important factor in the stability of emulsions.

Rheological properties of mayonnaise have been studied using different rheological techniques: steady shear rate–shear stress, time-dependent shear rate–shear stress, stress growth and decay at a constant shear rate, dynamic viscoelastic behavior, and creep-compliance viscoelastic behavior. More studies have been devoted to the study of rheological properties of mayonnaise than of salad dressings, probably because the former is a more stable emulsion and exhibits complex viscous and viscoelastic rheological behavior.

Because of the existence of yield stress as well as time-dependent rheological behavior of mayonnaise, it would seem reasonable to expect that traditional relationships between steady shear properties on one hand and small amplitude dynamic properties on the other that were found for polymeric liquids will not hold for mayonnaise. Bistany and Kokini (1983) showed that the Cox–Merz rule and other relationships at low shear rates and frequencies: did not hold for mayonnaise and other semisolid foods. Further, one must also be careful when using dynamic tests on emulsions, such as mayonnaise, in view of the potential for aiding coalescence of the oil droplets and consequently altering the structure of the sample.

Stress Overshoot Data of Mayonnaise

Several studies were conducted on the stress overshoot and/or decay at a constant shear rate. Kokini and Dickie (1981) obtained stress growth and decay data on mayonnaise and other foods at 0.1, 1.0, 10.0, and 100 s^{-1}. As expected from studies on polymers, shear stresses for mayonnaise and other food materials displayed increasing degrees of overshoot with increasing shear rates. The Bird–Leider empirical equation was used to model the transient shear stresses.

$$\sigma = -K\dot{\gamma}^n\left[1+\left(b\dot{\gamma}-1\right)\exp\left(-t/an\lambda\right)\right] \tag{5.30}$$

where, $\dot{\gamma}$ is the shear rate, t is time, λ is time constant, a and b are constants, and K and n are power law constants. At very small times for a constant shear rate, the model reduces to:

$$\sigma = \text{constant}\left(b\dot{\gamma}t\right) \tag{5.31}$$

Equation (5.23) simulates the elastic response or sudden overshoot when a constant shear rate is applied. The peak shear stress is reached at a time t_{max} given by:

$$t_{max} = an\lambda + \frac{1}{b\dot{\gamma}} \tag{5.32}$$

The magnitude of σ_{max} can be calculated by substituting the above expression for t_{max} in Eq. (5.23). After σ_{max} is reached, the term exp − $(1/an\lambda)$ simulates the exponential decay; the product: as a whole controls the exponential decay.

Kokini and Dickie (1981) found that the Bird–Leider equation provided moderately good predictions for peak shear stresses (σ_{max}) and the corresponding times (t_{max}), but the prediction of shear stress decay was poor. They suggested that a series of relaxation times, as opposed to a single exponential term, would be needed for mayonnaise and other materials.

Campanella and Peleg (1987) presented stress growth and decay data on mayonnaise at shear rates of 1.8, 5.4, 9.9, and 14.4 s^{-1} with a controlled shear rate viscometer and a concentric cylinder geometry. They modeled the data by a three-constant model that was a modification of Larson's (1985) model that was successfully employed for polyethylene melts with a wide distribution of molecular weights. The model employed by Campanella and Peleg was:

$$\frac{\sigma(t)}{\sigma_\infty} = \frac{\Gamma(m_1)\sin(m_1\pi)}{\pi(m_2\dot{\gamma})^{m_1-1}}\left(\frac{t^{1-m_1}e^{-m_2\dot{\gamma}}}{m_1} + \int_0^t s^{-m_1}e^{-m_2\dot{\gamma}\,s}\,ds\right) \tag{5.33}$$

where, σ_∞ is the steady state stress Γ is the gamma function; m_1 and m_2 are constants; $\dot{\gamma}$ is the shear rate; $s = t - t'$; and the deformation history of the experiment is described by:

$$\gamma = \Big|_{\dot{\gamma}(t-t')}^{\dot{\gamma}t} \quad {}^{-\infty<t'<0}_{0<t'<t} \tag{5.34}$$

The constant m_1 has a value $0 < m_1 < 1$ and it is representative of the relaxation modulus, when it is expressed as:

$$G(t-t')=C(t-t')^{-m_1} \tag{5.35}$$

The parameter m_2 is obtained from the time at which the maximum shear stress is reached, t_{max}:

$$m_2=\frac{1}{\dot{\gamma}_{max}} \tag{5.36}$$

The above relationship implies that t_{max} is inversely proportional to the imposed shear rate and this was found to be true for the experimental data. From experimental data, the values of $m_1=0.3$ and $m_2=0.07$ were determined. Even though the damping constant (m_2) was assumed to be shear rate dependent, a single representative value of m_2 was found to be satisfactory because m_2 was a weak function of shear rate.

Figoni and Shoemaker (1983) also found that a single first-order function did not adequately describe the stress decay data of a mayonnaise sample. They suggested that two independent first-order processes, each with a different rate constant but acting simultaneously are major contributors to the stress decay data on mayonnaise. The first process was attributed to the reversible flocculation–deflocculation process of oil droplets that occurs during the steady shear process, while the second term describes the changes occurring in the continuous phase. The possibility that the two first-order processes were applicable to the flocculation and deflocculation of aggregates of two size groups of oil droplets was also suggested.

Creep-Compliance Behavior of Mayonnaise

Creep-compliance behavior of mayonnaise and other emulsions have been investigated by Sherman and coworkers. The chief advantage of the creep-compliance test is that data can be obtained with minimal disturbance to the test samples and under linear-viscoelastic regime. Nearly all, if not all, of the previous studies were conducted with commercial (supermarket) samples whose compositions were not determined. Therefore, the role of composition on the rheological parameters could not be determined. In contrast, Kiosseoglou and Sherman (1983) employed the formulations shown in Table A512 in order to study the effect of egg white on the rheological properties of mayonnaise. In addition, the effects of pH, salt, and sugar on rheological properties were studied.

The creep-compliance data on the mayonnaises were modeled by means of the six parameter model:

$$J(t)=J_0+J_1\left(1-e^{-t/\tau_1}\right)+J_2\left(1-e^{-t/\tau_2}\right)+t/\eta \tag{5.37}$$

We note that the creep-compliance parameters can be used to calculate the moduli $E_i = 1/J_i$ and viscosities $\eta_i = E_i\tau_i$. In the following, the magnitudes of the viscoelastic parameters, that is, the moduli and the viscosities, will be discussed.

The viscoelastic parameters reached maximum values within one hour after preparation of the emulsions and then they decreased continuously as the emulsions were aged. The E_0 values (Pa) of freshly prepared mayonnaise increased as the pH decreased: pH 6.2, $E_0 = 750$; pH 4.5, $E_0 = 885$; pH 3.9, $E_0 = 1220$; and pH 3.3, $E_0 = 1351$. There was a substantial decrease in the viscoelasticity parameters as the emulsions were aged at room temperature at all values of pH except at pH 6.2.

At a constant pH 3.9, salt increased the initial viscoelasticity, the effect being greater at the 0.25 % (w/w) level than at the 1.25 % level. However, the rate of decrease in E_0 was lower at the 1.25 % NaCI than at 0.25 %. Sugar had much smaller effect than salt on the viscoelasticity of mayonnaise. Replacing egg yolk by egg white, while keeping the total solids content constant, increased E_0. When the egg yolk content was kept constant and additional egg white was incorporated, all the viscoelastic parameters increased substantially. With an egg yolk/egg white ratio of 0.5, the viscoelasticity parameters decreased least when the mayonnaises were aged.

The addition of salt to the egg yolk solution during the preparation of the solutions disrupts the granules and provides additional surface active material for adsorption, such as α- and β-lipovitellins and phosvitin. In this regard, Chang et al. (1972) observed with an electron microscope a continuous speckled layer around oil drops and also a fibrous membrane on the surface of drops after they were washed with water. The electron-dense particles in the speckled layer were interpreted to be egg yolk lipoproteins and the fibrous material to be livetin-phosvitin complex.

The low viscoelasticity of mayonnaise in the presence of sugar was considered to be due to the shielding effect exerted by sugar molecules on the protein groups which were involved in interaction and network formation between the oil drops. The higher viscoelasticity of mayonnaise containing egg white was thought to be due to strengthening of the interlinked network structure by interaction with the egg yolk components of the interfacial membrane around the oil drops.

Rheology of Salad Dressings

Rheological properties of salad dressings were also studied using the techniques: steady shear rate–shear stress, stress growth and decay at a constant shear rate, dynamic viscoelastic behavior, and creep-compliance viscoelastic behavior.

Tanaka and Fukuda (1976) studied the steady shear rate–shear stress behavior, thixotropic behavior, and dynamic viscoelastic behavior of salad dressing prepared with different stabilizing agents (0.6 %, w/w): guar gum, tragacanth gum, pectin, starch paste, tamarind seed extracts, and xanthan gum. The power law flow behavior index ranged from 0.20 when xanthan gum was used to 0.68 when corn starch was added. A thixotropy index (TI) was calculated as the sum of the difference between the magnitudes of the shear stress in increasing (S_{ui}) and decreasing (S_{di}) of shear rates:

$$TI = \sum_{1}^{4} \left(S_{\text{ui}} - S_{\text{di}} \right) \tag{5.38}$$

where, the magnitudes of the shear stresses: S_{ui} and S_{di} were read off at four unspecified shear rates. Magnitudes of *TI*, of apparent viscosity at a shear rate of 0.1 s^{-1}, of dynamic elasticity, G', and of dynamic viscosity, η', were highest for salad dressings with xanthan gum and lowest for dressings with corn starch.

Elliott and Ganz (1977) noted that steady shear and dynamic rheological properties of commercial salad dressings covered a wide range of magnitudes. Because the products were being sold successfully on the commercial market, they suggested that a strict control of rheological properties is not necessary. A modified Bingham body, shown in Fig. 3.37, was used to account for the observed response to dynamic tests of the samples that exhibited yield stresses. Stress growth and decay at a constant shear rate was analyzed as the excess work (W) of structure breakdown discussed in more detail in Chap. 3:

$$W = \dot{\gamma} \int_{A}^{B} \left(\sigma - \sigma_{\infty} \right) \mathrm{d}t \tag{5.39}$$

where, $\dot{\gamma}$ is the constant shear rate, σ is the time-dependent stress, σ_{∞} is the equilibrium shear stress value, and A and B are the times at which the stress curve overshot the equilibrium stress value and the time at which the equilibrium value was reached, respectively. The slopes of log η_a versus log $\dot{\gamma}$ data of the salad dressings were between −0.72 and −0.86.

Paredes et al. (1988) found commercial bottled salad dressings to be more viscous than dressings prepared from dry mixes. The Arrhenius activation energy of flow for the commercial dressings ranged between 10.5 and 20.5 kJ/mole. Thixotropic behavior at a constant shear rate was described better by the Weltman equation (Eq. 5.33) than the equation of Hahn et al. (Eq. 5.34).

$$\sigma = A_1 - B_1 \log t \tag{5.40}$$

$$\log\left(\sigma - \sigma_e \right) = A_2 - B_2 t \tag{5.41}$$

where, σ is the shear stress, t is time in seconds, σ_e is the equilibrium shear stress, and A_1, A_2, B_1, and B_2 are constants.

Commercial bottled salad dressings showed higher viscoelasticity than dressings prepared from dry mixes, and bottled reduced calorie dressings exhibited intermediate viscoelasticity (Paredes et al. 1989). A four-parameter model described the creep-compliance behavior of the commercial salad dressings and the effect of storage on a prepared salad dressing containing xanthan gum:

$$J(t) = J_0 + J_1(1 - e^{-t/\tau_1}) + t/\eta \tag{5.42}$$

The viscoelastic parameters of the model salad dressing (Table A513) decreased with storage time until 21 days and then increased from the 28th to the 42nd day. Specifically, the dramatic changes in the magnitudes of the viscoelastic parameters during the first two weeks are noteworthy. Kiosseoglou and Sherman (1983) also reported decrease in viscoelasticity of o/w emulsions during storage up to the maximum storage time of 20 days. The observed increase in viscoelasticity after 21 days by Paredes et al. (1989) was most likely related to the hydration of xanthan gum in the salad dressings, and that an eventual decrease and stabilization in the viscoelasticity of the salad dressings would be inevitable. The study of Paredes et al. (1989) points out, among other things, that the role of hydrocolloids as salad dressing ingredients must be studied (Andon 1987).

Melting characteristics of margarines and table spreads are important for flavor release and consumer acceptance. Oscillatory measurements as a function of temperature and also drop points were used to quantify rheological changes accompanying melting and DSC measurements were used to quantify melting of fat crystals in these products (Borwankar et al. 1992). For low-fat spreads rheological changes characteristic of melting could occur at a temperature higher than when fat melting occurs and that the fat melting is largely unaffected by emulsification. The rheology of low-fat spreads was governed by emulsion characteristics such as the proportion of the aqueous phase and the size of the water droplets. Drop point measurements provide a simple method to determine melting in the rheological sense. Perceived meltability of these products, in general, was due to a combined perception of cooling sensation accompanying the melting of fat crystals and the sensation of flow accompanying the rheological transitions. The cooling sensation was significant only for the case of butter and high-fat margarines; in these cases the fat melting and rheological transitions both occurred at about the body temperature. In reduced fat products, the cooling sensation was not significant and perceived meltability correlated with the viscoelastic and drop point data.

Processed Foods as Soft Materials

Many soft materials, such as foams, emulsions, pastes, and slurries, have intriguing rheological properties. Experimentally, there is a well-developed phenomenology for such systems: Their nonlinear flow behavior often follows the Herschel–Bulkley model:

$$\sigma - \sigma_{0\mathrm{H}} = K_K \dot{\gamma}^{n_H}$$

For the same materials, linear viscoelastic measurements often reveal storage, G′(ω), and loss, G″ω), moduli in nearly constant ratio (G″/G′ is usually about 0.1) with a frequency dependence that is either a weak power law (clay slurries, paints, microgels) or negligible (tomato paste, dense emulsions, dense multilayer vesicles, colloidal glasses). This behavior persists down to lower frequencies (about 10^{-3} 1 Hz depending on the measuring equipment used), in apparent contradiction to linear response theory, which requires that G″(ω) should be an odd function of ω. That similar anomalous rheology should be seen in such a wide range of soft materials suggests a common cause. Indeed, the frequency dependence indicated above points strongly to the generic presence of slow "glassy" dynamics persisting to arbitrarily small frequencies (Sollich et al. 1997).

Structural Analyses of Food Dispersions

Role of Suspended Particles in Soy Milk

While application of structure-based models to rheological data, such as the Casson model, provides useful information, structure-based analysis can provide valuable insight in to the role of the structure of a dispersed system. Bodenstab et al. (2003) estimated the contributions to flow shear stress of soy milk by the suspended particles and the suspending fluid.

$$\sigma = \sigma_s + \sigma_p \tag{5.43}$$

where, σ_s is the shear stress caused by the viscous forces generated by the suspending fluid (continuous phase) and σ_p is the shear stress caused by interaction between suspended particles. The former is temperature dependent and in the latter the temperature dependency can be neglected. In the direct interparticle interactions, Coulomb's mechanical friction forces, hydrogen bonds, electrostatic attraction, and hydrophobic attraction may be important. As pointed out earlier, in some food suspensions, it may not be easy to separate the suspending fluid from the suspended particles because part of the former being enclosed by the latter.

$$\sigma = (1-\phi_p)\eta_s \frac{\dot{\gamma}_{eq}^2}{\dot{\gamma}} + \frac{\dot{E}_p}{V\dot{\gamma}} \tag{5.44}$$

where, ϕ_p is the volume fraction of solids in the dispersion, η_s is the dynamic viscosity of the suspending fluid (continuous phase), $\dot{\gamma}$ is the apparent shear rate, $\dot{\gamma}_{eq}$ is the equivalent shear rate; $\dot{\gamma}_{eq}$ is defined as the shear rate that would be necessary to shear the suspending fluid in the absence of dispersed solids so that the same volumetric energy would be dissipated as in the dispersion, and $\dot{E}_p$ is the energy dissipation by direct interparticle forces. The term, $\dot{E}_p / V\dot{\gamma}$, has the dimensions of shear stress and it represents σ_p. In the above equation (Eq. 5.37), there are two unknown parameters: $\dot{\gamma}_{eq}$ and $\dot{E}_p$, that may be estimated from measurements at two temperatures on a specific dispersion. Here, the implicit assumptions are: (1) there is negligible change in the flow pattern of a dispersion when the temperature is varied over a moderate range, (2) the composition of the dispersion does not change over the temperature range, and (3) the interparticle forces depend only weakly on temperature.

Based on an earlier study of Sommer (1975) on synthetic-particle suspensions, the relationship for $\dot{\gamma}_{eq}$ is:

$$\dot{\gamma}_{eq} = \sqrt{\frac{\dot{\gamma}(\sigma_1 - \sigma_2)}{(1-\phi_p)(\eta_1 - \eta_2)}} \tag{5.45}$$

where the subscripts 1 and 2 refer to the values of stress and viscosity of the dispersion at temperatures 1 and 2. Typical values of temperature that were used were 10 and 25 °C, and 20 and 40 °C (Bodenstab et al. 2003).

Using the expression for $\dot{\gamma}_{eq}$, Eq. 5.37 becomes:

$$\sigma = \sigma_s + \sigma_p = \eta_s \frac{\sigma_1 - \sigma_2}{\eta_1 - \eta_2} \sigma_p \tag{5.46}$$

For the soy milks studied, direct interparticle interactions were found to be significant at particle concentrations above about 20 g/100 g. Between about 15.0 and 26.3 g/100 g concentration, direct interactions between suspended particles became greater than the viscous flow of the suspending fluid and dominated the overall properties of the dispersions. It appears that shear stresses resulting from interparticle forces were fairly independent of shear. However, σ_s, the shear stress caused by the suspending fluid, was much more shear sensitive.

Structural Components of Yield Stress

In the kinetic or structural approach to rheology of dispersions (Michaels and Bolger 1962) the basic flow units are assumed to be small clusters or floes that at low shear rates give the dispersion a finite yield stress. The clusters associate randomly to form weakly bonded aggregates and tenuous networks, giving rise to plastic and structural properties. In Chap. 3, the concept of static, σ_{0s}, and dynamic, σ_{0d}, yield stress was introduced. Based on the work of Michaels and Bolger (1962), from an energy balance at the point of maximum deformation (yield point) in the vane test, the contributions of different structural components to the total yield stress, σ_{0s}, may be estimated (Genovese and Rao 2003):

$$\sigma_{0s} = \sigma_b + \sigma_v + \sigma_n \tag{5.47}$$

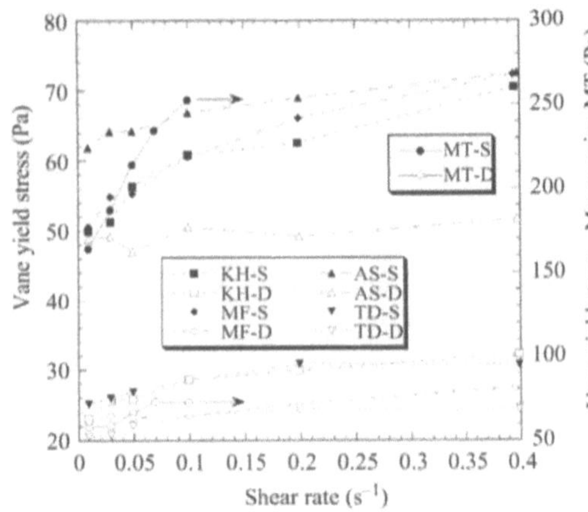

Fig. 5.19 Static (S) and dynamic (D) yield stresses of structured foods determined using the vane method at different shear rates. Products studied: apple sauce—AS, ketchup—KH, mustard—MF, tomato concentrate—TD, and mayonnaise MT (Genovese and Rao 2005)

The stress to break the bonds between the floes may be calculated as the difference between the static, σ_{0s}, and the dynamic, σ_{0d}, yield stresses of the samples with undisrupted and disrupted structure, respectively.

$$\sigma_b = \sigma_{0s} - \sigma_{0d} \tag{5.48}$$

where, σ_b is the stress required bonds between the floes, σ_v is the stress dissipated due to purely viscous drag, and σ_n is the stress required to break the aggregate network. Given that $\sigma_v = \eta_\infty \dot{\gamma}$ is very small in most dispersions, one can estimate the two contributions σ_b and σ_n.

Genovese and Rao (2005) determined the vane yield stress of commercial samples of apple sauce, tomato concentrate, mustard, mayonnaise, and tomato ketchup. For each sample, they determined first the static yield stress, σ_{0s}, of a sample with undisrupted structure, followed by a flow test that served the dual purposes of breaking down structure and also providing shear rate versus shear stress data, and finally determined the dynamic yield stress, σ_{0d}. Magnitudes of the static and dynamic yield stresses are shown in Fig. 5.19. For the foods studied, the viscous component of yield stress was small, so that the contributions of bonding and network components were of major interest. For the homogenized products, mayonnaise, mustard, and ketchup $\sigma_b > \sigma_n$ while for products processed in a finisher, applesauce and tomato puree $\sigma_n > \sigma_b$. The contribution of bonding to the static yield stress of the products that were homogenized: mayonnaise, mustard, and ketchup ranged between 53 and 65 %, while that of finished, nonhomogenized products: apple sauce and tomato puree was about 20 %.

Tárrega et al. (2006) found that the addition of a small amount of λ-carrageenan to starch–milk systems resulted in a large increase in the structure's strength. The contribution of bonding increased due to the presence of λ-carrageenan-casein network for both native and cross-linked maize starch dispersions. Achayuthacan et al. (2006) found that values of σ_b xmaize, cross-linked waxy maize, and cold water swelling decreased and those of σ_n increased with increase in xanthan gum concentration.

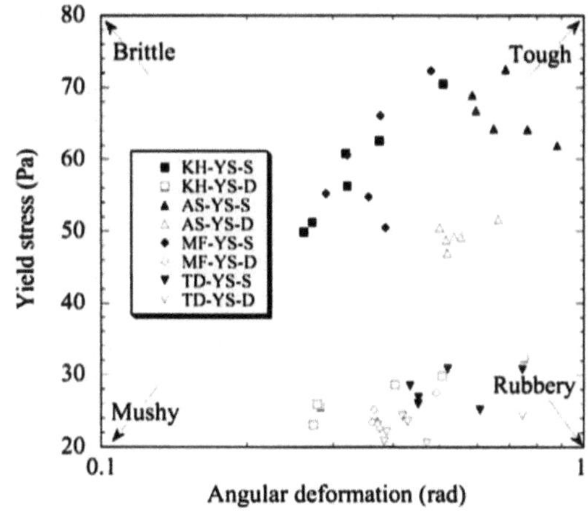

Fig. 5.20 Texture map of structured foods in which values of the static (S) and dynamic (D) yield stresses determined using the vane method at different shear rates were plotted against the corresponding values of the angular deformations. Products studied: apple sauce—AS, ketchup—KH, mustard—MF, and tomato concentrate—TD. Based on unpublished data Genovese and Rao (2005)

Texture Map Based on Vane Yield Stress

As stated earlier, values of the static and dynamic yield stresses are determined on samples that are undisturbed and disturbed, respectively. A texture map can be created when those yield values are plotted against the corresponding values of the deformation. The texture map of three heated starch dispersions based on yield stress versus deformation was shown in Fig. 4.27 (Genovese and Rao 2003). Figure 5.20 is a texture map of the structured foods: apple sauce, ketchup, mustard, and tomato concentrate (Genovese and Rao 2005). In the figure, values of the static and dynamic yield stresses determined using the vane method at different shear rates were plotted against the corresponding values of the angular deformations. Unlike a traditional texture map, a map based on static and dynamic yield stresses indicates the behavior of a food with both undisturbed and disturbed structure; it should be useful in understanding the role of processes that may alter a food's structure.

High Pressure Processing of Milk and Milk Proteins

Tattiyakul and Rao (2009) reviewed the effects of high pressure (HP) and ultrasonic processing of foods on their rheological properties. Because numerous studies have been conducted on HP processing of milk and its proteins (López-Fandiño 2006), some of the quantitative studies are reviewed. Milk processing at 150–400 MPa caused a certain amount of irreversible fragmentation of casein micelles, together with calcium release, increased milk viscosity, decreased milk turbidity, and decreased non casein nitrogen (Cheftel and Dumay 1996). The exact changes in individual caseins and whey proteins (WPs) are not known, but milk treatment accelerates subsequent casein coagulation by rennet or glucono-δ-lactone, and enhances the strength and water-holding capacity (WHC) of acid-set gels. Processing solutions of β-lactoglobulin (β-Lg) in the same HP range causes partial structure denaturation. β-Lg unfolding is more extensive and irreversible at neutral than at acid pH. In addition to the reduction of particle size, micelles in HP-treated milk are irregularly shaped and of enhanced voluminosity, due to higher hydration, and all these factors contribute to an increased viscosity in skimmed milk with increasing pressure and treatment time (López-Fandiño 2006).

As a consequence of micellar disruption and protein solubilization, the amount of milk protein associated with the milk fat globules also increases by HP treatments and this could, at least partially, influence the creaming properties of HP-treated milk. The milk fat globules also increases by HP treatments and this could, at least partially, influence the creaming properties of HP-treated milk (López-Fandiño 2006).

Kinetics of Thermal and Pressure-induced Changes

Hinrichs and Rademacher (2005) studied the kinetics of the combined effects of thermal and pressure-induced WP denaturation in bovine skim milk. Estimation of kinetic parameters for pressure-induced denaturation was based on isobaric experiments at different pressures. In addition, the pressure treatment was carried out at different temperatures because the rate of pressure-induced reactions could be accelerated or decelerated due to synergistic or antagonistic effects of temperature. Specifically, skim milk was treated under isobaric conditions (200–800 MPa) in combination with isothermal conditions (1–70 °C), and the levels of denaturation of β-lactoglobulin A and B, and α-lactalbumin were analyzed.

The kinetic equations applicable for the change in a protein are (Hinrichs and Rademacher 2005):

$$\frac{dC}{dt} = -k_{p,T} . C^n \tag{5.49}$$

$$\text{For n = 1,} \quad \frac{C_t}{C_0} = exp\,(-k_{p,T}\ t) \tag{5.50}$$

The rate constant, $k_{p,T}$, in Eq. 5.43 depends on both temperature and pressure. The isothermal activation volume, $\Delta V^{\#}$, can be calculated from experiments at a constant temperature and using a reference pressure, p_{ref}, say 500 MPa. For isothermal experiments, the Eqs. 3 and 4 are applicable when $n=1$ and when $n \neq 1$, respectively:

$$C_t = C_0\ exp\left(k_{ref,T}\ exp\left\{-\frac{\Delta V^{\#}}{RT}\left(p - p_{ref}\right)\right\} \cdot t\right) \tag{5.51}$$

$$C_t = C_0\left(1+(n-1)k_{ref,T}\ exp\left\{-\frac{\Delta V^{\#}}{RT}(p-p_{ref})\right\} C_0^{n-1}.t\right)^{(1/1-n)} \quad (5.52)$$

The kinetic parameters (order of reaction, rate constant at reference pressure and temperature and activation volume) in Eqs. (5.44) and (5.45) were estimated by means of nonlinear regression of the experimental data with an overall fit. Depending on the temperature, the determined formal reaction order, n, decreased from about $n=3$ at 1–10 °C to a value of 2 at 60–70 °C, for both β-Lg fractions (β-Lg A, β-Lg B). An order of about 2.5 was estimated in the measured temperature range for the denaturation of α-La.

The isobaric activation energy, E_a/RT , was incorporated to obtain the equation:

$$C_t = C_0\left(1+(n-1)k_{ref,T}\ exp\left\{-\frac{E_a}{RT}\left(\frac{1}{T}-\frac{1}{T_{ref}}\right)\right\} C_0^{n-1}.t\right)^{(1/1-n)} \quad (5.53)$$

The activation volume, $\Delta V^{\#}$, for the β-Lg fraction decreased from around 0 at 1 °C to less than −80 ml mol^{-1} at 70 °C. A similar behavior was found for the temperature dependence of the activation volume of α-La. The activation volume, $\Delta V^{\#}$, of about 0 at 1 °C indicated that denaturation showed no pressure dependence at this temperature. In addition, the rate constant, $k_{500,T}$ increased with increasing temperature, resulting in an accelerated denaturation reaction whereby pressure and temperature act synergistically.

Thus, the higher the applied temperature the higher the pressure-dependence of the denaturation reaction. The results suggest a synergistic mechanism, where the molecule unfolds more easily under pressure at elevated temperatures. However, alpha lactalbumin (α-La) showed a much higher pressure resistance than β-Lg. The former molecule is much more stable, as α-La is stabilized by four intramolecular disulfide bonds compared to two stabilizing β-Lg. Additionally, α-La possesses no free SH-groups as β-Lg does. In analogy to thermal treatment, it can be assumed that α-La is only denatured if free SH-groups are available. Due to the fast pressure-induced aggregation of β-Lg among each other and with aS2caseins and k-caseins, not enough free SH-groups are available for the oligomerization with a-La, resulting in the observed low rate of denaturation.

Mechanisms of Heat-induced and Pressure-induced Changes

Considine et al. (2007) noted that in general terms heat treatment and pressure treatment have similar effects: denaturing and aggregating the WPs and diminishing the number of viable microorganisms. However, it was pointed out that there are significant differences between the effects of the two treatments on protein unfolding and the subsequent thiol-catalyzed disulfide-bond interchanges that lead to different structures and product characteristics.

It was also noted that β-LG is one of the most pressure-sensitive proteins and α-lactalbumin (α-LA) is one of the most pressure resistant. In a heated WPC system, bovine serum albumin (BSA) is very sensitive and β-LG is more resistant. In a heated milk system, β-LG reacts with κ-casein (κ-CN) and not with α_S2-CN, but, in pressure-treated milk, β-LG forms adducts with either κ-CN or α_S2-CN. In both treatments, the role of β-LG is central to the ongoing reactions, involving α-LA and κ-CN in heated systems but involving κ-CN, α_S2-CN, and α-LA in pressurized systems.

Structural Changes in Milk Proteins

Kanno et al. (2002) studied the pressure-induced denaturation and gelation of β-Lg, α-La, and BSA at various concentrations at 200–800 MPa and 30 °C for 10 min. The α-helix content of β-Lg was found to decrease with increasing pressure and the random structure was increased; however, the secondary structure of α-La and of BSA was not affected. The microstructure of the β-Lg gels showed a porous network, while the BSA gels had a plate-like structure. The addition of β-Lg or cysteine to α-La induced a stronger gel at 800 MPa with a similar microstructure to that of the gel from BSA.

Effect of Pressure Release Rates

Fertsch et al. (2003) studied the effects of HP on the structure and firmness of whey protein isolate (WPI) and MC at 600 MPa and 30 °C, using a pressure build up rate of 200 MPa min^{-1} and holding times of 0, 15, or 30 min. In addition, pressure release rate was varied among 20, 200, and 600 MPa min^{-1}. After 3 min of pressure build up and a holding time of 0 min, gelling was only detected with a release rate of 20 MPa/min for WPI. Increasing holding time to 15 or 30 min increased gel firmness. Firmness for samples with 15 or 30 min holding times and different rates of pressure release did not differ significantly. As holding time increased, so did the number of disulfide bonds and therefore the strength of cross-linking in the gel structure. MC solution gelled under all treatment conditions, but firmness was less than for WPI gels with the same protein content. Firmness was not significantly affected by holding time but was determined by pressure release rate; the faster the pressure release, the firmer the gel. This was attributed to changes in microstructure; electron micrographs showed a fine microstructure of small particles with a pressure release time of 1 min, but several irregular aggregates of ~1 μm diameter with a release time of 3 min. Thus the firmness of pressure-induced gels depends on the type of protein present as well as the protein content.

Ultrahigh Pressure Homogenization of Milk

Milk acid gels are widely consumed as healthy food and there is a growing interest in producing gels that do not show syneresis during storage without added stabilizers (Lucey 2001). This is reflected in the great effort dedicated to the study of novel technologies for the processing of yogurt milk base including: enzymatic cross-linking by transglutaminase, extending shelf-life involving the use of carbon dioxide, and on improving several characteristics of yogurt, such as fermentation time, WHC and postacidification, which may include the use of high hydrostatic pressure.

Thus, in addition to using static HP, the effect of ultrahigh pressure homogenization (UHPH) is of interest. UHPH is based on the same principle as conventional homogenization but which works at significantly higher pressures (up to 350 MPa) (Serra et al. 2007).

UHPH samples were compared to both: HT milk with and without skim milk powder (SMP) added, since the increase in nonfat solids content (to E14 %) of the milk base using SMP is the most common practice in the current dairy industry (Serra et al. 2008a). Set-type yogurts prepared from milk UHPH treated at 200 or 300 MPa presented higher gel firmness in texture analysis, less syneresis, and lower titratable acidity compared with conventionally treated milk, even fortified with 3 % SMP. In view of these previous results and with the objective of giving a practical approach of this new technology, the study of the evolution of UHPH samples during storage was made by comparison with a product manufactured following the normal process applied in industry.

To obtain stirred yogurt, the coagulum is mechanically broken before cooling and packaging, which induces considerable changes in the rheological properties, although the physical properties of stirred yogurts are also affected by the original gel characteristics. However, in order to maintain the stability of the product during the shelf-life, some of the approaches adopted by manufacturers include: the use of exopolysacharide (EPS)-producing cultures is a common practice (Tamime and Robinson 2007) for improving the final texture of the product, because EPS contribute to a polymer-like behavior of the serum phase, which might have the ability to bind water and increase yogurt viscosity. The addition of pectin, gelatin, starch, and/or blends of these stabilizers (Tamime and Robinson 2007) is also a usual way to proceed due to the ease of use in the already disrupted coagulum.

Structural Changes due to UHPH Treatment

During the UHPH treatment, partial disintegration of casein micelles has been reported (Sandra and Dalgleish 2005), resulting in an increase in the number of casein fragments in solution, and hence in a higher solvation of these proteins. In addition, the partial disintegration of casein micelles is accompanied by the solubilization of colloidal calcium-phosphate (CCP) (Serra et al. 2008a). As a result, it is likely that the structure of UHPH gels is dominated by casein–casein interactions rather than by WP–casein interactions, which is in line with the thicker strands observed in the microstructure of the protein matrix and with the G′ values obtained. Dynamic moduli (G′ and G″) are related to the number and strength of interactions responsible for the network structure (Lucey et al. 1998). Therefore, the higher values of G′ obtained in HT+SMP gels could be due to the branched microstructure composed of WP and casein via disulfide bonds (Lucey et al. 1999).

In UHPH gels the backbone of the matrix is likely to be composed mainly of casein–casein unions by combinations of electrostatic and hydrophobic interactions. Hence, in UHPH gels the intensity of the linkages present in the network was considerably lower than those covalent bonds that constituted the matrix of HT+SMP yogurts. The solid-like behavior of the gel (G′) not only depends on the magnitude of interacting forces but also on the number of them. In both type of yogurts, UHPH and HT+SMP, there was an increase in the number of interacting particles compared with nonenriched milk. In UHPH gels, this increase was due to the reduction in particle size (Serra et al. 2008a), whereas in HT+SMP gels it was because of the higher concentration of protein. In this way, the reduction in fat droplet size demonstrated in earlier studies (Serra et al. 2008b), due to the UHPH treatment, could also explain some of the differences observed. In HT+SMP gels, the lower number of fat droplets resulted in the formation of a continuous protein network occasionally interrupted by embedded fat globules, whereas in UHPH gels the high number of fat droplets resulted in a discontinuous protein–fat droplets network. Hence, considering the same surface in contact with the rheometer probe, less protein–protein interactions were detected in UHPH gels compared with HT+SMP gels due to the different dispersion of fat; this fact could also explain the lower G′ values of UHPH gels.

Textural analysis showed higher values of firmness of UHPH gels (Table A520). It is important to note that in puncture tests, not only the linkages in the structure are important but the global distribution of particles also has a big effect. Microstructure of UHPH gels was much more homogeneous and compact, with large protein clusters, and less porous than that of HT+SMP gels, in agreement with results of gel density obtained in previous work (Serra et al. 2007). Therefore, the network of UHPH yogurts is more resistant to breakage than conventional yogurts that have a more branched matrix with a big number of pores. Firmness values hardly increased during cold storage, probably because the test applied was not sensitive enough to reflect the rearrangements that occurred at the particle–particle level.

Stirred gels from HT+SMP milk gave finer and less grainy structures than those from UHPH-treated milk; the main structure of the latter is hypothesized based on casein–casein unions. The heterogeneity in the structure of stirred gels from UHPH-treated milk could explain the higher values of G′ obtained, since the presence of microclusters implies more junctions between larger particles in dispersion. During cold storage, G′ values increased until day 14 in all cases (Table A520), probably due to overacidification and a tendency to recover the gel structure. It is noted that acidification during cold storage enhances casein–casein unions by means of the intensification of hydrophobic interactions and solubilization of colloidal calcium phosphate. In addition, in stirred gels, one can expect the gels to regain their structure after destructive treatments.

Myofibrillar Proteins

Iwasaki et al. (2006) reported that the microstructure of combined pressure-heat-induced chicken myofibrillar gel was composed of three-dimensional fine strands. In addition, pressurization at 200 MPa prior to heating increased the apparent elasticity of chicken myofibrillar gel and pork patty; however, pressure treatment above 200 MPa decreased it. The apparent elasticity of the pressure-treated (200 MPa) thermal myofibrillar gel was three times higher, and that of pork patty was twice higher than those of the unpressurized ones. The rheological properties of the low salt (1 % NaCl) pork sausage can be improved by pressure treatment at 200 MPa prior to heating.

Zamri et al. (2006) studied simultaneous heat and pressure treatments ranging from ambient temperature to 70 °C and from 0.1 to 800 MPa, respectively, in various combinations. Texture profile analysis (TPA) of treated samples was performed to determine changes in hardness. At treatment temperatures up to and including 50 °C, heat and pressure acted synergistically to increase meat hardness. However, at 60 and 70 °C, hardness decreased following pressure treatments of >200 MPa.

TPA performed on extracted myofibrillar protein gels showed similar effects of heat and pressure after treatment under similar conditions. DSC analysis of whole chicken meat samples revealed that at ambient pressure the unfolding of myosin was completed at 60 °C, unlike actin, which completely denatured only above 70 °C (Zamri et al. 2006).

Effects of HP processing on isolated myofibrillar proteins and myosin from cod were investigated, and results were compared with those using myofibrillar proteins from turkey meat (breast) (Angsupanich et al. 1999). When turkey breast muscle and isolated myofibrillar protein and myosin of cod or turkey (pH $\sim$7) were subjected to pressures up to 800 MPa for 20 min, DSC and SDS-PAGE indicated that HP-induced denaturation of myosin led to formation of structures that contained hydrogen bonds and were additionally stabilized by disulphide bonds. Disulphide bonds were also important in heat-induced myosin gels. Hardness of whole cod muscle estimated by TPA showed pressure-treated samples (400 MPa) to be harder than cooked (50 °C) or cooked and then pressure-treated or pressure-treated and then cooked samples. The results support the suggestion that pressure induces formation of heat labile hydrogen-bonded structures while heat treatment gives rise to structures that are primarily stabilized by disulphide bonds and hydrophobic interactions. As expected, turkey myosin was more stable than that of cod; however, the pressure-induced gelation mechanisms of both myosins were similar.

In the review of Rastogi et al. 2007, it was noted that high-pressure treatment of freshwater fish (carp, Cyprinus carpio) resulted in the gelling of fish paste, which is useful for product development. Breaking strength of pressure-induced carp gels was much lower than that of heat-induced carp gels or lizardfish gels. The gel-forming ability of myofibrillar proteins was increased by addition of transglutaminase. The gel strength and brightness of the pork paste gel was found to increase with an increase in pressure. The higher the temperature or the sodium chloride concentration used for gelation, the lower was the pressure needed to denature the protein. Changes in rheological properties of bovine myofibrillar proteins in solution were influenced by structural changes caused by high-pressure treatment. Further, increase in process pressure and holding time resulted in decrease in viscosity and shift toward the Newtonian flow behavior.

Chapleau and Lamballerie-Anton (2003) used response surface methodology to study the effect of pressure (0–600 MPa) and time (0–1800s) on the surface hydrophobicity, reactive sulfhydryl groups content, and the flowing properties of bovine myofibrillar proteins in 10 g L^{-1} solution. High-pressure treatment induced a threefold increase in the surface hydrophobicity of myofibrillar proteins between 0 and 450 MPa. The same upward trend was observed with the reactive sulfhydryl groups, whose content increased from 40 to 69%. Concerning rheological properties of solutions, the flow behavior index tended toward a maximum value close to Newtonian behavior ($n = 1$), whereas the viscosity decreased with the increase of pressure.

Soy Proteins

The minimum pressure required for inducing gelation of soy proteins was reported to be 300 MPa for 10–30 min and the gels formed were softer with lower elastic modulus in comparison with heat-treated gels (Okamoto et al. 1990). The treatment of soy milk at 500 MPa for 30 min changed it from a liquid state to a solid state, whereas at lower pressures and at 500 MPa for 10 min, the milk remained in a liquid state, but indicated improved emulsifying activity and stability (Kajiyama et al. 1995). The hardness of tofu gels produced by high-pressure treatment at 300 MPa for 10 min was comparable to heat-induced gels.

Dynamic viscoelastic behavior of soy protein isolate (SPI) dispersions with concentrations (10, 15, and 20%) that were subjected to HP treatment at (350, 450, 550, and 650 MPa) for 15 min at 23 ± 1.5 °C was studied (Ahmed et al. 2007). The frequency sweep rheological data (0.1–10.0 Hz showed that the elastic modulus (G') predominated over the viscous component (G'') for all concentration. Gel rigidity of pressurized samples showed no systematic pattern with applied pressure. However, concentration significantly increased the mechanical strength of the gels. Calorimetric studies confirmed denaturation of the SPI dispersions at 350 MPa irrespective of concentration. Electrophoresis results (both native and SDS) showed insignificant conformational changes in the protein subunits of the processed samples. At similar concentrations, the firmness of pressure-treated gels was lower than that of thermally induced gels.

Molina and Ledward (2003) studied the textural properties: adhesiveness, springiness (elasticity), and hardness, as well as the water-binding capacity of gels from 12 % dispersions of SPI and 11S and 7S globulin fractions subjected to heat (90 °C for 15 min) before or after pressurization at 300–700 MPa. Gelation only occurred for the 11S fraction after either heating or pressurization. Pressurization followed by heating (PHT) produced gels in all the dispersions; however, heating followed by pressurization (HP) produced a gel only in the 11S fraction. Adhesiveness of 11S PHT gels was not much affected by pressure, whereas elasticity tended to decrease and hardness to increase as pressure increased. Water loss increased with pressure applied to 11S PHT gels. Adhesiveness of SPI gels followed the trends of 11S gels, but values were higher; the same trend was observed for water-binding capacity. The 7S PHT gel had no adhesiveness at any applied pressure, while the elasticity decreased as pressure increased to 600 MPa and then 700 MPa; hardness decreased as pressure increased from 300 to 500 MPa, then remained stable. Water-binding capacity increased with pressure applied in 7S PHT gels.

Apichartsrangkoon (2003) studied the effects of HP processing (at 200–800 MPa at 20 or 60 °C) on the rheological properties, protein solubility, and SDS-PAGE profiles of hydrated (moisture content of 80 %) commercial soy protein concentrates (SPC). Shapes of the storage (G′) and loss (G″) modulus curves changed only slightly with temperature/pressure treatments. Overall, increasing temperature gave rise to greater rheological changes than did pressure. Chemical analysis of gel systems suggested limited disulphide bonding occurred as a result of pressure and temperature.

Puppo et al. (2004) demonstrated that the application of HP (200–600 MPa) on SPI at pH 8.0 resulted in an increase in a protein hydrophobicity and aggregation, a reduction in free sulfhydryl content, and a partial unfolding of the 7S and 11S fractions. A change in the secondary structure leading to a more disordered structure was also reported. At pH 3.0, the protein was partially denatured and insoluble aggregates were formed, the major molecular unfolding resulted in decreased thermal stability, increased protein solubility, and hydrophobicity.

Puppo et al. (2008) studied the effect of the combined T/HP (T 20–60 °C/HP 0.1–600 MPa) treatment on physicochemical and rheological properties of emulsions prepared with native SPIs at 7 % (w/v). The size and aggregation of oil droplets of emulsions prepared with SPI 7 % (w/v) solutions were not altered by the combined T/HP treatment and the emulsions did not flocculate or coalesce. Simultaneously, a significant increase in the apparent viscosity with increase in pressure was observed, which was reinforced by temperature. The phenomenon was attributed to the gelation of nonadsorbed soybean proteins. Temperature seemed to improve the gelation process but only up to 400 MPa. At higher pressures, the combined effect of temperature and pressure resulted in dissociation of the protein aggregates decreasing gelation.

Egg Proteins

Rastogi et al. (2007) summarized the effect of HP treatment on egg proteins. The conformation of the main protein component, ovalbumin, of egg white remained fairly stable when pressurized at 400 MPa, probably due to the four disulfide bonds and noncovalent interactions stabilizing the three-dimensional structure of ovalbumin. Egg yolk formed a gel when subjected to a pressure of 400 MPa for 30 min at 25°C, kept its original color, and was soft and adhesive. The hardness of the pressure-treated gel increased and adhesiveness decreased with an increase in pressure.

Appendix 5–A

Literature Values of Rheological Properties of Foods

Paul E. Okechukwu and M. A. Rao

Literature values of rheological properties of fluid foods are useful in design and handling applications. Earlier, Holdsworth (1971, 1993) presented compilations of the literature values of rheological properties of fluid foods. Magnitudes of the flow properties of several foods from the literature that were selected after examination of the measurement technique used are listed in several tables: (1) viscosity of water in Table A5.1, (2) fruit juices and purees in Table A5.2, (3) tomato pastes (Rao and Cooley 1992) in Table A5.3, (4) the effect of temperature on apparent viscosity and consistency index of fruit juices and purees in Table A5.4, (5) creep-compliance parameters of tomato concentrates (Yoo and Rao 1996) in Table A5.6, (6) values of yield stress of prepared apple sauce samples using the vane method and extrapolated Bingham yield stress in 5–F (Qiu and Rao 1988), (7) flow properties of chocolate in Table A5.8, (8) flow properties of chocolate before and after degasification in Table A5.9 (Fang et al. 1996), (9) flow properties of mustard, mayonnaise, fats, and oils in Table A5.10, (10) effect of storage on the viscoelastic properties of a model salad dressing (Paredes et al. 1989) in Table A5.12, (11) the effect of temperature on flow properties of oils as reported by Lang et al. (1992) in Table A5.13; an unusual model: In $\eta = A + (B/T + C)$ was used to describe the effect of temperature, (12) flow properties of dairy products in Table A5.14, (13) flow properties of meat batters in Table A5.15, (14) flow properties of egg products in Table A5.16, (15) flow properties of syrups and honeys in Table A5.17, and (16) viscosity of wines in Table A5.18. Some of the tables listed above have not been referred to in Chap. 5.

Table A5.1 Viscosity of water at different temperatures (Perry 1950)

	Temperature (°C)	Viscosity (mPa s)
Water	0	1.7921
	1	1.7313
	2	1.6728
	3	1.6191
	4	1.5674
	5	1.5188
	6	1.4728
	7	1.4284
	8	1.3860
	9	1.3462
	10	1.3077
	11	1.2713
	12	1.2363
	13	1.2028
	14	1.1709
	15	1.1404
	16	1.1111
	17	1.0828
	18	1.0559
	19	1.0299
	20	1.0050
	20.2	1.0000
	21	0.9810
	22	0.9579
	23	0.9358
	24	0.9142
	25	0.8937
	26	0.8737
	27	0.8545
	28	0.8360
	29	0.8180
	30	0.8007
	31	0.7840
	32	0.7679
	33	0.7523
	34	0.7371
	35	0.7225
	36	0.7085
	37	0.6947
	38	0.6814
	39	0.6685
	40	0.6560
	41	0.6439
	42	0.6321

Table A5.1 (continued)

Temperature (°C)	Viscosity (mPa s)
43	0.6207
44	0.6097
45	0.5988
46	0.5883
47	0.5782
48	0.5683
49	0.5588
50	0.5494
51	0.5404
52	0.5315
53	0.5229
54	0.5146
55	0.5064
56	0.4985
57	0.4907
58	0.4832
59	0.4759
60	0.4688
61	0.4618
62	0.4550
63	0.4483
64	0.4418
65	0.4355
66	0.4293
67	0.4233
68	0.4174
69	0.4117
70	0.4061
71	0.4006
72	0.3952
73	0.3900
74	0.3849
75	0.3799
76	0.3750
77	0.3702
78	0.3655
79	0.3610
80	0.3565
81	0.3521
82	0.3478
83	0.3436
84	0.3395
85	0.3355
86	0.3315

Table A5.1 (continued)

Temperature (°C)	Viscosity (mPa s)
87	0.3276
88	0.3239
89	0.3202
90	0.3165
91	0.3130
92	0.3095
93	0.3060
94	0.3027
95	0.2994
96	0.2962
97	0.2930
98	0.2899
99	0.2868
100	0.2838

Table A5.2 Flow properties of fruit and vegetable products

Product	Conc (% solids)	Method	Temp (°C)	Shear Rate (s^{-1})	K^a (Pa s^n)	n	Yield Stress(Pa)	Reference
Tomato juice (pH 4.3)	5.80	Cone cylinder	32.2	500–800	0.22	0.22		Harper and El-Sahrigi (1965)
			48.9		0.27	0.54		
			65.6		0.37	0.47		
	12.80		32.3		2.10	0.43		
			48.9		1.18	0.43		
			65.6		2.28	0.34		
			82.2		2.12	0.35		
	16.00		32.3		3.16	0.45		
			48.9		2.27	0.45		
			65.6		3.18	0.40		
			82.2		3.27	0.38		
	25.00		32.3		12.9	0.41		
			48.9		10.5	0.42		
			65.6		8.0	0.43		
			82.2		6.1	0.44		
Tomato juice	30.00	Cone cylinder	32.3	500–800	18.7	0.40		Harper and Lieberman (1962)
			48.9		15.1	0.42		
			65.6		11.7	0.43		
			82.2		7.9	0.44		

Table A5.2 (continued)

Product	Conc (% solids)	Method	Temp (°C)	Shear Rate (s^{-1})	K[a] (Pa s^n)	n	Yield Stress(Pa)	Reference
Tomato conc.	35.74	Parallel plate carimed	25.0	4–576	297.0	0.38		Alviar and Reid (1990)
	24.16				65.7	0.27		
	17.86				26.1	0.26		
	11.78				6.0	0.38		
	35.74	Back extrusion	25.0	4–576	249.0	0.29		
	24.16				58.5	0.21		
	17.86				24.3	0.23		
	11.78				5.82	0.27		
Tomato paste	29.7°Brix	Haake mixer	32.0		208.0	0.27	206	Rao et al. (1993)
	29.7°Brix		39.0		179.0	0.31	180	
	23.8°Brix		33.0		48.0	0.47	40	
	23.8°Brix		39.0		34.0	0.52	29	
	16.3°Brix		25.0		24.0	0.23	38	
Tomato sauce	7.2°Brix	Conc cylinder Haake RV20.	20.0	0–10	4.0	0.40		McCarthy and Seymour (1994b)
	8.9°Brix		20.0	0–10	7.3	0.33		
Pears (Bartlett)	18.30	Conc cylinder	32.3	0.1–2600	2.3	0.49	3.5	
			48.9		1.9	0.48		
			65.6		1.6	0.48		
			82.2		1.5	0.48		
	26.13		32.3		6.2	0.45	8.1	
			48.9		5.0	0.45		
			65.6		4.2	0.46		
			82.2		3.6	0.46		
	31.00		32.3		10.9	0.45	19.1	
			48.9		8.80	0.45		
			65.6		6.8	0.46		
			82.2		5.6	0.46		
	37.16		32.3		17.0	0.46		
			48.9		13.5	0.46		
			65.6		10.4	0.46		
			82.2		9.40	0.46		
	45.75		32.3		35.5	0.48	33.9	
			48.9		26.0	0.48		
			65.6		20.0	0.48		
			82.2		16.0	0.48		
Pear Puree	14.60	Tube visc	27.0	100–2000	5.3	0.38		Saravaeos (1968)

Table A5.2 (continued)

Product	Conc (% solids)	Method	Temp (°C)	Shear Rate (s^{-1})	K^a (Pa s^n)	n	Yield Stress(Pa)	Reference
	16.00		30.0		5.6	0.35		Harper (1960)
	24.30	Conc cylinder	26.6	1–1000	5.8	0.39		
	33.40				38.5	0.38		
	37.60				49.7	0.38		
	37.50				64.8	0.38		
	47.60				120.0	0.33		
	51.30				205.0	0.34		
	49.30				170.0	0.34		
	48.80				150.0	0.34		
	15.20				4.25	0.35		
Apple sauce	11.00	Brook-field RTV	30.0	5–50	11.6	0.34		Saravacos (1970)
	11.00		82.0	5–50	9.0	0.34	11.6	Sarava-cos and Moyer (1967)
		Tube visc	27.0	100–2000	12.7	0.28		Saravacos (1968)
Apple sauce		Haake mixer	32.0		200.0	0.42	240	Rao et al. (1993)
Depectin.	75°Brix		20–70			1.00		Saravacos (1970)
Apple juice	50°Brix		20–70			1.00		
	30°Brix		20–70			1.00		
	15°Brix		20–70			1.00		
Cloudy apple juice	65.5°Brix		20–70			0.65		
	50°Brix		20–70			0.85		
	40°Brix		20–70			1.00		
	30°Brix		20–70			1.00		
	15°Brix		20–70			1.00		
Apple juice	69.8°Brix	Haake RV2/ Deer Rheom.	25.0		0.24	1.00		Rao et al. (1993)
	65.3°Brix		23.0		0.09	1.00		
	69.8°Brix		43.0		0.03	1.00		
	51.1°Brix		25.0		0.02	1.00		
Apricot puree	16	Brook-field RTV	30.0	5–50	6.80	0.30		Sarava-cos and Moyer (1967)
	13.8	Tube visc.	27.0		7.2	0.41		Saravacos (1970)

Table A5.2 (continued)

Product	Conc (% solids)	Method	Temp (°C)	Shear Rate (s^{-1})	K[a] (Pa s^n)	n	Yield Stress(Pa)	Reference
Apricot	17.7	Conc cylinder	26.7	1–1000	5.4	0.29		Harper (1960)
	44.3				56.0	0.39		
	51.4				108.0	0.36		
	55.2				152.0	0.34		
	59.3				300.0	0.32		
	41.4				540.0	0.35		
	23				11.2	0.35		
Peach	40.1				58.5	0.35		
	49.8				85.5	0.34		
	49.6				152.0	0.34		
	58.4				440.0	0.34		
	17				13.8	0.35		
	21.9				21.1	0.55		
	26				13.4	0.40		
	29.6				18.0	0.40		
	37.5				44.0	0.38		
	10.9				9.40	0.44		
Peach puree	11.7	Brook-field RTV tube visc.	30.0	5–50	7.20	0.28		Saravacos (1970)
Plum	14		30.0	5–50	2.2	0.34		Saravacos (1970)
Plum	14		82.2	5–50	2.0	0.34		
Concord	64		20–70			0.90		Saravacos (1970)
	50					1.00		
	30					1.00		
	15					1.00		
Passion fruit	15.6°Brix	Tube visc	21.6		0.07	0.74		Rao et al. (1974)
	20.7°Brix		21.4		0.29	0.53		
	25.3°Brix		20.0		0.80	0.52		
	30.6°Brix		23.2		1.62	0.49		
	33.4°Brix		20.0		2.62	0.45		
Papaya	7.3°Brix		26.0		0.91	0.53		Rao (1977)
Mango	9.3°Brix		24.2		2.06	0.33		
Guava	10.3°Brix		23.4		3.90	0.49		
Guava puree	7.2°Brix	Tube Visco	24.0	100–10000	0.26	0.68		Brekke and Myers (1978)
	7.4°Brix		24.0		0.28	0.68		

Table A5.2 (continued)

Product	Conc (% solids)	Method	Temp (°C)	Shear Rate (s^{-1})	K^a (Pa s^n)	n	Yield Stress(Pa)	Reference
Guava concentrate	12.3°Brix		24.0		1.60	0.55		
	15.9°Brix		24.0		5.00	0.47		
	22.7°Brix		24.0		41.00	0.32		
Depectin. guava puree	8.8°Brix		24.0		<0.1	1.65		
	8.9°Brix		24.0		<0.1	1.78		
Depectin. guava conc	15.1°Brix		24.0		0.30	0.71		
	24.3°Brix		24.0		1.60	0.62		
	31.4°Brix		24.0		2.30	0.59		
Conc. orange juice hamlin early	42.5°Brix	Haake RV12 conc cylinder	25.0	0–500	0.41	0.59		Crandall et al. (1982)
	15.0			0–500	0.60	0.60	0.0	
				0–500	0.92	0.68		
			−10.0	0–500	1.43	0.71		
Hamlin late	41.4°Brix		25.0	0–500	0.19	0.735		
			15.0	0–500	0.81	0.56		
			0.0	0–500	0.18	0.62		
			10.0	0–500	1.39	0.71		
Pineapple early	41.4°Brix		25.0	0–500	0.26	0.64		
			15.0	0–500	0.59	0.59		
			0.0	0–500	0.89	0.68		
			−10.0	0–500	1.22	0.71		
Pineapple late	40.3°Brix		25.0	0–500	0.86	0.53		
			15.0	0–500	1.34	0.54		
			0.0	0–500	1.86	0.64		
			−10.0	0–500	3.64	0.63		
Valencia early	43.1°Brix		25.0	0–500	0.51	0.54		
			15.0	0–500	0.67	0.61		
			0.0	0–500	1.40	0.62		
			−10.0	0–500	2.72	0.61		
Valencia late	41.9°Brix		25.0	0–500	0.84	0.54		
			15.0	0–500	1.18	0.57		
			0.0	0–500	1.86	0.64		
			−10.0	0–500	4.14	0.63		
Naval	65.1°Brix		−18.5		29.20	0.71		Rozema and Beverloo (1974)

Table A5.2 (continued)

Product	Conc (% solids)	Method	Temp (°C)	Shear Rate (s^{-1})	K[a] (Pa s^n)	n	Yield Stress(Pa)	Reference
			14.1		14.60	0.76		
			–9.3		10.80	0.74		
			–5.0		7.90	0.72		
			–0.7		5.90	0.71		
			10.1		2.70	0.73		
			19.9		1.60	0.72		
			29.5		0.9	0.74		
Orange juice	64.9°Brix	Haake RV2 conc cylinder	–18.0		18.3	0.80		Vitali and Rao (1984)
Pera orange, 3.4 % pulp			–14.0		10.6	0.81		
			–10.0		7.1	0.79		
			–5.0		4.9	0.78		
			0.0		3.2	0.79		
			10.0		1.6	0.79		
			20.0		0.7	0.83		
			30.0		0.4	0.82		
Pera orange, 5.7 % pulp	65°Brix		–18.0		24.5	0.76		
			–14.0		15.6	0.76		
			–10.0		8.8	0.79		
			5.0		6.5	0.77		
			0.0		4.5	0.76		
			10.0		2.1	0.78		
			20.0		1.3	0.77		
			30.0		0.7	0.80		
Pera orange, 8.6 % pulp	65.3°Brix		–18.0		42.8	0.70		
			–14.0		25.3	0.73		
			–10.0		15.8	0.74		
			–5.0		11.1	0.73		
			0.0		7.3	0.74		
			10.0		3.9	0.73		
			20.0		1.9	0.76		
			30.0		1.4	0.73		
Pera orange, 11.1 % pulp	64.8°Brix		–18.0		58.9	0.67		
			–14.0		34.6	0.70		
			–10.0		23.9	0.69		
			–5.0		16.0	0.69		
			0.0		11.5	0.69		
			10.0		5.6	0.70		

Table A5.2 (continued)

Product	Conc (% solids)	Method	Temp (°C)	Shear Rate (s^{-1})	K^a (Pa s^n)	n	Yield Stress(Pa)	Reference
			20.0		2.9	0.72		
			30.0		2.1	0.70		
Valencia orange, 21.2% pulp	65.3°Brix		–18.3		109.9	0.55		
			–14.2		59.7	0.61		
			–9.7		40.6	0.60		
			–5.1		24.5	0.63		
			–0.4		18.5	0.62		
			10.2		8.3	0.65		
			19.7		6.1	0.61		
			29.6		2.5	0.68		
Mango pulp 3.49% db pectin (pH 3.56)	16° Brix	Cone cylinder	30	20–250	5.1	0.97	Manohar et al. (1990)	
		$r_1/r_2 = 0.94$	40		4.4	0.98		
		Rheotest2	50		3.8	0.99		
			60		3.3	0.99		
			70		2.8	0.99		
	20°Brix		30	20–250	7.9	0.28		
			40		6.8	0.28		
			50		5.4	0.27		
			60		4.5	0.28		
			70		3.6	0.27		
	25°Brix		30	20–250	9.8	0.29		
			40		8.2	0.29		
			50		7.5	0.28		
			60		6.4	0.28		
			70		5.6	0.28		
	30°Brix		30	20–250	19.3	0.28		
			40		14.7	0.98		
			50		12.3	0.28		
			60		10.3	0.28		
			70		8.5	0.28		
3.38% db pectin	16°Brix		30	20–250	2.9	0.29		
			40		2.5	0.29		
			50		2.3	0.29		
			60		1.9	0.29		
			70		1.7	0.28		
3.29% db pectin	16°Brix		30	20–250	2.7	0.29		
			40		2.4	0.28		

Table A5.2 (continued)

Product	Conc (% solids)	Method	Temp (°C)	Shear Rate (s^{-1})	K[a] (Pa s^n)	n	Yield Stress(Pa)	Reference
			50		2.1	0.29		
			60		1.8	0.28		
			70		1.5	0.29		
3.23% db pectin	16°Brix		30	20–250	1.7	0.28		
			40		1.4	0.28		
			50		1.1	0.29		
			60		0.9	0.29		
			70		0.8	0.29		
			30	20–250	2.4	0.29		
			40		2.0	0.29		
			50		1.6	0.29		
			60		1.4	0.29		
			70		1.2	0.29		
			30	20–250	3.0	0.29		
			40		2.6	0.29		
			50		2.2	0.29		
			60		1.9	0.29		
			70		1.7	0.29		
			30	20–250	5.3	0.28		
			40		4.4	0.29		
			50		3.7	0.28		
			60		3.1	0.29		
			70		2.5	0.29		
Mango pulp juice	16°Brix	Cone cylinder Rheotest2	25		10.0	0.32		Rao et al. (1985)
Mango pulp juice	17°Brix	Rotational RVT Brookfield	25		27.8	0.33		Gunjal and Waghmare (1987)
Mango pulp juice	16.0°Brix	Rotational LVF Brookfield	25		12.30	0.280		Garcia et al. (1974)
Tamarind juice	7°°Brix	Cone cylinder Rheotest2 $R_1/R_{2=}0.98$	35.0	10–400	0.002	0.99		Manohar et al. (1991)
			45.0		0.001	1.0		
			55.0		0.001	1.00		
	9°Brix		25.0		0.006	1.00		
			45.0		0.004	1.00		

Table A5.2 (continued)

Product	Conc (% solids)	Method	Temp (°C)	Shear Rate (s^{-1})	K^a (Pa s^n)	n	Yield Stress(Pa)	Reference
			55.0		0.003	1.00		
	23°Brix		25.0		0.087	0.76		
			35.0		0.062	0.75		
			45.0		0.050	0.73		
			55.0		0.039	0.75		
	28°Brix		25.0		0.35	0.63		
			35.0		0.26	0.62		
			45.0		0.19	0.63		
			55.0		0.16	0.63		
			65.0		0.14	0.63		
	36°Brix		25.0		0.46	0.63		
			35.0		0.35	0.61		
			45.0		0.26	0.63		
			55.0		0.23	0.62		
			65.0		0.21	0.61		
	44°Brix		30.0		1.23	0.63		
			40.0		0.83	0.63		
			50.0		0.63	0.64		
			60.0		0.53	0.61		
			70.0		0.43	0.63		
	50°Brix		35.0		3.40	0.62		
			45.0		2.65	0.63		
			55.0		1.92	0.63		
			65.0		1.53	0.63		
			70.0		1.38	0.63		
	62°Brix		27.0		9.07	0.63		
			35.0		6.28	0.63		
			45.0		4.46	0.63		
			55.0		3.32	0.62		
			70.0		2.23	0.63		
Pineapple juice	7°Brix	Conc cylinder narrow gap viscometer	30.0	100–600	0.03	1.00		Varshney and Kumbhar (1978)
	22°Brix		30.0		0.06	0.91		
	36°Brix		30.0		0.13	0.81		
	7°Brix	Conc cylinder narrow gap viscometer	45.0		0.02	1.03		
	22°Brix		45.0		0.05	0.90		
	36°Brix		45.0		0.12	0.81		

Table A5.2 (continued)

Product	Conc (% solids)	Method	Temp (°C)	Shear Rate (s^{-1})	K[a] (Pa s^n)	n	Yield Stress(Pa)	Reference
Orange juice	13°Brix	Conc cylinder narrow gap viscometer	30.0	100–600	0.06	0.86		
	22°Brix		30.0		0.12	0.79		
	33°Brix		30.0		0.14	0.78		
	13°Brix	Conc cylinder narrow gap viscometer	45.0		0.02	1.00		
	22°Brix		45.0		0.08	0.79		
	33°Brix		45.0		0.12	0.98		
Pulp free rorange juice % pectin (galacturonic acid)								Mizrahi (1979)
		Brookfield LVT cone cylinder						
0.047	60°Brix			1–1,142	0.016	0.55		
0.019	65°Brix				0.003	0.91		
0.034	60°Brix				0.005	0.73		
Raspberry	15°Brix	Cone cylinder	5.0	3–2,000	0.029	0.90		Ibarz and Pagan (1987)
juice		Haake RV12 NV						
			10.0		0.022	0.92		
			15.0		0.017	0.93		
			20.0		0.014	0.94		
			25.0		0.010	0.96		
			30.0		0.009	0.96		
			35.0		0.006	0.99		
			45.0		0.004	1.01		
			55.0		0.003	1.00		
			60.0		0.002	1.02		
	20°Brix	Cone cylinder Haake RV12 NV	5.0	3–2,000	0.058	0.90		

Table A5.2 (continued)

Product	Conc (% solids)	Method	Temp (°C)	Shear Rate (s^{-1})	K^a (Pa s^n)	n	Yield Stress(Pa)	Reference
			10.0		0.045	0.92		
			15.0		0.032	0.94		
			20.0		0.030	0.94		
			25.0		0.023	0.96		
			30.0		0.019	0.96		
			35.0		0.014	0.99		
			45.0		0.011	0.99		
			55.0		0.010	1.01		
			60.0		0.008	1.00		
	25°Brix	Cone cylinder Haake RV12 NV	5.0	3–2,000	0.344	0.76		
			10.0		0.169	0.86		
			15.0		0.111	0.91		
			20.0		0.072	0.94		
			25.0		0.064	0.94		
			30.0		0.050	0.95		
			35.0		0.033	1.00		
			45.0		0.024	1.00		
			55.0		0.021	1.00		
			60.0		0.019	1.01		
	30°Brix	Cone cylinder Haake RV12 NV	5.0	3–2,000	0.739	0.74		
			10.0		0.437	0.82		
			15.0		0.365	0.81		
			20.0		0.202	0.90		
			25.0		0.142	0.91		
			30.0		0.124	0.91		
			35.0		0.101	0.92		
			45.0		0.067	0.95		
			55.0		0.049	0.95		
			60.0		0.040	0.96		
	35°Brix	Cone cylinder Haake RV12 NV	5.0	3–2,000	1.86	0.69		
			10.0		1.30	0.72		
			15.0		0.81	0.80		
			20.0		0.67	0.79		

Table A5.2 (continued)

Product	Conc (% solids)	Method	Temp (°C)	Shear Rate (s^{-1})	K[a] (Pa s^n)	n	Yield Stress(Pa)	Reference
			25.0		0.56	0.79		
			30.0		0.34	0.86		
			35.0		0.28	0.87		
			45.0		0.17	0.91		
			55.0		0.12	0.92		
			60.0		0.08	0.95		
	41°Brix	Cone cylinder Haake RV12 NV	5.0	3–2,000	6.25	0.59		
			10.0		4.30	0.64		
			15.0		3.36	0.66		
			20.0		2.57	0.68		
			25.0		1.61	0.73		
			30.0		1.34	0.75		
			35.0		0.95	0.79		
			45.0		0.46	0.88		
			55.0		0.35	0.87		
			60.0		0.30	0.88		
Apple juice	71°Brix	Haake RV12 cone cylinder	5.0		1.26	1.00		Ibarz et al. (1987)
			15.0		0.26	1.00		
			25.0		0.22	1.00		
			35.0		0.13	1.00		
			40.0		0.088	1.00		
			45.0		0.050	1.00		
			60.0		0.027	1.00		
Apple juice	35°Brix		25.0		0.001	1.00		
	45°Brix		25.0		0.008	1.00		
	50°Brix		25.0		0.018	1.00		
	60°Brix		25.0		0.066	1.00		
	71°Brix		25.0		0.217	1.00		
Pear juice	70°Brix		5.0		1.46	1.00		Ibarz et al. (1987)
			10.0		0.876	1.00		
			15.0		0.527	1.00		
			25.0		0.233	1.00		
			35.0		0.110	1.00		
			45.0		0.061	1.00		
			60.0		0.031	1.00		
Pear juice	30°Brix		25.0		0.003	1.00		

Table A5.2 (continued)

Product	Conc (% solids)	Method	Temp (°C)	Shear Rate (s^{-1})	K[a] (Pa s^n)	n	Yield Stress(Pa)	Reference
	40°Brix		25.0		0.005	1.00		
	45°Brix		25.0		0.008	1.00		
	50°Brix		25.0		0.013	1.00		
	55°Brix		25.0		0.019	1.00		
	60°Brix		25.0		0.041	1.00		
	65°Brix		25.0		0.074	1.00		
	70°Brix		25.0		0.233	1.00		
Black currant juice clarified	35°Brix	Haake RV12 conc cylinder	5.0		0.0096	0.0096		Ibarz et al. (1992a)
			10.0		0.0077	1.00		
			15.0		0.0064	1.00		
			20.0		0.0056	1.00		
			25.0		0.0049	1.00		
			30.0		0.0044	1.00		
			35.0		0.0038	1.00		
			40.0		0.0035	1.00		
			45.0		0.0031	1.00		
			50.0		0.0027	1.00		
			55.0		0.0026	1.00		
			60.0		0.0024	1.00		
	40°Brix		5.0		0.0138	1.00		
			10.0		0.0105	1.00		
			15.0		0.0090	1.00		
			20.0		0.0075	1.00		
			25.0		0.0065	1.00		
			30.0		0.0057	1.00		
			35.0		0.0050	1.00		
			40.0		0.0046	1.00		
			45.0		0.0041	1.00		
			50.0		0.0038	1.00		
			55.0		0.0033	1.00		
			60.0		0.0031	1.00		
	45°Brix		5.0		0.0234	1.00		
			10.0		0.0174	1.00		
			15.0		0.0136	1.00		
			20.0		0.0112	1.00		
			25.0		0.0095	1.00		
			30.0		0.0079	1.00		
			35.0		0.0069	1.00		
			40.0		0.0061	1.00		
			45.0		0.0055	1.00		

Table A5.2 (continued)

Product	Conc (% solids)	Method	Temp (°C)	Shear Rate (s^{-1})	K^a (Pa s^n)	n	Yield Stress(Pa)	Reference
			50.0		0.0050	1.00		
			55.0		0.0047	1.00		
			60.0		0.0042	1.00		
	50°Brix		5.0		0.0371	1.00		
			10.0		0.0284	1.00		
			15.0		0.0215	1.00		
			20.0		0.0175	1.00		
			25.0		0.0143	1.00		
			30.0		0.0116	1.00		
			35.0		0.0100	1.00		
			40.0		0.0084	1.00		
			45.0		0.0077	1.00		
			50.0		0.0067	1.00		
			55.0		0.0061	1.00		
			60.0		0.0056	1.00		
	55°Brix		5.0		0.0744	1.00		
			10.0		0.0549	1.00		
			15.0		0.0400	1.00		
			20.0		0.0325	1.00		
			25.0		0.0245	1.00		
			30.0		0.0195	1.00		
			35.0		0.0161	1.00		
			40.0		0.0135	1.00		
			45.0		0.0130	1.00		
			50.0		0.0113	1.00		
			55.0		0.0099	1.00		
			60.0		0.0091	1.00		
	60°Brix		5.0		0.1589	1.00		
			10.0		0.1093	1.00		
			15.0		0.0802	1.00		
			20.0		0.0558	1.00		
			25.0		0.0449	1.00		
			30.0		0.034	1.00		
			35.0		0.027	1.00		
			40.0		0.022	1.00		
			45.0		0.020	1.00		
			50.0		0.018	1.00		
			55.0		0.015	1.00		
			60.0		0.013	1.00		
	64.5°Brix		5.0		0.500	1.00		
			10.0		0.298	1.00		
			15.0		0.208	1.00		
			20.0		0.138	1.00		

Table A5.2 (continued)

Product	Conc (% solids)	Method	Temp (°C)	Shear Rate (s^{-1})	K^a (Pa s^n)	n	Yield Stress(Pa)	Reference
			25.0		0.101	1.00		
			30.0		0.073	1.00		
			35.0		0.055	1.00		
			40.0		0.044	1.00		
			45.0		0.035	1.00		
			50.0		0.029	1.00		
			55.0		0.024	1.00		
			60.0		0.020	1.00		
Clarified peach	40(°Brix)	Cone cylinder Haake Rotovisco	5.0	50–700	0.016	1.00		Ibarz et al. (1992b)
juice	45				0.025	1.00		
	50				0.044	1.00		
	55				0.085	1.00		
	60				0.181	1.00		
	65				0.526	1.00		
	69				2.087	1.00		
	40		10.0		0.013	1.00		
	45				0.019	1.00		
	50				0.033	1.00		
	55				0.063	1.00		
	60				0.123	1.00		
	65				0.404	1.00		
	69				1.288	1.00		
	40		15.0		0.011	1.00		
	45				0.016	1.00		
	50				0.027	1.00		
	55				0.048	1.00		
	60				0.092	1.00		
	65				0.249	1.00		
	69				0.743	1.00		
	40		20.0		0.009	1.00		
	45				0.013	1.00		
	50				0.021	1.00		
	55				0.037	1.00		
	60				0.062	1.00		
	65				0.175	1.00		
	69				0.485	1.00		
	40		25.0		0.008	1.00		
	45				0.011	1.00		
	50				0.017	1.00		
	55				0.029	1.00		

Table A5.2 (continued)

Product	Conc (% solids)	Method	Temp (°C)	Shear Rate (s^{-1})	K^a (Pa s^n)	n	Yield Stress(Pa)	Reference
	60				0.049	1.00		
	65				0.126	1.00		
	69				0.324	1.00		
	40		35.0		0.006	1.00		
	45				0.008	1.00		
	50				0.012	1.00		
	55				0.020	1.00		
	60				0.033	1.00		
	65				0.070	1.00		
	69				0.155	1.00		
	40		45.0		0.005	1.00		
	45				0.006	1.00		
	50				0.009	1.00		
	55				0.013	1.00		
	60				0.021	1.00		
	65				0.041	1.00		
	69				0.084	1.00		
	40		60.0		0.004	1.00		
	45				0.005	1.00		
	50				0.006	1.00		
	55				0.009	1.00		
	60				0.013	1.00		
	65				0.023	1.00		
	69				0.042	1.00		
Tomato conc. (hot break samples), 0.857 % pectin (as galactu ronic acid anhydride)	10°Brix		15.0		6.98	0.37		Fita et al. (1983)
		Contraves Rheomat 15	20.0		7.00	0.36		
			30.0		0.240	0.36		
			40.0		6.03	0.37		
			50.0		4.84	0.38		
			60.0		4.75	0.37		
			70.0		4.49	0.35		
	14°Brix		15.0		14.6	0.38		
			20.0		13.9	0.37		
			30.0		12.6	0.34		
			40.0		12.3	0.34		
			50.0		11.7	0.32		

Table A5.2 (continued)

Product	Conc (% solids)	Method	Temp (°C)	Shear Rate (s^{-1})	K^a (Pa s^n)	n	Yield Stress(Pa)	Reference
			60.0		12.2	0.32		
			70.0		12.2	0.24		
	15°Brix		15.0		19.7	0.33		
			20.0		21.1	0.33		
			30.0		18.9	0.33		
			40.0		17.0	0.33		
			50.0		14.2	0.34		
			60.0		12.0	0.33		
			70.0		11.3	0.32		
	20°Brix		15.0		40.0	0.31		
			20.0		34.4	0.31		
			30.0		33.6	0.32		
			40.0		28.8	0.33		
			50.0		25.2	0.33		
			60.0		23.2	0.29		
			70.0		17.2	0.29		
	25°Brix		15.0		76.9	0.28		
			20.0		61.5	0.29		
			30.0		76.7	0.28		
			40.0		60.3	0.27		
			50.0		57.7	0.26		
			60.0		47.2	0.31		
			70.0		39.4	0.32		
	28°Brix		15.0		65.1	0.40		
			20.0		81.3	0.33		
			30.0		10.5	0.25		
			40.0		86.8	0.27		
			50.0		67.7	0.28		
			60.0		87.6	0.22		
			70.0		43.6	0.26		
Tomato conc CB 0.857% pectin (as galact-uronic acid anhydride)	10°Brix	Con-traves Rheomat 15	15.0		0.45	0.43		
			20.0		0.38	0.43		
			30.0		0.56	0.38		
			40.0		0.27	0.44		
			50.0		0.40	0.41		
			60.0		0.26	0.42		
			70.0		0.12	0.45		
	14°Brix		15.0		4.1	0.35		
			20.0		3.2	0.34		

Table A5.2 (continued)

Product	Conc (% solids)	Method	Temp (°C)	Shear Rate (s^{-1})	K[a] (Pa s^n)	n	Yield Stress(Pa)	Reference
			30.0		3.0	0.38		
			40.0		2.8	0.33		
			50.0		2.3	0.39		
			60.0		2.3	0.36		
			70.0		2.0	0.35		
	15°Brix		15.0		6.8	0.34		
			20.0		7.4	0.36		
			30.0		6.6	0.38		
			40.0		5.8	0.39		
			50.0		5.1	0.38		
			60.0		4.3	0.36		
			70.0		4.1	0.38		
	20°Brix		15.0		8.6	0.35		
			20.0		6.9	0.39		
			30.0		7.1	0.36		
			40.0		6.4	0.40		
			50.0		5.4	0.37		
			60.0		5.5	0.37		
			70.0		4.9	0.36		
	25°Brix		15.0		17.2	0.34		
			20.0		15.4	0.35		
			30.0		14.0	0.35		
			40.0		12.3	0.36		
			50.0		11.2	0.35		
			60.0		8.9	0.38		
			70.0		9.4	0.34		
	28°Brix		15.0		29.3	0.34		
			20.0		27.5	0.35		
			30.0		25.9	0.34		
			40.0		23.0	0.33		
			50.0		19.4	0.35		
			60.0		17.9	0.33		
			70.0		15.1	0.35		
Apple sauce (Mott's)	18.2°Brix	Haake RV3	25.0		34.0	0.42	34	
Apple sauce (Stop and shop)	18.1				6.4	0.43	30	
Baby food banana (Beech-Nut)	15		25.0		5.4	0.62	21	
Baby food banana (Gerber)	15.1		25.0		28.0	0.59	28	

Table A5.2 (continued)

Product	Conc (% solids)	Method	Temp (°C)	Shear Rate (s^{-1})	K^a (Pa s^n)	n	Yield Stress(Pa)	Reference
Baby food peach (Beech-Nut)	10.8		25.0		0.600	0.83	14	
Baby food peach (Gerber)	16		25.0		1.40	0.60	13	
Mustard (Golden's)			25.0		5.50	0.52	35	
Mustard (Stop & shop)					3.40	0.56	20	
Tomato conc.	%TS25.74	Parallel plates	25.0		297.0	0.38		Alviar and Reid (1990)
	24.16	Carri-Med	25.0		65.7	0.27		
	17.86		25.0		26.1	0.26		
	11.78		25.0		6.00	0.33		
	%TS25.74	Back extrusion	25.0		249.0	0.29		Alviar and Reid (1990)
	24.16		25.0		58.5	0.21		
	17.86		25.0		24.3	0.23		
	11.78		25.0		5.8	0.27		
Tomato ketchup		Bohlin VOR conc cylinder	25.0	50–2000	6.1	0.40		Autio (1991)
		Haake RV3 conc cylinder	25.0	50–2000	5.6	0.41		
Conc. orange juice		Haake RV12 conc cylinder						
Hamlin early	42.5°Brix		25.0	0–500	0.08	0.86	1.36	Crandall et al. (1982)
			15.0	0–500	0.13	0.86	1.88	
			0.0	0–500	0.17	0.97	3.72	
			–10.0	0–500	0.35	0.95	5.64	
Hamlin late	41.4°Brix		25.0	0–500	0.031	1.06	0.94	
			15.0	0–500	0.229	0.77	2.07	
			0.0	0–500	0.370	0.83	3.72	
			–10.0	0–500	0.339	0.96	5.64	
Pineapple early	40.3°Brix		25	0–500	0.06	0.91	0.90	

Table A5.2 (continued)

Product	Conc (% solids)	Method	Temp (°C)	Shear Rate (s^{-1})	K[a] (Pa s^n)	n	Yield Stress(Pa)	Reference
			15	0–500	0.15	0.83	1.7	
			0	0–500	0.26	0.90	3.1	
			–10	0–500	0.08	1.19	7.5	
Pineapple late	41.8°Brix		25	0–500	0.10	0.91	2.8	
			15	0–500	0.18	0.88	4.3	
			0	0–500	0.46	0.88	6.2	
			–10	0–500	0.47	0.99	15.0	
Valencia early	43.1°Brix		25	0–500	0.12	0.84	1.5	
			15	0–500	0.19	0.82	1.9	
			0	0–500	0.36	0.86	4.3	
			–10	0–500	0.89	0.81	7.5	
Valencia late	41.9°Brix		25	0–500	0.15	0.84	24.4	
			15	0–500	0.26	0.82	3.4	
			0	0–500	0.48	0.88	6.2	
			–10	0–500	0.77	0.92	15.0	
Clarified banana juice	20.0	Conc cylinder Rheo-test-2	30.0	50–750	0.001			Lopez et al. (1989)
	30.0		30.0		0.003			
	39.6		30.0		0.006			
	49.0		30.0		0.010			
Tamarind kernel flour, pH=7, suspension								
	2	Conc cylinder Rheo-test-2	24.0	3–1,300	0.00045	0.87	0.003	Bhattacharya, et al. (1991)
	4		24.0		0.00056	0.84	0.004	
	6		24.0		0.00167	0.82	0.015	
	8		24.0		0.00488	0.78	0.019	
	10		24.0		0.00681	0.75	0.021	

[a] Some values were converted from cp or mPa s units

Table A5.3 Solids fraction and steady shear rheological properties of tomato pastes and serum (Rao and Cooley 1992)

Solids[a] Fraction %	Flow Behavior Index, n (–)	Consistency Index, K (Pa s^n)	Yield Stress σ_{0c} (Pa)	Casson Visosity, η_{Ca} (Pa s)	Serum Visosity (mPa s)
0.81	0.25	252.4	176.9	7.8	43.3
0.78	0.18	206.4	196.3	1.1	96.5
0.73	0.29	222.6	133.5	10.2	62.7
0.72	0.30	188.8	109.3	9.6	36.5
0.54	0.22	164.5	157.9	1.4	4.3
0.62	0.24	168.7	123.0	3.8	–
0.71	0.27	220.5	138.1	8.8	56.4
0.74	0.29	240.2	129.3	15.0	29.0
0.71	0.29	231.2	130.6	12.8	25.4
0.72	0.29	207.6	125.5	9.5	53.8
0.82	0.30	245.1	149.3	13.0	140.3
0.77	0.35	234.2	127.6	17.3	76.5
0.73	0.34	245.7	113.9	22.2	61.8
0.63	0.32	205.3	99.5	16.3	5.5
0.69	0.32	204.7	105.0	14.2	53.6
0.72	0.40	324.5	77.9	85.6	76.0
0.66	0.31	238.4	144.2	13.1	125.8
0.65	0.29	200.5	140.5	7.0	59.0
0.69	0.28	197.3	120.0	8.3	48.2
0.66	0.25	179.2	131.2	3.8	5.0
0.66	0.33	273.1	112.4	32.5	–
0.71	0.33	174.0	121.0	7.1	48.2
0.53	0.16	139.0	170.2	0.1	29.1
0.50	0.13	177.1	211.6	0.1	5.7
0.53	0.22	77.8	92.7	0.2	32.3

[a] Solids content of pastes; serum was obtained by removing insoluble solids by centrifugation

Table A5.4 Effect of temperature on flow properties of fruit and vegetable products Effect of temperature model based on viscosity, $\eta_a = \eta_\infty \exp(E_{av}/RT)$ (1) Effect of temperature model based on consistency index, $K = K_\infty \exp(E_{ak}/RT)$ (2) where, T is temperature (°K), η_∞ and K_∞ are collision factors or values at infinite temperature, R is gas constant, and E_{av} and E_{ak} are activation energies based in viscosity and consistency index, respectively

Sample	η_∞(Pa s)	E_{av}(kJ mol^{-1})	Shear Rate (s^{-1})	K_∞ Pa s^n	E_{ak}(kJ mol^{-1})	Temp. Range (°C)	Reference
Pera orange juice, 65°Brix							
0% Pulp	1.299E–10	51.5	100	7.672E–11	54.0	30–(–18)	Vitali and Rao (1984)
3.4%	7.755E–10	48.2		9.881 E–10	49.8		
5.7%	4.400E–09	43.1		3.130E–08	43.1		

Table A5.4 (continued)

Sample	η_∞(Pa s)	E_{av}(*kJ mol*$^{-1}$)	*Shear Rate* (*s*$^{-1}$)	K_∞ *Pa s*^{n}	E_{ak}(*kJ mol*$^{-1}$)	*Temp. Range* (*°C*)	Reference
8.6%	9.526E–09	44.0		1.742E–08	45.2		
11.1%	2.202E–08	44.0		4.195E–08	44.4		
Pera orange juice, 5.7% Pulp							
65°Brix	4.410E–09	44.8	100	3.13E–08	43.1	30–(–18)	
62°Brix	2.670E–08	39.4		1.39E–07	40.6		
58°Brix	9.340E–08	34.8		1.31E–07	36.8		
56°Brix	1 280E–07	33.5		8.95E–08	37.3		
52°Brix	1.640E–07	31.4		6.62E–08	35.6		
Pera orange juice, 4.6% Pulp							
65°Brix	3.45E–09	45.6	100	1.86E–08	10.6	30–(–18)	
62°Brix	2.48E–08	39.8		3.77E–08	40.2		
58°Brix	9.56E–08	35.6		2.30E–07	36.0		
56°Brix	1.29E–07	33.9		1.09E–07	36.8		
52°Brix	1.83E–07	31.8		1.66E–07	34.8		
Concentrated apple juice from Mclntosh apples							
45.1°Brix	3.394E–07	25.1					
50.4°Brix	1.182E–07	28.9					
55.2°Brix	2.703E–09	39.4					
60.1°Brix	3.935E–10	45.6					
64.9°Brix	7.917E–12	56.9					
68.3°Brix	1.156E–12	64.1					
Concentrated grape juice from concord grapes							
43.1°Brix	8.147E–08	29.3					
49.2	1.074E–08	35.6					
54	9.169E–10	43.1					
59.2	1.243E–10	49–4					
64.5	1.340E–12	51.5					
68.3	6.086E–12	60.7					
Mango juice							
16°Brix,3.49% pectin			0.0145	15.0	30–70		Manohar et al. (1990)
20°Brix,3.49% pectin			0.01677	15.5	30–70		
25°Brix,3.49% pectin			0.03275	14.6	30–70		
30°Brix,3.49% pectin			0.0339	15.8	30–70		
16° Brix,3.38% pectin			0.00894	14.9	30–70		
16°Brix,3.29% pectin			0.00855	14.7	30–70		
16°Brix,3.23% pectin			0.004	15.2	30–70		
21°Brix,3.23% pectin			0.00614	15.1	30–70		
25°Brix,3.23% pectin			0.00799	15.1	30–70		
28°Brix,3.23% pectin			0.0122	15.3	30–70		
Depectinized apple juice							Saravacos (1970)

Table A5.4 (continued)

Sample	η_∞(Pa s)	E_{av}*(kJ mol⁻¹)*	*Shear Rate* (s^{-1})	K_∞ *Pa* s^n	E_{ak}*(kJ mol⁻¹)*	*Temp. Range (°C)*	Reference
75°Brix	59.5						
50°Brix	35.2						
30°Brix	26.4						
15°Brix	22.2						
Cloudy apple juice							Saravacos (1970)
65.5°Brix	38.1						
50°Brix	25.5						
40°Brix	24.3						
30°Brix	21.4						
15°Brix	14.7						
Peach puree, 11.7°Brix				7.1			Saravacos (1970)
Cocord grape juice							Saravacos (1970)
64°Brix	46.9						
50°Brix	28.9						
30°Brix	26.0						
15°Brix	22.2						
Passion fruit juice-concentrated							Rao et al. (1974)
Fruit juice-temperature effect							
15.6°Brix	18.8						
20.7°Brix	17.2						
25.3°Brix	16.7						
30.6°Brix	15.9						
33.4°Brix	13.4						
Mango pulp juice, 16°Brix					16.0		Rao et al., (1985)
Mango pulp juice, 17°Brix					11.0		Gunjal and Waghmere (1987)
Combined Concentration & Temperature Models & Data							
Tamarind Juice, 7–62°Brix					18.7	30–70	Manohar et al. (1991)
		$K_0 = AC^b$: $A = 1.091E-11$, $b = 4.815$, C is concentration in°Brix $K = K_0\, exp(E_{ak}/RT)$; $E_{ak} = 18.7$ kJ mol⁻¹					
Raspberry juice, 15–41°Brix		,			37.9	5–60	Ibarz and Pagan, (1987)

Table A5.4 (continued)

Sample	η_∞(Pa s)	E_{av}(*kJ mol*$^{-1}$)	*Shear Rate* (*s*$^{-1}$)	K_∞ *Pa s*^{n}	E_{ak}(*kJ mol*$^{-1}$)	*Temp. Range* (°*C*)	Reference
		$K=K_\infty\ exp(E^{ak}/RT)$; $E_{ak}=37.9$ kJ mol^{-1} $K_\infty=AC^b$: $A=2.076E-15$ and $b=5.018$					
Concentrated apple juice, 71°Brix							Ibarz et al., (1987)
		η_∞:$=3.385E-10\ E_{ar}50.2$ $K=AC^b$, $A=9.304E-14$, $b=4.892$ $K=K_1\exp(BC)$, $K_{1=}7.9E-5$, $B=0.106$					
Clarified banana juice					25.1	50–750	Lopez et al. (1989)

Table A5.5 Viscoelastic parameters of tomato concentrates[a] (Yoo and Rao 1996) A. Power relationship between G' G'' and oscillatory frequency, rad s^{-1}(Yoo and Rao 1996)

Sample °Brix	Intercept	G' Slope	R^2	Intercept	G'' Slope	R^2
Concentrates from juice using 0.69 mm (0.027 in) finisher screen						
21.0	5,384	0.16	0.99	1,223	0.20	0.99
22.7	6.970	0.17	0.94	1,643	0.21	0.99
25.9	9,403	0.16	0.94	2,208	0.21	0.99
29.8	11,227	0.17	0.96	2,869	0.20	0.99
32.2	13,737	0.18	0.97	3,808	0.19	0.99
Concentrates from juice using a 0.84 mm (0.033 in) finisher screen						
21.0	8,400	0.16	0.99	2,047	0.19	0.99
24.4	11,617	0.17	0.96	2,907	0.19	0.99
30.4	17,141	0.18	0.99	4,754	0.19	0.99

[a] Numbers in parentheses are standard deviations of parameters

Table A5.6 Creep-compliance parameters (Yoo and Rao 1996)

Sample (°Brix)	J_0 (Pa^{-1})	J_1 (Pa^{-1})	τ_1 (s)	J_2 (Pa^{-1})	τ_2 (s)	η (Pa s)
Concentrates from 0.69 mm (0.027 in) finisher screen						
21.0	1.93×10^{-4}	1.37×10^{-4}	3.96	2.05×10^{-4}	24.1	1.11×10^{6}
	(5.73×10^{-5})	(1.28×10^{-4})	(1.08)	(8.99×10^{-5})	(4.07)	(5.34×10^{5})
22.7	2.04×10^{-4}	1.64×10^{-4}	2.84	1.90×10^{-4}	26.6	8.83×10^{5}
	(8.03×10^{-6})	(3.67×10^{-5})	(0.95)	(2.23×10^{-5})	(2.20)	(6.42×10^{4})
25.9	9.58×10^{-5}	4.51×10^{-5}	3.23	1.14×10^{-4}	23.2	1.76×10^{6}
	(1.26×10^{-5})	(1.42×10^{-5})	(1.10)	(3.38×10^{-5})	(4.24)	(3.67×10^{5})
29.8	8.42×10^{-5}	4.38×10^{-5}	2.83	9.56×10^{-5}	20.6	1.88×10^{6}
	(2.43×10^{-6})	(1.11×10^{-5})	(1.10)	(2.92×10^{-5})	(2.56)	(2.91×10^{5})
32.2	7.77×10^{-5}	4.60×10^{-5}	3.13	1.06×10^{-4}	21.0	1.96×10^{6}

Table A5.6 (continued)

Sample (°Brix)	J_0 (Pa^{-1})	J_1 (Pa^{-1})	τ_1 (s)	J_2 (Pa^{-1})	τ_2 (s)	η (Pa s)
	(7.72×10^{-6})	(1.46×10^{-5})	(0.47)	(3.31×10^{-5})	(1.67)	(2.21×10^{5})
Concentrates from 0.84 mm (0.033 in) finisher screen						
21.0	1.68×10^{-4}	7.22×10^{-5}	3.17	1.87×10^{-4}	20.6	9.66×10^{5}
	(3.15×10^{-5})	(1.97×10^{-5})	(0.78)	(4.31×10^{-5})	(3.94)	68(2.99×10^{5})
24.4	1.25×10^{-4}	5.27×10^{-5}	2.61	1.12×10^{-4}	19.4	1.40×10^{6}
	(2.48×10^{-5})	(1.04×10^{-5})	(0.91)	(3.00×10^{-5})	(1.12)	(2.34×10^{5})
30.4	7.51×10^{-5}	4.54×10^{-5}	3.68	7.04×10^{-5}	22.44	2.84×10^{6}
	(7.29×10^{-6})	(1.16×10^{-5})	(0.73)	(2.09×10^{-5})	(3.33)	(3.58×10^{5})

Table A5.7 Magnitudes of yield stress of apple sauce determined using the vane method and extrapolopated (Bingham) values (Qiu and Rao 1988)

	Vane[b] and RPM				
					Bingham
Sample[a]	11,1.0 rpm	11,0.4 rpm	12, 1.0 rpm	12, 0.4 rpm	Yield Value
RGA25	63.9	55.8	66.1	58.0	70.6
RGA27	58.9	53.1	55.9	51.8	66.2
RGA29	87.1	76.0	84.1	74.2	81.6
RGA45	83.7	73.7	80.9	76.9	81.0
RGA47	86.0	72.1	87.6	81.4	82.2
RGA49	55.8	52.0	54.5	51.3	73.1
RGB25	56.6	46.5	56.1	47.2	70.8
RGB27	57.4	48.1	54.0	45.9	69.2
RGB29	67.0	57.0	68.8	62.0	71.6
RGB45	51.5	45.8	52.6	47.5	62.6
RGB47	58.9	49.3	54.0	48.0	69.4
RGB49	76.0	65.1	72.3	64.2	74.4

[a] Sample code: RG stands for Rhode Island greening cultivar; A indicates firmness of 37.4 N, B for 74.3 N; the first number indicates the finisher screen opening size: 2 for 1.6 mm screen, 4 for 3.2 mm screen; the last number indicates the finisher speed (500, 700, or 900 rpm)
[b] 11 is an eight-blade vane described by Rao and Cooley (1984), 12 is a six-blade vane described in Qiu and Rao (1988)

Table A5.8 Casson model parameters of commercial chocolate samples (Chevalley 1991)

Brand Name	Method	Temp. (°C)	Shear Rate (s^{-1})	Plastic Viscosity η_∞(Pa s)	Yield Stress (Pa)
ArniCoop	Haake RV20	40	5–60	2 5	10.9
			60–5	2.5	9.1
Cailler (Nestle)			5–60	3.7	16.5
Cadbury dairy			5–60	4.4	33.7
Milk GB			60–5	4.8	20.5
Galaxy GB			5–60	4.7	78.0

Table A5.8 (continued)

Brand Name	Method	Temp. (°C)	Shear Rate (s^{-1})	Plastic Viscosity η_∞(Pa s)	Yield Stress (Pa)
			60–5	4.9	49.2
Lindt F			5–60	1.8	10.3
			60–5	1.9	8.9
Nestle F			5–60	2.6	19.8
			60–5	2.6	15.9
Poulain F			5–60	2.7	19.4
Suchard Milka			5–60	1.9	12.2
			60–5	1.9	9.4
Yorkie GB Dark chocolate			5–60	4.2	10.0
Bournville			5–60	3.9	42.9
Black magic			5–60	2.4	10.1
Cote d'Or F			5–60	4.0	23.8
Cremant			5–60	2.70	25.8
Terry plain GB			5–60	1.90	20.4
			60–5	1.90	19.0
Nestle white			5–60	4.70	14.2
Rowntree white			5–60	6.20	9.6
Suchard white			5–60	2.40	38.5

Table A5.9 Casson model parameters: Yield stress (σ_{0C}) and plastic viscosity (η_∞) of cocoa mass as a function of temperature (Fang et al. 1996)

	Predegasified sample		Degasified sample	
Temp	σ_{0C}	η_∞	σ_{0C}	η_∞
40	19.0	4.55	2.45	4.64
60	19.6	1.77	2.36	1.94
80	14.8	1.13	2.36	0.84
93	7.31	1.12	2.12	0.573
110			1.35	0.331
130			0.97	0.204

Table A5.10 Flow properties of dairy products

Product	Conc. (% solids)	Method	Temp (°C)	Shear Rate (s−1).	K[a] (Pa s^n)	n	Yield Stress (Pa)	Reference
Milk			20		0.0020	1.00		Kokini (1992)
			30		0.0015	1.00		
			40		0.0011	1.00		
			50		0.0010	1.00		
			60		0.0008	1.00		

Table A5.10 (continued)

Product	Conc. (% solids)	Method	Temp (°C)	Shear Rate (s−1).	K[a] (Pa s^n)	n	Yield Stress (Pa)	Reference
			70		0.0007	1.00		
			80		0.0006	1.00		
			0		0.0034	1.00		
			5		0.0031	1.00		
			10		0.0026	1.00		
			15		0.0023	1.00		
			20		0.0020	1.00		
			25		0.0017	1.00		
			30		0.0015	1.00		
			35		0.0013	1.00		
			40		0.0012	1.00		
Skim milk	Normal	Unknown	25		0.0014	1.00		Hayes (1987)
Whole milk	12.3							Wayne and Shoemaker (1988)
Whole milk	Normal	Unknown	20		0.0212	1.00		
Whole milk	12.3	Rotational	25		0.0018	1.00		
0.05482%	11.9	Rheometer	25		0.0018	1.00	0.021	
Butter fat milk	12.1	Weissenberg	25		0.0018	1.00	0.0211	
Skim milk	9.7	1°cone & plate	25		0.0016	1.00	0.0465	
	15.5		25		0.0024	1.00	0.0747	
	16.3		25		0.0026	1.00	0.0357	
	17.7		25		0.0028	1.00	0.0268	
	18.9		25		0.0031	1.00	0.0341	
	20.2		25		0.0034	1.00	0.0371	
Stirred yoghurts								
0.05% fat								
	12.5	Rheomat 30 & Carrimed CS 100 (cone & plate)	10	15–300	8.7000	0.250	2	Benezech and Maingonnat (1993)
	11.6		10	15–300	13.4000	0.160	2	
	12.0		10	15–300	8.9000	0.250	2	
1.6% fat	16.0		10	15–300	14.3000	0.195	2	
Yoplait	Unknown	Haake	10	0–500	1.4300	0.531	6.94	Ramaswamy and Basak (1991)
			15	0–500	1.2100	0.532	7.16	

Table A5.10 (continued)

Product	Conc. (% solids)	Method	Temp (°C)	Shear Rate (s−1).	K[a] (Pa s[n])	n	Yield Stress (Pa)	Reference
			20	0–500	0.8600	0.568	6.34	
			25	0–500	0.6100	0.596	5.45	
Delisle	Unknown		10	0–500	3.6200	0.418	13.7	
Whey protein solution								
	10	Bohtin VOR						
	20	(cone and plate)	22	0–230	2.7700	1.00		
	30				12.2	0.960		
					62.1	0.950		
Whole Milk Retentates								Sierzant and Smith (1993)
6.29 % protein, 6.88 % fat, 1.03 % ash, 6.08 % lactose 20.27 Haake Rotovisco			5	1–1170	0.01084	0.97		
6.44 % protein, 7.06 % fat, 1.03 % ash, 5.96 lactose 20.49			5		0.01109	0.96		
6.49 % protein, 7.03 % fat, 1.06 % ash, 6.45 % lactose 21			5		0.01862	0.90		
10.19 % protein, 10.12 % fat, 1.36 % ash, 6.46 % lactose 28.13			5		0.04074	0.95		
10.70 % protein, 10.60 % fat, 1.42 % ash, 6.90 % lactose 29.62			5		0.04841	0.93		
9.83 % protein, 10.33 % fat, 1.06 % ash, 6.45 % lactose 28.82			5		0.0389	0.94		
10.19 % protein, 10.12 % fat, 1.36 % ash, 6.46 % lactose 28.13			50		0.00385	1.03		
10.70 % protein, 10.60 % fat, 1.42 % ash, 6.90 % lactose 29.62			50		0.00359	1.04		
9.83 % protein, 10.33 % fat, 1.06 % ash, 6.45 % lactose 28.82 50				0.004416	1.01			
10.19 % protein, 10.12 % fat, 1.36 % ash, 6.46 % lactose 28.13 75				0.00385	1.18			

Table A5.10 (continued)

Product	Conc. (% solids)	Method	Temp (°C)	Shear Rate (s−1).	K[a] (Pa s[n])	n	Yield Stress (Pa)	Reference
10.70 % protein, 10.60 % fat, 1.42 % ash, 6.90 % lactose 29.62 75				0.00359	1.090			
9.83 % protein, 10.33 % fat, 1.06 % ash, 6.45 % lactose 28.82 75				0.00441	1.130			
11.63 % protein, 14.4 % fat, 1.67 % ash, 8.28 % lactose 35.99 5				0.6607	0.810			
11.07 % protein, 13.02 % fat, 1.60 % ash, 8.50 % lactose 34.19 5				0.3715	0.810			
10.48 % protein, 13.10 % fat, 1.61 % ash, 9.62 % lactose 34.61 5				0.3758	0.810			
11.63 % protein, 14.4 % fat, 1.67 % ash, 8.28 % lactose 35.99 50				0.0254	0.900			
11.07 % protein, 13.02 % fat, 1.60 % ash, 8.50 % lactose 34.19 50				0.01428	0.950			
10.48 % protein, 13.10 % fat, 1.61 % ash, 9.62 % lactose 34.61 50				0.008128	1.040			
11.63 % protein, 14.4 % fat, 1.67 % ash, 8.28 % lactose 35.99 75				0.1778	0.690			
11.07 % protein, 13.02 % fat, 1.60 % ash, 8.50 % lactose 34.19 75				0.1884	0.650			
10.48 % protein, 13.10 % fat, 1.61 % ash, 9.62 % lactose 34.61 75				0.6918	0.540			
Dulce de leche milk household sample								
71, pH=6.1 Haake 25			1.35–28.35	28.38	0.6	45.45		Rovedo et al. (1991)
71 Conc cyl. 40				15.69	0.61	18.28		
71 MV11 55				5.7	0.65	9.65		
Confectionery sample								
71, pH=6.1 Haake 25				78.3	0.43	4.57		
71 Conc cyl. 40				36.7	0.47	4.85		
71 MV11 55				15.76	0.47	0.0		

Effect of temperature on concentrated freeze skim milk (Chang and Hartel 1997) viscosity data obtained over the shear rate range, 0.1–14 s^{-1} and temperature range 20 to −3.4 °C

Total Solids (%)	η_∞(Pa. s)	E_a (kJ mol^{-1})
9.56	5.62*E*–8	25.1
15	1.92 *E*–9	28.9
20	5.93*E*–8	32.9
25	2 83*E*–8	30.7
30	4.68*E*–10	42.0
35	1.03*E*–11	52.8
40	1.68*E*–12	60.3

Table A5.11 The effect of temperature and concentration on the flow behavior of milk concentrates (Velez-Ruiz and Barbosa-Canovas. 1998)

Concentration (% w/w)	Flow Behavior Index (dimensionless)			Consistency Coefficient (Pa s^n)			Yield Stress (Pa)		
	5 °C	15 °C	25 °C	5 °C	15 °C	25 °C	5 °C	15 °C	25 °C
12.6	1.0	1.0	10	0.003	0.002	0.002			
17.2	1.0	1.0	1.0	0.005	0.003	0.003			
19.6	1.0	1.0	1.0	0.006	0.004	0.003			
22.3	1.0	1.0	1.0	0.007	0.005	0.004			
24.9	0.97	0.97	0.95	0.013	0.008	0.007			
30.5	0.96	0.94	0.94	0.026	0.018	0.011			
42.4	0.91	0.93	0.89	0.171	0.077	0.069	0.097	0.072	0.018
48.6	0.10	0.10	0.54	41.3	40.1	2.09	0.204	0.218	0.061

The combined effect of temperature and concentration on the consistency coefficient of milk concentrates: In $K = \beta_0 + \beta_1(\tau) - \beta_2 x$, where X is concentration of milk (w/w, %) (Velez-Ruiz and Barbosa-Canovas 1998)

Storage Time (weeks)	β_0	β_1	β_2	R^2
1	–18.22	2.973	0.1278	0.98
2	–18.68	2.992	0.1502	0.97
3	–21.18	3.206	0.2178	0.95
4	–24.77	3.982	0.2622	0.88

[a] Some values were converted from cp or mPa s units

Table A5.12 Formulations of three mayonnaises by Kiosseoglu and Sherman (1983)

	% wt/wt		
Ingredient	1	2	3
Egg yolk	6.00	3.99	6.00
Egg white	—	8.28	12.00
Water	12.00	5.73	—
Acetic acid	0.20	0.20	0.20
Sugar	1.55	1.55	1.55
Salt	0.25	0.25	0.25
Oil	80.00	80.00	80.00

Table A5.13 Viscoelastic parameters of model salad dressing during storage at 2.8 °C (Paredes et al. 1989)

Storage Time (Days)	E_0 (dyne $cm^{-2} \times 10^{-3}$)	E_1 (dyne $cm^{-2} \times 10^{-3}$)	η_1 (poise $\times 10^{-3}$)	η_N (poise $\times 10^{-3}$)
1	4.17	1.08	7.5	74.4
3	1.64	1.92	16.7	114.6
7	1.49	2.38	31.3	51.6
14	0.72	0.89	52.8	154.2
21	0.65	0.43	132.8	492.4
28	0.63	1.00	180.0	493.8
42	4.17	1.45	15.0	303.0

Table A5.14 Rheological properties of mustard, mayonnaise, oils, and fats

Item	Temp. (°C)	Shear Rate (s^{-1})	Consistency[a] Index (Pa S^n)	Flow Behavior Index	Reference
Mustard	25	30–1300	18.50	0.39	Kokini (1992)
	25	30–1300	19.10	0.39	
	25	40–1100	27.00	0.34	
	25	40–1100	33.00	0.28	
Castor	10		2.42	1.0	Kokini (1992)
	30		0.451	1.0	
	40		0.231	1.0	
	100		0.017	1.0	
Corn	38		0.317	1.0	Kokini (1992)
	25		0.057	1.0	
Cottonseed	20		0.070	1.0	Kokini (1992)
	38		0.039	1.0	
Linseed	50		0.018	1.0	Kokini (1992)
	90		0.007	1.0	
Olive	10		0.138	1.0	Kokini (1992)
	40		0.036	1.0	
	70		0.012	1.0	
Peanut	25		0.066	1.0	
	38		0.025	1.0	

Table A5.14 (continued)

Item	Temp. (°C)	Shear Rate (s^{-1})	Consistency[a] Index (Pa S^n)	Flow Behavior Index	Reference
	21.1	0.32–64	0.065	1.0	
	37.8	0.32–64	0.039	1.0	
	54.4	0.32–64	0.027	1.0	
Rapeseed	0		2.530	1.0	
	20		0.163	1.0	
	30		0.096	1.0	
Safflower	38		0.029	1.0	
	25		0.092	1.0	
Sesame	38		0.032	1.0	
Soybean	30		0.041	1.0	
	50		0.021	1.0	
	90		0.008	1.0	
Sunflower	38		0.031	1.0	
Mayonnaise	25	30–1300	6.40	0.55	
	25	30–1300	6.60	0.54	
	25	40–1100	4.20	0.60	
	25	40–1100	4.70	0.59	
Cream (10% fat)	40		0.00148	1.0	Kokini (1992)
	60		0.00107	1.0	
	80		0.00083	1.0	
Cream (20% fat)	40		0.00238	1.0	
	60		0.00171	1.0	
	80		0.00129	1.0	
Cream (30% fat)	40		0.00395	1.0	
	60		0.00289	1.0	
	80		0.0022	1.0	
Cream (40% fat)	40		0.0069	1.0	
	60		0.0051	1.0	
	80		0.00395	1.0	
Canola	4		0.1627	1.0	Lang et al. (1992)
	10		0.1172	1.0	
	25		0.057	1.0	
	40		0.0334	1.0	
Refined bleached	4		0.1568	1.0	
winterized	10		0.1197	1.0	
	25		0.0589	1.0	
	40		0.0333	1.0	
Crambe	23.9		0.0867	1.0	Noureddini et al. (1992)
	37.8		0.0487	1.0	
	48.9		0.0330	1.0	
	60.0		0.0234	1.0	

Table A5.14 (continued)

Item	Temp. (°C)	Shear Rate (s^{-1})	Consistency[a] Index (Pa S^n)	Flow Behavior Index	Reference
	82.2		0.0132	1.0	
	100		0.00906	1.0	
	110		0.00755	1.0	
Rapeseed	23.9		0.0788	1.0	
	37.8		0.0449	1.0	
	48.9		0.0303	1.0	
	600		0.0214	1.0	
	82.2		0.0121	1.0	
	100		0.00841	1.0	
	110		0.00715	1.0	
Corn	23.9		0.0523	1.0	
	37.8		0.0308	1.0	
	48.9		0.0227	1.0	
	60		0.0157	1.0	
	82.2		0.00928	1.0	
	100		0.00657	1.0	
	110		0.0056	1.0	
Soybean	23.9		0.0543	1.0	
	37.8		0.0318	1.0	
	48.9		0.0233	1.0	
	60		0.00161	1.0	
	82.2		0.0951	1.0	
	100		0.00673	1.0	
	110		0.00571	1.0	
Milkweed	23.9		0.0563	1.0	
	37.8		0.0327	1.0	
	48.9		0.0238	1.0	
	60.0		0.0164	1.0	
	82.2		0.0096	1.0	
	100		0.00679	1.0	
	110		0.00571	1.0	
Coconut	37.8		0.028	1.0	
	48.9		0.0198	1.0	
	60		0.0133	1.0	
	82.2		0.00755	1.0	
	100		0.00522	1.0	
	110		0.00436	1.0	
Lesquerella	23.9		0.275	1.0	
	37.8		0.121	1.0	
	48.9		0.0698	1.0	
	60		0.0435	1.0	
	82.2		0.0204	1.0	
	100		0.0127	1.0	
	110		0.0102	1.0	
Nonanoic	23.9		0.00725	1.0	
	37.8		0.00497	1.0	
	48.9		0.00392	1.0	
	60		0.00303	1.0	

Table A5.14 (continued)

Item	Temp. (°C)	Shear Rate (s^{-1})	Consistency[a] Index (Pa S^n)	Flow Behavior Index	Reference
	71.1		0.00248	1.0	
	82.2		0.00202	1.0	
	98.9		0.00156	1.0	
	110		0.00137	1.0	
Capric	37.8		0.00633	1.0	
	48.9		0.00496	1.0	
	60		0.00372	1.0	
	71.1		0.00302	1.0	
	82.2		0.00242	1.0	
	98.9		0.00184	1.0	
	110		0.0016	1.0	
Laurie	48.9		0.00747	1.0	
	60		0.00542	1.0	
	71.1		0.00428	1.0	
	82.2		0.00338	1.0	
	98.9		0.0025	1.0	
	110		0.0021	1.0	
Myristic	60		0.00748	1.0	
	71.1		0.00583	1.0	
	82.2		0.00450	1.0	
	98.9		0.00326	1.0	
	110		0.00278	1.0	
Palmitic	71.1		0.00764	1.0	
	82.2		0.00580	1.0	
	98.9		0.00413	1.0	
	110		0.00347	1.0	
Stearic	82.2		0.00731	1.0	
	98.9		0.00508	1.0	
	110		0.00424	1.0	
	121.1		0.00350	1.0	
Oleic	23.9		0.02900	1.0	
	37.8		0.01770	1.0	
	48.9		0.01320	1.0	
	60		0.00932	1.0	
	71.1		0.00727	1.0	
	82.2		0.00561	1.0	
	98.9		0.00403	1.0	
	110		0.00341	1.0	
Erucic	37.8		0.03230	1.0	
	48.9		0.02240	1.0	
	60		0.01590	1.0	
	71.1		0.01190	1.0	
	82.2		0.00890	1.0	
	98.9		0.00618	1.0	
	110		0.00516	1.0	
Palm oil	35.0		0.0963	1.0	Timms (1985)
	40.0		0.0404	1.0	
	45.0		0.0333	1.0	
	50.0		0.0270	1.0	

Table A5.14 (continued)

Item	Temp. (°C)	Shear Rate (s^{-1})	Consistency[a] Index (Pa S^n)	Flow Behavior Index	Reference
Palm olein	20.0		0.0853	1.0	
	25.0		0.0692	1.0	
	30.0		0.0563	1.0	
	35.0		0.0434	1.0	
	40.0		0.0386	1.0	
	45.0		0.0326	1.0	
	50.0		0.0265	1.0	
Palm kernel	30.0		0.0440	1.0	
	35.0		0.0356	1.0	
	40.0		0.0300	1.0	
	45.0		0.0254	1.0	
	50.0		0.0210	1.0	
Coconut	25.0		0.0481	1.0	
	30.0		0.0398	1.0	
	35.0		0.0329	1.0	
	40.0		0.0274	1.0	
	45.0		0.0233	1.0	
	50.0		0.0190	1.0	
Soybean	20.0		0.0636	1.0	
	25.0		0.0516	1.0	
	30.0		0.0431	1.0	
	35.0		0.0368	1.0	
	40.0		0.0317	1.0	
	45.0		0.0268	1.0	
	50.0		0.0225	1.0	
Peanut oil	30.0		0.0345	1.0	Krishna (1993)
Peanut +6.8 % free fatty acids	30.0		0.0360	1.0	
Peanut +3 % rice bran wax	30.0		0.0626	1.0	
Peanut +1 % oryzanol	30.0		0.0374	1.0	
Peanut +1 % glyceride monostearate	30.0		0.0375	1.0	
Peanut +2 % soy lecithin	30.0		0.0414	1.0	
Rice bran oil triglyceride	30.0		0.0366	1.0	
Comm. rice bran oil	30.0		0.0370	1.0	
Cocoa butter using Ubbelohde type glass					Sterling and Wuhrmann,

Table A5.14 (continued)

Item	Temp. (°C)	Shear Rate (s^{-1})	Consistency[a] Index (Pa S^n)	Flow Behavior Index	Reference
capillary viscometer					1960
	30.0		0.0722	1.0	
	28.0		0.0864	1.0	
	27.0		0.0986	1.0	
	26.0		0.1114	1.0	
	25.0		0.1241	1.0	
	24.0		0.1380	1.0	

[a] Some values were converted from cp or mPa s units

Table A5.15 Effect of temperature on rheological properties of oils and fats (Lang et al. 1992)

Oil	Rheol. Method	Temp Range	Temperature Model	Model Constants
Canola	Cannon-Fenske	4–100 °C	$\vartheta = \exp(C_0 + C_1 T - C_2 T^2)$ C_0 = 0.0519, C_1 = 0.0002	
(refined)			ϑ = kinematic viscosity, [$(mm)^2 s^{-1}$]	
Crambe	Cannon-Fenske	23.9–110 °C	$\ln \eta = A - \frac{B}{T+C}$ η = viscosity (m Pa s), T(K)	A = –2.0277E + 00, B = 9.2659E + 02, C = 1.5427E + 02
Rapeseed	Cannon-Fenske	23.9–110 °C	same	A = –2.44621 E + 01, B = 6.21021 E + 03, C = 2.6671 E – 02
Corn	Cannon-Fenske	23.9–110 °C	same	A = –1.95604E + 01, B = 5.22544E + 03, C = 1.9943E – 02
Soybean	Cannon-Fenske	23.9–110 °C	same	A = –2.00596E + 01, B = 5.32374E + 03, C = 2.0634E – 02
Milkweed	Cannon-Fenske	23.9–110 °C	same	A = –2.03845E + 01, B = 5.40303E + 03, C = 2.0948E – 02
Coconut	Cannon-Fenske	23.9–110 °C	same	A = 1.0945E + 01, B = –2.25154E + 03, C = 9.1778E + 05
Lesquerella	Cannon-Fenske	23.9–110 °C	same	A = 6.6123E + 00, B = –6.28877E + 03, C = 1.7800E + 06
Nonanoic	Cannon-Fenske	23.9–110 °C	same	A = –1.28399E + 01, B = 3.44593E + 03, C = 1.0839E – 02
Capric	Cannon-Fenske	37.8–110 °C	same	A = –2.6280E + 00, B = 7.1762E + 02, C = –1.5129E + 02

Table A5.15 (continued)

Oil	Rheol. Method	Temp Range	Temperature Model	Model Constants
Lauric	Cannon-Fenske	48.9–110 °C	same	A=–2.5030E+00, B=7.0888E+02, C=–1.6475E+02
Myristic	Cannon-Fenske	60.0–110 °C	same	A=6.1590E – 01, B=–1.94571 E+03, C=8.0399E+05
Palmitic	Cannon-Fenske	71.1–110 °C	same	A=–3.43507E+01, B=7.86679E+03, C=3.9312E – 02
Stearic	Cannon-Fenske	82.2–110 °C	same	A=2.6780E+00, B=–3.43749E+03, C=1.1344E+06
Oleic	Cannon-Fenske	23.9–110 °C	same	A=–1.64441 E+01, B=4.55414E+03, C=1.5075E – 02
Erucic	Cannon-Fenske	37.8–110 °C	same	A=–1.82905E+01, B=5.10332E+03, C=1.7229E – 02

Table A5.16 Rheological properties of egg products, and activation energies of flow (Rao 1977)

Name of Product	Conc. (% solids)	pH	Method	Temp (°C)	K (Pa s^n)	n	σ_0 (Pa)	Ea (kJ mol^{-1})
Whole egg	25.0	7.5	capillary visc	30	0.0064	1.0		24.7
Stabilized egg white	12.0	7.0	capillay visc	30	0.0027	1.0		16.2
Plain yolk	45.0	6.8	capillary visc	30	0.0920	1.0		26.8
Sugared yolk (10 % sucrose)	50	6.3	capillary visc	30	0.0560	1.0		27.5
Salted yolk (10 % salt)	50.0	6.0	capillary visc	30	0.0400	1.0		34.9

Table A5.17 Rheological properties of meats

Product	Conc. (% solids)	Method	Temp (°C)	Shear rate (s^{-1})	K (Pa s^n)	n	Yield Stress (Pa)	Reference
Minced beef	26.8	Haake MV1	20	4.7–1694	4.67	0.33	4.47	Bianchi et al. (1985)
21.7 % protein								
4.7 % fat								
73.2 % water								

Table A5.17 (continued)

Product	Conc. (% solids)	Method	Temp (°C)	Shear rate (s^{-1})	K (Pa s^n)	n	Yield Stress (Pa)	Reference
Minced beef/soy protein mixtures								
9: 1	26.8				2.1	0.49	4.4	
7:3	26.8				9.2	0.30	7.3	
5:5	26.8				12.0	0.31	18	

Table A5.18 Rheological properties of foods containing sugars

Product	Conc. (% solids)	Method	Temp. (°C)	Shear Rate (s^{-1})	K^a (Pa s^n)	n	Reference
Corn syrup	48.4	Cone. cyl.	26.6	1–1000	0.0053	1.0	Harper (1960)
	54.5				0.0106	1.0	
	63.6				0.0338	1.0	
	72.5				0.1600	1.0	
	29.6				0.0205	1.0	
	36.4				0.0228	1.0	
	40.8				0.0290	1.0	
	36.9				0.0245	1.0	
	41.5				0.0228	1.0	
	23.0				0.0122	1.0	
	25.0				0.0146	1.0	
	29.0				0.0151	1.0	
	31.0				0.0149	1.0	
	30.0				0.0148	1.0	
	36.2				0.0186	1.0	
	39.4				0.0218	1.0	
	42.0				0.0240	1.0	
Corn syrup	77.4° Brix	Cone cyl.	20	1–10	5.46	1.0	McCarthy and Seymour (1994b)
Honey	18.2		8.9		52.8	1.0	Munro (1943)
			9.8		46.8	1.0	
			10.9		40.3	1.0	
			12.1		34.08	1.0	
			12.8		30.00	1.0	
			13.4		27.12	1.0	
			14.2		23.76	1.0	
			14.8		20.88	1.0	
			15.8		17.04	1.0	
			16.8		14.4	1.0	
			17.6		12.24	1.0	
			18.3		10.8	1.0	
			19.1		9.6	1.0	
			20.1		8.4	1.0	

Table A5.18 (continued)

Product	Conc. (% solids)	Method	Temp. (°C)	Shear Rate (s^{-1})	K^a (Pa s^n)	n	Reference
			22.0		6.72	1.0	
			23.3		5.76	1.0	
			25.0		4.8	1.0	
			28.6		3.24	1.0	
			30.8		2.4	1.0	
			34.5		1.87	1.0	
			36.8		1.51	1.0	
			40.0		1.20	1.0	
			41.6		0.98	1.0	
			42.8		0.86	1.0	
			43.3		0.82	1.0	
			44.5		0.80	1.0	
			45.7		0.75	1.0	
			47.1		0.68	1.0	
			49.7		0.58	1.0	
			51.8		0.52	1.0	
			56.5		0.37	1.0	
			60.1		0.30	1.0	
			63.9		0.24	1.0	
			67.4		0.19	1.0	
			71.6		0.15	1.0	
			75.9		0.11	1.0	
			78.5		0.10	1.0	
			80.0		0.09	1.0	
Buckwheat	18.6		3.9		70.56	1.0	
			5.2		61.20	1.0	
			7.0		52.32	1.0	
			8.1		46.56	1.0	
			9.3		40.80	1.0	
			10.3		36.96	1.0	
			11.1		33.12	1.0	
			12.2		27.60	1.0	
			12.7		25.20	1.0	
			13.8		20.16	1.0	
			15		16.20	1.0	
			16		12.82	1.0	
			17.2		10.32	1.0	
			18.1		8.88	1.0	
			18.5		8.16	1.0	
			19.2		7.44	1.0	
			20.5		6.24	1.0	
			21.9		5.28	1.0	
			22.9		4.68	1.0	
			24.8		3.86	1.0	

Table A5.18 (continued)

Product	Conc. (% solids)	Method	Temp. (°C)	Shear Rate (s^{-1})	K^a (Pa s^n)	n	Reference
			26.5		3.26	1.0	
			28.1		2.83	1.0	
			31		2.16	1.0	
			36.6		1.44	1.0	
			46.9		0.72	1.0	
			48.7		0.61	1.0	
			54		0.44	1.0	
			58		0.34	1.0	
			65.6		0.23	1.0	
			66.9		0.20	1.0	
			69.4		0.17	1.0	
			72.2		0.15	1.0	
			74.1		0.14	1.0	
			76.2		0.14	1.0	
			77.8		0.14	1.0	
			78.9		0.15	1.0	
			79.6		0.15	1.0	
Goldenrod	19.4		6.6		33.6	1.0	
			7.7		30.0	1.0	
			8.8		26.18	1.0	
			9.5		23.52	1.0	
			10.6		19.92	1.0	
			10.9		15.60	1.0	
			12.7		12.96	1.0	
			13.5		11.28	1.0	
			14		10.32	1.0	
			14.4		9.36	1.0	
			15.7		7.92	1.0	
			16.1		7.44	1.0	
			16.9		6.48	1.0	
			18.0		6.0	1.0	
			19.9		5.04	1.0	
			21.6		3.96	1.0	
			22.9		3.38	1.0	
			24.3		2.93	1.0	
			26.7		2.35	1.0	
			29.8		1.82	1.0	
			32.4		1.51	1.0	
			34.8		1.20	1.0	
			37.0		0.93	1.0	
			39.0		0.82	1.0	
			40.5		0.75	1.0	
			43.4		0.60	1.0	

Table A5.18 (continued)

Product	Conc. (% solids)	Method	Temp. (°C)	Shear Rate (s^{-1})	K^a (Pa s^n)	n	Reference
			46.7		0.49	1.0	
			48.7		0.41	1.0	
			52.4		0.32	1.0	
			55.8		0.26	1.0	
			59.8		0.21	1.0	
			64.0		0.17	1.0	
			68.0		0.14	1.0	
			71.4		0.12	1.0	
			74.9		0.10	1.0	
			78.2		0.09	1.0	
			80.4		0.07	1.0	
Sweetclover	17		12.0		43.20	1.0	
			13.0		37.44	1.0	
			14.8		30.24	1.0	
			15.7		25.80	1.0	
			16.8		21.12	1.0	
			17.3		18.96	1.0	
			18.6		15.60	1.0	
			19.6		13.68	1.0	
			21.1		11.28	1.0	
			23.0		8.88	1.0	
			24.7		7.20	1.0	
			26.5		6.00	1.0	
			28.2		4.80	1.0	
			30.0		4.08	1.0	
			31.8		3.36	1.0	
			34.6		2.52	1.0	
			35.8		2.16	1.0	
			37.6		1.97	1.0	
			38.7		1.87	1.0	
			40.4		1.56	1.0	
			43.1		1.25	1.0	
			45.9		1.01	1.0	
			48.3		0.86	1.0	
			50.4		0.77	1.0	
			54.5		0.63	1.0	
			58.4		0.52	1.0	
			62.0		0.44	1.0	
			65.9		0.34	1.0	
			68.2		0.28	1.0	
			70.6		0.25	1.0	
			73.4		0.19	1.0	
			75.6		0.17	1.0	

Table A5.18 (continued)

Product	Conc. (% solids)	Method	Temp. (°C)	Shear Rate (s^{-1})	K^a (Pa s^n)	n	Reference
			77.9		0.14	1.0	
			80.1		0.12	1.0	
Sage	18.6		11.7		72.96	1.0	
			12.0		66.72	1.0	
			12.2		63.12	1.0	
			12.5		59.52	1.0	
			12.8		56.64	1.0	
			12.9		54.24	1.0	
			13.5		49.44	1.0	
			13.9		44.88	1.0	
			14.2		42.00	1.0	
			14.6		39.36	1.0	
			15.1		35.76	1.0	
			15.8		31.56	1.0	
			16.2		29.52	1.0	
			16.8		27.12	1.0	
			17.3		25.20	1.0	
			17.9		23.52	1.0	
			18.8		21.60	1.0	
			20.2		18.48	1.0	
			21.6		15.36	1.0	
			23.4		12.24	1.0	
			25.9		8.88	1.0	
			27.8		6.96	1.0	
			30.7		5.52	1.0	
			32.1		4.56	1.0	
			34.0		3.84	1.0	
			36.2		3.12	1.0	
			38.8		2.40	1.0	
			40.9		1.92	1.0	
			42.0		1.64	1.0	
			44.1		1.48	1.0	
			45.7		1.30	1.0	
			48.0		1.15	1.0	
			50.7		0.93	1.0	
			53.4		0.76	1.0	
			56.0		0.62	1.0	
			60.1		0.48	1.0	
			63.6		0.39	1.0	
			67.4		0.30	1.0	
			72.0		0.23	1.0	
			73.6		0.21	1.0	

Table A5.18 (continued)

Product	Conc. (% solids)	Method	Temp. (°C)	Shear Rate (s^{-1})	K^a (Pa s^n)	n	Reference
			75.5		0.19	1.0	
			76.8		0.17	1.0	
			77.6		0.16	1.0	
			78.7		0.15	1.0	
			80.5		0.14	1.0	
Sucrose solution	20	Unknown	0.0		0.003818	1.0	Perry (1950)
			5.0		0.003166	1.0	
			10.0		0.002662	1.0	
			15.0		0.002275	1.0	
			20.0		0.001967	1.0	
			25.0		0.001710	1.0	
			30.0		0.001510	1.0	
			35.0		0.001336	1.0	
			40.0		0.001197	1.0	
			45.0		0.001074	1.0	
			50.0		0.000974	1.0	
			55.0		0.000887	1.0	
			60.0		0.000811	1.0	
			65.0		0.000745	1.0	
			70.0		0.000688	1.0	
			75.0		0.000637	1.0	
			80.0		0.000892	1.0	
			85.0		0.000552	1.0	
	40		0.0		0.014820	1.0	
			5.0		0.011600	1.0	
			10.0		0.009830	1.0	
			15.0		0.007496	1.0	
			20.0		0.006223	1.0	
			25.0		0.005206	1.0	
			30.0		0.004398	1.0	
			35.0		0.003776	1.0	
			40.0		0.003261	1.0	
			45.0		0.002858	1.0	
			50.0		0.002506	1.0	
			55.0		0.002277	1.0	
			60.0		0.001989	1.0	
			65.0		0.001785	1.0	
			70.0		0.001614	1.0	
			75.0		0.001467	1.0	
			80.0		0.001339	1.0	
			85.0		0.001226	1.0	
			90.0		0.001127	1.0	
			95.0		0.001041	1.0	

Table A5.18 (continued)

Product	Conc. (% solids)	Method	Temp. (°C)	Shear Rate (s^{-1})	K[a] (Pa s^n)	n	Reference
	60		10.0		0.113900	1.0	
			15.0		0.074900	1.0	
			20.0		0.056700	1.0	
			25.0		0.044020	1.0	
			30.0		0.034010	1.0	
			35.0		0.026620	1.0	
			40.0		0.021300	1.0	
			45.0		0.017240	1.0	
			50.0		0.014060	1.0	
			55.0		0.011710	1.0	
			60.0		0.009870	1.0	
			65.0		0.008370	1.0	
			70.0		0.007180	1.0	
			75.0		0.006220	1.0	
			80.0		0.005420	1.0	
			85.0		0.004750	1.0	
			90.0		0.004170	1.0	
			95.0		0.003730	1.0	
Molasses			21.0		6.600	1.0	Hayes (1987)
			37.8		1.872	1.0	
			49.0		0.920	1.0	
			66.0		0.374	1.0	

[a] Some values were converted from cp or mPa s units

Table A5.19 Rheological properties of alcohols and wines, and effect of temperature (activation energy, E_a)

Name of product	Sugar (%)	Alcohol (%)	Method	Temp. (°C)	Shear Rate (s^{-1})	K[a] (mPa s^n)	n	Reference
Ethanol density = 760kg/m^3				20		1.2	1.0	Hayes (1987)
Methyl alcohol				0		0.81	1.0	Hayes (1987)
density = 810 kg/m^3				20		0.59	1.0	
WinesModel solutions Model 1, %m/v glycerin = 0								
	0.0	12.0	Rheomat RM-15	25		1.36	1.0	Rao et al. (1977)
E_a = 20.2 kJ mol^{-1} Model 2, %m/v glycerin = 1.3								
	1.0	12.0				1.32	1.0	

Table A5.19 (continued)

Name of product	Sugar (%)	Alcohol (%)	Method	Temp. (°C)	Shear Rate (s^{-1})	K[a] (mPa s^n)	n	Reference
E_a = 23.1 kJ mol^{-1} Model 3, %m/v glycerin = 1.3	5.0	12.0				1.30	1.0	
E_a = 25.0 kJ mol^{-1} Model 4, %m/v glycerin = 1.3	10.0	12.0				2.06	1.0	
E_a = 26.7 kJ mol^{-1} White Wines Rhine, %m/v glycerin = 0.78	2.7	12.6				1.55	1.0	
E_a = 22.0 kJ mol^{-1} Aurora Blanc, %m/v glycerin = 1.06	1.2	11.5				1.41	1.0	
E_a = 23.3 kJ mol^{-1} Elvira, %m/v glycerin = 0.76	0.33	12.6				1.44	1.0	
E_a = 23.8 kJ mol^{-1} Emerald Riesling, %m/v glycerin = 0.87	0.64	10.9				1.46	1.0	
E_a = 24.7 kJ mol^{-1} Chablis, %m/v glycerin = 0.71	0.21	11.1				1.43	1.0	
E_a = 18.8 kJ mol^{-1} Red Wines Ives, %m/v glycerin = 0.76	0.4	11.20				1.37	1.0	
E_a = 23.7 kJ mol^{-1} Baco Noir, %m/v glycerin = 0.90	0.05	12.40				1.52	1.0	
E_a = 21.5 kJ mol^{-1} Rubicon, %m/v glycerin = 0.69	3.35	12.61				1.48	1.0	
E_a = 22.3 kJ mol^{-1} Hybrid chancellor, %m/v glycerin = 0.72	0.05	12.20				1.45	1.0	

Table A5.19 (continued)

Name of product	Sugar (%)	Alcohol (%)	Method	Temp. (°C)	Shear Rate (s^{-1})	K[a] (mPa s^n)	n	Reference
$E_a = 19.4$ kJ mol^{-1} Ruby Cabernet, %m/v glycerin = 0.64								
	0.83	11.30				1.48	1.0	
$E_a = 19.9$ kJ mol^{-1} Burgundy, %m/v glycerin = 1.32								
	1.53	12.77				1.43	1.0	
$E_a = 19.1$ kJ mol^{-1} Rosé Wines Catawba, %m/v glycerin = 0.85								
	6.7	12.2				1.67	1.0	
$E_a = 24.7$ kJ mol^{-1} Fortified Wines Cream sherry, %m/v glycerin = 0.67								
		13.4	16.6			2.45	1.0	
$E_a = 23.5$ kJ mol^{-1} Cream sherry, %m/v glycerin = 0.37								
	11.63	17.1				1.95	1.0	
$E_a = 28.2$ kJ mol^{-1}								
Beer, density = 1000 kg/m^3					0.0	1.30	1.0	Hayes (1987)
Chardonnay wine, density = 1010					14	3.90	1.0	Lopez et al. (1989)
$E_a = 14.2$ kJ mol^{-1}					16	3.74	1.0	
					20	3.45	1.0	
					22	3.32	1.0	
Chardonnay wine, density = 1000					14	3.82	1.0	
					16	3.67	1.0	
					20	3.38	1.0	
					22	3.25	1.0	
Chardonnay wine, density = 995					14	3.78	1.0	
					16	3.63	1.0	
					20	3.35	1.0	
					22	3.22	1.0	

[a] Some values were converled from cp or mPa s units

Table A5.20 Storage modulus (G′, Pa) and firmness (N) of set-type and stirred yogurts obtained from ultra-high pressure homogenization or heat-treated (HT) milk. (*Source:* Serra et al., 2009)

Parameters	Storage time (days)				
	1	7	14	21	28
Modulus, set-type yogurt					
HT + SMP[1]	1481$^{a/E}$	1633$^{a/C}$	1754.5$^{a/A}$	1699$^{a/B}$	1552.1$^{a/D}$
200 MPa	1121$^{b/C}$	1293$^{b/B}$	1342$^{b/B}$	1312.3$^{b/B}$	1417$^{b/A}$
300 MPa	992$^{c/C}$	1181$^{c/B}$	1212$^{c/B}$	1179.3$^{c/B}$	1255.3$^{c/A}$
Modulus, stirred yogurt					
HT +SMP[1]	177.1$^{c/C}$	189.85$^{c/B}$	202.96$^{c/A}$	180.91$^{c/BC}$	184.5$^{c/BC}$
200 MPa	246.48$^{b/B}$	264.36$^{b/AB}$	274.13$^{b/A}$	264.16$^{b/AB}$	260.16$^{b/AB}$
300 MPa	295.95$^{a/C}$	337.85$^{a/A}$	315.73$^{a/B}$	339.4$^{a/A}$	356.31$^{a/A}$
Firmness, set-type yogurt					
HT +SMP[1]	1.42$^{a/A}$	1.42$^{a/A}$	1.43$^{a/A}$	1.37$^{a/B}$	1.42$^{a/A}$
200 MPa	1.83$^{b/AB}$	1.79$^{b/B}$	1.81$^{b/B}$	1.86$^{b/A}$	1.87$^{b/A}$
300 MPa	1.99$^{c/D}$	2.13$^{c/C}$	2.27$^{c/AB}$	2.33$^{c/A}$	2.23$^{c/B}$
Strain (stirred yogurt)					
HT +SMP	3.37$^{a/B}$	3.76$^{a/B}$	3.69$^{a/B}$	3.98$^{a/A}$	3.58$^{a/B}$
200 MPa	4.92$^{b/A}$	4.52$^{b/B}$	4.52$^{b/B}$	4.47$^{b/B}$	4.87$^{b/A}$
300 MPa	5.73$^{c/A}$	5.18$^{c/BC}$	4.87$^{c/C}$	5.41$^{c/AB}$	5.43$^{c/AB}$

$^{a-c}$ Different superscripts denote differences ($P<0.05$) between different treatment within the same parameter

$^{A-E}$ Different superscripts denote differences ($P<0.05$) between different days of storage within the same treatment

[1] Heat-treated milk at 90 °C, 90s, homogenized at 15 MPa and fortified with 3 % SMP

References

Ahmed, J., A. Ayad, H. S. Ramaswamy, I. Alli, and Y. Shao. 2007. Dynamic viscoelastic behavior of high pressure treated soybean protein isolate dispersions. *International Journal of Food Properties* 10 (2):397–411.

Alviar, M. and Reid, D. S. 1990. Determination of rheological behavior of tomato concentrates using back extrusion. *J Food Sci.* 55: 554–555.

Andon, S. A. 1987. Interaction of salad dressing ingredients and hydrocolloids. *Food Eng.* 59: 47–48.

Angsupanich, K., M. Edde, and D. A. Ledward. 1999. Effects of High Pressure on the Myofibrillar Proteins of Cod and Turkey Muscle. *Journal of Agricultural and Food Chemistry* 47 (1):92–99.

Apichartsrangkoon, A. 2003. Effects of on rheological properties of soy protein gels. *Food Chemistry* 80 (1):55–60.

Autio, K. 1991. Measurement of flow curves for model liquids and real food systems with two commercial viscometers. *J. Food Eng.* 13: 57–66.

Bayindirli, L. 1992. Mathematical analysis of variation of density and viscosity of apple juice with temperature and concentration. *J. Food Proc. Preser.* 16: 23–28.

Bayindirli, L. 1993. Density and viscosity of grape juice as a function of concentration and temperature. *J. Food Proc. Preser.* 17: 147–151.

Benezech, T. and Maingonnat, J. F.1993. Flow properties of stirred yoghurt: structured parameter approach in describing time dependency. *J. Texture Stud.* 24: 455–473.
Bhattacharya, S., Bal, S., Mukherjee, R. K., and Battacharya, S. 1991. Rheological behaviour of Tamarind Tamarindus indica kernel suspension. *J Food Eng.* 13: 151–158
Bianchi, M. A., Pilosof, A. M. R., and Bartholomai, G. B. 1985. Rheological behaviour of comminuted meat systems containing soy protein isolates. *J Texture Stud.* 16: 193–206.
Bistany, K. L. and Kokini, J. L. 1983. Dynamic viscoelastic properties of foods in texture control. *J. Rheol.* 27: 605–620.
Bodenstab. S., Juillert, M., Bauer, W., and Sommer, K. 2003. Separating the role of particles and the suspending fluid for the flow of soy milks. *J. Food Sci.* 68(5): 1722–1730.
Borwankar, R. P., Frye, L. A., Blaurock, A. E., and Sasevich, F. J. 1992. Rheological characterization of melting of margarines and tablespreads. *J. Food Eng.* 16: 55–74
Brekke, J. E. and Myers, A. L. 1978. Viscometric behavior of guava purees and concentrates. *J. Food Sci.* 43:272–273.
Campanella, O. H. and Peleg, M. 1987. Determination of yield stress of semi-liquid foods from squeezing flow data. *J. Food Sci.* 52: 214–215,217.
Caradec, R L. and Nelson, P. E. 1985. Effect of temperature on the serum viscosity of tomato juice. *J. Food Sci.* 50: 1497–1498.
Carter, R. D. and Buslig, B. S. 1977. Viscosity and particle size distribution in commercial Florida frozen concentrated orange juice. *Proc. Fla. State Hort. Soc.* 90: 130–132.
Chang, C. M., Powrie, W. D., and Fennema, O. 1972. Electron microscopy of mayonnaise. *Can. Inst. Food Sci. Technol. J.* 5: 134–137.
Chang, Y. and Hartel, R. W. 1997. Flow properties of freeze concentrated skim milk. *J. Food Eng.* 31:375–386.
Chapleau, N. J., and M. I. Lamballerie-Anton. 2003. Changes in myofibrillar proteins interactions and rheological properties induced by high-pressure processing. *European Food Research and Technology* 216 (6):470–476.
Cheftel, J. C., and E. Dumay. 1996. Effects of high pressure on dairy proteins: a review. In *Progress in Biotechnology*, edited by R. Hayashi and C. Balny: Elsevier Science B.V.
Chevalley, J. 1975. Rheology of chocolate. *J. Texture Stud.* 6: 177–196.
Chevalley, J. 1991. An adaptation of the Casson equation for the rheology of chocolate. *J. Texture Stud.* 22: 219–229.
Considine, T., H. A. Patel, S. G. Anema, H. Singh, and L. K. Creamer. 2007. Interactions of milk proteins during heat and high hydrostatic pressure treatments – A Review. *Innovative Food Science & Emerging Technologies* 8 (1):1–23.
Crandall, P. G., Chen, C. S., and Carter, R. D. 1982. Models for predicting viscosity of orange juice concentrate. *Food Technol.* 36(5): 245–252.
den Ouden, F. W. C. and van Vliet, T. 1993. Determination of particle size distribution in tomato concentrate, in *Food Colloids and Polymers: Stability and Mechanical Properties*, eds. E. Dickinson and P. Walstra, pp. 285–288, The Royal Society of Chemistry, Cambridge, England.
Dickinson, E. and Stainsby, G. 1982. *Colloids in Food*, Applied Science Publishers, New York.
Dickinson, E. and Stainsby, G. 1987. Progress in the formulation of food emulsions and foams. *Food Technol.* 41: 74, 76–81, 116.
Durán, L. and Costell, E. 1982. Rheology of apricot puree: characterization of flow. *J. Texture Stud.* 13: 43–58.
Durán, L. and Costell, E. 1985. Influence of homogenization on the rheological behaviour of apricot puree. *Acta Alimentaria* 14: 201–210.
Elliott, J. H. and Ganz, A. J. 1977. Salad dressings—preliminary rheological characterization. *J. Texture Stud.* 8: 359–377.
Fang, T. N., Tiu, C., Wu, X., and Dong, S. 1996. Rheological behaviour of cocoa dispersions. *J. Texture Stud.* 26: 203–215.
Fang, T., Zhang, H., Hsieh, T. T., and Tiu, C. 1997. Rheological behavior of cocoa dispersions with cocoa butter replacers. *J. Texture Stud.* 27: 11–26.

Fernandez-Martin, F. 1972. Influence of emperature and composition on some physical properties of milk and milk concentrates. II. Viscosity. *J. Dairy Res.* 39: 75–82.

Fertsch, B., M. Müller, and J. Hinrichs. 2003. Firmness of pressure-induced casein and whey protein gels modulated by holding time and rate of pressure release. *Innovative Food Science & Emerging Technologies* 4 (2):143–150.

Figoni, P. I. and Shoemaker, C. F. 1983. Characterization of time dependent flow properties of mayonnaise under steady shear. *J. Texture Stud.* 14: 431–442.

Fito, P. J., demente, G., and Sanz, F. J. 1983. Rheological behaviour of tomato concentrates hot break and cold break. *J. Food Eng.* 2: 51–62.

Ganani, E. and Powell, R. E. 1985. Suspensions of rodlike particles: literature review and data correlations. *J. Compos. Mater.* 19: 194–215.

Garcia, R., Rivera, J., and Rolz, C. 1974. Rheological properties of tropical products and their enzymic classification. *Proc. IV. Int. cong. Fd. Sci. Technol.* 2: 18–26.

Genovese, D. B. and Rao, M. A. 2003. Vane yield stress of starch dispersions. *J. Food Sci.* 68(7): 2295–2301.

Genovese, D. B. and Rao, M. A. 2005. Components of vane yield stress of structured food dispersions. *J. Food Sci.* 70(8): E498–E504.

Giboreau, A., Cuvelier, G., and Launay, B. 1994. Rheological behavior of three biopolymer/water systems with emphasis on yield stress and viscoelastic properties. J. Texture Stud. 25: 119–137.

Gunjal, B. B. and Waghmare, N. J. 1987. Flow characteristics of pulp, juice and nectar of "Baneshan" and "Neelum" mangoes. *J. Food Sci. Technol. India* 24: 20–23.

Harper, J. C. 1960. Viscometric behavior in relation to evaporation of fruit purees. *Food Technol.* 14: 557–560.

Harper, J. C. and Lieberman, K. W. 1962. Rheological behavior of pear purees. *Proc. I Int. Cong. Food Sci. Technol.* 1: 719–728.

Harper, J. C. and El-Sahrigi, A. F. 1965. Viscometric behavior of tomato concentrates. *J. Food Sci.* 30: 470–476.

Harrison, L. J. and Cunningham, F. E. 1985. Factors influencing the quality of mayonnaise: a review. *J. Food Qual.* 8: 1–20.

Harrison, L. J. and Cunningham, F. E. 1986. Influence of frozen storage time on properties of salted yolk and its functionality in mayonnaise. *J. Food Qual.* 9: 167–174.

Hartel, R. W. 1998. Phase transitions in chocolate and coatings, in *Phase/State Transitions in Foods: Chemical, Structural, and Rheological Changes*, eds. M. A. Rao and R. W., pp. 217–251, Marcel Dekker, Inc., New York.

Hayes, G. D. 1987. *Food Engineering Data Handbook*, John Wiley and Sons, Inc., New York.

Hinrichs, J., and B. Rademacher. 2005. Kinetics of combined thermal and pressure-induced whey protein denaturation in bovine skim milk. *International Dairy Journal* 15 (4):315–323.

Higgs, S. J. 1974. An investigation into the flow behaviour of complex non-Newtonian foodstuffs. *J. Phys. D: Appl. Phys.* 7: 1184–1191.

Holdsworth, S. D. 1971. Applicability of rheological models to the interpretation of flow and processing behavior of fluid food products. *J. Texture Stud.* 2: 393–418.

Holdsworth, S. D. 1993. Rheological models used for the prediction of the flow properties of food products: a literature review. *Trans. IChemE* 71, Part C: 139–179.

Hunter, R. J. and Nicol, S. K. 1968. The dependence of plastic flow behavior of clay suspensions on surface properties. *J. Colloid and Interface Sci.* 28: 250–259.

Ibarz, A. and Pagan, J. 1987. Rheology of raspberry juices. *J. Food Eng.* 6: 269–289.

Ibarz, A., Vicente, M., and Graell, J. 1987. Rheological behavior of apple juice and pear juice and their concentrates. *J. Food Eng.* 6: 257–267.

Ibarz, A., Pagan, J., and Miguelsanz, R. 1992a. Rheology of clarified Fruit Juices. II, blackcurrant juices. *J. Food Eng.* 15: 63–73.

Ibarz, A., Gonzalez, C., Esplugas, S., and Vicente, M. 1992b. Rhoelogy of clarified juices. 1: Peach juices. *J. Food Eng.* 15: 49–61.

Iwasaki, T., K. Noshiroya, N. Saitoh, K. Okano, and K. Yamamoto. 2006. Studies of the effect of hydrostatic pressure pretreatment on thermal gelation of chicken myofibrils and pork meat patty. *Food Chemistry* 95 (3):474–483.

Jeffrey, D. J. and Acrivos, A. 1976. The rheological properties of suspensions of rigid particles. *Am. Inst. Chem. Engineers. J.* 22: 417–432.

Jinescu, V. V. 1974. The rheology of suspensions. *Int. Chem. Eng.* 14: 397–120.

Kajiyama, N., S. Isobe, K. Uemura, and A. Noguchi. 1995. Changes of soya protein under ultra high hydraulic pressure. *International Journal of Food Science and Technology* 30 (2):147–158.

Kanno, C., T.-H. Mu, and H. Rikimaru. 2002. Gel formation of individual milk whey proteins under hydrostatic pressure. In *Progress in Biotechnology*: Elsevier Science B.V.

Kiosseoglou, V. D. and Sherman, P. 1983. Influence of egg yolk lipoproteins on the rheology and stability of o/w emulsions and mayonnaise 1. Viscoelasticity of groundnut oil-in-water emulsions and mayonnaise. *J. Texture Stud.* 14: 397–417.

Kokini, J. L. 1992. Rheological properties of Foods, in *Handbook of Food Engineering*, eds. D. R. Heldman and D. B. Lund, pp. 1–38. Marcel Dekker Inc., New York.

Kokini, J. L. and Dickie, A. 1981. An attempt to identify and model transient viscoelastic flow in foods. *J. Texture Stud.* 12: 539–557.

Kokini, J. L. and Plutchok, G. J. 1987. Viscoelastic properties of semisolid foods and their biopolymeric components. *Food Technol.* 41(3): 89–95.

Krishna, A. G. G. 1993. Influence of viscosity on wax settling and refining loss in rice bran oil. *J. Am. Oil Chem. Soc.* 70: 895–898.

Lang, W., Sokhansanj, S., and Sosulski, F. W. 1992. Modelling the temperature dependence of kinematic viscosity for refined canola oil. *J. Am. Oil Chem. Soc.* 69: 1054–1055.

Larson, R. G. 1985. Constitutive relationships for polymeric materials with power-law distributions of relaxation times. *Rheol. Acta* 24: 327–334.

Lopez, A., Ibarz, A., Pagan, J., and Vilavella, M. 1989. Rheology of wine musts during fermentation. *J. Food Eng.* 10: 155–161.

López-Fandiño, R. 2006. High pressure-induced changes in milk proteins and possible applications in dairy technology. *International Dairy Journal* 16 (10):1119–1131.

Lucey, J. A. 2001. The relationship between rheological parameters and whey separation in milk gels. *Food Hydrocolloids* 15 (4–6):603–608.

Lucey, J. A., C. T. Teo, P. A. Munro, and H. Singh. 1998. Microstructure, permeability and appearance of acid gels made from heated skim milk. *Food Hydrocolloids* 12 (2):159–165.

Lucey, J. A., P. A. Munro, and H. Singh. 1999. Effects of heat treatment and whey protein addition on the rheological properties and structure of acid skim milk gels. *International Dairy Journal* 9 (3–6):275–279.

Manohar, B., Ramakrishna, P., and Ramteke, R.S. 1990. Effect of pectin content on flow properties of mango pulp concentrates. *J. Texture Stud.* 21: 179–190.

Manohar, B., Ramakrishna, P., and Udayashankar, K. 1991. Some physical properties of tamarind Tamarindus indica L. juice concentrates. *J. Food Eng.* 13: 241–258.

Maschmeyer, R. O. and Hill, C. T. 1977. Rheology of concentrated suspensions of fibers in tube flow. *Trans. Soc. Rheol.* 21: 183–194.

McCarthy, K. L. and Seymour, J. D. 1994a. A fundamental approach for the relationship between the Bostwick measurement and Newtonian fluid viscosity. *J. Texture Stud.* 24: 1–10.

McCarthy, K. L. and Seymour, J. D. 1994b. Gravity current analysis of the Bostwick consistometer for power law foods. *J. Texture Stud.* 25: 207–220.

Metz B., Kossen, N. W. F., and van Suijdam, J. C. 1979. The rheology of mould suspensions, in *Advances in Biochemical Engineering*, eds. T. K. Ghose, A. Fiechter, and N. Blakebrough, Vol. 2. 103–156, Springer Verlag, New York.

Metzner, A. B. 1985. Rheology of suspensions in polymeric liquids. *J. Rheol.* 29: 739–775.

Michaels, A. S. and Bolger, J. C. 1962. The plastic behavior of flocculated kaolin suspensions. *Ind. Eng. Chem. Fundam.* 1: 153–162.

Mizrahi, S. and Firstenberg, R. 1975. Effect of orange juice composition on flow behaviour of six-fold concentrate. *J. Texture Stud.* 6: 523–532.

Mizrahi, S. 1979. A review of the physicochemical approach to the analysis of the structural viscosity of fluid food products. J. *Texture Stud.* 10: 67–82.
Mizrahi, S. and Berk, Z. 1970. Flow behavior of concentrated orange juice. *J. Texture Stud.* 1: 342–355.
Mizrahi, S. and Berk, Z. 1972. Flow behaviour of concentrated orange juice: mathematical treatment. *J. Texture Stud.* 3: 69–79.
Molina, E., and D. A. Ledward. 2003. Effects of combined high-pressure and heat treatment on the textural properties of soya gels. *Food Chemistry* 80 (3):367–370.
Munro, J. A. 1943. The viscosity and thixotropy of honey. *J. Econ. Entomol.* 36: 769–777.
Musser, J. C. 1973. Gloss on chocolate and confectionery coatings, in proceedings of the 27th Pennsylvania Manuf. Confect. Assoc. Production Conference, Lancaster, PA, pp. 46–50.
Noureddini, H., Teoh, B. C., and Clements, L. D. 1992. Viscosities of vegetable oils and fatty acids. *J. Am. Oil Chem. Soc.* 69: 1189–1191.
Okamoto, M., Y. Kawamura, and R. Hayashi. 1990. Application of high pressure to food processing: textural comparison of pressure- and heat-induced gels of food proteins. *Agricultural and Biological Chemistry* 54 (1):183–189.
Okechukwu, P. E. and Rao M. A. 1995. Influence of granule size on viscosity of corn starch suspension. *J. Texture Stud.* 26: 501–516.
Paredes, M. D. C., Rao, M. A., and Bourne, M. C. 1988. Rheological characterization of salad dressings. Steady shear, thixotropy and effect of temperature. *J. Texture Stud.* 19: 247–258.
Paredes, M. D. C., Rao, M. A., and Bourne, M. C. 1989. Rheological characterization of salad dressings. Effect of storage. *J. Texture Stud.* 20: 235–250.
Perona, P. 2005, Bostwick degree and rheological properties: an up-to-date viewpoint, Appl. Rheol. 15: 218–229.
Perry, J. H. 1950. *Chemical Engineer's Handbook*, 3rd ed., p. 374. McGraw-Hill, New York.
Praschan, V. C. 1981. *Quality Control Manual for Citrus Processing Plants*, Safety Harbor, Florida: Intercit.
Puppo, C., N. Chapleau, F. Speroni, M. deLamballerie-Anton, F. Michel, C. Anon, and M. Anton. 2004. Physicochemical Modifications of High-Pressure-Treated Soybean Protein Isolates. *Journal of Agricultural and Food Chemistry* 52 (6):1564–1571.
Puppo, M. C., V. Beaumal, N. Chapleau, F. Speroni, M. de Lamballerie, M. C. Anon, and M. Anton. 2008. Physicochemical and rheological properties of soybean protein emulsions processed with a combined temperature/high-pressure treatment. *Food Hydrocolloids* 22 (6):1079–1089.
Qiu, C.-G. and Rao, M. A. 1988. Role of pulp content and particle size in yield stress of apple sauce. *J. Food Sci.* 53: 1165–1170.
Qiu, C.-G. and Rao, M. A. 1989. Effect of dispersed phase on the slip coefficient of apple sauce in a concentric cylinder viscometer. J. Texture Stud. 20: 57–70.
Quemada, D., Fland, R, and Jezequel, R H. 1985. Rheological properties and flow of concentrated diperse media. *Chem. Eng. Comm.* 32: 61–83.
Ramaswamy, H. S. and Basak, S. 1991. Rheology of stirred yoghurts. *J. Texture Stud.* 22: 231–241.
Rao, M. A. 1977. Rheology of liquid foods-a review. *J. Texture Stud.* 8: 135–168.
Rao, M. A. 1987. Predicting the flow properties of food suspensions of plant origin. Mathematical models help clarify the relationship between composition and rheological behavior. *Food Technol.* 41(3): 85–88.
Rao, M. A. 1992. The structural approach to rheology of plant food dispersions. Spanish J. Food Sci. & Technol. 32: 3–17.
Rao, M.A. and Bourne, M. C. 1977. Analysis of the plastometer and correlation of Bostwick consistometer data. *J. Food Sci.* 42: 261–264.
Rao, M. A. and Cooley, H. J. 1984. Determination of effective shear rates of complex geometries. *J. Texture Stud.* 15: 327–335.
Rao, M. A. and Cooley, H. J. 1992. Rheology of tomato pastes in steady and dynamic shear. *J. Texture Stud.* 23: 415–425.
Rao, M. A., Otoya Palomi, L. N., and Bernhardt, L. W. 1974. Flow properties of tropical fruit purees. *J. Food Sci.* 39: 160–161.

Rao, M. A., Kenny, J. F., and Nelson, R. R. 1977. Viscosity of American Wines as a function of temperature (In German). *Mitteilungen-Klosterneuberg*. 27: 223–226.

Rao, M. A., Bourne, M. C., and Cooley, H. J. 1981. Flow properties of tomato concentrates. *J. Texture Stud.* 12: 521–538.

Rao, M. A., Cooley, H. J., and Vitali, A. A. 1984. Flow properties of concentrated fruit juices at low temperatures. *Food Technol.* 38(3): 113–119.

Rao, K., L,, Eipeson, W. E., Rao, P. N. S., Patwardhan, M. V., and Ramanathan, P. K. 1985. Rheological properties of mango pulp and concentrate. *J. Food Sci. Technol. India* 22: 30–33.

Rao, M. A., Cooley, H. J., Nogueira, J. N., and McLellan, M. R. 1986. Rheology of apple sauce: effect of apple cultivar, firmness, and processing parameters. *J. Food Sci.* 51:176–179.

Rao, M. A., Cooley, H. J., Ortloff, C., Chang, K,, and Wijts, S. C. 1993. Influence of rheological properties of fluid and semi solid foods on the performance of a filler. *J. Food Process Eng.* 16: 289–304.

Rastogi, N. K., K. S. M. S. Raghavarao, V. M. Balasubramaniam, K. Niranjan, and D. Knorr. 2007. Opportunities and challenges in high pressure processing of foods. *Critical Reviews in Food Science and Nutrition* 47:69–112.

Ross-Murphy, S. B. 1984. Rheological methods, in *Biophysical Methods in Food Research*, ed. H. W.-S. Chan, pp. 138–199, Blackwell Scientific Publications, London.

Rovedo, C. O., Viollaz, P. E., and Suarez, C. 1991. The effect of pH and temperature on the rheological behavior of Dulce de Leche. A typical dairy Argentine product. *J. Dairy Sci.* 74: 1497–1502.

Rozema, H. and Beverloo, W. A. 1974. Laminar isothermal flow of non Newtonian fluids in a circular pips. *Lebensmittel Wissenschaft und Technologie*. 7: 223–228.

Sandra, S., and D. G. Dalgleish. 2005. Effects of ultra-high-pressure homogenization and heating on structural properties of casein micelles in reconstituted skim milk powder. *International Dairy Journal* 15 (11):1095–1104.

Saravacos, G. D. 1968. Tube viscometry of fruit purees and juices. *Food Technol.* 22: 1585–1968.

Saravacos, G. D. 1970. Effect of temperature on viscosity of fruit juices and purees. *J. Food Sci.* 35: 122–125.

Saravacos, G. D. and Moyer, J. C. 1967. Heating rates of fruit products in an agitated kettle. *Food Technol.* 27: 372–376.

Serra, M., A. J. Trujillo, J. M. Quevedo, B. Guamis, and V. Ferragut. 2007. Acid coagulation properties and suitability for yogurt production of cows' milk treated by high-pressure homogenisation. *International Dairy Journal* 17 (7):782–790.

Serra, M., A. J. Trujillo, P. D. Jaramillo, B. Guamis, and V. Ferragut. 2008a. Ultra-High Pressure Homogenization-Induced Changes in Skim Milk: Impact on Acid Coagulation Properties. *Journal of Dairy Research* 75:69–75.

Serra, M., A. J. Trujillo, B. Guamis, and V. Ferragut. 2008b. Evaluation of physical properties during storage of set and stirred yogurts made from ultra-high pressure homogenization-treated milk. *Food Hydrocolloids In Press,* Corrected Proof.

Serra, M., A. J. Trujillo, B. Guamis, and V. Ferragut. 2009. Evaluation of physical properties during storage of set and stirred yogurts made from ultra-high pressure homogenization-treated milk. *Food Hydrocolloids* 23 (1):82–91.

Servais, C., Ranc, H., and Roberts, I. D. 2004. Determination of chocolate viscosity. *J. Texture Stud.* 34(5–6): 467–498

Sherman, P. 1970. Industrial Rheology, Academic Press, New York.

Sierzant, R. and Smith, D. E. 1993. Flow behavior properties and density of whole milk retentates as affected by temperature. *Milchwissenschaft*. 48(1): 6–10.

Sollich, P., Lequeux, F., Hébraud P. and Cates M. F. 1997. Rheology of soft glassy materials. Physical Review Letters 78:2020–2023.

Sommer, K. 1975. Bestimmung des strömungs-bzw. Festkörperreibungsanteils der viscosität kozentrierter suspensionen. *Rheol. Acta* 14: 347–351. (Cited in Bodenstab et al. (2003).

Sornsrivichai, T. 1986. A study on rheological properties of tomato concentrates as affected by concentration methods, processing conditions, and pulp content, Ph.D. thesis, Cornell University, Ithaca, NY.

Sterling, C. and Wuhrmann, J. J. 1960. Rheology of cocoa butter. I. Effect of contained fat crystals on flow properties. *Food Res.* 25: 460–463.

Takada, N. and Nelson, P. E. 1983. A new consistency method for tomato products: the precipitate weight ratio. *J. Food Sci.* 48: 1460–1462.

Tamime, A. Y., and R. K. Robinson. 2007. *Yoghurt Science and Technology*. Cambridge (UK): Woodhead Publishing Ltd.

Tanaka, M. and Fukuda, H. 1976. Studies on the texture of salad dressings containing xanthan gum. *Can. Inst. Food Sci. Technol. J.* 9: 130–134.

Tanglertpaibul, T. and Rao, M. A. 1987a. Rheological properties of tomato concentrates as affected by particle size and methods of concentration. *J. Food Sci.* 52: 141–145.

Tanglertpaibul, T. and Rao, M. A. 1987b. Flow properties of tomato concentrates: effect of serum viscosity and pulp content. *J. Food Sci.* 52: 318–321.

Tanglertpaibul, T. and M. A. Rao. 1987c. Intrinsic viscosity of tomato serum as affected by methods of determination and methods of processing concentrates. J. Food Sci. 52:1642–1645 & 1688.

Tárrega, A., Costell, E., and Rao M. A. 2006. Vane yield stress of native and cross-linked starch dispersions in skim milk: effect of starch concentration and λ-carrageenan addition. *Food Sci. & Tech. Int.* 12(3): 253–60.

Tattiyakul, J. & Rao, M. A. 2009. Effect of high pressure and ultrasonic processing of foods on rheological properties, p. 207–224, in *Novel Food Processing – Effects on Rheological and Functional Properties*, eds. J. Ahmed, H. S. Ramaswamy, S. Kasapis and J. Boye, CRC Press, Inc., Boca Raton and New York.

Timms, R. E. 1985. Physical properties of oils and mixtures of oils. *J. Am. Oil Chem. Soc.* 62(2): 241–248.

Trifiro, A., Saccani, G., Gherardi, S., and Bigliardi, D. 1987. Effect of content and sizes of suspended particles on the rheological behavior of apricot purees. *Industria Conserve* 62: 97–104.

Varshney, N. N. and Kumbhar, B. K. 1978. Effect of temperature and concentration on rheological properties of pineapple and orange juice. *J. Food Sci. Technol. India* 15(2): 53–55.

Velez-Ruiz, J. F. and Barbosa-Canovas, G. V. 1998. Rheological Properties of concentrated milk as a function of concentration, temperature and storage time. *J. Food Eng.* 35: 177–190.

Vitali, A. A. 1983. Rheological behavior of frozen concentrated orange juice at low temperatures (in Portuguese), Ph.D. thesis, University of São Paulo, São Paulo, Brazil.

Vitali, A. A. and Rao, M. A. 1984. Flow properties of low-pulp concentrated orange juice: Effect of temperature and concentration. J. Food Sci. 49(3): 882–888.

Vitali, A. A. and Rao, M. A. 1984a. Flow Properties of low-pulp concentrated orange juice: serum viscosity and effect of pulp content. *J. Food Sci.* 49: 876–880.

Vitali, A. A. and Rao, M. A. 1984b. Flow Properties of low-pulp concentrated orange juice: Effect of temperature and concentration. *J. Food Sci.* 49: 882–888.

Wayne, J. E. B. and Shoemaker, C. F. 1988. Rheological characterization of commercially processed fluid milks. *J. Texture Stud.* 19: 143–152.

Wayland, M. 1994. Functional effects of emulsifiers in chocolate. *Manuf. Confect.* 745: 111–117.

Wildemuth, C. R. and Williams, M. C. 1984. Viscosity of suspensions modeled with a shear-dependent maximum packing fraction. *Rheol. Acta* 23: 627–635.

Yoo, B. and Rao, M. A. 1994. Effect of unimodal particle size and pulp content on rheological properties of tomato puree. *J. Texture Stud.* 25: 421–436.

Yoo, B. and Rao, M. A. 1996. Creep and dynamic rheological behavior of tomato concentrates: effect of concentration and finisher screen size. *J. Texture Stud.* 27: 451–459.

Zamri, A. I., D. A. Ledward, and R. A. Frazier. 2006. Effect of Combined Heat and High-Pressure Treatments on the Texture of Chicken Breast Muscle (Pectoralis Fundus). *Journal of Agricultural and Food Chemistry* 54 (8):2992–2996.

Chapter 6
Rheological Behavior of Food Gels

Rheological studies can provide much useful information on sol-gel and gel-sol transition, as well as on the characteristics of a gel. There are several definitions of what a gel is that are based on either phenomenological and/or molecular criteria. Flory (1953) defined a gel to consist of polymeric molecules cross-linked to form a tangled interconnected network immersed in a liquid medium. Hermans (1949) defined it as a two-component system (e.g., gelling polymer and the solvent, water or aqueous solution in foods) formed by a solid finely dispersed or dissolved in a liquid phase, exhibiting solid-like behavior under deformation; in addition, both components extend continuously throughout the entire system, each phase being interconnected. At the molecular level, gelation is the formation of a continuous network of polymer molecules, in which the stress-resisting bulk properties (solid-like behavior) are imparted by a framework of polymer chains that extends throughout the gel phase. Furthermore, gel setting involves formation of cross-links, while softening or liquefaction (often called melting) involves their destruction.

In many food products, gelation of polysaccharides is critical to the formation of desired texture. Most biopolymers form physical gels, structured by weak interactions (hydrogen, electrostatic, hydrophobic), and, thus, at a gross level belong to Flory's third mechanism (Flory 1974): "Polymer networks formed through physical aggregation, predominantly disordered, but with regions of local order." They are often thermoreversible and almost invariably occur in the presence of excess of solvent, usually water or aqueous electrolyte. Therefore, they are called solvated networks. In most biopolymer gels, the polymer chains form extended junction zones by means of side-by-side associations of a physical nature, in contrast to the typical single covalent bonds found in chemically cross-linked networks. Consequently, in physical gelation the formation and breakdown of the junction zones are usually reversible, the cross-link functionality is very high, and the junction zones have a finite lifetime. The formation of these kinds of transient networks is determined by the chemical composition of both the polymer and the solvent which constitute the gelling system, and by temperature and time.

In this chapter, the formation of gels and their softening (melting) are reviewed. Furthermore, some of the contents of the chapter have been discussed by the authors in a recent review of the subject (Lopes da Silva et al. 1998). Although most food

M. A. Rao, *Rheology of Fluid, Semisolid, and Solid Foods*, Food Engineering Series, DOI 10.1007/978-1-4614-9230-6_6, © Springer Science+Business Media New York 2014

gels are formed by first dissolving a gelling polymer in water and can be studied by means of traditional treatments, starch gels are composites of starch granules in a matrix of gelled amylose and they are not thermoreversible. Therefore, their rheological behavior is treated separately in this chapter. Additionally, information on rheological behavior of dispersions of starches alone and protein-starch mixtures can be found in Chap. 4

Rheological Tests to Evaluate Properties of Gel Systems

As mentioned in Chap. 3, dynamic (oscillatory) rheological tests provide valuable information on the viscoelastic nature of foods. Because gels are viscoelastic materials, dynamic rheological tests are well suited for studying the characteristics of gels as well as of gelation and melting. From dynamic rheological tests in the linear viscoelastic range, the storage modulus, G', and the loss modulus, G'', and tan $\delta = (G'/G'')$ is the loss tangent can be obtained (Rao 1992). In addition, the complex viscosity, $\eta^* = (G^*/\omega)$ is another useful parameter; where ω is the frequency of oscillation (rad s^{-1}), and $G^* = \sqrt{(G')^2 + (G'')^2}$. Three types of dynamic tests can be conducted to obtain useful properties of gels, and of gelation and melting: (1) frequency sweep studies in which G' and G'' are determined as a function of frequency (ω) at fixed temperatures, (2) temperature sweep in which G' and G'' are determined as a function of temperature at fixed ω, and (3) time sweep in which G' and G'' are determined as a function of time at fixed ω and temperature. As a reminder, all the above tests should be conducted in the linear viscoelastic range.

Figure 3.33 illustrates a time sweep in which G' and G'' were determined as a function of time at fixed ω and temperature on low-methoxyl (LM) pectin + Ca^{2+} gels containing sucrose. This type of test, often called a gel cure experiment, is well suited for studying structure development in physical gels. Figure 3.34 shows values of $\Delta G'$ of LM pectin + Ca^{2+} gels containing 20 % sucrose due to increasing the calcium content from 0.10 to 0.15 % (Grosso and Rao 1998).

In addition, other measurement techniques in the linear viscoelastic range, such as stress relaxation, as well as static tests that determine the modulus are also useful to characterize gels. For food applications, tests that deal with failure, such as the dynamic stress/strain sweep to detect the critical properties at structure failure, the torsional gelometer, and the vane yield stress test that encompasses both small and large strains are very useful.

Mechanisms of Gelation

Because different biopolymer gel systems can be encountered, different gelation mechanisms also can be encountered. Because of variations in the number and nature of the cross-links, framework flexibility, attractions and repulsions between framework elements, and interactions with solvent, different properties of the formed gels are to be expected (Rees 1972). The range of systems is quite extensive, from systems formed purely by topological entanglements to the complex networks formed by ordered fibrous assemblies, such as by actin and fibrin filaments in vivo. Some of the characteristics of the gelation processes found in food polymers are given in Table 6.1 (Lopes da Silva et al. 1998).

Polysaccharides of the galactan group (carrageenans, agarose) form thermoreversible gels through a transition from a disordered to a helical macromolecular conformation (Rees 1969; Morris et al. 1980a). Alginates and LM pectins form gels by chain stacking through the interaction between divalent ions and the carboxyl groups of the polysaccharide macromolecule (Grant et al. 1973; Gidley et al. 1979). In gels of methyl and hydroxypropyl derivatives of cellulose, the gel network is formed by hydrophobic association of chains, showing very distinct properties, like gel setting when heated and gel melting when cooled (Haque and Morris 1993). A phase separation process was suggested to interpret the gelation of amylose (Miles et al. 1985; Doublier and Choplin 1989). Gels of bovine serum albumin are fibrillar in nature, first unfolding by heating and then by aggregation, leading to gels of different properties depending on pH and ionic strength (Richardson and Ross-Murphy 1981a). Gelatin networks are formed by triple helices randomly distributed in space and separated by chain segments of random-coil conformation (Pezron et al. 1990).

Classification of Gels

Clark and Ross-Murphy (1987) classified biopolymer gels based on the level of order of the macromolecule both before and during the network formation: (1) gels formed from disordered biopolymers, such as carrageenans, pectins, starch, gelatin, and (2) gel networks that involve specific interactions between denser and less flexible particles, such as thermally denatured globular proteins and aggregated proteins from enzymatic or chemical action. Gel-forming biopolymers can also be divided into "cold-setting" or "heat-setting," based on the two main gelation mechanisms. In the former, gelation is induced by cooling and includes those biopolymer gels which occur in nature that provide structures in biological systems. In the latter, gelation occurs by heating and includes those systems where gelation involves extensive denaturation of the biopolymer, for example, thermally unfolded globular proteins.

Many foods contain proteins and, hence, protein gels have been studied extensively. When a protein's concentration is equal to or above the critical minimum concentration required to form a gel network (c_0), aggregation will ultimately produce a gel network that is most commonly detected by a shift from a fluid to a solid state (Foegeding 2006). Based on a model proposed by Dobson (2003), Foegeding (2006) proposed that starting with a native protein in the monomer or oligomer form, unfolding/aggregation proceeds through an intermediate state prior to reaching the unfolded state. The intermediate state can form disordered aggregates or prefibrillar species that eventually form amyloid fibrils. The unfolded molecule can also form disordered aggregates. Based on this model, there is the potential to form

Table 6.1 Gelation characteristics of important food polymers

Polymer	Important structural features affecting gelation	Main factors promoting gelation	Mechanism of gelation
Carrageenans	Alternating β-1,3- and α-1,4-linked galactose residues; presence of 3,6-anhydride residues; presence of sulfated residues	Cooling; presence of potassium or other gel-promoting cations	Association of the molecular chains into double helices followed by aggregation of the ordered "domains"
Agarose	Alternating 1,3-linked β-D-galactopyranose and 1, 4-linked 3,6-anhydro-α-L-galactopyranose	Cooling	Helix formation followed by aggregation
Alginates	Alternating blocks of β-1,4-linked D-mannuronic acid and α-1, 4-linked L-guluronic acid residues	Presence of divalent cations (usually Ca^{2-})	Specific site binding of calcium with the carboxyl groups of the polyuronic acid residues (mainly the polyguluronic), "egg-box" model
Pectins	α(1,4)-D-galacturonic acid residues, partially methyl esterified; presence of 1,2-linked L-rhamnosyl residues		
High-methoxyl pectins	Degree of methylation higher than 50 %	Cooling; low pH and low water activity	Junction zones stabilized by hydrogen bonds and hydrophobic interactions between the ester methyl groups
Low-methoxyl pectins	Degree of methylation lower than 50 %	Presence of divalent cations (usually Ca^{2-})	Specific site binding of calcium with the carboxyl groups of the polyuronic acid residues, "egg-box" model
Gellan gum	Glucose, glucuronic acid, and rhamnose residues (2:1:1); presence of O-acetyl and O-L-glyceryl substituents	Cooling; presence of gel-promoting cations	Ion-mediated aggregation of double helices
Starch	Linear fraction (amylose): α-1,4-linked D-glucose residues branched fraction (amylopectin): α-1,4- and α-1:6-linked D-glucose residues; higher DP	Cooling	Composite gels: amylopectin granules threaded by a gelled amylose matrix
Gelatin	High content of pyrolidine residues (pro + hypro): presence of glycine as every third residue; isoelectric point affected by acid or alkali treatment	Heating solutions followed by cooling	Formation of triple helices randomly distributed in space and separated by chain segments of random-coil conformation

Table 6.1 (continued)

Polymer	Important structural features affecting gelation	Main factors promoting gelation	Mechanism of gelation
Bovine serum albumin	Globular protein (M.W. ~66,500 Da); isoelectric point ~5.1; characteristic secondary structure, with a specific content of α-helix, β-sheet and disordered peptide chain conformations; presence of disulfide bonds and one free sulfydryl group	Heating solutions followed by cooling	Denaturation followed by aggregation, likely by formation of *β*-sheet regions and other less specific protein-protein associations
β-lactoglobulin	Globular protein (M.W. ~18,500 Da); characteristic secondary structure, with a specific content of α-helix, β-sheet and disordered peptide chain conformations; isoelectric point ~5.4–5.5; presence of two disulfide bridges and a free thiol group	Heating solutions followed by cooling	Denaturation followed by aggregation
Caseins	Different proteins: variation in the degree of phosphorylation, glycosylation, hydrophobicity and amphipathic structures. Presence of disulfide groups small molecules, c. 20,000–24,000 Da isoelectric points: pH 4.5–4.9	Renneted gels: *enzymatic* action followed by precipitation by Ca^{2+} Acid coagulation: *acidification* and instability of the casein micelles	Renneted gels: k-casein hydrolysis, instability of the casein micelles followed by coagulation, mainly by electrostatic interactions involving Ca^{2+} Acid coagulation: *solubilization* of the colloidal calcium phosphate and aggregation when the i.p. is approached (charge neutralization)

three different types of aggregates (A), designated as: A-Intermediate, A-Unfolded, and A-Amyloid (Dobson 2003; Foegeding 2006). It is noted that the designation of the "unfolded" state does not imply a complete loss of secondary structure.

Globular proteins can form gels varying in appearance from opaque to clear, and having either high or low water holding capacity (Foegeding 2006). As a first approximation, the network formed may be designated as being either fine-stranded or particulate. Increased ionic strength or a pH ~ pI, that decrease electrostatic repulsion, promote the formation of a particulate gel network. Gelation under low ionic strength and pi<pH<pI produces fine-stranded gels. Particulate gels are opaque and have low water holding capacity, while fine-stranded gels are transparent or translucent and have high water holding capacity. Depending on the conditions at gelation, the microstructure of particulate gels consists of large aggregates, ranging

in size from several hundred to >1,000 nm. Fine-stranded gel networks are composed of flexible (curved) strands or more rigid fibrils depending on the pH and ionic strength. Strand diameters are about the diameter of one or several protein molecules. The linear fibrils formed at pH 2.0 and low ionic strength are considered to be amyloid fibrils (Gosal et al. 2002; Foegeding 2006).

Because the proteins often studied are: β-lactoglobulin, bovine serum albumin, egg white ovalbumin, egg white (protein mixture), or whey protein isolate (protein mixture), it is of interest to note that their gels fall into one of the following types (Foegeding 2006):

- *Fine-stranded at neutral pH* gels are formed at or close to pH 7.0 and contain little, if any, added salts,
- *Fine-stranded at low pH* gels are formed at or close to pH 2.0 and contain little, if any, added salts,
- *Particulate at neutral pH* gels are formed at or close to pH 7.0 and generally NaCI concentration is increased to shift the network from fine-stranded to particulate.
- *Particulate at the isoelectric point* gels are formed around the isoelectric point and, thus, the pH will vary with protein type.

Based on the macroscopic behavior of gel systems, a practical and useful distinction can be made between those systems that are free-standing as a consequence of the development of the three-dimensional network, called "true gels," and those characterized by a tenuous gel-like network, which is easily broken when submitted to a high enough stress, called "weak gels" (Doublier et al. 1992). Clark and Ross-Murphy (1987) distinguished between these two kinds of gelled systems based on the operational definition of a gel from mechanical spectroscopy, designating "strong gels" as those networks that have "finite energy," and "weak gels" as those systems which are transient in time and that even well above co show some of the properties usually attributed to the presence of hyperentanglements. A traditional example of weak gel behavior is the viscoelastic behavior of xanthan gum dispersions.

The difference between the entanglement network, the "strong (true) gel" (Figs. 6.1a and 6.2a) and the "weak gel" (Figs. 6.1b and 6.2b), can be easily established using mechanical spectroscopy or by a strain dependence of the moduli (Clark and Ross-Murphy 1987). In the former, the molecular rearrangements within the network are very reduced over the time scales analyzed, G' is higher than G'' throughout the frequency range, and is almost independent of frequency, ω. In "weak gels," there is a higher dependence on frequency for the dynamic moduli, suggesting the existence of relaxation processes occurring even at short time scales, and lower difference between moduli values, indicating that a lower percentage of the stored energy is recovered.

High Pressure Effect

Although in most studies protein gels are formed by heating and subsequent cooling, they can also be created by using high pressure. Heat treatment of surimi gel

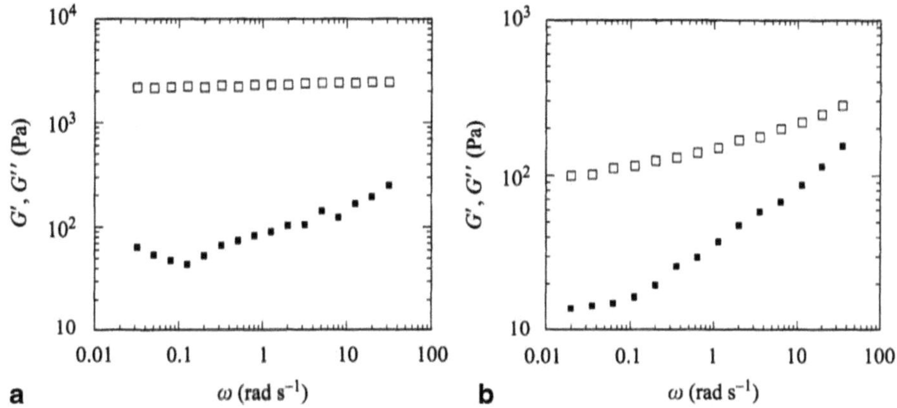

Fig. 6.1 Mechanical spectra (□ G', ■ G'') of a 1 % high-methoxyl pectin gel, pH 3.0, 60 % (w/w) sucrose measured after aging for 36 h at two different aging temperatures: (**a**) 5 °C, (**b**) 30 °C

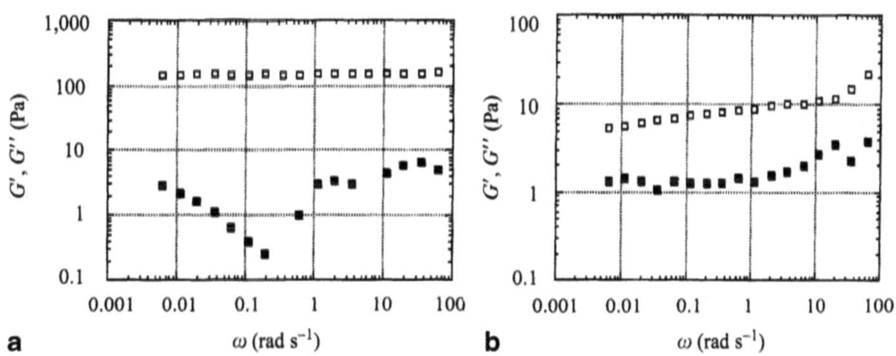

Fig. 6.2 Mechanical spectra (□ G', ■ G'') of low-methoxyl pectin/calcium gels: 8.1 g L^{-1}, $[Ca^{2+}]=6.2\times10^{-3}$ mol L^{-1}, 0.1 mol L^{-1} NaCl, measured at 20 °C after a quenching time of 21 h, for two different pH values: **a** pH 3.0; **b** pH 6.8

induced denaturation and reaggregation of protein resulting in the formation of a strong network structure. However, pressure treatment contributed to the formation of weak cross-links which were formed by aggregation of protein in its native form (Ahmad et al. 2004). Isobaric isothermal denaturation of β-lactoglobulin and α-lactalbumin was found to follow third- and second-order kinetics, respectively. Isothermal pressure denaturation of both β-lactoglobulin fractions did not differ significantly and the denaturation rate accelerated with increasing temperature. The activation energy ~70–100 kJ mol^{-1} for β-lactoglobulin A and B was not dependent to a great extent on the pressure. This result was interpreted as at that pressures >200 MPa, the denaturation rate is limited by the aggregation rate, while pressure forces the molecules to unfold (Hinrichs and Rademacher 2004).

Theoretical Treatment of Gels

Theoretical and experimental treatments of gels go hand-in-hand. The former are covered first because they will help us understand gel point and other concepts. Two main theories have been used to interpret results of experimental studies on gels: the classical theory based on branching models developed by Flory and Stockmayer, and the percolation model credited to de Gennes. Gelation theories predict a critical point at which an infinite cluster first appears. As with other critical points, the sol-gel transition can be in general characterized in terms of a set of generally applicable (universal) critical exponents.

Assuming high gel homogeneity for single biopolymer systems and reversibility of the cross-linking process to explain the critical concentration aspect, the classical theory and its extension and modifications, including the cascade theory (Gordon and Ross-Murphy 1975), were used to describe the gelling process of several biopolymers. It was used to describe the gel growth process after the gel point and to explain the role of important parameters in the gelation of physical gels, such as the critical gel concentration, and the concentration dependence of the gel modulus. The classical theory was used to describe gelation of, among several others biopolymer systems, bovine serum albumin (Richardson and Ross-Murphy 1981a), gelatin (Peniche-Covas et al. 1974), agarose (Watase et al. 1989), egg albumen and ovalbumin (Hsieh and Regenstein 1992a), and maltodextrins (Kasapis et al. 1993).

Rubber Elasticity

An equation for the modulus of ideal rubber was derived from statistical theory that can be credited to several scientists, including Flory, and Guth and James (Sperling 1986). A key assumption in derivation of the equation is that the networks are Gaussian. For the same Gaussian networks, Treloar (1975) derived an equation for the shear modulus, G

$$G = \left(\frac{gcRT}{M_c}\right) \tag{6.1}$$

where T is temperature (K), R is gas constant, and M_c is the mean molecular weight of chains joining adjacent cross-links. Equation 6.1 can be valid only under the following assumptions (Treloar 1975; Fu 1998): (1) the network contains N chains per unit volume, a chain being defined as the segment of molecule between successive points of cross-linkage; (2) the mean-square end-to-end distance for the whole assembly of chains in the unstrained state is the same as for a corresponding set of free chains; (3) there is no change of volume on deformation; (4) the junction points between chains move on deformation as if they were embedded in an elastic continuum; and (5) the entropy of the network is the sum of the entropies of the individual chains.

In a solvent-swollen polymer, the volume fraction of swollen polymer, υ_2, in the mixture of the polymer and solvent needs to be considered. Therefore, the relationship between G' and M_c is given by (Treloar 1975):

$$G = \left(\frac{\rho RT}{M_{\mathrm{c}}} \right) (\upsilon_2)^{1/3} \tag{6.2}$$

where ρ(g cm^{-3}) is the density of the polymer in the unswollen state. Most biopolymer networks have structures that do not conform to the main assumptions of the theory of rubber elasticity, because the cross-links are of low-energy physical interactions and they are not limited to single points on the chains but correspond to more or less extended junction zones. In addition, the polymer chains between junction zones can be characterized by several degrees of rigidity and deviate from Gaussian behavior. Stiff chains increase their flexibility with increasing temperature due to the increased number of allowed torsion angles. Consequently, changes in the rigidity of the molecular chains can also explain the decrease in the gel modulus when the temperature increases. Therefore, the theory is not applicable to most biopolymer gels, since the assumption of a purely entropy contribution to the elasticity of a network chain is likely invalid in most cases. For thermoreversible gels, the decrease in the number of junction zones with temperature is usually more pronounced than any increase from the entropy contribution.

In spite of the above drawbacks, Eqs. 6.1 and 6.2 are popular mainly due to their simple form and have been used to estimate M_c of protein (e.g., egg, gelatin) and other gels from shear modulus-concentration data (Table 6.2; Fu 1998). Its successful application to protein gels initially has been attributed to the greater flexibility of polypeptide chains in comparison to polysaccharide chains that are relatively stiff. In addition, the inability to employ low strain rates in early experimental studies on polysaccharide gels could be another reason.

Oakenfull's Modification

Oakenfull (1984) developed an extension of Eq. 6.1 for estimating the size of junction zones in noncovalently cross-linked gels subject to the assumptions (Oakenfull 1987): (1) the shear modulus can be obtained for very weak gels whose polymer concentration is very low and close to the gel threshold, that is, the polymer chains are at or near to maximum Gaussian behavior and (2) the formation of junction zones is an equilibrium process that is subject to the law of mass action. Oakenfull's expression for the modulus is (Oakenfull 1984)

$$G = -\frac{RT_{\mathrm{c}}}{M} \frac{M[J] - c}{M_{\mathrm{J}}[J] - c} \tag{6.3}$$

where M is the number average molecular weight of the polymer, R is the gas constant, T is the absolute temperature, c is the weight concentration of the polymer, M_j

Table 6.2 Examples of application of the rubber elasticity theory to biopolymer gels

System	Measurement and results	Reference
Collagen from epimysial connective tissues	Using stress-strain tests to calculate molecular weight between covalent cross-links (M_c)	McClain et al. (1969)
Wheat gluten and dough	Using extension tests to calculate the number of cross-links per unit volume	Muller (1969)
Kamaboko	Using extension and compression tests to develop the relationship between stress and strain and then to conclude that orientation of segments of kamaboko's network was negligible, and similar to the behavior of the vulcanized rubber in small deformation	Takagi and Simidu (1972)
Wheat flour doughs	Using shear modulus obtained at different temperature to conclude that model of rubber cannot be applied to dough elasticity	Bloksma and Nieman (1975)
Calcium, potassium, and sodium iota-carrageenate	Using rigidity modulus to calculate molecular weight of polymer chains between the network junctions (M_c)	Morris and Chilvers (1981)
HMP/sucrose and gelatin	Using shear modulus obtained from compression tests for noncovalent gels to calculate the association constant for the formation of junction zones, the number average molecular weight of the junction zones, and the number of cross-linking loci per junction zone	Oakenfull (1984)
Surimi	Using elastic moduli obtained from stress relaxation tests to calculate the number of cross-links per unit volume	Iso et al. (1984)
Fish meat	Using elastic modulus (Eo) obtained from compression tests to calculate the number of chains per unit volume	Mochizuki et al. (1987)
Egg white	Using stress relaxation tests to calculate the density of permanent cross-links	Ziegler and Rizvi (1989)
Egg white and ovalbumin	Using compression tests to calculate molecular weight of the interchain (M_c)	Hsieh and Regenstein (1992b)
Soy protein isolate	Using storage modulus (G') obtained from dynamic oscillatory tests to calculate the number of disulfide bridges per molecule	Chronakis et al. (1996)
Xanthan/gaiactomannans	Using shear modulus (G) obtained from creep tests to calculate molecular weight between cross-links (M_c)	Kloek et al. (1996)
Waxy maize starches	Using plateau storage modulus (G') obtained from small amplitude oscillatory tests to calculate molecular weight between cross-links (M_c)	Gluck-Hirsch and Kokini (1997)
Covalently cross-linked alginate gels	Results not consistent with prediction based on the rubber elasticity theory, since the model does not consider cross-linking with a second macromolecule	Eiselt et al. (1999)
Low-methoxyl pectin	M_c from plateau G' obtained from oscillatory tests	Fu and Rao (2001)
Egg albumen protein	M_c from plateau G' obtained from oscillatory tests	Christ et al. (2005)
Methylcellulose	Using plateau G' obtained from oscillatory tests to interpret M_c	Kuang et al. (2006)

is the number average molecular weight of the junction zones, and *[J]* is the molar concentration of junction zones. From assumption 2, it follows:

$$K_J = [J]M_J^n\{n(c - M_J[J])\}^{-n} \tag{6.4}$$

where K_j is the association constant for the formation of junction zones and n is the number of segments from the different polymer molecules involved. Numerical methods have been used to solve f or *M*, M_j, K_j, and *n* after substituting for *[J]* to obtain a relationship between shear modulus and concentration (Oakenfull 1984, 1987).

Cascade Theory

The cascade theory, developed by Gordon, is an extension of the classical Flory–Stockmayer concepts of gelation. An expression for the shear modulus, *G*, can be written as

$$G = \mathrm{Ne}\, aRT \tag{6.5}$$

where Ne is the number of elastically active network chains per primary chain (moles V^{-1}), and *a* is generalized front factor. From cascade theory

$$G = \{Nf\alpha(1 - \upsilon^2)(1 - \beta)/2\}aRT \tag{6.6}$$

where *f* is the number of sites (functionalities) along each molecule's length, α is the fraction of such sites that have reacted, and υ and β are derived from cascade approach to network formation, and initially $N = (c/M)$

$$\upsilon = (1 + \alpha + \alpha\upsilon)^{f-1} \tag{6.7}$$

$$\beta = (f - 1)\alpha\upsilon/(1 - \alpha + \alpha\upsilon) \tag{6.8}$$

The influence of temperature is taken into account for the cross-linking equilibrium constant *K*

$$K = \exp(\Delta S^\circ/R)\exp(-\Delta H^\circ/RT) \tag{6.9}$$

The critical gelling concentration, c_c, (the symbol c_c is used here to denote a concentration different than c_0 at initiation of gelling) at of polymer the melting temperature T_m can be calculated from

$$c_c = M(f - 1)/2K(T_m)f(f - 2)^2 \tag{6.10}$$

$$T_m = \Delta H^\circ / \{\Delta S^\circ + R\ln(2cf(f-2)^2/M(f-1))\} \tag{6.11}$$

Clark (1994) modeled the variation of the modulus of κ-carrageenan gels as a function of polymer's molecular weight using the cascade approach. From modulus-temperature data, thermodynamic and functional parameters were determined for gelatin (Clark et al. 1994): high ΔH°–600–1,600 kJ mol^{-1}, $T_m \sim 30\,°C$, a ~ 1, $f \sim 3$–100, and for pectin: low ΔH°–0.8 kJ mol^{-1}, $T_m \sim 130\,°C$, $a \sim 1{,}000$–5,000, $f \sim 3$–100. The relative insensitivity of the parameters a and f, and the high value of $T_m \sim 130\,°C$ noted for pectin gels point out the need for additional investigations on the applicability of the cascade theory to other gels. Galactomannan gelation was analyzed by cascade approach (Richardson et al. 1999). Recently, Mao and Rwei (2006) applied the cascade concept to describe the temperature and concentration dependence of gel modulus for mixed gels of xanthan and locust bean gum (LBG). Despite these successful efforts at modeling of biopolymer gelation according to a branching process by cascade theory, drawbacks have been reported mainly for processes that approach colloidal aggregation (e.g., maltodextrin gelation; Loret et al. 2004).

Percolation Theory

Percolation describes the geometrical transition between disconnected and connected phases as the concentration of bonds in a lattice increases. It is the foundation for the physical properties of many disordered systems and has been applied to gelation phenomena (de Gennes 1979; Stauffer et al. 1982). At just above gelation threshold, denoting the fraction of reacted bonds as p and $p = p_c + \Delta p$, p_c the critical concentration (infinite cluster), the scaling laws (critical exponents) for gel fraction (S_∞), and modulus (E) are

$$S_\infty \cong (p - p_c)^\beta; \quad E \cong (p - p_c)^t \tag{6.12}$$

A number of authors have used concentration of the polymer, c, in place of p. Theoretical values for the exponents from the percolation theory are: $\beta = 0.39$, $t \sim 1.7$–1.9; also, $t > \beta$ because of dangling chains. In contrast, the classical theory's predictions for the same quantities are

$$\beta_{cl} = 1, \quad t_{cl} = 3 \tag{6.13}$$

According to the percolation model, the chains in a swollen gel need not be Gaussian. For a true gel when $\sigma >> E$, the stress (σ) versus strain (λ) relationships from the percolation model (Eq. 6.14) and the classical theory (Eq. 6.15) are, respectively

$$\sigma = E\lambda^{5/2} \tag{6.14}$$

$$\sigma = E\lambda^2 \tag{6.15}$$

Verification of the values of the exponents in Eqs. 6.14 and 6.15 continue to be the subjects of considerable debate between proponents of the two theories. The viscoelastic properties of the system that characterize the sol-gel transition are also important features of the percolation model

$$\eta_0 \sim \varepsilon^{-s} \tag{6.16}$$

$$G_0 \sim \varepsilon^{t} \tag{6.17}$$

$$G'(\omega) \sim G''(\omega) \sim \omega^{\Delta} \tag{6.18}$$

$$\eta^{*}(\omega) \sim \omega^{\Delta-1} \tag{6.19}$$

where ε is the relative distance to the gel point, $\varepsilon = (|p - p_c|/p_c)$ and $\Delta = t/(s+t)$. From Eq. 6.18, it follows that tan δ is independent of frequency at the gel point

$$\tan\delta = G''/G'' = \text{constant} = \tan(\Delta\pi/2) \tag{6.20}$$

The predictions of the scalar percolation theory have been experimentally confirmed for chemical gelation (Durand et al. 1989; Adam 1991) and physical gelation (Tokita et al. 1984; Axelos and Kolb 1990; Wang et al. 1991). For ovalbumin gels at pH 3.0 and NaCl 30 mM, the exponent t was estimated to be 1.79 ± 0.25 (van der Linden and Sagis 2001). However, for amyloid fibrillar β-lactoglobulin gels, prepared at pH 2, plotting values of G' extrapolated to infinite time, G'_{∞} versus $(c/c_0) - 1$, Gosal et al. (2004) obtained values of the exponent, t, in the range 2.2–2.8. The same authors reported a value of 3.1 for the data of Pouzot et al. (2004). They suggest that for heat-set β-lactoglobulin gels, prepared at pH 2, the exponent appears to be between 2 and 3. The higher values could be consistent with the Bethe lattice value of 3 and the values smaller than 3 could also be explained by the classical approach if network wastage is taken into account.

It should be noted that the exponents are defined to be critical only when data are accessed within the critical region. A practical guide for the critical region is given by Stauffer et al. (1982) who suggest $10^{-2} < (p/p_c) - 1 < 10^{-1}$. However, it appears that very few of the data of Gosal et al. (2004) satisfied this condition. Also, other theoretical studies assuming the same isotropic force between neighboring sites predicted a value of the exponent, t, to be 2.0. Even higher values are predicted for macroscopically nonhomogeneous materials assuming a central force and a bending energy term contributing to the elastic energy. This "universality" class has a lower bound value of 2.85 and an upper bound of 3.55 (van der Linden and Sagis 2001).

Fractal Model

Wu and Morbidelli (2001) extended the model of Shih et al. (1990) discussed in Chap. 2 to include gels that are intermediate between the strong-link and the weak-link regimes. In addition to the modulus, they also considered the critical strain, γ_c, at which the linear viscoelastic region ends

$$G' \alpha \phi^A \tag{6.21}$$

$$\gamma_c = \phi^B \tag{6.22}$$

The exponents A and B are given by

$$A = \beta/(d - D_f); \quad \beta = (d-2) + (2+x)(1-\alpha) \tag{6.23}$$

$$B = (d - \beta - 1)/(d - D_f) \tag{6.24}$$

where the parameter $\alpha = 0$ for a strong-link gel, $\alpha = 1$ for a weak-link gel, and $0 < \alpha < 1$ for intermediate gels; x is the fractal dimension of the gel's backbone, and d is the Euclidian dimension of the system (usually taken as 3). One obtains for weak links, $B = 1/(d - D_f)$, and for strong links, $B = -(1+x)/(d - D_f)$. Thus, one must have reliable data on modulus and critical strain versus protein concentration in order to calculate values of D_f and β.

In the strong-link regime, where the interfloc links are stronger than the intrafloc links, the modulus of a gel is given by that of intralinks. In the weak-link regime, where the flocs are more rigid than the interfloc links, the elasticity of the interfloc links determines the elasticity of the gel. In the weak-link regime, the limit of linearity increases with increasing volume fraction; in contrast, in the strong-link regime it decreases. The exponent x, which represents the backbone fractal dimension or tortuosity of the network, has a value of 1.0–1.3 for a colloidal gel. A value of $x = 1.3$ was used by Wu and Morbidelli (2001) to calculate values of D_f, ranging between 1.73 and 2.82, and β, ranging between 1.0 and 4.4, of several protein gels. It should be noted that for $x = 1.3$, the magnitude of β should be between 1.0 and 4.3.

Pouzot et al. (2004) reported applicability of the fractal model to β-lactoglobulin gels prepared by heating at 80 °C and pH 7 and 0.1M NaCl. They suggested that the gels may be considered as collections of randomly close packed "blobs" with a self-similar structure characterized by a fractal dimension D_f 2.0 ± 0.1.

One concern with applying the fractal model to gels is that the critical concentration, c_0, is assumed to be zero. In addition, for amyloid fibrillar networks derived from β-lactoglobulin, Gosal et al. (2004) suggested that the assumption of self-similar structures would not be valid. Nevertheless, when applicable, the fractal model provides a means of understanding the structure of gels based on purely rheological properties.

Gel Point and Sol-Gel Transition by Rheological Measurements

During gelation, a polymer undergoes a phase transition from a liquid to a gel. The sol-gel transition is a critical point where one characteristic length scale (viz. the size of the largest molecule) diverges. It is a transition in connectivity between a sol where the "monomers" are not connected to a gel where they are connected. It is not a thermodynamic transition, nothing special is happening to the free energy, and it is not associated with any singularity in concentration fluctuations (Joanny 1989). Thus, independent of the system studied or the mechanism involved, gelation is a critical phenomena where the transition variable will be the connectivity of the physical or chemical bonds linking the basic structural units of the material. Therefore, rheological properties are very sensitive indicators of the critical gel point.

Rheological Definition of "Gel Point"

The idea of gel point has received much attention in synthetic and biopolymer gels. One can talk about gel point as an instant in time or as a specific temperature. In this chapter, the symbols t_c and T_{gel} will be used for the gel time and gel temperature (Tobitani and Ross-Murphy 1997a; Foegeding et al. 1998), respectively. One can imagine many qualitative definitions of gel point, especially based on visual observation of individuals. Here, however, methods based on rheological measurements are emphasized.

Before the gel point, the connectivity is small and the material typically relaxes rapidly. Near the gel point the relaxation time rises sharply and at the gel point it diverges to infinity (or at least to very long times for a finite sample); in addition, the relaxation spectrum does not contain a characteristic time anymore. After the gel point, if the network has reached a high degree of development, the maximum relaxation time of the final network is also very short.

Classical Concept

Experimental detection of the gel point is not always easy since the equilibrium shear modulus is technically zero at the gel point and any applied stress will eventually relax, but only at infinite time. From the classical theory, the attributes of the gel point are an infinite steady shear viscosity and a zero equilibrium modulus at zero frequency limit (Fig. 6.3; Flory 1953). These criteria have been widely employed to detect the gel point of chemical gels. However, because continuous shearing affects gel formation, accurate information from viscosity measurement is not possible in the close vicinity of the gel point. Furthermore, information regarding the transition

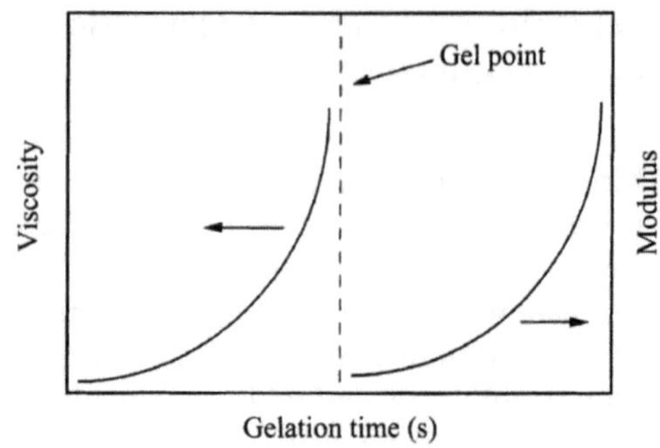

Fig. 6.3 Schematic diagram of the classical definition of changes in viscosity and modulus at the gel point

itself could only be obtained by extrapolation, thereby introducing uncertainties in the determination of the gelation moment.

Recently, focus has been placed on small amplitude oscillatory shear technique to measure the dynamic moduli during the gelation process to identify the gel point. By this method, the continuous evolution of the viscoelastic properties throughout the gelation process can be followed. Because the strain is kept small, modification of molecular structure caused by shear is minimized that, in turn, is an important advantage over the classical methods based on the diverging viscosity. Some of the magnitudes of the critical exponents obtained for the steady-state viscoelastic properties of different biopolymers are shown in Table 6.3 (Lopes da Silva et al. 1998) and they are in agreement with different approaches to the percolation problem. More important is the fact that irrespective of the mechanism or complexity of the gelation process, the changes in the viscoelastic parameters near the gel point obey similar scaling laws.

G′ and G″ Crossover

Studies on thermosetting resins (Tung and Dynes 1982) suggested that the gel point might occur at the time at which G' and G'' cross each other at a given frequency. This criterion was applied to the gelation of some biopolymers, such as gelatin (Djabourov et al. 1988b), β-lactoglobulin (Stading and Hermansson 1990), maltodextrins (Kasapis et al. 1993), and κ-carrageenan gels (Stading and Hermansson 1993). However, the time of G'-G'' crossover was found to be dependent on the oscillatory frequency in the case of both synthetic polymers (Winter et al. 1988; Muller et al. 1991) and biopolymers (Djabourov et al. 1988b). As the gel time is an intrinsic property of the material, it cannot be dependent on the frequency of the dynamic rheological experiment. Another concern is that in low-concentration gelling systems, the viscoelastic moduli may be too low to give a measurable signal by a conventional rheometer. However, it is likely that the G'-G'' crossover time might be close to the sol/gel transition time.

Table 6.3 Some values of the critical exponents (S and T) for biopolymer gelation

Polymer systems	S	T	Reference
Casein	–	1.8–2.6[a]	Tokita et al. (1984)
Gelatin	1.48	1.82	Djabourov et al. (1988b)
	1.1	1.70–1.9	Borchard and Burg (1989)
–	0.95	–	Dumas and Bacri (1980)
LMP/calcium	0.82	1.93	Axelos and Kolb (1990)
Alginate/calcium	0.88	–	Wang et al. (1991)
Alginate/cobalt	1.14	–	
LMP/calcium	0.70	1.70	Audebrand et al. (2006)

[a] Depending on frequency

Power Law Behavior of the Shear Modulus

Winter and coworkers (Winter and Chambon 1986; Chambon and Winter 1987; Winter et al. 1988) gave the first experimental demonstration of the power law behavior of the relaxation modulus at gel point, and that the scaling of the dynamic moduli and dynamic viscosity according to power laws in oscillatory frequency. Power law relaxation at gel point seems to be a general property of both chemical gelation and physical gelation involving either synthetic polymers or biopolymers, respectively. However, the different values of the relaxation exponent (Δ) shown in Table 6.4 reflect the fact that it is not a universal parameter for gelation. Even for the same polymeric material, relatively wide range of values have been obtained for this critical exponent and for some physical gels it was found to be dependent on polymer concentration and thermal history (Michon et al. 1993; Lopes da Silva and Gonçalves 1994), as well as chain stiffness (Matsumoto et al. 1992) and molecular weight (Lu et al. 2005). As the exponent Δ depends on molecular structure, it may be useful to distinguish between different gel structures (Lopes da Silva et al. 1998).

Threshold G′ Value

Equation 6.20 is the most generally valid gel point criterion and it has been successfully applied for the detection of the gel point of a wide variety of polymers (te Nijenhuis and Winter 1989; Lin et al. 1991; Muller et al. 1991; Scanlan and Winter 1991; Michon et al. 1993). Equations 6.16–6.20 still provide the most generally valid gel point criterion and they have been successfully applied for the detection of the gel point of a wide variety of polymers (te Nijenhuis 1997; Winter and Mours 1997). Nevertheless, arguments have been formulated (Kavanagh and Ross-Murphy 1998) that are not in favor of the general application of the assumptions implicit in the Winter's approach. One reason is that because entanglements are neglected, Winter's approach may be of limited usefulness for physical networks that exhibit highly ordered structures before gelation, that is, when the lowest detected value

Table 6.4 Some values of the critical relaxation exponent (Δ) obtained for physical gels

Physical gels	Δ	Reference
Poly (vinyl chloride) in plasticizers	0.8	te Nijenhuis and Winter (1989)
Thermoplastic elastomeric poly (propylene)	0.125	Lin et al. (1991)
LM pectin/calcium systems	0.71	Axelos and Kolb (1990)
Xanthan/carob and iota-carrageenan	0.50	Cuvelier et al. (1990)
Xanthan/chromium	0.35	Nolte et al. (1992)
Sodium alginates	0.53–0.84	Matsumoto et al. (1992)
Gelatin	0.22–0.71	Michon et al. (1993)
HM pectin/sucrose systems	0.41–0.65	da Silva and Gonçalves (1994)
Schizophyllan/sorbitol	0.50	Fuchs et al. (1997)
Xanthan/aluminum	0.16–0.22	Richteret al. (2004)
Xanthan/carob gum mixtures	0.62–0.67	Richter et al. (2004)
Alginate/calcium	0.37–0.72	Lu et al. (2005)
Chitosan/acetic acid-water-propanediol	≈ 0.5	Montembault et al. (2005)

of G' may already be greater than G'', as was the case with some heat-set globular protein gels (Gosal et al. 2004).

Also, some practical limitations may occur when applying these criteria. In fact, for many biopolymer systems, gelation begins from a sol state characterized by a very low viscosity that is often below the resolution of the rheometer in small strain oscillatory experiments. On the other hand, the lowest detected value of G' may already be greater than G'' (Gosal et al. 2004). Therefore, other criteria have been used for the gel point. Many authors have assessed the gel point as when G' has increased to a value greater than the experimental noise level, an approach that is not very rigorous. Nevertheless, this criterion was used for amylose (Doublier and Choplin 1989), β-lactoglobulin (Stading and Hermansson 1990), gelatin (Ross-Murphy 1991b), and BSA (Richardson and Ross-Murphy 1981a; Tobitani and Ross-Murphy 1997). In the latter, it was shown that this criterion systematically indicated a gel time less than that measured on the basis of viscosity changes.

Extrapolation of G' Value

Several methods have been used to estimate either the gel temperature or the gel time. Extrapolation of the rapidly rising values of G' to the time axis was suggested as an accurate estimate of the gel point, t_c (Clark 1991). Hsieh et al. (1993) determined T_{gel} of whey and egg proteins by extrapolation of the rapidly rising values of G' to the temperature axis. In addition, the gel time was calculated for the gelation of bovine serum albumin by monitoring the increasing viscosity as a function of time (Richardson and Ross-Murphy 1981a) and taking the time corresponding to the asymptotic value of viscosity. In this method, although small shear rates were used, the high strains necessarily imposed probably affected the gel critical point.

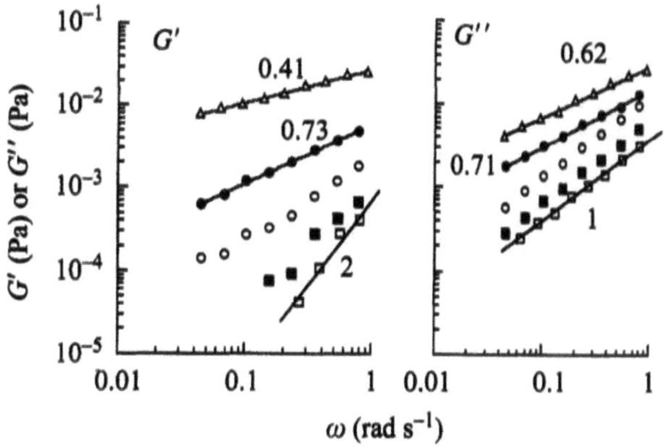

Fig. 6.4 Mechanical spectra recorded at 20 °C of 6.1 g L^{-1} low-methoxyl pectin (pH 7, 0.1 mol L^{-1} NaCl) with different calcium contents aged for 20 h at 20 °C: □ no added calcium, ■ 1.4×10^{-3} mol L^{-1}, O 1.9×10^{-3} mol L^{-1}, • 2.1×10^{-3} mol L^{-1}, Δ 2.4×10^{-3} mol L^{-1}. Numbers are values of the slopes of lines

The time at which the modulus versus time plot showed a sudden increase in slope on a log-log plot, called the log discontinuity method, was used to determine the gel time (Gosal et al. 2004). It was claimed that the log discontinuity method most closely reflects the expected percolation threshold behavior.

Critical Viscoelastic Behavior at the Gelation Threshold

Based on the above defined criteria, dynamic rheological experiments have been widely used to follow the aggregation process before the gel point and throughout the sol-gel transition for several biopolymer systems (refer to "Rheological behavior of food gel systems," First edition). Due to its well-defined viscoelastic changes near the gel point and the easy experimental assessment of these changes by small-deformation rheology, calcium-induced LM pectin gelation is a useful example to illustrate the critical viscoelastic behavior generally observed at the gelation threshold.

Dynamic rheological experiments have been used to follow the aggregation process before the gel point and throughout the sol-gel transition for calcium-induced LM pectin gelation as a function of the calcium added to the system and as a function of the aging time (da Silva et al. 1996). Figure 6.4 shows G' and G'' at 20 °C plotted against angular frequency (ω) at various calcium concentrations added to a pectin solution (c_p=6.1 g L^{-1}, pH 7, 0.1 mol L^{-1} NaCl). The LM pectin dispersion without added calcium shows the classical viscoelastic behavior of a macromolecular solution in the terminal zone at low frequencies with $G'(\omega) \propto \omega^2$ and $G''(\omega) \propto \omega$. From this low-frequency behavior, the system smoothly evolves toward a plateau value with increasing calcium concentration, with both moduli increasing in magnitude and decreasing their frequency dependence. It was very close to the gel point when the calcium concentration equaled 0.083 g L^{-1}, for which $G'(\omega) \propto \omega^{0\ 73}$ and $G''(\omega) \propto \omega^{0.71}$. Thus, this critical behavior at gel point was in satisfactory agreement with the percolation model prediction (2/3) based on Rouse-like dynamics (Martin et al. 1989).

Fig. 6.5 Dynamic moduli (□ G', ■ G'') as a function of frequency at 20 °C, measured at different aging times ([low-methoxyl pectin] = 6.1 g L^{-1}, [Ca^{2+}] = 2.4 × 10^{-3} mol^{-1}, pH 7, 0.1 mol^{-1} NaCl). To avoid superposition of data, values (A) shown were added to G'

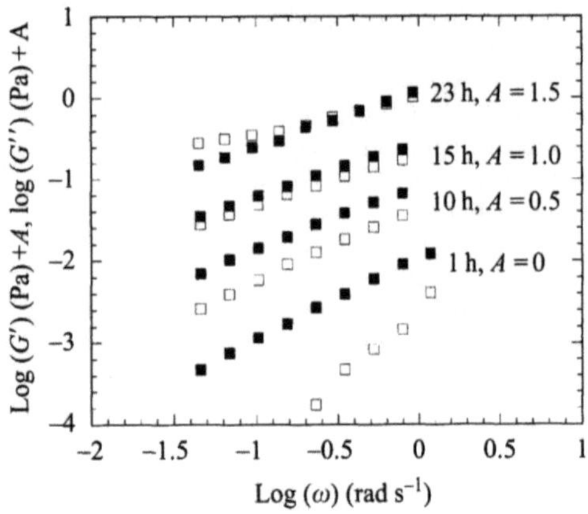

Figure 6.5 shows the evolution of the dynamic moduli for an LM pectin/calcium system in the vicinity of the gel point as a function of the aging time. The evolution of the dynamic moduli was similar to that one observed as a function of the calcium concentration. In the initial period of aging, the system showed the typical liquid-like behavior. Then, both moduli increased with time, G' increasing more rapidly than G'' and with lower dependence on frequency. After 15 h of aging, the system was just above the critical gel point, with a viscoelastic exponent Δ in the range of 0.65–0.68. After the gel point, G' passed beyond G'', first in the lower frequency range where one can observe the initial formation of the elastic plateau.

Influence of Concentration and Temperature on Gel Time

The influence of concentration on gel time of gelatin gels was described (te Nijenhuis, 1981) by the equation

$$\ln(t_c) = a_1 + a_2 c \tag{6.25}$$

Based on gelation of high-methoxyl (HM) pectins, Oakenfull (1984) suggested the equation

$$t_c = \frac{1}{kc^n} \tag{6.26}$$

where k is a reaction rate constant and n is the order of reaction. It is to be understood that reaction orders higher than 4 are not encountered often and a minimum (critical) concentration is necessary for gelation. However, gel time versus concentration data at 75 and 80 °C for β-lactoglobulin at pH 2 followed simple power law relationships with slopes (power law indices) lying between −5.5 and −6.4. In contrast, the simple cascade model predicts a power law index of 1 for gel times at higher concentrations and a sudden divergent increase in the gel time at a critical concentration c_0 (Gosal et al. 2004). Also, of relevance is the work of Clark (1994) and Clark et al. (1994) that is based on kinetics and the cascade theory with the limitations of this approach mentioned earlier.

Tobitani and Ross-Murphy (1997) used kinetic and scaling concepts to arrive at the concentration dependence as

$$t_c \approx \frac{nN_{b0}}{kN_{a0}^n} \tag{6.27}$$

$$t_c \sim \frac{1}{N_{a0}^n} s \sim \frac{1}{c^n} \tag{6.28}$$

where N_{a0}^n is the initial number of active sites of the polymer per unit volume, N_{b0}^n is the initial number of junctions composed of n reacted cites, and the symbol ~ is used to indicate that the two terms on either side can be equated by scaling, but not by absolute values. The influence of temperature, T(K), was taken into account in the reaction rate constant, k

$$k = A\exp\left(\frac{-E_a}{RT}\right) \tag{6.29}$$

where A is the preexponential factor, E_a is the apparent activation energy, and R is the gas constant. da Silva et al. (1995) also found that the effect of temperature on gelation rates followed the Arrhenius equation. Tobitani and Ross-Murphy (1997) combined the above equations and then modified the resulting expression to take into account the need for the cross term of (ln c) and temperature

$$\ln(t_c) = a_1 + a_2\ln(c) + a_3T + a_4T\ln(c) \tag{6.30}$$

where a_1, a_2, a_3, and a_4 are coefficients to be determined from experimental data.

Experimental gel threshold times, indicated by symbols, of bovine serum albumin (Tobitani and Ross-Murphy 1997) are shown as a function of temperature and the threshold times 100, 1,000, and 10,000 s in Fig. 6.6. The threshold values of time were determined as the time at which G' of a sample reached a threshold value of between 2 and 3 Pa. Furthermore, values calculated using the above equation are shown as lines in the figure. In the diagram, the upper region represents the gel state and the lower region the sol state of a sample. For a given set of concentration and temperature values, the threshold time (e.g., 100 s) for gel formation can be found from the figure; the diagram can also be used to predict the possibility of gel formation for any given concentration and temperature.

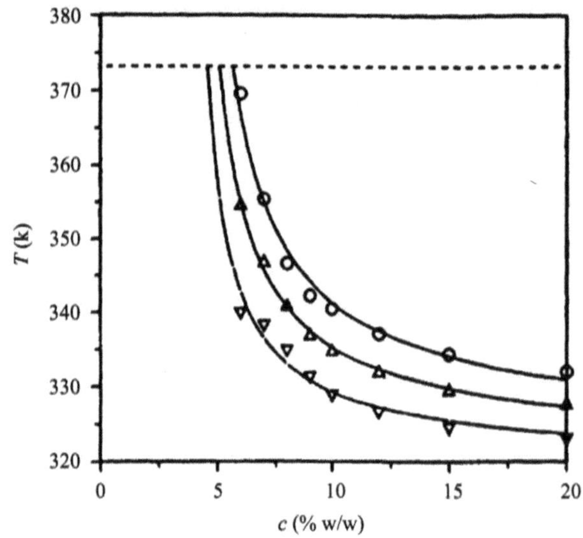

Fig. 6.6 Temperature versus concentration state diagram illustrating the sol-gel boundary for BSA in which threshold times for gelation are shown: circles, 100 s; ▲ triangles up, 1,000 s; ▼ triangles down, 10,000 s. Horizontal broken line is boiling point of solvent. Curves shown are based on Eq. 6.26 (Tobitani and Ross-Murphy 1997)

Evaluation of Structure Development during Biopolymer Gelation

Aqueous biopolymer gels exhibit a general characteristic behavior when the structure development is evaluated by dynamic rheological measurements (Clark and Ross-Murphy 1987). The gradual development of the network structure is reflected by a progressive increase in storage modulus (G'). The beginning of the gelation process is dominated by the viscous behavior of the system ($G''> G'$) and the elastic behavior dominates the final stages of the experiment ($G'>>G''$). Both moduli increase as a result of the increasing density of junction zones, but G' rises more sharply until it intercepts and then exceeds G''. After an initial rapid increase, resulting from the rapid formation of junction zones into the polymer network, G' keeps increasing continuously as a result of the slower formation and rearrangement of junction zones resulting in a net decrease in the length of elastically active chains; values of G' eventually reach a pseudoplateau region. It is emphasized that in many biopolymer gelation processes, for example, gelatin (te Nijenhuis 1981; Ross-Murphy 1991b) and HM pectin/fructose and sucrose systems (Rao and Cooley 1993; da Silva and Gongalves 1994), because of the nonequilibrium behavior, even after aging for long periods, G' continues to increase steadily as a result of a continuous reorganization of the network. As expected from a network that has attained a state close to equilibrium, the kinetics applicable at longer times are quite different from those observed close to the gel point.

Kinetic measurements close to the gel point can provide important information about the gelation process of biopolymers, since it is under these conditions that the mechanism of gel formation can be reasonably related to its mechanical properties. As an example, for HM pectin gelation, the power law dependence of the gelation

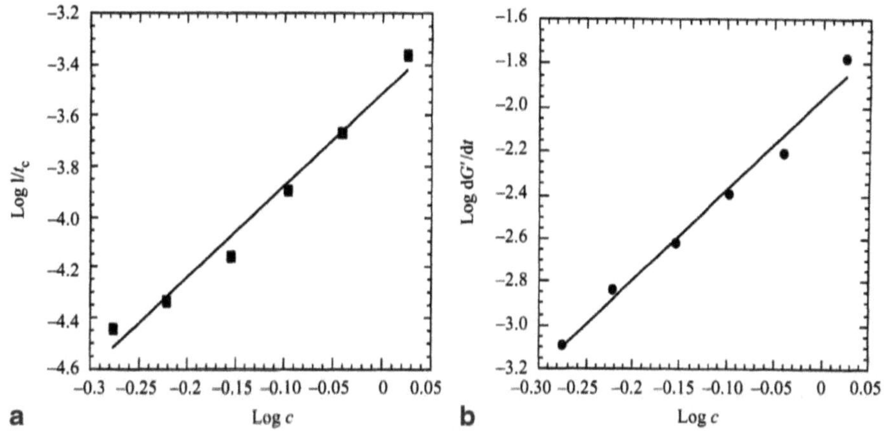

Fig. 6.7 Influence of high-methoxyl pectin concentration on the gelation rate (inverse of the apparent gel time), t_c (**a**) and on the rate of change in G' at the gelation threshold (**b**). The slope of the straight line in (**a**) was 3.6 (R^2 =0.981); least squares fit of data in (**b**) gave a slope of 4.1 and R^2 =0.986

rate (inverse of the gel time, defined as the time at which $G'=G''$) upon concentration was found to have a slope of 3.6 (Fig. 6.7a). Taking into consideration the kinetic model of Ross-Murphy (1991a), using an apparent second-order reaction ($n=2$) and assuming $p=1.8$, the approximate critical value for the percolation exponent (Stauffer et al. 1982), the limiting value of the exponent of a log-log plot of $(1/t_c)$ versus c at high (c/c_0) was equal to np, in close agreement with the slope of 3.6 obtained in Fig. 6.7a (da Silva and Gonçalves 1994).

$$t_c \approx K/[((c/c_0)^n - 1)^p] \tag{6.31}$$

The rate of change in G', dG'/dt, and the modulus G' at the gelation threshold are also sensitive to changes in pectin concentration. The rate of change in G' increased with pectin concentration (Fig. 6.7b). The usual high power dependence obtained for dG'/dt with polymer concentration, was probably related to the complex behavior typical at low biopolymer concentration close to the gel threshold. Pronounced concentration dependence has been reported for a wide range of systems. For BSA gelation, the slope of the modulus growth near the gel point was found to be proportional to c^{27} for low concentrations, and even at higher concentrations dG'/dt was proportional to c^6 (Richardson and Ross-Murphy 1981a). For physical gels of poly(acrylonitrile) in dimethylformamide close to the critical gel concentration, Bisschops (1955) found dG'/d$t \sim c^{22}$. In contrast, Nolte et al. (1992) found a much lower power dependence of dG'/dt on polymer concentration: dG'/d$t \sim c^{1.25}$, for the gelation of xanthan with chromium ions, even close to the critical gelling concentration. Close to the critical concentration the interpretation of these exponents is complex. It is likely that the different mechanisms of macromolecular association involved account for this wide range of values observed.

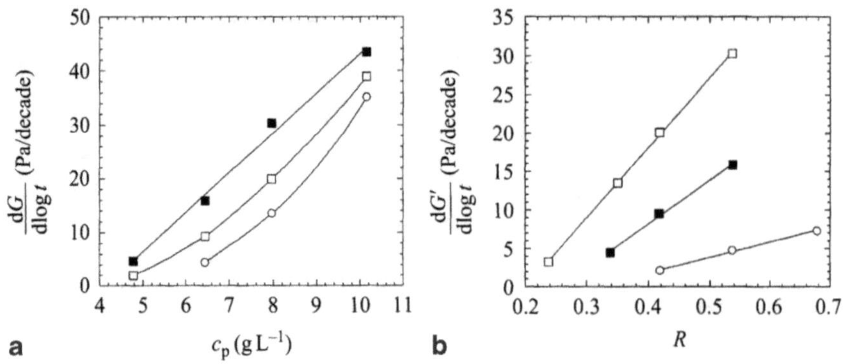

Fig. 6.8 Rate of aging for low-methoxyl pectin gelled systems studied near the pseudoplateau region: (**a**) dG'/d log t, against pectin concentration (0.1 mol L^{-1} NaCl, pH 7) for different calcium:pectin ratios: O R=0.34; ■ R=0.42; □ R=0.54; (**b**) dG'/d log t, against the stoichiometric calcium:pectin ratio, R, for different pectin concentrations: O 4.78 g L^{-1}; ■ 6.45 g L^{-1}; □ 7.96 g L^{-1}

When the storage modulus is plotted against log aging time, typically the curves obtained are approximately linear after some time. The slope of these curves (dG'/dt) can be interpreted as a rate of aging in this long time range, a criterion used by te Nijenhuis (1981) to analyze the rate of aging of gelatin and polyvinyl chloride gels. Examples of the application of this approach are shown in Fig. 6.8 for LM pectin/Ca^{2+} gelation. Magnitudes of the rates of aging increased when the pectin concentration was increased (Fig. 6.8a), or when the calcium/pectin ratio was increased (Fig. 6.8b), suggesting that the annealing processes, such as rearrangement of junction zones and lateral chain aggregation resulting in the net increase in the length of elastically active chains, became more prevalent at higher pectin concentration and/or higher calcium/pectin ratio.

Effect of Temperature on the Rate of Structure Formation and Kinetic Data

Most gelation processes involving biopolymers can be seen as thermal effects, because the gelation mechanism basically is related to the change of only one thermodynamic parameter: the temperature. Depending on the molecular organization of the polymer network, temperature can have different effects on the viscoelasticity of the gel. The study of the influence of temperature on structure formation in biopolymer systems used as gelling agents in food formulations could be useful in understanding the thermal behavior of food commodities where they are incorporated and can provide additional insights into the nature of the gelation process of these polymers.

Fig. 6.9 Arrhenius treatment for the influence of temperature on the gelation rate of 1 % high-methoxyl pectin dispersion (60 % sucrose, pH 3), measured at different aging temperatures in the range 15–60 °C

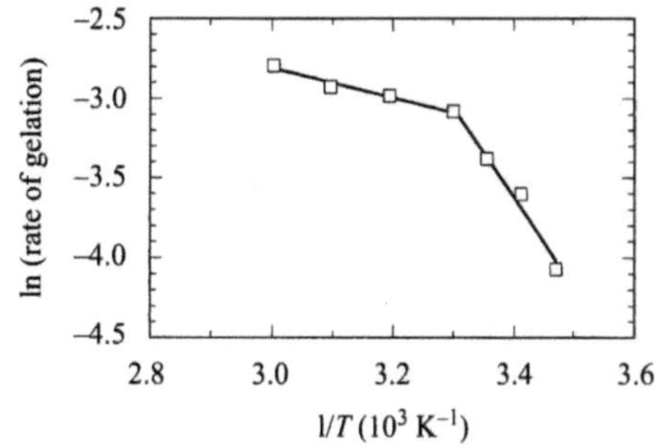

Structure Development Rate

Structure development rate (SDR), defined as dG'/dt, can be measured under isothermal conditions, at different aging temperatures (cure curves), and also during the gelation process promoted by decreasing the temperature of the biopolymer dispersion (non-isothermal conditions; Rao and Cooley 1993). The analysis of SDRs under isothermal conditions, as opposed to non-isothermal conditions, can be very useful to avoid the nonequilibrium effects and the effect of thermal history of the system.

Structure Development under Isothermal Conditions

The evolution of dG'/dt with time during a cure experiment has been studied during the gelation of HM pectin/sucrose systems (da Silva et al. 1995). The gelation rate of HM pectin/sucrose systems was strongly dependent upon the temperature, showing a complex behavior due to several kinds of intermolecular interactions involved. The influence of temperature on these intermolecular interactions, mainly hydrogen bonding and hydrophobic interactions, could be described by an Arrhenius relationship (Fig. 6.9). Two different temperature ranges could be identified, one between 15 and 30 °C and the other between 30 and 60 °C, yielding apparent activation energies of 46.4 and 7.7 kJ mol^{-1}, respectively. It is clear that the gelation rate was energetically favored above 30 °C.

Two different processes were proposed to explain the discontinuity observed, each one having rates with different temperature dependence. One of them was the initial intermolecular local contacts, essentially by hydrogen bonding and involving short chain segments. This stage can have similarities with the nucleation step found in the gelation of biopolymers that undergo a coil-to-helix transition. This process was considered the rate determining process above 30 °C. The second different rate process was attributed to the subsequent lateral aggregation of chains and enlargement of junction zones, essentially governed by hydrophobic interactions, favored by increasing temperature, probably stabilized or strengthened by van der Waals forces. This was considered to be the rate limiting process below 30 °C. Hydrogen

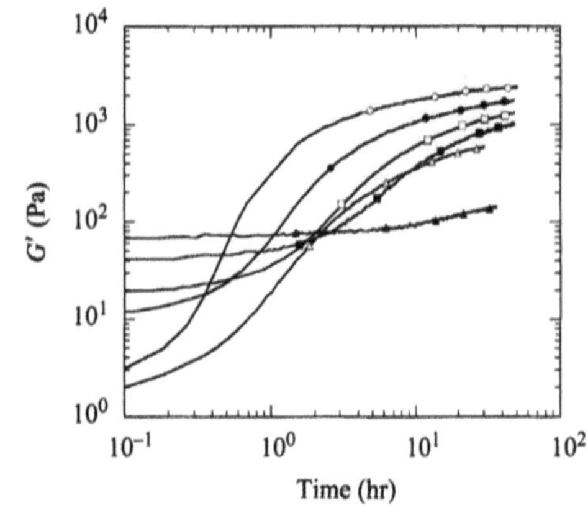

Fig. 6.10 Storage modulus (G') as a function of the aging time for 1 % high-methoxyl pectin (60 % sucrose, ph 3), at several aging temperatures: (▲) 5 °C, (■) 10 °C, (□) 15 °C, (•) 20 °C, (O) 30 °C, (Δ) 50 °C. Data were obtained at 3 % strain and 0.5 Hz, every 3 min during the first 3 h, and then every 10 min. For clarity, only some of the experimental points are shown

bonds are less dependent upon the separation of molecules and have enough energy to be stable even when they are few in numbers. Hydrogen bonds acting at larger intermolecular distances can be essential to the subsequent development of hydrophobic interactions and van der Waals forces, through clustering of nonpolar groups involving larger portions of the pectin chains.

The kinetics at longer times were quite different. As shown in Fig. 6.10, the initial period of gelation was characterized by concave cure curves, after which they became convex and then almost linear. At 5 °C, the convex region was not perceptible. This different behavior, and the effect of temperature on the rate of aging of the HM pectin gels, could be clearly illustrated when the SDRs were analyzed at different aging temperatures.

Figure 6.11 shows SDRs as a function of the aging time at different temperatures. Significant differences in SDRs were observed during the first 4 h of aging. The general behavior is an initial increase in SDR, followed by a decrease in the SDR values, which continue to decrease steadily until the end of the aging period. The sharpest initial increase in the aging process was observed at 30 °C, the maximum in SDR was reached at a lower aging time than at the other aging temperatures. For the aging at 5 °C, a different behavior was observed, with an initial rapid decrease during the first hour, followed by an almost constant SDR value until the end of the aging period. Between 15 and 50 °C, the initial increase in SDR was due to the initial rapid increase in the storage modulus during the aging of the pectin gel. This exponential time dependence of the increase in elasticity of the system should correspond to the rapid formation of junction zones between pectin chains. After this stage, the SDR decreased until it reached a value, always higher than zero, but almost independent of the aging time or of the aging temperature, that corresponded to a slow increase in storage modulus and to the almost linear region of the cure curves. This latter stage should correspond to a slow reorganization of the network allowing for creation of new junction zones, or increase in the extension of the junctions between molecular chains.

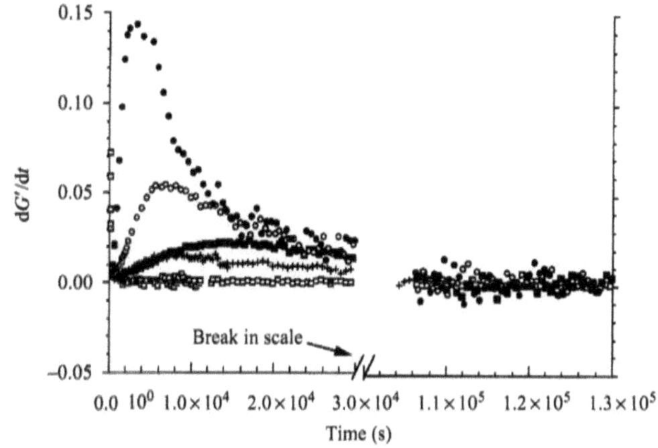

Fig. 6.11 Structure development rates (SDRs) during aging of a 1 % high-methoxyl pectin/60 % sucrose (pH 3.0) dispersion under isothermal conditions at: (□) 5 °C, (■) 15 °C, (O) 20 °C, (•) 30 °C, and (+) 50 °C. SDR was calculated by numerical differentiation followed by smoothing of the derivative values

Structure Development Rate under Non-Isothermal Conditions

Based on rheological data obtained from temperature sweep experiments, SDRs can also be measured during the gelation process promoted by decreasing the temperature of the biopolymer dispersion (non-isothermal conditions) as illustrated for the HM pectin/sucrose system in Fig. 6.12. During gelation induced by cooling, the SDR increased with decreasing temperature, showing a higher variation at the higher and lower temperature ranges. A temperature range where the SDR increases with decrease in temperature can be found during the gelation of biopolymers that undergo coil-helix transitions, for example, gelatin (Djabourov et al. 1988a) or agarose (Braudo et al. 1991). However, in these cases, the effect of temperature on the formation of ordered structures typically follows a sigmoidal shape, instead of the logarithmic shape observed, and consequently, the effect of temperature on the rate of structure development shows a bell-shaped curve, with a maximum value at an intermediate temperature.

Kinetics of Gelation

Kinetic data can also be obtained for the structure development process during cooling of biopolymer dispersions. This kind of analysis is difficult due to the combined effects of time and temperature. Usually, the data are obtained using time as an independent variable and changing the temperature and test sample for each run.

A non-isothermal kinetic model was used to describe the gelation process of the HM pectin/sucrose system during cooling (da Silva et al. 1995), taking into

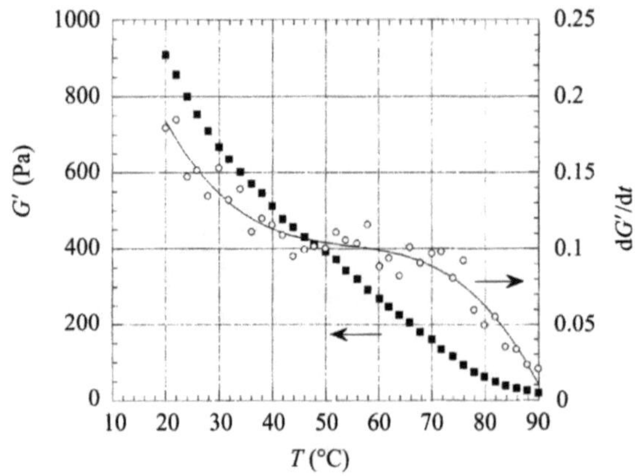

Fig. 6.12 Influence of temperature on storage modulus (G') and dG'/dt (SDR) for 1 2 high-methoxyl pectin/60 h sucrose dispersion during cooling at 0.5 °C min^{-1} from 90 to 20 °C at 3 a strain and 0.5 Hz; (■) G'; (O) dG'/dt calculated after cubic spline fit; (- - -) dG′/dt calculated after polynomial equation fit

consideration kinetic effects as well as thermal effects on biopolymer gelation. It was adapted from the "linear increasing temperature method" (Rhim et al. 1989) to changes in structure under non-isothermal conditions (temperature sweeps) by linearly decreasing the temperature at a constant rate. This non-isothermal kinetic model was based on the combination of the classical rate equation, the Arrhenius equation, and the time-temperature relationship, yielding

$$\ln\left(\frac{1}{G'^n}\frac{dG'}{\mathrm{d}t}\right) = \ln k_0 - \left(\frac{E_a}{RT}\right) \tag{6.32}$$

where G' is storage modulus, taken as a measure of the number of junction zones in the network, n is reaction rate order, t is time, k_o is the Arrhenius frequency factor, E_a is the activation energy of the process, R is universal gas constant, and T is the absolute temperature.

An important step in using this method to describe the gelation process of biopolymers is that the order of the reaction rate can be evaluated, a parameter that can be difficult to determine for this kind of process. It was assumed that at least in the incipient gelation process of the HM pectin/sucrose aqueous system, macromolecular aggregation could he described by a bimolecular association equilibrium, that is, $n=2$. This was also supported by the results obtained for the concentration dependence of the gelation rate. However, that a more complex mechanism due to the change in temperature or in the number of preformed junction zones during the cooling-induced gelation of our system can also occur should not be excluded.

Assuming the reaction order $n=2$, Eq. 6.28 was used to estimate the activation energy of the process, by using a linear least squares routine (Fig. 6.13). The process could be satisfactorily described by assuming a two-step process, corresponding to two different temperature ranges. Between 90 and 62 °C, the least squares linear regression yielded $E_a = 132.0$ kJ mol^{-1} and $k_o = 5.09 \times 10^{14}$ Pa^{-1} s^{-1}, and between 50 and 20 °C, $E_a = 27.6$ kJ mol^{-1} and $k_o = 1.77 \times 10^{-2} \cdot$ Pa^{-1} s^{-1}

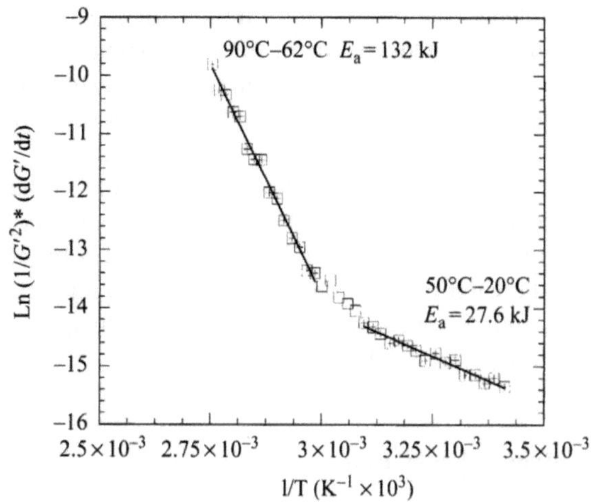

Fig. 6.13 Arrhenius plot for the change in the storage modulus (G') during cooling of a 1 % high-methoxyl pectin/60 % sucrose dispersion (pH 3.0). Apparent second-order kinetics and a two-stage gelation process were assumed

Since data points were estimated in a continuous manner during the initial period of gelation induced by cooling, the kinetic parameters obtained were likely affected by the previous thermal history of the material. The higher activation energy in the higher temperature range may reflect the higher energy barrier between the unaggregated and aggregated macromolecular states, under temperature conditions not favorable to the development of intermolecular interactions that could promote the formation of the pectin network. Under thermal conditions that favor interchain aggregation, as was the case below 50 °C, the structure development process was energetically easier, and the activation energy was lower.

Evaluation of Structure Loss during Melting/Softening of Biopolymer Gels

Temperature Dependence of Gel Modulus

Many biopolymer gels follow the general behavior of viscoelastic amorphous polymers: a decrease in viscoelastic parameters with increasing temperature (Ferry 1980). Such biopolymer gels include: alginates (Andresen and Smidsrød 1977), κ-carrageenan (Watase and Nishinari 1987b), agarose/κ-carrageenan mixed gels (Zhang and Rochas 1990), globular proteins (Richardson and Ross-Murphy 1981a, 1981b), and LM pectin/calcium gels (Garnier 1992). This behavior is typical of polysaccharide gels where hydrogen bonding or electrostatic interactions are the only significant interactions that stabilize the polymer network. Consequently, for most biopolymer gels the rupture of junction zones stabilized by thermolabile interactions may explain the decrease in modulus with increasing temperature, in addition to the increase in flexibility of the polymer chains. Nishinari et al. (1985)

introduced the "reel-chain" model for the molecular mechanism underlying the temperature dependence of the viscoelastic behavior of thermoreversible biopolymer gels, that was further theoretically treated by Higgs and Ball (1990). SLRs of only a few binary gels, such as HM pectin/sodium alginate (Rao and Cooley 1995), have been studied.

Clearly, temperature plays a crucial role in the formation and disruption of the typical physical cross-links found in biopolymer gels. Activation energies associated with the breakage of junction zones in biopolymer gels have been estimated from the temperature dependence of different rheological parameters. For both protein and polysaccharide gels, activation energies ranging from 10 to 275 kJ mol^{-1} have been reported (Table 6.5).

Eldridge-Ferry Plot

Eldridge and Ferry (1954) studied the relationship between the melting point (T_m), and concentration (c) and molecular weight (M) of gelatin gels. They found linear relationships between ln (c) and $1/T_m$, and between ln (M) and T_m. Assuming an equilibrium binary association of polymer chains and using the van't Hoff law, they suggested the following relationship from which it is possible to calculate the enthalpy (ΔH_m) of the cross-linking process and to get some information about the stability of the junction zones into the gel:

$$-\left[\frac{\mathrm{d}\ln c}{\mathrm{d}T_m}\right]_{MW} = \frac{\Delta H_m}{RT^2} \tag{6.33}$$

This approach was satisfactorily applied to other biopolymer gelled systems, including agarose (Watase and Nishinari 1987a), HM pectin/dimethyl sulfoxide (DMSO) gels (Watase and Nishinari 1993), κ-carrageenan (Watase and Nishinari 1987b), and LM pectin/Ca^{2+} and LM pectin/Ca^{2+}/sucrose systems (Fu and Rao 1999).

Melting temperatures (T_m) were determined for several LM pectin/calcium gels at different pectin and calcium concentrations. During the heating process of the gel, the storage modulus G' decreased with increasing temperature. T_m was selected to be either the temperature at which G' became too small to be accurately measured or that corresponding to the inflection point of the plot of G' versus temperature. T_m increased as either the pectin or the calcium concentration increased, due to the increase in cross-linking junctions between the pectin chains. For the LM pectin + calcium, plots of ln c_p versus T_m^{-1} (Fig. 6.14) showed good linearity. Similar plots for LM pectin + calcium + sucrose system also showed good linearity (Fu and Rao 1999).

The dissociation enthalpies of junction zones (ΔH_m) for melting of the LM pectin/Ca^{2+} and LM pectin/Ca^{2+}/sucrose gels that reflect the thermal stability of the junction zones were estimated from the slopes of the straight lines of the Eldridge–Ferry plots shown in Fig. 6.14. No significant effects of the pectin and calcium/pectin ratio (R) on the ΔH_m values were observed. The enthalpy change obtained in

Table 6.5 Magnitudes of activation energy for viscoelastic properties of biopolymer gels

Sample	E_a (kJ mol^{-1})	Technique	Reference
κ-carrageenan gels (0.1 mol L^{-1} KCl) 15 g L^{-1}	87	Temperature dependence of the shift factor required for superposition of creep curves (9–50 °C)	Braudo et al. (1984)
κ-carrageenan gels 15 g L^{-1}	10.0	Arrhenius equation applied to the elastic modulus measured at different temperatures by compression tests	Zhang and Rochas (1990)
Xanthan/chromium gels	42	Temperature dependence of the gelation rate measured at different temperatures	Nolte et al. (1992)
1.5 % HM pectin gels (70 % sucrose, pH 3)	155–160	Temperature dependence of the shift factor required for superposition of creep curves (25–50 °C)	Plashchina et al. (1979)
2 % HM pectin gels (65 % sucrose, pH 3)	130	Temperature dependence of the shift factor required for superposition of creep curves (11–49 °C)	Kawabata (1977)
2.5 % LM pectin gels (0.05 % calcium)	273	Temperature dependence of the shift factor required for superposition of creep curves (5.5–24.5 °C)	Kawabata (1977)
LM pectin/calcium gels (c_p = 4–8.5 g/ L^{-1}; R = 0.3–0.5)	11–86	Temperature dependence of the shift factor required for time-temperature superposition of mechanical spectra (5–40 °C).	Garnier (1992)
Whey protein gels (protein 14 %, pH 8)	87–143	Time-temperature superposition of creep curves and WLF equation (15–75 °C)	Katsuta and Kinsella (1990)

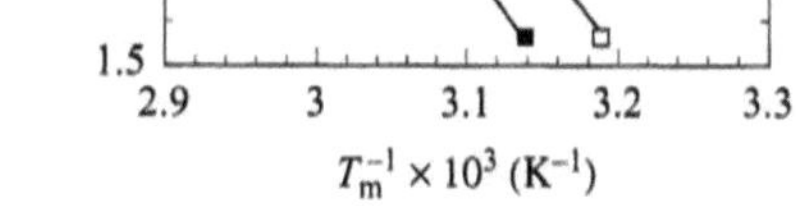

Fig. 6.14 Eldridge-Ferry plots ln c_p versus $1/T_m$ for low-methoxyl pectin + calcium gels. No sugar, 0.1 mol L^{-1} NaCl, pH 7.0, for two different R ratios: (□) R = 0.42 and (■) R = 0.54

this way characterizes the enthalpy associated with the strongest network junctions averaged over the concentration range, that is, the junction zones that remain stable until the melting temperature is reached (Braudo et al. 1984). The temperature dependence of the viscoelastic behavior of the gel is another method to measure the breakdown enthalpy of network junctions (ΔH_{bT}). The enthalpy change obtained in this way characterizes gels of a certain concentration and relates to the temperature interval for which the viscoelastic properties were measured. Its magnitude can be

Table 6.6 Magnitudes of enthalpy (ΔH_m) of melting of network junctions of biopolymer gels

Sample	ΔH_m (kJ mol^{-1})	Technique[a]	Reference
Gelatin gels (c=2–6%)			
chilled at 0°C	205–305[b]	1	Eldridge and Ferry (1954)
chilled at 15°C	502–920[b]		
HM pectin gels (c=2%)			
(70% sucrose, pH 3.0)	20	2	Plashchina et al. (1979)
HM pectin gels	126	1	Braudo et al. (1984)
(50% glycerol, pH 2.7)	172		
Ca-salt of furcellaran	121	1	Braudo et al. (1984)
(70% sucrose, pH 7.6)	192		
k-furcellaran gels	54	1	Braudo et al. (1984)
(0.15M KCl, pH 7.6)	40	2	
Agarose gels	$1.5–1.6 \times 10^3$	1	Watase and Nishinari (1987a)
(c=4–18%)			
k -carrageenan gels	60	1	Watase and Nishinari (1987b)
(c=2–12%)			
LM pectin/calcium gels	38–45[e]	1	da Silva and Rao (unpublished results)
	50–75[c]	2	Garnier (1992)
LM pectin/calcium/sucrose gels	43–50[e]	1	Fu and Rao (1999)
HM pectin/DMSO gels	115–440[d]	1	Watase and Nishinari (1993)
pH 2.4, pH 6.9	45–125[e]		

[a] Method 1—Eldridge–Ferry approach (Eldridge and Ferry 1954); Method 2—superposition of viscoelastic data at different temperatures, and plot of the slope of Ln bT versus 1/T (Schultz and Myers 1969)
[b] Depending on molecular weight
[c] Depending on R and pectin concentration
[d, e] Depending on DMSO content

regarded as a measure of the variation in the cross-link strength (i.e., in the number of cross-links) within the gel (Braudo et al. 1984). Both parameters can serve as measures of stability of the gel junction zones, but each of them reflects different manifestation of the breakdown process, and, thus, their magnitudes are not necessarily the same for all gelled systems. In fact, as can be seen in Table 6.6, it seems that the difference between ΔH_{bT} and ΔH_m is larger for gels with low water activity, for which an important contribution from hydrophobic interactions to the stability of the gel may be expected. Therefore, junction zones with different stability might be present and the hydrophobic interactions contribute to the relatively high values of ΔH_m.

Values of ΔH_m obtained for the LM pectin/calcium gels were close to those found for κ-carrageenan and κ-furcellaran gels, but much lower than those of either gelatin or agarose gels. The κ-carrageenan and κ-furcellaran gels are formed by helical structures aggregated by potassium ions (Morris et al. 1980a), a mechanism that energetically is similar to the gelation of LM pectin via cross-linking by calcium ions. In the case of gelatin and agarose, gelation is associated to helix formation which results in a large entropy decrease, which must be compensated by a large enthalpy decrease. The difference between agarose and κ-carrageenan gels may be due to the different sulfate content (Watase and Nishinari 1987a). The smaller values of ΔH_m obtained for the LM pectin gels, if compared with these gels and even with the HM pectin/glycerol or HM pectin/DMSO gels, suggest that the junction zones in LM pectin/calcium are less heat-resistant. In contrast, the enthalpy ΔH_{bT} obtained for LM pectin/calcium gels is higher than that obtained for HM pectin gels, which suggests a cooperative process in the formation or disruption of effective junction zones in the pectin/calcium system (Garnier 1992).

Application of Thermal Cycles

Application of thermal cycles is one of the methods that has been used to study the thermal behavior of biopolymer gels. For thermoreversible gels, important hysteresis effects, related to the nonequilibrium nature of the gel phase can be identified by different methods, including dynamic rheological methods, for systems such as gelatin (Djabourov et al. 1988a), agarose (Watase and Nishinari 1987a; Braudo et al. 1991) *κ*-carrageenan gels (Morris et al. 1980a), and methylcellulose gels (Haque and Morris 1993). Thermal hysteresis have their origin in stabilization of the ordered form by further aggregation (Liang et al. 1979; Morris et al. 1980b): when the conformational ordering on cooling is accompanied by aggregation of the ordered structures, the melting of the aggregates usually occurs over a high temperature range.

Nishinari et al. (1990) described the thermoreversible gel-sol transition by a zipper model approach. The junction zones were assumed to be made up of association of molecular zippers standing for a rigid ordered molecular structure such as helices or extended molecules. The gel-sol transition for thermoreversible gels was compared with a zipper opening from both ends. In the case of heating, the gel-sol transition caused by the opening of molecular zippers will start as soon as the temperature arrives at the point where the segments begin to be released from a zipper. On cooling, in contrast, the pair wise coupling cannot start as easily because of the difficulty for a long molecule to find its partner in appropriate positions for zipper construction. Hence, a state that is like super cooling may occur. This explains why T_m is far higher than T_{gel} and that the rise of G' in cooling is far steeper than the decay of G' on heating, as observed for several thermoreversible biopolymer gels.

Structure Loss Rate

Rates of structure disruption or loss rates (SLR), defined as $-dG'/dt$, can also be measured when the gels are subjected to an increasing temperature in order to generate kinetic data for the melting process. The changes in the HM pectin network

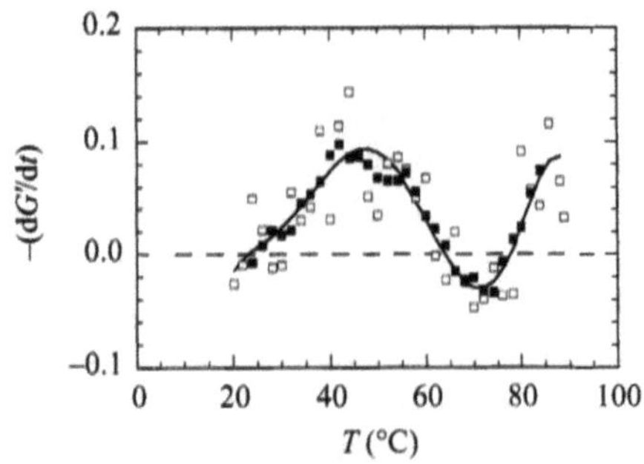

Fig. 6.15 Effect of temperature on SLR (−d*G*′/d*t*) of a 1 % high-methoxyl pectin/60 % sucrose gel during heating from 20 to 90 °C, right after a cooling scan from 90 to 20 °C. □ d*G*′/d*t* calculated after cubic spline fit; the line is for dG′/d*t* calculated after fitting a polynomial equation to the data; ■ smoothed d*G*′/d*t* values calculated after cubic spline fit

Fig. 6.16 Effect of temperature on SLR of a cured gel (20 °C, 48 h) during heating from 20 to 90 °C. Values of d*G*′/d*t* calculated by derivation of a polynomial equation fitted to the data (continuous line), or after cubic spline fit and smoothed using a "moving average" procedure (■); (□)*G*′ values

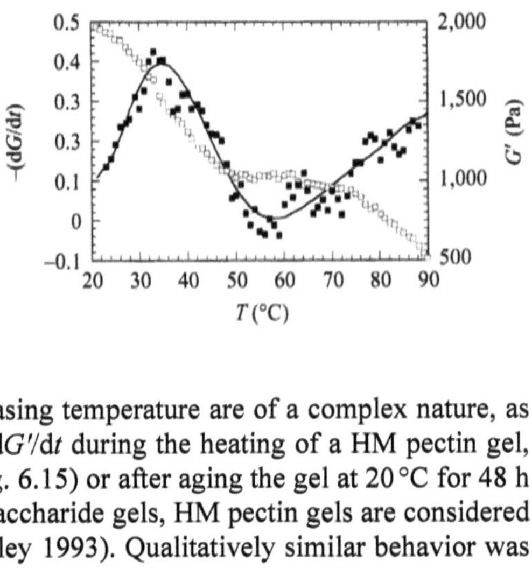

structure when subjected to an increasing temperature are of a complex nature, as seen in plots of temperature versus d*G*′/d*t* during the heating of a HM pectin gel, immediately after a cooling scan (Fig. 6.15) or after aging the gel at 20 °C for 48 h (Fig. 6.16). In contrast to most polysaccharide gels, HM pectin gels are considered to be heat-irreversible (Rao and Cooley 1993). Qualitatively similar behavior was obtained for the two experiments. Three different zones can be identified: an initial increase in rate of structure loss, then a decrease in SLR, and finally another temperature range, above 60 °C, where SLR increased again. Some quantitative differences can be found between the behavior of these two networks. The gel analyzed, whose behavior is in Fig. 6.15, was far from an equilibrium state, and a higher mobility of the pectin chains in the network can be expected. For some temperatures, it can be seen that values of SLR were negative, meaning that an increase in temperature even contributed to structure formation. In the previously aged gel (Fig. 6.16), the higher number and/or the higher extension of the junction zones also lead to higher SLR values.

A different behavior is expected with thermoreversible gels. Using the LM pectin/calcium network as an example, the influence of temperature on SLR was

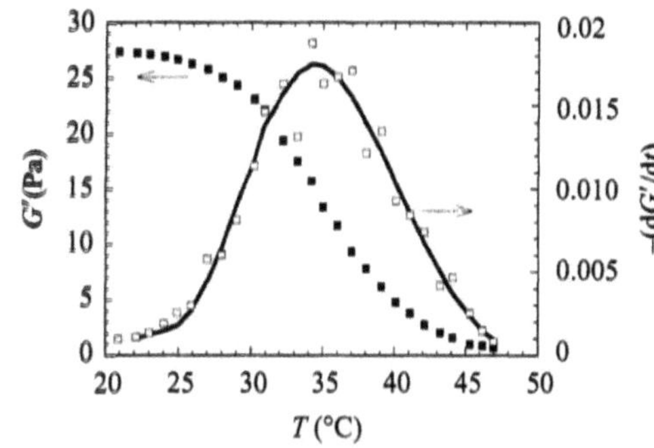

Fig. 6.17 Dependence on temperature of SLR for a cured gel (24 h) under heating from 20 to 50 °C ([low-methoxyl pectin] = 6.4 g L^{-1}, [Ca^{2+}] = 0.18 g L^{-1}, 0.1 mol L^{-1} NaCl, pH 7.0). Values of dG'/dt calculated after cubic spline fit (G); also shown for comparison are the dG'/dt values calculated by derivation of a polynomial equation fitted to the data (line), and the $G'(\omega)$ values

examined (Fig. 6.17). The influence of temperature on the rate of structure disruption followed a bell-shaped curve, with a maximum value at an intermediate temperature.

Kinetic data can also be obtained for the structure loss process during heating of biopolymer gels. The non-isothermal kinetic model previously described (Eq. 6.28) can be used to describe the melting process of thermoreversible gels and to obtain apparent activation energies for this non-isothermal process, and was applied to the melting of LM pectin/calcium gels. A multiple linear regression of ln(−dG'/dt) versus ln G' and $1/T$ was conducted in order to evaluate the order of reaction (n). This treatment showed that invariably the order of reaction (n) for the process under analysis should be close to 2. Exceptions were found for the higher pectin concentration analyzed, and for high R values, where a general tendency for a higher reaction order was observed. Taking the reaction order $n=2$, Eq. 6.18 was used to estimate the activation energy of the process, by using a linear least squares routine. As shown in Fig. 6.18 for an LM pectin/calcium gel ($c_p=6.4$ g L^{-1}, $R=0.42$), the melting process could be satisfactorily described by this non-isothermal kinetic approach.

Mixed Polymer Gels

The behavior in systems of two or more polymers is important because it is different than that each would exhibit in the absence of the other and the knowledge should help formulate new foods as well as to innovative methods to generate attractive food structures and textures. Polysaccharides and proteins are the main structure and texture-forming macromolecules in food. Thus, mixtures of these biopolymers are good models to understand structure-property relationships in real food systems. Due to the high physical, chemical, and structural diversity among polysaccharides

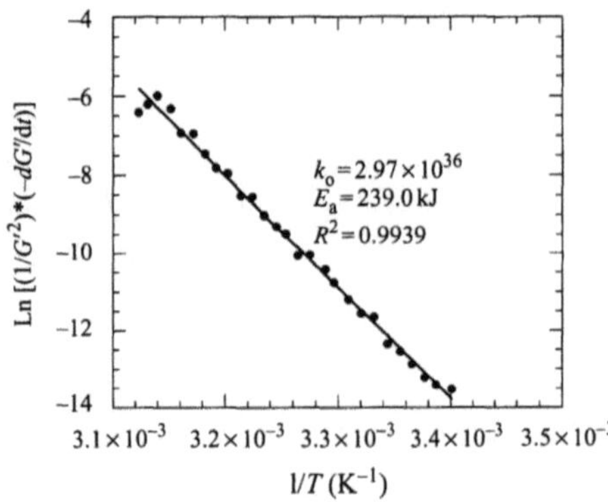

Fig. 6.18 Arrhenius type plot for change in the storage modulus (G') during heating for the low-methoxyl pectin system shown in Fig. 6.17

and proteins, a wide range of different combinations and resulting properties are possible in such multicomponent systems. Segregative interactions, arising from thermodynamic incompatibility, either by phase separation (which raises the effective concentration of both components) or by exclusion effects promoting self-association within a single phase, are the general rule observed among different biopolymers. Nonetheless, in a few systems gelation may occur by association of the two different polymers into conformational ordered cooperative junction zones that are analogous to those in single-component polysaccharide gels. A few examples of these systems are discussed in the following section, with special emphasis on their rheological behavior.

Polysaccharide-Polysaccharide Mixtures

One possible combination to create a simple two-component gelled system is by blending a gelling and a nongelling biopolymer. Typical examples of such multicomponent gels are polysaccharide mixtures involving galactomannans, which are typical random-coil nongelling polysaccharides, on the one hand and a gelling biopolymer such as κ-carrageenan (Fernandes et al. 1991), agar (Turquois et al. 1993), or LM pectin gels formed by cooling in the presence of Ca^{2+} (Lopes da Silva et al. 1996) on the other. The combination of these polysaccharides has been shown to offer the possibility of having structural and textural diversity, and consequently a broader spectrum of application in food systems. In general, the presence of even a small amount of galactomannan leads to an increase in the final rigidity of the gel or promotes gelation, and introduces important changes in the viscoelastic properties of the gelling biopolymer. These changes can be observed during gel formation and curing, commonly associated to lower gel times and faster gelation processes, and also in the cured gels. The synergistic effects involved are not all clearly understood and the mechanism of gelation of these binary systems is still a matter of debate. Two antagonistic models have been proposed involving either a cooperative association into mixed junction zones, or a mutual exclusion with a liquid-liquid phase separation.

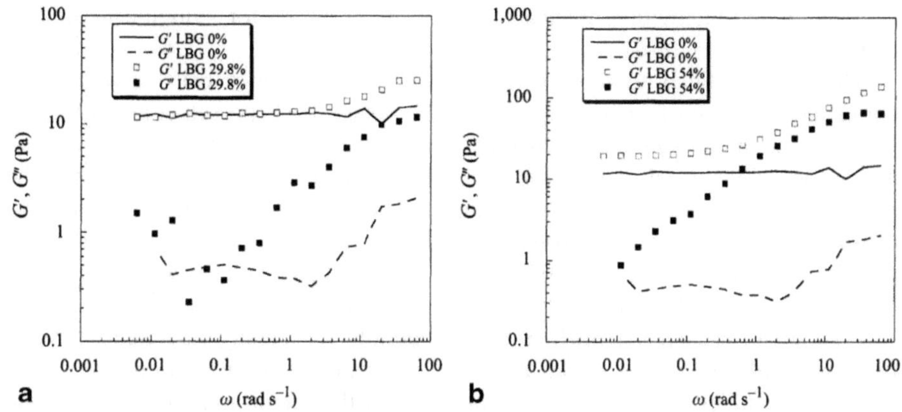

Fig. 6.19 Effect of locust bean gum addition on the mechanical spectra of an LMP/Ca system (c_p=6.4 g L^{-1}, [Ca^{2+}]=3.7×10^{-3} mol L^{-1}). a LBG 2.7 g L^{-1} (30 %), b LBG 7.6 g L^{-1} (54 %; Reproduced with permission from Lopes da Silva et al. 1996)

For the LM pectin/galactomannan systems, changes observed in the rheological properties of the pectin gels were related to limited galactomannan microphase separation, and the entrapment of the nongelling polysaccharide within the pectin-calcium network. The addition of the galactomannan increased both the storage (G') and the loss moduli (G''), especially at short time scales. However, a greater frequency dependence and a smaller separation of G' and G'' was also observed, meaning a general decrease in the gel-like character. Examples of small-deformation rheological data (mechanical spectra) obtained for these mixtures and for the gelling system alone are shown in Fig. 6.19.

Various multicomponent systems, including mixed gels of LBG and Na- and Ca-κ-carrageenan (Stading and Hermansson 1993) and some composites obtained when both components gel by themselves (Doublier et al. 1992), also showed a clear increase of the elastic modulus in the high frequency range or even the presence of two elastic plateaus. This indicates the presence of additional relaxation mechanisms when compared with the behavior of one gel component alone, and has been taken as an indication of the heterogeneity of the gelled multicomponent systems. Some of these systems may, thus, resemble the structural organization of a gelled emulsion using the terminology of Morris (1990).

Other interesting two-component gelling systems are obtained from mixtures of xanthan gum and polysaccharides, especially from the β (1–4) glycan family, namely galactomannans or glucomannans (Williams et al. 1991; Mannion et al. 1992; Annable et al. 1994; Goycoolea et al. 1995; Schorsch et al. 1997; Paradossi et al. 2002). These mixtures form thermoreversible gels despite the fact that, under similar conditions, neither of the components forms gels independently. The synergistic effects observed lead to gels with increased elasticity and strength. These systems are among the few examples where association appears to occur by formation of specific heterotypic junctions (Morris 1990). Recently, similar synergistic association and interaction mechanism has been proposed for xyloglucan-xanthan admixtures (Kim et al. 2006).

The viscoelastic properties of the mixed systems change considerably when compared with that of each component. As an example (Schorsch et al. 1997), the galactomannan and xanthan mechanical spectra were typical of a viscous solution and the so-called "weak-gel," respectively. However, the mechanical spectra of the mixed systems showed structured systems that behaved as gels, with a higher value of storage modulus and higher solid-like character than each component. The ability of galactomannans to form synergistic gels with xanthan increased with the proportion of unsubstituted mannose residues in the chain backbone (not shown here).

A variety of viscoelastic profiles may be obtained from many other admixtures of polysaccharides involving either associative or segregative interactions by changing the type of biopolymers, their proportion in the blend, ionic conditions, and temperature. A class of synergistic gels, also described on the basis of formation of heterotypic junctions between components, is the one resulting from interactions among glucomannans and disordered gelling polysaccharides such as κ-carrageenan (Kohyama et al. 1993) and gellan gum (Miyoshi et al. 1996; Nishinari et al. 1996). Gellan gum, a gelling extracellular polysaccharide produced by aerobic fermentation of *Sphyngomonas elodea,* has also received increasing attention regarding its interaction with other polysaccharides. It exhibits both synergistic and nonsynergistic interactions with polysaccharides, such as κ-carrageenan (Nishinari et al. 1996), iota-carrageenan (Rodriguez-Hernandez and Tecante 1999), xanthan (Rodriguez-Hernandez and Tecante 1999), and xyloglucans (Nitta et al. 2003; Ikeda et al. 2004). Recently, the rheological behavior observed for the nonsynergistic admixture of κ-carrageenan-agar gels has been interpreted based on a segregative phase separation mechanism, where agar-rich phase formed a continuous phase and κ-carrageenan-rich phase formed a discontinuous gelled phase (Norziah et al. 2006).

Many other studies can be found in the literature that contain more details on structure, mechanisms and rheology of mixed biopolymer gels, including reviews (Morris 1990, 2000; Zasypkin et al. 1997). Also, recently, Lopes da Silva and Rao (2006) discussed several mixed biopolymer systems containing pectin.

Protein-Polysaccharide Mixtures

Segregative demixing, due to thermodynamic incompatibility, is also widely accepted as a general phenomenon in aqueous protein-polysaccharide (Grinberg and Tolstoguzov 1997; Doublier et al. 2000; Wang et al. 2001) especially with high molecular weight biopolymers and in the absence of complex coacervation. This well-illustrated thermodynamic incompatibility arises mainly from the low entropy of mixing and results in segregation and phase separation, with both macromolecules tending to demix into different regions each being enriched in one of the biopolymers. The occurrence and extension of demixing is influenced by a wide range of intrinsic and extrinsic factors, including biopolymer concentration, molecular weight, conformation, charge density, and, therefore, pH and ionic strength of the solvent (Tolstoguzov 2001), and it is expected to greatly affect the rheological behavior of the multicomponent system. Nevertheless, proteins and polysaccharides with oppositely charged groups show net attraction, resulting in complex coacervation or associative phase separation. The mixture separates into two phases: a lower phase containing the protein/polysaccharide complex and an upper phase containing mostly the solvent. The term coacervation is used with liquid-liquid phase separation (Weinbreck 2004). It was pointed out in Chap. 4 that in protein-polysaccharide systems, the assymmetry of modulus versus concentration plots is due to phase separation and inversion as first shown by Tolstoguzov (1985). Such systems exist widely in living organisms and can be naturally associated to maintain cell integrity or induce cell division, but can be incompatible and participate in cell partition.

Usually, the rheological behavior of mixed systems differs markedly from the pure biopolymer system. Several examples can be found in literature for the significant rheological changes that accompany phase separation in mixed biopolymer solutions or gels. We will briefly focus on the rheological behavior of selected polysaccharide-protein mixtures, considering three main cases regarding the protein network: (1) mixtures containing the disordered gelatin, which form thermoreversible gels on cooling, (2) mixtures containing globular proteins, essentially whey proteins that form thermally induced gels, and (3) the effect of polysaccharides on casein aggregation and gelation.

Gelatin-Polysaccharide Systems

Addition of polysaccharides to a gelatin solution generally accelerates and increases the process of structure formation during gelation. The elastic modulus of filled gelatin gels increased with filler concentration in the single-phase region but decreased in the two-phase region because of the increase in imperfection of the gel network (Marrs 1982). Results on gelatin/maltodextrin mixed systems (Kasapis et al. 1993a) showed an initial increase in gelation time with increasing maltodextrin concentration followed by a decrease at higher concentrations, associated to phase inversion from a gelatin-continuous network at low concentrations of maltodextrin to a maltodextrin-continuous network at higher concentrations. The storage moduli (G') of the phase-separated gelatin/maltodextrin cogels have been related quantitatively to the experimentally determined concentration dependence of G' for the constituent polymers. Distribution of water between the phases was examined by using polymer blending laws to derive calculated moduli for gelatin-continuous and maltodextrin-continuous networks over the entire range of solvent partition (Kasapis et al. 1993b).

Papageorgiou et al. (1994) studied gellan/gelatin systems under conditions where electrostatic interactions between the two polymers were suppressed. They found an increase in the gel strength of 5 % gelatin gel upon the addition of 0.5 % gellan, and they suggested that the two components behave as thermodynamically incompatible polymers, which gel independently in their respective phases, as evidenced by mechanical spectroscopy, differential scanning calorimetry, and compression testing. Changing polymer concentration and ionic strength also promotes phase inversion from a gellan-continuous phase to a system where gelatin forms the supporting matrix. On the other hand, electrostatic effects were suggested to explain the synergistic effects observed for gelatin/iota-carrageenan systems, corresponding to the increase in the storage and loss moduli of the mixtures, compared with the values for individual components (Michon et al. 1996). The effect of segregative phase separation on the rheology and microstructure of gelatin/dextran systems was studied by Owen and Jones (1998). They attributed the effect of thermal history on the properties of the gelatin/dextran mixture to the relative occurrence of phase separation and gelation. The concept of kinetic entrapment, associated to that biopolymer that gels first in a mixture and changes on thermal regime, is known to exert a determinant role on the structure of the final gel (Alevisopoulos et al. 1996). Recently, an interesting study on mixtures offish gelatin/κ-carrageenan (Haug et al. 2003) showed the possibility of the coexistence of both associative and segregative phase separation, although the associative phase separation is most likely to be kinetically trapped.

Whey Protein-Polysaccharides

Whey protein concentrates (WPC) and isolates (WPI) are increasingly used as food ingredients due to their unique nutritional and functional properties. Therefore, not surprisingly, the gelation properties of these complex protein systems have been studied extensively, and several studies have been made on the gel formation of whey proteins in combination with polysaccharides. These studies have focused on the effect of ionic gelling (e.g., Ndi et al. 1996; Capron et al. 1999; Eleya and Turgeon 2000; Beaulieu et al. 2001; Donato et al. 2005) or nongelling polysaccharides (Zasypkin et al. 1996; Walkenstrom et al. 1998; Bryant and McClements 2000), on the effect of starch and starch-derived polysaccharides (Manoj et al. 1997; Shim and Mulvaney 2001), and also on the influence of neutral nongelling polysaccharides (Syrbe et al. 1995; Wang et al. 2001; Monteiro et al. 2005; Tavares and Lopes da Silva 2003; Tavares et al. 2005) on the whey protein gelation. The results obtained so far are important contributions to the adequate manipulation of the rheological properties of these biopolymer mixtures and to the attractive concept of engineered food structure development.

Rheological techniques have been widely used to describe the properties of these systems, and have been successful tools to illustrate the effect of phase behavior on the viscoelasticity of the mixed systems. Capron et al. (1999) studied mixed β-lactoglobulin/κ-carrageenan systems using rheological measurements and microscopy (TEM). They showed that the first step of the aggregation process of β-lactoglobulin was not changed by the presence of κ-carrageenan, whereas an increase in gelation rate was observed in the second step. Other rheological studies confirmed the synergistic effects in κ-carrageenan/denaturated whey protein systems (Tziboula and Home 1999; Eleya and Turgeon 2000). The synergistic interaction was related to the formation of two cocontinuous networks as a result of segregative phase separation before gelation occurred. A segregative phase separation phenomenon also appears to have occurred in whey protein/xanthan mixtures (Zasypkin et al. 1996; Sanchez et al. 1997; Bryant and McClements 2000). The limiting concentration for the gelation of proteins decreases due to the presence of the polysaccharide. Synergistic effects were observed at low xanthan concentrations while antagonistic effects were reported at high xanthan content. Bryant and McClements (2000) reached similar conclusions by mixing heat-denatured whey proteins and xanthan. Rheological methods have also been used to study the BSA/Na-alginate systems (Neiser et al. 1998). The results obtained showed trends similar to those reported for whey protein/xanthan systems, with a remarkable effect of pH and ionic strength on the final gel properties.

A few studies have also reported on the effects of nongelling neutral polysaccharides on whey protein gelation (Olsson et al. 2000, 2002; Monteiro et al. 2005; Tavares et al. 2005), revealing interesting relationships between the microstructure of the biphasic mixed networks and the observed viscoelastic behavior and pointing out the importance of the structural characteristics of the polysaccharide.

By combining results from rheological and microstructural studies, Olsson et al. (2002) demonstrated the complexity of structure-rheology relationships in phase-separated networks of nongelling amylopectins and particulate β-lactoglobulin gels. They found that the cluster size and the pore size between the strands of clusters were related to storage modulus and stress at fracture as long as the connectivity of the network strands was good; however, a decrease in gel strength occurred at high polysaccharide contents due to decreased network connectivity. The results also demonstrated the importance of a full rheological characterization, since small and large deformation tests responded differently to the gel microstructure (Olsson et al. 2000). The low sensitivity of linear viscoelastic properties to some microstructural changes in biphasic polysaccharide/whey protein gels was also pointed out by Tavares et al. (2005). Despite these limitations, combination of microstructural analysis and small-deformation oscillatory rheology has proved to be useful to discriminate the effect of degree of branching and molecular weight of galactomannans on the properties of fine-stranded whey protein gels (Monteiro et al. 2005; Tavares et al. 2005).

Some examples of the rheology-microstructure relationships obtained are illustrated in Fig. 6.20a–c; specifically, the rheological data are shown in Figs. 6.20a, b and the microstructure data in 6.20c. The LBG with the lowest molecular mass, 168 kDa, exhibited rheological behavior shown in Fig. 6.20a. However, it did not change the microstructural arrangement of the protein network at this length scale (top left, Fig. 6.20c). However, for the mixtures with galactomannan samples with higher molecular weights: 425, 786, and 1,930 kDa, phase separation occurred with irregular masses of polysaccharide solution dispersed in the continuous protein matrix. The size of the polysaccharide-rich areas and the degree of connection between them increase with increasing galactomannan concentration or molecular weight. Interconnection between the polysaccharide-rich droplets occurred for the higher LBG molecular weight and concentration analyzed, originating bicontinuous phase-separated systems. The viscoelastic profile of the gels in the presence of the LBG fractions changed dramatically. Increasing the molecular weight from 168 to 786 kDa resulted in higher moduli with G' less dependent on frequency (e.g.,

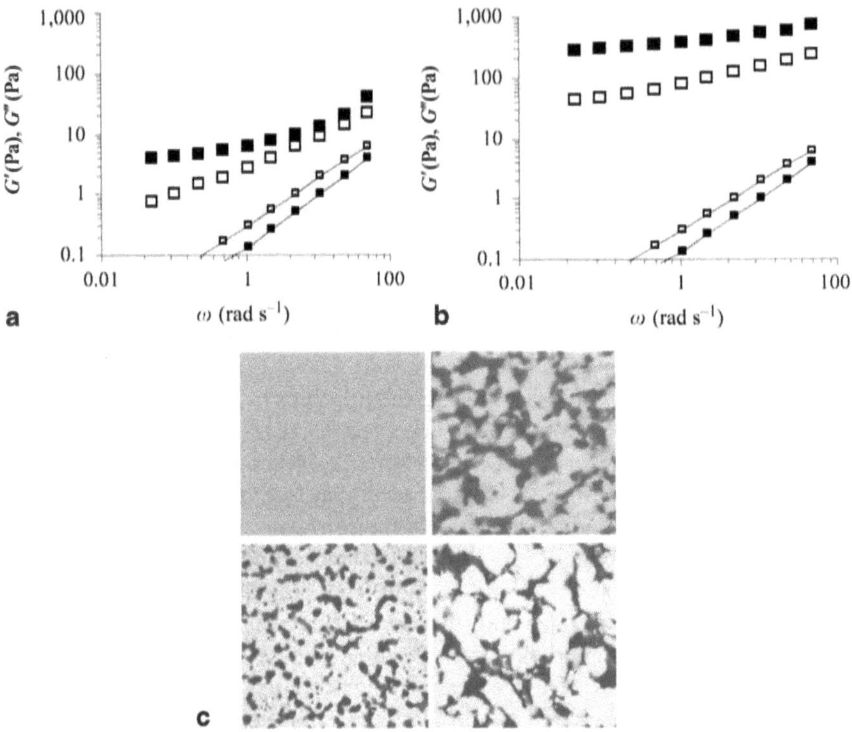

Fig. 6.20 Mechanical spectra showing the frequency dependence of G′ (■) and G″ (□) for 11 % whey protein alone and in two mixtures with locust bean gum (20 °C, pH 7): a LBG: 168 kDa and b LBG: 786 kDa. Small *symbols* and *lines* denote the behavior of whey protein alone. In (c) microstructure of mixtures, molecular weight of LBG counterclockwise from *top left*: 168, 425, 786, and 1930 kDa (Reproduced with permission from Monteiro et al. 2005)

Fig. 6.20b for LBG with 786 kDa); the corresponding microstructure is shown at top right, Fig. 6.20c. However, for the highest molecular weight LBG studied (not shown here), G' decreased slightly, its dependence on frequency increased, and a smaller separation between G' and G'' was observed, that is, an elastically less perfect network was formed, probably related to the formation of a bicontinuous network with lower connectivity within the protein-rich gelled phase.

Casein-Polysaccharides

Segregative phase separation, probably mediated by depletion/flocculation mechanism, was proposed for admixtures of micellar casein and galactomannans (Bourriot et al. 1999a; Tuinier et al. 2000) or anionic exopolysaccharide (Tuinier et al. 1999). The presence of a monophasic or biphasic domains was dependent on biopolymer concentration. Modifications of the flow and the viscoelastic properties with respect to the individual components were clearly evident in the two-phase domain (not shown here). The phase separation process yields a continuous network mostly composed of the aggregated micellar casein, where the two phases coexist, one containing mainly casein micelles forming a continuous network (white areas), whereas the other one is enriched with guar gum. The mechanical spectrum obtained for the biphasic system, compared with the pure biopolymers or the monophasic mixed system, exhibited higher viscoelastic moduli and was less dependent upon frequency, and the G' curve tended to level off and cross the G'' curve, suggesting a solid-like behavior despite the low moduli values.

Studies on casein/carrageenan and casein/pectin interactions (Dickinson 1998; Bourriot et al. 1999b; Langendorff et al. 2000; Maroziere and de Kruif 2000; Matia-Merino et al. 2004) have revealed different phase behaviors dependent on biopolymer charge density and concentration, and consequently a diversity of rheological profiles. As for many other biopolymer blends, the presence of charged groups has a profound impact on the phase behavior; in the presence of charged biopolymers, with no added salts, the entropy-driving force for phase separation is often suppressed and gelled interpenetrating networks can form. Complex ionic effects have been often reported. For example, a rheological and microstructural study (Matia-Merino et al. 2004) on mixed gels obtained from LM amidated pectin and sodium caseinate showed that both the amount of pectin and the presence of calcium have a major influence on the interaction in that system. A general reduction of the gel modulus and an increase in gelation time was observed for calcium-free systems and for systems with calcium but low pectin content; under those conditions, increasing the pectin content prevented aggregation of the caseinate particles into a coherent network. However, a large increase in the modulus was observed at higher pectin contents and in the presence of added calcium. The combination of a gradual release of calcium ions from caseinate aggregates as the pH was lowered and the electrostatic adsorption of pectin onto the casein particles were suggested as the main factors influencing the interactions between the two biopolymers and the final gel structure.

Biopolymer Mixtures under Flow Fields

When liquid-liquid phase separation occurs, resulting in water-in-water emulsions where the different biopolymers are concentrated in the different phases, and when one or both phases can gel, these systems can be used to produce anisotropic microgel particles and/or gel composites with anisotropic inclusions, resulting in a variety of interesting microstructures and morphologies. An interesting concept to generate alternative gel structures with basis on phase-separated biopolymer blends has been studied and developed, essentially related to flow-induced generation of anisotropic gel structures. Different biopolymer mixtures, such as aqueous mixtures of gellan/κ-carrageenan, gellan/sodium alginate, gelatin/dextran, and gelatin/guar, submitted to certain flow conditions during gelation are able to produce a broad range of microstructures and textures (Wolf et al. 2000; Antonov et al. 2004; Simeone et al. 2002, 2005). The stresses developed in the dispersion medium of a water-inwater emulsion can deform and orient the liquid-dispersed particles. This occurs if the viscosity of the dispersed particles is lower than that of the continuous medium. The size and shape of the dispersed particles can be fixed by gelation. Morphology and polydispersity of the dispersed particles may be controlled through defined flow conditions and reasonable quantitative predictions about the system's structure may be achieved based on the knowledge of the rheology of the phases, interfacial tension, flow field, and the gelation kinetics (Stokes et al. 2001; Wolf et al. 2001; Antonov et al. 2004; Lundell et al. 2004).

Phase Diagrams and Modeling

Phase separation in solution can be detected by the immediate development of turbidity on mixing, eventually resulting in two separate layers whose compositions fall on a "cloud-point curve" or "binodal" that define the boundary between the single-phase and two-phase states of the system. In Fig. 6.21, the curve represents the binodal of a polysaccharide (alginate, [A]) and protein (casein, [C]). The thin straight lines are tie lines from which the concentrations of coexisting protein-rich and polysaccharide-rich phases can be determined as shown in the filled circles. Below the binodal, the polymer concentrations of the single homogeneous phase solution can be determined. In particular, the open circle represents the critical point below which the system remains as a single homogeneous phase.

In Chap. 2, simple Takayanagi models for the modulus of two-component blends were discussed that are applicable to the shear modulus G_c of a composite formed from components (polymers) x and y with shear moduli G_x and G_y, respectively. If ϕ_x and $\phi_y = 1 - \phi_x$ are the respective volume fractions of the two components, the equations for the upper and lower bound values are

$$G_c = \phi_x G_x + \phi_y G_y \quad \text{(upper bound, isostrain, if } G_x > G_y) \tag{6.34}$$

$$(1/G_c) = (\phi_x/G_x) + (\phi_y/G_y) \quad \text{(lower bound, isosterss, if } G_x < G_y) \tag{6.35}$$

The former approximates the weighted-average of the individual moduli, while the latter approximates the weighted average of the individual compliances. If the filler x is either much softer or much harder than the surrounding matrix, the above equations can be reduced, respectively, to

$$G_c \approx G_x \phi_x \; or \; (G_c/G_x) = \phi_x, \quad \text{when } G_x \gg G_y \tag{6.36}$$

$$(1/G_c) \approx (\phi_x/G_x) \text{ or } (G_c/G_x) \approx (1/\phi_x), \quad \text{when } G_x \ll G_y \tag{6.37}$$

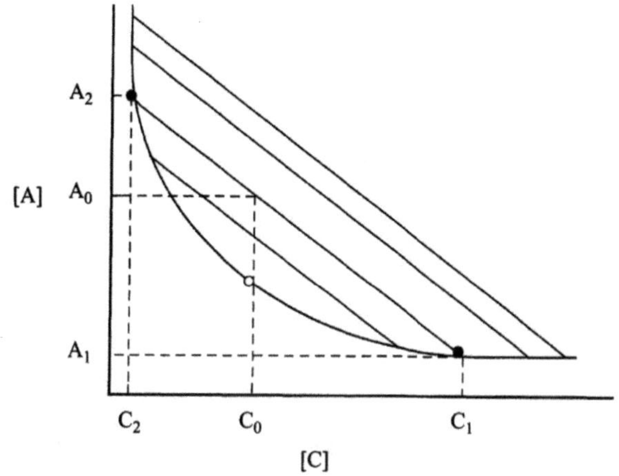

Fig. 6.21 Phase diagram of a mixed solution of two biopolymers, for example, sodium alginate and casein. The curve represents the binodal and the *thin straight lines* (- - -) are tie lines from which the concentrations of coexisting protein-rich and polysaccharide-rich phases can be determined as shown at the *filled circles* (•). The *open circle* (○) represents the critical point below which the system remains as a single homogeneous phase

It is reasonable to expect the major component to form the continuous phase, that is, at $\phi_x > 0.5$. More complex Takayanagi models have been reviewed by Manson and Sperling (1976).

Morris (1992) explored the manner in which values of G_x, G_y, and the upper ($G_u = G_c$) and lower bound values ($G_l = G_c$) vary with the fraction of the component x. It was assumed that the two polymers x and y are confined entirely to their respective phases, the modulus varies as the square of the polymer concentration, and that the volume of the polymer chains is negligible compared with the volume of the solvent, so that

$$\phi_x = x \quad \text{and} \quad \phi_y = 1 - x \tag{6.38}$$

Furthermore, it was also assumed that $M_y = (n')^2 M_x$, where n' is an integer. The results obtained with $n'=2$ (the symbol n' is used to avoid confusion with the symbol n used for the reaction rate order) are shown in Fig. 6.22 where data reported in Morris (1992) were replotted. The values of the modulus G_x are high at low values of x and decrease with increasing x reaching a final value $G_x = M_x$ at $x=1$. Conversely, G_y is high at high values of x and decreases to $G_y = M_y$ at $x=0$. At a critical point, $G_x = G_y$, that is, the moduli of both phases are identical. In addition, at the critical point, G_u and G_l reach minimum and maximum values, respectively. Also, of interest are the general results for the critical concentration and modulus

$$x_{\text{crit}} = (1/[n'+1]) \quad \text{and} \quad G_{\text{crit}} = M_x (n+1)^2 \tag{6.39}$$

Phase-separated polymer mixtures are important in many practical applications, such as confectionery and low-fat foods. Values of the fraction (x) of a component can be replaced by those of the fraction of solvent (S_x) associated with polymer (Morris 1998) so that Fig. 6.22 may also be used to understand the role of solvent partitioning. A key assumption is that there is complete separation of the two

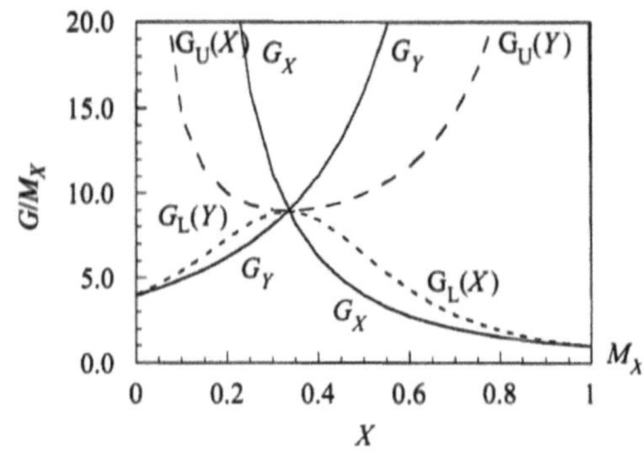

Fig. 6.22 Distribution of the moduli of components x, y, upper (G_u), and lower bound (G_l) moduli from the Takayanagi isostress and isostrain blending laws; the continuous phase is indicated in parenthesis (plotted from data in Morris 1992)

polymers into their respective phases, so that the relative phase volumes (hence, the polymer concentration in each phase) may be characterized by the "*p-factor,*" defined as

$$\text{p-factor} = \frac{(\text{water/polymer})\,\text{in phase 1}}{(\text{whater/polymer})\,\text{in phase 2}} \tag{6.40}$$

Using calibration curves of modulus versus concentration of each polymer, experimental moduli of mixed gel networks, and the Takayanagi blending laws discussed earlier, one can calculate the best value of the *p-factor* by using numerical methods. Such data then provide information on the distribution of the two polymers in a mixed gel system. Several examples of application of the earlier described concepts can be found in literature: gelatin-maltodextrin cogels (Kasapis et al. 1993c), milk proteins-maltodextrin gels (Chronakis et al. 1996a), whey protein-maltodextrin (Manoj et al. 1997), agarose-starch composite gels (Mohammed et al. 1998), and gelatin-low methoxyl pectin cogels (Gilsenan et al. 2003).

Low-Fat Spreads

Low-fat spreads are mixed gel systems of fat, water, emulsifiers, and hydrocolloids. Because they often undergo emulsion breakdown, starch derivatives and gelatin are used to provide the aqueous phase with acceptable rheological and melting characteristics. In a recent study, creep-compliance studies were used to examine the rheological characteristics of butter, margarine, and a low-fat spread (Chronakis and Kasapis 1995; Evageliou et al. 1997). It will be recalled from Chap. 3 that an experimental creep-compliance curve is described by a general equation of the form

$$J(t) = J_0 + J_{\mathrm{m}}\,(1 - e^{-t/\tau_{\mathrm{m}}}) + (t/\eta_{\mathrm{N}}) \tag{6.41}$$

where J_0 describes the instantaneous compliance, η_N is the viscosity, and $J_m[1-\exp(-t/\tau_m)]$ describes the time-dependent retarded elastic region, where m

Table 6.7 Creep-compliance parameters of low-fat spreads at 5 °C (Evageliou et al. 1997; Stanley et al. 1998)

Sample	Applied stress ($Pa \times 10^3$)	Jo($Pa^{-1} \times 10^4$)	Voigt units	η_N ($Pa^{-1} \times 10^6$)
Gelatin gel,	4.52	1	0	51
3.6% 0.2	3.0	4.25	3	62
Margarine,	1.8	0.10	2	473
80% fat	2.2	0.14	3	69
Spread, 3%	0.3	0.69	2	172
Fat + Gelatin	1.4	0.98	4	199
Spread,	0.53	0.35	2	173
22% fat + Gelatin	0.75	0.47	3	45

is the number of Kelvin-Voigt units required to describe the experimental data precisely.

In Table 6.7, obvious differences can be seen among the parameters required to describe the creep-compliance data obtained during 1 h on a gelatin gel, margarine, and two experimental low-fat spreads shown in Fig. 6.23 (Evageliou et al. 1997; Stanley et al. 1998). For example, the instantaneous compliance of margarine was lower, increasing the applied stress had more of an effect on the margarine sample than on the gelatin sample, and at the higher stress level the network was disrupted. The effect of increasing the applied stress resulted in more Voigt units. The two low-fat spreads reflected the combination of a gelatin-structured aqueous phase and a fat lattice phase; with the sample containing 3.6% fat, the instantaneous compliance was affected by the gelatin-structured aqueous phase and the sample containing 22.5% fat had high values of viscosity reflecting that only a small deformation caused terminal flow.

STARCH GELS

Most solid foods can be described as composites. Sometimes, they can be regarded to be simple filler-matrix composites. The filler may be starch granules, fat globules, or any cellular material embedded in a matrix of cross-linked protein or polysaccharide base. Usually, both components possess a degree of deformability giving rise to a complex mechanical response to deformation.

In Chap. 4, the important role of starch granule swelling and release of amylose was discussed. Furthermore, it was shown that the volume fraction of starch granules, denoted as *cQ*, plays a major role in the rheological behavior of starch dispersions. Based on vane yield stress measurements, it was also shown that dispersions of different starches have different adhesion characteristics (interparticle bonding) between granules and the continuous medium. These two characteristics play important roles in the rheology of starch gels that form when a heated dispersion of

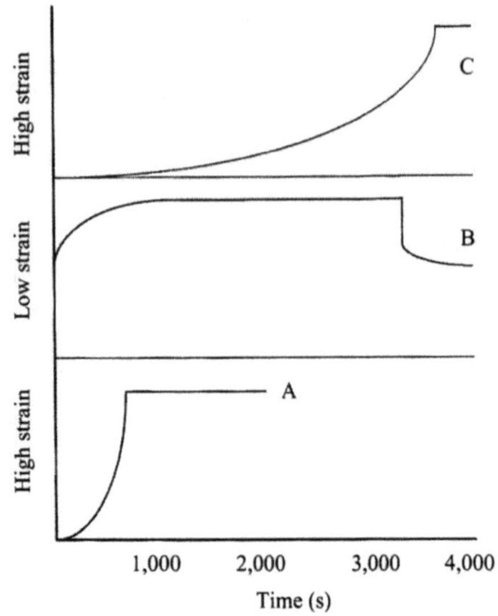

Fig. 6.23 Creep-compliance curves at 5 °C of butter (**a**), a commercial low-fat spread containing gelatin (**b**), and an experimental product with 5.2 % fat content and containing maltodextrin (**c**); Adapted from Evageliou et al. 1997; Stanley et al. 1998)

starch above a critical concentration is cooled after being heated above the starch gelatinization temperature.

Earlier studies on polymer composites have shown that the modulus (rigidity) of these composites is dependent on the rigidity of the matrix and the volume fraction and rigidity of the filler particles. As was shown for dispersions in Chap. 2, particle shape has been shown to be very important in composites. In addition, in foods, adhesion between the filler and the continuous phase among the different biopolymers for each other will differ and hence the strength of interaction between filler particles and matrix will also differ (Brownsey et al. 1987).

We denote the dynamic rheological modulus of a composite as G', that of the amylose matrix as G_0' and of the starch granule as G'_1, and the volume fraction of the granules as ϕ. The various parameters that affect the rigidity of a composite may be seen in Eq. 6.42 of Kerner (1956) and its modification, Eq. 6.45, by Lewis and Nielsen (1970) for reinforcement (G'/G_0') by spherical particles

$$G'/G_0' = (1 + AB\phi)/(1 - B\phi) \tag{6.42}$$

where ϕ is the volume fraction of the granules and the parameter A is given by

$$A = \frac{7 - 5V}{8 - 10V} \tag{6.43}$$

For spherical particles with Poisson's ratio, $V = 0.5$, the magnitude of $A = 1.5$. The parameter B is given by

$$B = \{(G_1'/G_0') - 1\}/\{(G_1'/G_0') + A\} \tag{6.44}$$

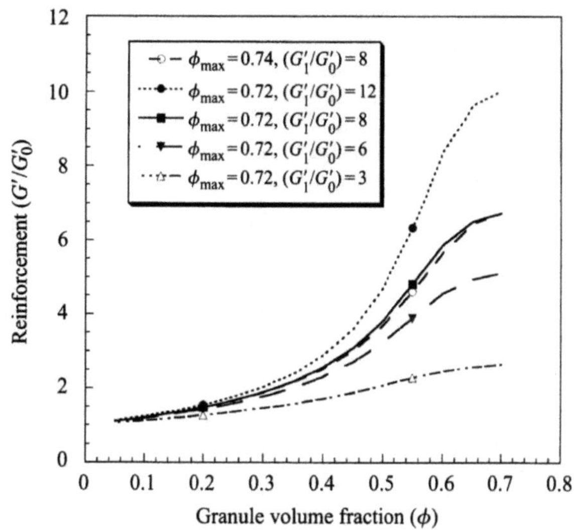

Fig. 6.24 Estimated reinforcement (G'/G_0') by spherical particles in a matrix using the Kerner-Nielsen equation: G′ is the modulus of the composite, G_0' of the amylose matrix, and G_1' of the starch granule

Nielsen's modification, that takes into account the maximum packing fraction of filler particles, is given by

$$G'/G_0' = (1 + AB\phi)/(1 - B\psi\phi) \tag{6.45}$$

The term $\psi\phi$ is given by

$$\psi\phi = 1 - \exp\{-\phi/[1 - (\phi/\phi_m)]\} \tag{6.46}$$

where ϕ_m is the maximum volume fraction.

Although ϕ, ϕ_m, and G_0' can be determined by conducting experiments carefully, there does not appear to be a reliable method for determination of G_1', the modulus of the granule. For corn starch granules, values of $G_1' = 800$ and 500 Pa were estimated when the dispersions were heated at 70 and 85 °C, respectively (Carnali and Zhou, 1996). Other reported values in the literature were 475 and 650 Pa.

Values of reinforcement as a function of granule volume fraction predicted by Nielsen's equation (Eq. 6.45) for two values of ϕ_m and different values of G_1'/G_0' are shown in Fig. 6.24. The combination of values: $\phi_m = 0.72$ and $G_1'/G_0' = 12$, and $\phi_m = 0.74$ and $G_1'/G_0' = 8$ were those used by Carnali and Zhou (1996) for defatted corn starch granules dispersed in amylose matrix. Reasonable agreement with values predicted by Nelsen's equation was also reported by Brownsey et al. (1987) for a gelatin gel filled with Sephadex G-50 spheres. Two useful results in Fig. 6.24 are: (1) that the ratio G_1'/G_1' plays an important role which can be seen at values of ϕ>0.3 and (2) that below ϕ=3, reinforcement is not significantly affected by the various parameters.

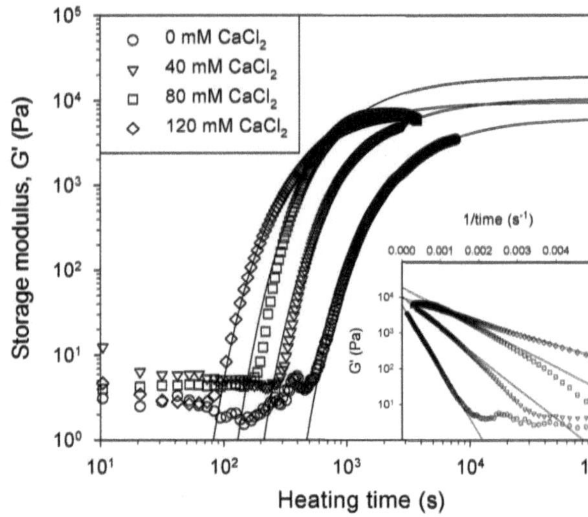

Fig. 6.25 Gel cure data for 10% w/w WPI with 0–120 mM $CaCl_2$ heated at 80 °C on the rheometer. Lines show the fit of Eq. 6.47 to data, and the inset shows the linear fits of the equation

Table 6.8 Effect of $CaCl_2$ on gel times, t_{gel} (s), for 10% whey protein isolate heated at 80 °C on the rheometer. Three different methods were used for estimating t_{gel}

$CaCl_2$	0 mM	40 mM	80 mM	120 mM
t_{gel} (s) when G'>10 Pa	637	334	209	115
t_{gel} (s) from manual selection	491	271	156	73
t_{gel} (s) from Eq. 6.47	471	208	126	83

Kerner's equation is based on the assumption that there is good adhesion between the granules and the gel around them, and that the granules are uniformly distributed. Under conditions of high strain, deviations from the mechanical behavior predicted by the low-strain analysis may occur, due to slip at the matrix-filler interface. Furthermore, due to nonuniform distribution of stress and strain throughout the material may result in a more complex mechanical response to deformation at high strains.

Effect of calcium on the morphology and functionality of whey protein nanofibrils

Loveday et al. (2011) examined the effect of calcium on the morphology and rheology of whey protein nanometer-scale fibril gels that formed spontaneously due to the self-assembly of the fibrils. They fitted the gel cure data to the model of Kavanagh et al. (2000; Fig. 6.25):

$$G' = G'_{\text{inf}} \exp(-B/t) \tag{6.47}$$

They found that G'_{inf} increased to a maximum with 80 mM $CaCl_2$ then decreased with more $CaCl_2$, so that G'_{inf} with 40 mM and G'_{inf} with 120 mM $CaCl_2$ were approximately equal. Values of $(dG'/dt)_{max}$ also increased to a maximum with 80 mM $CaCl_2$ and decrease with more $CaCl_2$, although in a subsequent experiment $(dG'/dt)_{max}$ increased by another 35% when $CaCl_2$ was increased to 120 mM. The time at which the maximum in (dG'/dt) occurred was between $4t_{gel}$ and $5t_{gel}$ in all cases. G'_{inf} was higher with 80 mM $CaCl_2$ than at any other level. This may be because the rate of fibril nucleation with 80 mM $CaCl_2$ was optimal for the formation of a strong elastic network.

References

Adam, M. 1991. Growth process of polymers near the gelation threshold. Die Makromolekulare Chemie, Macromol. Symp. 45: 1–9.

Ahmad, M. U., Tashiro, Y., Matsukawa, S., and Ogawa, H. 2004. Comparison of gelation mechanism of surimi between heat and pressure treatment by using rheological and NMR relaxation measurements. J. Food Sci. 69(9): E497–E501.

Alevisopoulos, S., Kasapis, S., and Abeysekera, R. 1996. Formation of kinetically trapped gels in the maltodextrin-gelatin system. Carbohydr. Res. 293: 79–99.

Annable, P., Williams, P. A., and Nishinari, K. 1994. Interaction in xanthan-glucomannan mixtures and the influence of electrolyte. Macromolecules 27: 4204–4211.

Andresen, I. L. and Smidsrød, O. 1977. Temperature dependence of the elastic properties of alginate gels. Carbohydr. Res. 58: 271–279.

Antonov, Y. A., Van Puyvelde, P., Moldenaers, P., and Leuven, K. U. 2004. Effect of shear flow on the phase behavior of an aqueous gelatin-dextran emulsion. Biomaeromoleeules 5: 276–283.

Audebrand, M., Kolb, M., and Axelos, M. A.V. 2006. Combined rheological and ultrasonic study of alginate and pectin gels near the sol-gel transition. Biomaeromoleeules 7(10): 2811–2817.

Axelos, M. A. V. and Kolb, M. 1990. Crosslinked biopolymers: experimental evidence for scalar percolation theory. Phys. Rev. Lett. 64: 1457–1460.

Beaulieu, M., Turgeon, S. L., and Doublier, J. L. 2001. Rheology, texture and microstructure of whey proteins/low methoxyl pectins mixed gels with added calcium. Int. Dairy J. 11: 961–967.

Bisschops, J. 1955. Gelation of concentrated polyacrylonitrile. J. Poly. Sci. 17: 89–98.

Bloksma, A. H. and Nieman, W. 1975. The effect of temperature on some rheological properties of wheat flour doughs. J. Texture Stud. 6: 343–361.

Borchard, W. and Burg, B. 1989. Investigations of the complex shear modulus and the optical rotation in the system gelatin-water during the thermoreversible gelation process, in Molecular Basis of Polymer Networks, eds. A. Baumgärtner and C. E. Picot, pp. 162–168, Springer-Verlag, Berlin.

Bourriot, S., Garnier, C., and Doublier, J. L. 1999a. Phase separation, rheology and microstructure of micellar casein-guar gum mixtures. Food Hydrocolloids 13: 43–49.

Bourriot, S., Garnier, C., and Doublier, J. L. 1999b. Micellar casein/kappa-carrageenan mixtures. 1. Phase separation and ultrastructure. Carbohydr. Polym. 40: 145–157.

Braudo, E. E., Muratalieva, I. R., Plashchina, I. G., and Tolstoguzov, V. B. 1991. Correlation between the temperatures of formation/breakdown of the gel network and conformational transitions of agarose macromolecules. Carbohydr. Polym. 15: 317–321.

Braudo, E. E., Plashchina, I. G., and Tolstoguzov, V. B. 1984. Structural characterisation of thermoreversible anionic polysaccharide gels by their elastoviscous properties. Carbohydr. Polym. 4: 23–48.

Brownsey, G. J., Ellis, H. S., Ridout, M. J., and Ring, S. G. 1987. Elasticity and failure in composite gels. J. Rheol. 31: 635–649.

Bryant, C. M. and McClements, D. J. 2000. Influence of xanthan gum on physical characteristics of heat-denaturated whey protein solutions and gels. Food Hydrocolloids 14: 383–390.

Capron, I., Nicolai, T., and Smit, C. 1999. Effect of addition of κ-carrageenan on the mechanical and structural properties of β-lactoglobulin gels. Carbohydr. Poly. 40: 233–238.

Carnali, J. O. and Zhou, Y. 1996. An examination of the composite model for starch gels. J. Rheol. 40(2): 221–234.

Chambon, F. and Winter, H. H. 1987. Linear viscoelasticity at the gel point of a crosslinking PDMS with imbalanced stoichiometry. J. Rheol. 31: 683–697.

Chronakis, I. S. and Kasapis, S. 1995. A rheological study on the application of carbohydrate-protein incompatibility of the development of low fat commercial spreads. Carbohydr. Polym. 28: 367–373.

Christ, D., Takeuchi, K. P., and Cunha, R. L. 2005. Effect of sucrose addition and heat treatment on egg albumen protein gelation. J. Food Sci. 70(3): E230–E238.

Chronakis, I. S., Kasapis, S., and Richardson, R. K. 1996a. Small deformation rheological properties of maltodextrin-milk protein systems. Carbohydr. Polym. 29: 137–148.

Chronakis, I. S., Kasapis, S., Richardon, R. K., and Doxastakis, G. 1996b. Characterisation of a commercial soy isolate by physical techniques. J. Texture Stud. 26: 371–389.

Clark, A. H. 1991. Structural and mechanical properties of biopolymer gels, in Food Polymers, Gels and Colloids, Dickinson, E. ed., pp. 322–338, The Royal Society of Chemistry, Cambridge, UK.

Clark, A. H. 1994. Rationalisation of the elastic modulus-molecular weight relationship for κ-carrageenan gels using cascade theory. Carbohydr. Polym. 23: 247–251.

Clark, A. H. and Ross-Murphy, S. B. 1987. Structural and mechanical properties of biopolymer gels. Adv. in Polym. Sci. 83: 57–192.

Clark, A. H., Evans, K. T., and Farrer, D. B. 1994. Shear modulus-temperature meltdown profiles of gelatin and pectin gels. Int. J. Bio. Macromol. 16: 125–130.

Cuvelier, G., Peigney-Noury, C., and Launay, B. 1990. Viscoelastic properties of physical gels: critical behaviour at the gel point, in Gums and Stabilisers for the Food Industry 5, eds. G. O. Phillips, D. J. Wedlock and P. A. Williams, pp. 549–552, IRL Press, Oxford, UK.

da Silva, J. A. L. and Gonçalves, M. P. 1994. Rheological study into the ageing process of high methoxyl pectin/sucrose aqueous gels. Carbohydr. Polym. 24: 235–245.

da Silva, J. A. L., Gonçalves, M. P., and Rao, M. A. 1995. Kinetics and thermal behaviour of the structure formation process in HMP/sucrose gelation. Int. J. Biol. Macromol. 17: 25–32.

da Silva, J. A. L., Gonçalves, M. P., Doublier, J. L., and Axelos, M. A. V. 1996. Effect of galactomannans on the viscoelastic behaviour of pectin/calcium networks. Carbohydr. Polym. 24: 235–245.

Dickinson, E. 1998. Stability and rheological implications of electrostatic milk protein/polysaccharide interactions. Trends Food Sci. Technol. 9: 347–354.

de Gennes, P. G. 1979. Scaling Concepts in Polymer Physics, Cornell University Press, Ithaca, New York.

Djabourov, M., Leblond, J., and Papon, P. 1988a. Gelation of aqueous gelatin solutions. I. Structural investigation. J. de Physique 49: 319–332.

Djabourov, M., Leblond, J., and Papon, P. 1988b. Gelation of aqueous gelatin solutions. II. Rheology of the sol-gel transition. J. *de* Physique 49: 333–343.

Dobson, C. M. 2003. Protein folding and misfolding. Nature 426: 884–890.

Donato, L., Garnier, C., Novales, B., Durand, S., and Doublier, J. L. 2005. Heat-induced gelation of bovine serum albumin/low-methoxyl pectin systems and the effect of calcium ions. Biomacromolecules 6: 374–385.

Doublier, J. L. and Choplin, L. 1989. A rheological description of amylose gelation. Carbohydr. Res. 193: 215–226.

Doublier, J. L., Garnier, C., Renard, D., and Sanchez, C. 2000. Protein-polysaccharide interactions. Curr. Opin. Colloid and Interface Sci. 5: 202–214.

Doublier, J. L., Launay, B., and Cuvelier, G. 1992. Viscoelastic properties of food gels, in Viscoelastic Properties of Foods, eds. M. A. Rao and J. F. Steffe, Chapter 14, Elsevier Science Publishers, Barking, England.

Dumas, J. and Bacri, J.C.1980. New method of viscosity measurement near the gelatin sol-gel transition. Le Journal de Physique—Letters 41: 279–282.

Durand, D., Naveau, F., Busnel, J. P., Delsanti, M., and Adam, M. 1989. Evolution of polyurethane gel fraction near the gelation threshold. Macromolecules 22: 2011–2012.

Eiselt, P., Lee, K. Y., and Mooney, D. J. 1999. Rigidity of two-component hydrogels prepared from alginate and poly(ethylene glycol)-diamines. Macromolecules 32(17): 5561–5566.

Eldridge, J. E. and Ferry, J. D. 1954. Studies of the cross-linking process in gelatin gels. III. Dependence of melting point on concentration and molecular weight. J. Phys. Chem. 58: 992–995.

Eleya, M. M. O. and Turgeon, S. L. 2000. Rheology of κ-carrageenan and β-lactoglobulin mixed gels. Food Hydrocolloids 14: 29–40.

Evageliou, V., Alevisopolous, S., and Kasapis, S. 1997. Application of stress-controlled analysis to the development of low-fat spreads. J. Texture Stud. 28: 319–335.

Ferry, J. D. 1980. Viscoelastic Properties of Polymers, 3rd ed., John Wiley and Sons Inc., New York, USA.

Flory, P. J. 1953. Principles of Polymer Chemistry, Cornell University, Ithaca, NY, USA.

Flory, P. J. 1974. Introductory lecture. Faraday Discuss. Chem. Soc. 57: 7–18.

Foegeding, E. A. 2006. Food Biophysics of Protein Gels: A challenge of nano and macroscopic proportions. Food Biophys. 1: 41–50.

Foegeding, E. A., Li, H., and Bottcher, S. R. 1998. Gelation of globular proteins, in Phase/State Transitions in Foods: Chemical, Rheological and Structural Changes, eds. M. A. Rao and R. W. Härtel, pp. 111–156, Marcel Dekker, Inc., NY.

Fu, J.-T. 1998. Rheology of sol-gel and gel-sol transition of low-methoxyl pectin + Ca^{2+} gels: the effect of sweeteners, Ph.D thesis, Cornell University, Ithaca, NY.

Fu, J.-T. and Rao, M. A. 1999. The influence of sucrose and sorbitol on gel-sol transition of low-methoxyl pectin + Ca^{2+} gels. Food Hydrocolloids 13: 371–380.

Fu, J.-T. and Rao, M. A. 2001. Rheology and structure development during gelation of low-methoxyl pectin gels: the effect of sucrose. Food Hydrocolloids 15: 93–100.

Fuchs, T., Richtering, W., Burchard, W., Kajiwara, K., and Kitamura, S. 1997. Gel point in physical gels: rheology and light scattering from thermoreversibly gelling schizophyllan. Polymer Gels and Networks 5(6): 541–559.

Garnier, C. 1992. Gelification des pectines en presence de calcium: Étude physico-chimique et rheologique. Ph.D thesis, Université de Nantes, Nantes, France.

Gidley, M. J., Morris, E. R., Murray, E. J., Powell, D. A., and Rees, D. A. 1979. Spectroscopic and stoichiometric characterisation of the calcium-mediated association of pectate chains in gels and in the solid state. J. Chem. Soc. Chem. Comm. (22): 990–992.

Gilsenan, P. M., Richardson, R. K., and Morris, E. R. 2003. Associative and segregative interactions between gelatin and low-methoxy pectin: part 3 quantitative analysis of co-gel moduli. Food Hydrocolloids 17: 751–761.

Gluck-Hirsch, J. B. and Kokini, J. L. 1997. Determination of the molecular weight between cross-links of waxy maize starches using the theory of rubber elasticity. J. Rheol. 41: 129–139.

Gordon, M. and Ross-Murphy, S. B. 1975. The structure and properties of molecular trees and networks. Pure Appl. Chem. 43: 1–26.

Gosal, W. S., Clark, A. H., and Ross-Murphy, S. B. 2002. Novel amyloid fibrillar networks derived from a globular protein: β-lactoglobulin. Langmuir 18: 7174–7181.

Gosal, W. S., Clark, A. H., and Ross-Murphy, S. B. 2004. Fibrillar β-lactoglobulin gels: part 2. dynamic mechanical characterization of heat-set systems. Biomaeromoleeules 5: 2420–2429.

Goycoolea, F. M., Richardson, R. K., Morris, E. R., and Gidley, M. J. 1995. Stoichiometry and conformation of xanthan in synergistic gelation with locust bean gum or konjac glucomannan—evidence for heterotypic binding. Macromolecules 28: 8308–8320.

Grant, G. T., Morris, E. R., Rees, D. A., Smith, P. J. C., and Thom, D. 1973. Biological interactions between polysaccharides and divalent cations: the egg-box model. FEBSLett. 32(1): 195–198.

Grinberg, V. Y., and Tolstoguzov, V. B. 1997. Thermodynamic incompatibility of proteins and polysaccharides in solutions. Food Hydrocolloids 11: 145–158.

Grosso, C. R. F. and Rao, M. A. 1998. Dynamic rheology of structure development in low-methoxyl pectin + Ca^{2+} + sugargels. Food Hydrocolloids 12: 357–363.

Haque, A. and Morris, E. R. 1993. Thermogelation of methylcellulose. Part I: molecular structures and processes. Carbohydr. Polym. 22: 161–173.

Haug, I., Williams, M. A. K., Lundin, L., Smidsrod, O., and Draget, K. I. 2003. Molecular interactions in, and rheological properties of, a mixed biopolymer system undergoing order/disorder transitions. Food Hydrocolloids 17: 439–444.

Hermans, P. H. 1949. Gels, in Colloid Science, ed. H. R. Kruyt, Vol. 2, pp. 483–651, Elsevier Publishing Company, Amsterdam, The Netherlands.

Higgs, P. G. and Ball, R. C. 1990. A "reel-chain" model for the elasticity of biopolymer gels, and its relationship to slip-link treatments of entanglements, in Physical Networks. Polymers and Gels, eds. W, Burchard and S. B. Ross-Murphy, Chapter 15, Elsevier Applied Science Publishers, Barking, England.

Hinrichs, J. and Rademacher, B. 2004. High pressure thermal denaturation kinetics of whey proteins. J. Dairy Res. 71(4): 480–488.

Hsieh, Y. L. and Regenstein, J. M. 1992a. Modeling gelation of egg albumen and ovalbumin. J. Food *Sci.* 57: 856–861.

Hsieh, Y. L. and Regenstein, J. M. 1992b. Elastic attributes of heated egg protein gels. J. Food Sci. 57(4): 862–868.

Hsieh, Y.-L., Regenstein, J. M., and Rao, M. A. 1993. The gel point of whey and egg proteins from dynamic rheological data. J. Food Sci. 58: 116–119.

Ikeda, S., Nitta, Y., Kim, B. S., Temsiripong, T., Pongsawatmanit, R., and Nishinari, K. 2004. Single-phase mixed gels of xyloglucan and gellan. Food Hydrocolloids 18: 669–675.

Iso, N., Mizuno, H., Saito, T., Ohzeki, F., and Kurihara. N. 1984. The change of the rheological properties of surimi (minced fish-meat) by heating. Bull. Japan. Soc. Sci. Fish. 50: 1045–1049.

Joanny, J. F. 1989. The sol-gel transition. Physica B 156, 157: 381–385.

Kasapis, S., Morris, E. R., Norton, I. T., and Brown, C. R. T. 1993a. Phase-equilibria and gelation in gelatin maltodextrin systems. 3. Phase-separation in mixed gels. Carbohydr. Polym. 21:261–268.

Kasapis, S., Morris, E. R., Norton, I. T., and Clark, A. H. 1993b. Phase equilibria and gelation in gelatin/maltodextrin systems—Part I: gelation of individual components. Carbohydr. Polym. 21: 243–248.

Kasapis, S., Morris, E. R., Norton, I. T., and Clark, A. H. 1993c. Phase-equilibria and gelation in gelatin maltodextrin systems. 4. Composition-dependence of mixed-gel moduli. Carbohydra. Polym. 21: 269–276.

Katsuta, K. and Kinsella, J. E. 1990. Effects of temperature on viscoelastic properties and activation energies of whey protein gels. J. Food Sci. 55: 1296–1302.

Kavanagh, G. M. and Ross-Murphy, S. B. 1998. Rheological characterisation of polymer gels. Prog. Polym. Sci. 23: 533–562.

Kavanagh, G. M., Clark, A. H. and Ross-Murphy, S. B. 2000. Heat-induced gelation of globular proteins: 4. Gelation kinetics of low pH β-lactoglobulin gels. Langmuir 16:9584–94.

Kawabata, A. 1977. Studies on chemical and physical properties of pectic substances from fruits. Mem. Tokyo Univ. Agric. 19: 115–200.

Kerner, E. H. 1956. The elastic and thermo-elastic properties of composite media. Proc. Phys. Soc. Sect. B 69: 808–813.

Kim, B. S., Takemasa, M., and Nishinari, K. 2006. Synergistic interaction of xyloglucan and xanthan investigated by rheology, differential scanning calorimetry, and NMR. Biomacromolecules 7: 1223–1230.

Kloek, W., Luyten, H., and van Vliet, T. 1996. Small and large deformation behaviour of mixtures of xanthan and enzyme modified galactomannans. Food Hydrocolloids 10: 123–129.

Kohyama, K., Iida, H., and Nishinari, K. 1993. A mixed system composed of different molecular weights konjac glucomannan and kappa-carrageenan: large deformation and dynamic viscoelastic study. Food Hydrocolloids 7: 213–226.

Kuang, Q. L., Cheng, G. X., Zhao, J., and Li, Y. J. 2006. Thermogelation hydrogels of methylcellulose and glycerol-methylcellulose systems. J. Appl. Polym. Sci. 100(5): 4120–4126.

Langendorff, V., Cuvelier, G., Michon, C., Launay, B., Parker, A., and de Kruif, C. G. 2000. Effects of carrageenan type on the behaviour of carrageenan/milk mixtures. Food Hydrocolloids 14: 273–280.

Lewis, T. B. and Nielsen, L. E. 1970. Dynamic mechanical properties of particulate-filled composites. J. Appl. Polym. Sci. 14: 1449–1471.

Liang, J. N., Stevens, E. S., Morris, E. R., and Rees, D. A. 1979. Spectroscopic origin of conformation-sensitive contributions to polysaccharide optical activity: vacuum-ultraviolet circular dichroism. Biopolymers 18: 327–333.

Lin, Y. G., Maliin, D. T., Chien, J. C. W., and Winter, H. H. 1991. Dynamic mechanical measurement of crystallization-induced gelation in thermoplastic elastomeric poly(propylene). Macromolecules 24: 850–854.

Lopes da Silva, J. A., Rao, M. A., and Fu, J-T. 1998. Rheology of structure development and loss during gelation and melting, in Phase/State Transitions in Foods: Chemical, Rheological and Structural Changes, eds. M. A. Rao and R. W. Hartel, pp. 111–156, Marcel Dekker, Inc., NY.

Lopes da Silva, J. A. and Rao, M. A. 2006. Pectins: Structure, functionality and uses, in Food Polysaccharides and Their Applications: Second Edition, Revised and Expanded, eds. A. M. Stephen, G. O. Phillips, and P. A. Williams, pp. 353–411, CRC Press, Inc., Boca Raton and New York.

Lopes da Silva, J. A. L., Gonçalves, M. P., Doublier, J. L., and Axelos, M. A.V. 1996. Effect of galactomannans on the viscoelastic behaviour of pectin/calcium networks. Polymer Gels and Networks 4: 65–83.

Loret, C., Meunier, V., Frith, W. J., and Fryer, P. J. 2004. Rheological characterisation of the gelation behaviour of maltodextrin aqueous solutions. Carbohydr. Polym. 57 (2): 153–163.

Loveday, S. M., Su, J., Rao, M. A., Anema, S. and Singh, H. 2011. Effect of calcium on the morphology and functionality of whey protein nanofibrils. Biomacromolecules 12:3780–3788.

Lu, L., Liu, X. X., Dai, L., and Tong, Z. 2005. Difference in concentration dependence of relaxation critical exponent n for alginate solutions at sol-gel transition induced by calcium cations. Biomaeromoleeules 6(4): 2150–2156.

Lundell, C., Walkenstrom, P., Loren, N., and Hermansson, A. M. 2004. Influence of elongational flow on phase separated inclusions within gelling biopolymer drops. Food Hydrocolloids 18: 805–815.

Mannion, R. O., Melia, C. D., Launay, B., Cuvelier, G., Hill, S. E., Harding, S. E., and Mitchell, J. R. 1992. Xanthan/locust bean gum interactions at room temperature. Carbohydr. Polym. 19: 91–97.

Manoj, P., Kasapis, S., and Hember, M. W. N. 1997. Sequence-dependent kinetic trapping of biphasic structures in maltodextrin-whey protein gels. Carbohydr. Polym. 32: 141–153.

Manson, J. A. and Sperling, L. H. 1976. Polymer Blends and Composites, Plenum Press, NY.

Mao, C. F. and Rwei, S. P. 2006. Cascade analysis of mixed gels of xanthan and locust bean gum. Polymer 41: 7980–7987.

Maroziene, A. and de Kruif, C. G. 2000. Interaction of pectin and casein micelles. Food Hydrocolloids 14: 391–394.

Marrs, W. M. 1982. Gelatin carbohydrate interactions and their effect on the structure and texture of confectionery gels. Prog. Food Nutr. Sci. 6: 259–268.

Martin, J. E., Adolf, D., and Wilcoxon, J. P. 1989. Viscoelasticity near the sol-gel transition. Phys. Rev. A Gen. Phys. 39: 1325–1332.

Matia-Merino, L., Lau, K., and Dickinson, E. 2004. Effects of low-methoxyl amidated pectin and ionic calcium on rheology and microstructure of acid-induced sodium caseinate gels. Food Hydrocolloids 18:271–281.

Matsumoto, T., Kawai, M., and Masuda, T. 1992. Influence of chain stiffness on the gelation and gel structure of alginate aqueous systems. J. Chem. Soc. Faraday Trans. 88(18): 2673–2676.
McClain, P. E., Kuntz, E., and Pearson, A. M. 1969. Application of stress-strain behaviour to thermally contracted collagen from epimysical connective tissues. J. Agric. Food Chem. 17: 629–632.
Michon, C., Cuvelier, G., and Launay, B. 1993. Concentration dependence of the critical viscoelastic properties of gelatin at the gel point. Rheol. Acta 32: 94–103.
Michon, C., Cuvelier, G., Launay, B., and Parker, A. 1996. Viscoelastic properties of λ-carrageenan/gelatin mixtures. Carbohydr. Polym. 31: 161–169.
Miles, M. J., Morris, V. J., and Ring, S. G. 1985. Gelation of amylose. Carbohyd. Res. 135: 257–269.
Miyoshi, E., Takaya, T., and Nishinari, K. 1998. Effects of glucose, mannose and konjac glucomannan on the gel-sol transition in gellan gum aqueous solutions by rheology and DSC. Polymer Gels and Networks 6: 273–290.
Mochizuki, Y., Saito, T., Iso, N., Mizuno, H., Aochi, A., and Noda, M. 1987. Effects of adding fat on rheological properties of fish meat gel. Bull. Japan. Soc. Sci. Fish. 53: 1471–1474.
Mohammed, Z. H., Hember, M. W. N., Richardson, R. K., and Morris, E. R. 1998. Application of polymer blending laws to composite gels of agarose and crosslinked waxy maize starch. Carbohydr. Polym. 36: 27–36.
Monteiro, S. R., Tavares, C. A., Evtuguin, D. V., Moreno, N., and Lopes da Silva, J. A. 2005. Influence of galactomannans with different molecular weights on the gelation of whey proteins at neutral pH. Biomacromolecules 6: 3291–3299.
Montembault, A., Viton, C., and Domard, A. 2005. Rheometric study of the gelation of chitosan in a hydroalcoholic medium. *Biomaterials* 26(14): 1633–1643.
Morris, E. R. 1990. Mixed polymer gels, in Food Gels, ed. R Harris, pp. 291–359, Elsevier Science Publishers, Barking, UK.
Morris, E. R. 1992. The effect of solvent partition on the mechanical properties of biphasic biopolymer gels: an approximate theoretical treatment. Carbohydr. Polym. 17: 65–70.
Morris, E. R. 1998. Segregative interactions in biopolymer co-gels, in Phase/State Transitions in Foods: Chemical, Rheological and Structural Changes, eds. M. A. Rao and R. W. Hartel, pp. 111–156, Marcel Dekker, Inc., NY.
Morris, E. R. 2000. Rheology of biopolymer co-gels, in Hydrocolloids, Part 2: Fundamentals and Applications in Food, Biology, and Medicine, ed. K. Nishinari, pp. 135–0146, Elsevier Science, Amsterdam, The Netherlands.
Morris, E. R., Rees, D. A., Norton, I. T., and Goodall, D. M. 1980a. Calorimetric and chiroptical evidence of aggregate-driven helix formation in carrageenan systems. Carbohydr. Res. 80: 317–323.
Morris, E. R., Gidley, M. J., Murray, E. J., Powell, D. A., and Rees, D. A. 1980b. Characterization of pectin gelation under conditions of low water activity, by circular dichroism, competitive inhibition and mechanical properties. Int. J. Biol. Macromol. 2: 327–330.
Morris, V. J. and Chilvers, G. R. 1981. Rheological studies on specific ion forms of iota-carrageenate gels. J. Sci. Food. Agric. 32: 1235–1241.
Muller, H. G. 1969. Application of the statistical theory of rubber elasticity to gluten and dough. Cereal Chem. 46: 443–446.
Muller, R., Gérard, E., Dugand, P., Rempp, P., and Gnanou, Y. 1991. Rheological characterization of the gel point: a new interpretation. Macromolecules 24: 1321–1326.
Ndi, E. E., Swanson, B. G., Barbosa-Canovas, G. V., and Luedecke, L. O. 1996. Rheology and micro-structure of β-lactoglobulin/sodium polypectate gels. J. Agric. Food Chem. 44: 86–92.
Neiser, S., Draget, K. I., and Smidsrød, O. 1998. Gel formation in heat treated bovine serum albuminsodium alginate systems. Food Hydrocolloids 12:127–132.
Nishinari, K., Koide, S., and Ogino, K. 1985. On the temperature dependence of elasticity of thermoreversible gels. J. *de* Physique 46: 793–797.
Nishinari, K., Koide, S., Williams, P. A., and Phillips, G. O. 1990. A zipper model approach to the thermoreversible gel-sol transition. *J. de* Physique 51: 1759–1768.

Nishinari, K., Miyoshi, E., Takaya, T., and Williams, P. A. 1996. Rheological and DSC studies on the interaction between gellan gum and konjac glucomannan. Carbohydr. Polym. 30: 193–207.

Nitta, Y., Kim, B. S., and Nishinari, K. 2003. Synergistic gel formation of xyloglucan/gellan mixtures as studied by rheology, DSC, and circular dichroism. Biomacromolecules 4: 1654–1660.

Nolte, H., John, S., Smidsrød, O., and Stokke, B. T. 1992. Gelation of xanthan with trivalent metal ions. Carbohydr. Polym. 18: 243–251.

Norziah, M. H., Foo, S. L., and Karim, A. A. 2006. Rheological studies on mixtures of agar (Gracilaria changii) and κ-carrageenan. Food Hydrocolloids 20: 204–217.

Oakenfull, D. G. 1984. A method for using measurements of shear modulus to estimate the size and thermodynamic stability of junction zones in noncovalently cross-linked gels. J. Food Sci. 49: 1103–1104, 1110.

Oakenfull, D. G. 1987. The chemistry of high-methoxyl pectins, in The Chemistry and Technology of Pectin, ed., R. H. Walter Chapter 5, Academic Press, New York.

Olsson, C., Stading, M., and Hermansson, A. M. 2000. Rheological influence of nongelling amylopectins on beta-lactoglobulin gel structures. Food Hydrocolloids 14: 473–483.

Olsson, C., Langton, M., and Hermansson, A. M. 2002. Microstructures of beta-lactoglobulin/amylopectin gels on different length scales and their significance for rheological properties. Food Hydrocolloids 16(2): 111–126.

Owen, A. J. and Jones, R. A. L. 1998. Rheology of simultaneously phase separating and gelling biopolymer mixtures. Macromolecules 31: 7336–7339.

Papageorgiou, M., Kasapis, S., and Richardson, R. K. 1994. Steric exclusion phenomena in gellan gelatin systems. 1. Physical-properties of single and binary gels. Food Hydrocolloids 8: 97–112.

Paradossi, G., Chiessi, E., Barbiroli, A., and Fessas, D. 2002. Xanthan and glucomannan mixtures: Synergistic interactions and gelation. Biomaeromoleeules 3: 498–504.

Peniche-Covas, C. A. L., Dev, S. B., Gordon, M., Judd, M., and Kajiwara, K. 1974. The critically branched state in a covalent synthetic system and in the reversible gelation of gelatin. Faraday Discuss. Chem. Soc. 57: 165–180.

Pezron, I., Herning, T., Djabourov, M., and Leblond, J. 1990. Scattering from a biopolymer solution in the sol and gel states: the gelatin example in Physical Networks. Polymers and Gels, eds. W. Burchard and S. B. Ross-Murphy, Chapter 18, Elsevier Applied Science Publishers, Barking, UK.

Plashchina, I. G., Fomina, O. A., Braudo, E. E., and Tolstoguzov, V. B. 1979. Creep study of high-esterified pectin gels. I. The creep of saccharose-containing gels. Colloid Poly. Sci. 257: 1180–1187.

Pouzot, M., Nicolai, T., Durand, D., and Benyahia, L. 2004. Structure factor and elasticity of a heat-set globular protein gel. Macromolecules 37:614–620.

Rao, M. A. 1992. Measurement of viscoelastic properties of fluid and semisolid Foods, in Viscoelastic Properties of Foods, eds. M. A. Rao, and Steffe, J. F., Chapter 8, Elsevier Applied Science Publishers, Barking, England.

Rao, M. A. and Cooley, H. J. 1993. Dynamic rheological measurement of structure development in high-methoxyl pectin/fructose gels. J. Food Sci. 58: 876–879.

Rao, M. A. and Cooley, H. J. 1995. Rates of structure development during gelation and softening of high-methoxyl pectin-sodium alginate-fructose mixtures. Food Hydrocolloids 9: 229–235.

Rees, D. A. 1969. Structure, conformation, and mechanism in the formation of polysaccharide gels and networks. Adv. Carbohydr. Chem. Biochem. 24: 267–332.

Rees, D. A. 1972. Polysaccharide gels—a molecular view. Chemistry and Industry, 19: 630–636.

Richardson, R. K. and Ross-Murphy, S. B. 1981a. Mechanical properties of globular protein gels: 1. Incipient gelation behaviour. Int. J. Biol. Macromol. 3: 315–322.

Richardson, R. K. and Ross-Murphy, S. B. 1981b. Mechanical properties of globular protein gels. II: Concentration, pH and ionic strength dependence. British Poly. J. 13: 11–16.

Richardson, P. H., Clark, A. H., Russell, A. L., Aymard, P. and Norton, I. T. 1999. Galactomannan gelation: A thermal and rheological investigation analyzed using the cascade model. Macromolecules 32(5): 1519–1527.

Richter, S., Boyko, V., Matzker, R., and Schroter, K. 2004. Gelation studies: Comparison of the critical exponents obtained by dynamic light scattering and rheology, 2(a)—A thermoreversible gelling system: mixtures of xanthan gum and locust-bean gum. Macromolecular Rapid Communications 25(16): 1504–1509.

Rhim, J. W., Nunes, R. V., Jones, V. A., and Swartzel, K. R. 1989. Determination of kinetic parameters using linearly increasing temperature. J. Food Sci. 54: 446–450.

Rodd, A. B., Cooper-White, J., Dunstan, D. E., and Boger, D. V. 2001. Gel point studies for chemically modified biopolymer networks using small amplitude oscillatory rheometry. Polymer 42(1): 185–198.

Rodriguez-Hernandez, A. I. and Tecante, A. 1999. Dynamic viscoelastic behavior of gellan-iota-carrageenan and gellan-xanthan gels. Food Hydrocolloids 13: 59–64.

Ross-Murphy, S. B. 1991a. The estimation ofjunction zone size from geltime measurements. Carbohydr. Polym. 14: 281–294.

Ross-Murphy, S. B. 1991b. Incipient behaviour of gelatin gels. Rheologica Acta 30: 401–411.

Sanchez, C., Schmitt, C., Babak, V. G., and Hardy, J. 1997. Rheology of whey protein isolate xanthan mixed solutions and gels. Effect of pH and xanthan concentration. *Nahrung* 41: 336–343.

Scanlan, J. C. and Winter, H. H. 1991. Composition dependence of the viscoelasticity of end-linking poly(dimethylsiloxane) at the gel point. Macromolecules 24: 47–54.

Schorsch, C., Garnier, C., and Doublier, J. L. 1997. Viscoelastic properties of xanthan/galactomannan mixtures: comparison of guar gum with locust bean gum. Carbohydr. Polym. 34: 165–175.

Schultz, R. K. and Myers, R. R. 1969. The chemorheology of poly(vinyl alcohol)-borate gels. Macromolecules 2: 281–285.

Shih W-H, Shih W. Y,, Kim, S-I, Liu, J., and Aksay, I. A. 1990. Scaling behavior of the elastic properties of colloidal gels. Physical Review A 42(8): 4772–4779.

Shim, J. and Mulvaney, S. J. 2001. Effect of heating temperature, pH, concentration and starch/whey protein ratio on the viscoelastic properties of com starch/whey protein mixed gels. J. Sci. Food Agric. 81: 706–717.

Simeone, M., Sibillo, V., Tassieri, M., and Guido, S. 2002. Shear-induced clustering of gelling droplets in aqueous biphasic mixtures of gelatin and dextran. J. Rheol. 46: 1263–1278.

Simeone, M., Tassieri, M., Sibillo, V., and Guido, S. 2005. Effect of sol-gel transition on shear-induced drop deformation in aqueous mixtures of gellan and kappa-carrageenan. J. Colloid Interface Sci. 281: 488–494.

Sperling, L. H. 1986. Introduction to Physical Polymer Science, John Wiley, New York.

Stading, M. and Hermansson, A. M. 1990. Viscoelastic behaviour of β-lactoglobulin gel structures. Food Hydrocolloids 4: 121–135.

Stading, M. and Hermansson, A. M. 1993. Rheological behaviour of mixed gels of κ-carrageenan-locust bean gum. Carbohydr. Polym. 22: 49–56.

Stanley, D. W., Aguilera, J. M., Baker, K. W., and Jackman, R. L. 1998. Structure/property relationships of foods as affected by processing and storage, in Phase/State Transitions in Foods: Chemical, Rheological and Structural Changes, eds. M. A. Rao and R. W. Hartel, pp. 1–56, Marcel Dekker, Inc., NY.

Stauffer, D., Coniglio, A., and Adam, M. 1982. Gelation and critical phenomena. Adv. Polymer Sci. 44: 103–158.

Stokes, J. R., Wolf, B., and Frith, W. J. 2001. Phase-separated biopolymer mixture rheology: prediction using a viscoelastic emulsion model. J. Rheol. 45: 1173–1191.

Syrbe, A., Femandes, P. B., Dannenberg, F., Bauer, W. J., and Klostermeyer, H. 1995. Whey protein-polysaccharide mixtures: polymer incompatibility and its application, in Food Macromolecules and Colloids, eds. E. Dickinson and D. Lorient, pp. 328–339, The Royal Society of Chemistry, London.

Takagi, I. and Simidu, W. 1972. On rheological properties and structure of kamaboko. I. Application of rubber elasticity theory to kamaboko. Bull. Japan. Soc. Sci. Fish. 38: 299–303.

Tavares, C. and Lopes da Silva, J. A. 2003. Rheology of galactomannan-whey protein mixed systems. Int. Dairy J. 13: 699–706.

Tavares, C., Monteiro, S. R., Moreno, N., and Lopes da Silva, J. A. 2005. Does the branching degree of galactomannans influences their effect on whey protein gelation? Colloids and Surfaces A. Physicochemical and Engineering Aspects, pp. 270–271: 213–219.

te Nijenhuis, K. 1981. Investigation into the ageing process in gels of gelatin/water systems by the measurement of their dynamic moduli. Part I—phenomenology. Colloid Polym. Sci. 259: 522–535.

te Nijenhuis, K. 1997. Thermoreversible networks. Adv. Polym. Sci. 130: 1–235.

te Nijenhuis, K. and Winter, H. H. 1989. Mechanical properties at the gel point of a crystallizing poly(vinyl chloride) solution. Macromolecules 22: 411–414.

Tobitani, A. and Ross-Murphy, S. B. 1997a. Heat-induced gelation of globular proteins. 1. Model for the effects of time and temperature on the gelation time of BSA. Macromolecules 30: 4845–4854.

Tobitani, A. and Ross-Murphy, S. B. 1997b. Heat-induced gelation of globular proteins. 2. Effect of environmental factors on single component and mixed protein gels. Macromolecules 30: 4855–4862.

Tokita, M., Niki, R., and Hikichi, K. 1984. Percolation theory and elastic modulus of gel. J. Phys. Soc. Japan 53: 480–482.

Tolstoguzov, V. B. 1985. Functional properties of protein-polysaccharide mixtures, in Functional Properties of Food Macromolecules, eds. J. Mitchell and D. A. Ledward, pp. 385–415, Elsevier Applied Science Publishers, London.

Tolstoguzov, V. B. 2001. Functional properties of food proteins and role of protein-polysaccharide interactions. Food Hydrocolloids 4: 429–468.

Treloar, L. R. G. 1975. The Physics of Rubber Elasticity, 3rd ed., Clarendon Press, Oxford, England.

Tuinier, R., ten Grotenhuis, E., Holt, C., Timmins, P. A., and de Krui, C. G. 1999. Depletion interaction of casein micelles and an exocellular polysaccharide. Phys. Rev. E 60: 848–856.

Tuinier, R., ten Grotenhuis, E., and de Kruif, C. G. 2000. The effect of depolymerised guar gum on the stability of skim milk. Food Hydrocolloids 14:1–7.

Tung, C.-Y. M. and Dynes, P. J. 1982. Relationship between viscoelastic properties and gelation in thermosetting systems. J. Appl. Polym. Sci. 27: 569–574.

Turquois, T., Taravel, F. R., and Rochas, C. 1993. Synergy of the agarose-carob galactomannan blend inferred from nmr and rheological studies. Carbohydr. Res. 238: 27–38.

Tziboula, A. and Horne, D. S. 1999. Influence of whey protein denaturation on κ-carrageenan gelation. Colloid *Surf.* B: *Biointerfaces* 12: 299–308.

Van der Linden, E. and Sagis, L. M. C. 2001. Isotropic force percolation in protein gels. Langmuir 17: 5821–5824.

Walkenstrom, P., Panighetti, N., Windhab, E., and Hermansson, A. M. 1998. Effects of fluid shear and temperature on whey protein gels, pure or mixed with xanthan. Food Hydrocolloids 12: 469–479.

Wang, Z.-Y., Zhang, Q.-Z., Konno, M., and Saito, S. 1991. Sol-gel transition of alginate solution by the additions of various divalent cations: critical behavior of relative viscosity. Chem. Phys. Lett. 186(4, 5): 463–466.

Wang, S., van Dijk, J. A. P. P., Odijk, T., and Smit, J. A. M. 2001. Depletion induced demixing in aqueous protein-polysaccharide solutions. Biomaeromoleeules 2: 1080–1088.

Watase, M. and Nishinari, K. 1987a. Dynamic viscoelasticity and anomalous thermal behaviour of concentrated agarose gels. Die *Makromolekulare Chemie* 188: 1177–1186.

Watase, M. and Nishinari, K. 1987b. Rheological and thermal properties of carrageenan gels—effect of sulfate content. Die *Makromolekulare Chemie* 188: 2213–2220.

Watase, M. and Nishinari, K. 1993. Effects of pH and DMSO content on the thermal and rheological properties of high methoxyl pectin-water gels. Carbohydr. Polym. 20: 175–181.

Watase, M., Nishinari, K., Clark, A. H., and Ross-Murphy, S. B. 1989. Differential scanning calorimetry, rheology, X-ray, and NMR of very concentrated agarose gels. Macromolecules 22: 1196–1201.

Weinbreck, F. 2004. Whey protein/Polysaccharide Coacervates: Structure and Dynamics. Ph.D thesis, Utrecht University, The Netherlands.
Williams, P.A., Day, D. A., Langdon, M. J., Phillips, O. G., and Nishinari, K. 1991. Synergistic interaction of xanthan gum with glucomannans and galactomannans. Food Hydrocolloids 6: 489–493.
Winter, H. H. and Chambon, F. 1986. Analysis of linear viscoelasticity of a crosslinking polymer at the gel point. J. Rheol. 30: 367–382.
Winter, H. H. and Mours, M. 1997. Rheology of polymers near liquid–solid transitions. Adv. Polym. Sci. 134: 165–234.
Winter, H. H., Morganelli, P., and Chambon, F. 1988. Stoichiometry effects on rheology of model polyurethanes at the gel point. Macromolecules 21: 532–535.
Wolf, B., Frith, W. J., Singleton, S., Tassieri, M., and Norton, I. T. 2001. Shear behaviour of biopolymer suspensions with spheroidal and cylindrical particles. Rheol. Acta 40: 238–247.
Wolf, B., Scirocco, R., Frith, W. J., and Norton, I. T. 2000. Shear-induced anisotropic microstructure in phase-separated biopolymer mixtures. Food Hydrocolloids 14: 217–225.
Wu, H. and Morbidelli, M. 2001. A model relating structure of colloidal gels to their elastic properties. Langmuir 17: 1030–1036.
Zasypkin, D. V., Braudo, E. E., and Tolstoguzov, V. B. 1997. Multicomponent biopolymer gels. Food Hydrocolloids 11: 159–170.
Zasypkin, D. V., Dumay, E., and Cheftel, J. C. 1996. Pressure- and heat-induced gelation of mixed β-lactoglobulin/xanthan solutions. Food Hydrocolloids 10: 203–211.
Zhang, J. and Rochas, C. 1990. Interactions between agarose and κ-carrageenans in aqueous solutions. Carbohydr. Polym. 13: 257–271.
Ziegler, G. R. and Rizvi, S. S. H. 1989. Determination of cross-link density in egg white gels from stress relaxation data. *J. Food. Sci.* 54: 218–219.

Chapter 7
Role of Rheological Behavior in Sensory Assessment of Foods and Swallowing

Sensory perception of foods is based on the integration of information about numerous aspects of a food, through a number of senses that reach the brain. Among these, the structural information plays an important role. The surface structure of a food product is first perceived by vision, and then the bulk structure is assessed by tactile and kinaesthetic senses combined with hearing while the food is chewed. In spite of the fact that texture is a perceived attribute, which is dynamically evaluated during consumption, many attempts have been done to gain insights into the texture of foods through rheological and structural studies. Several reviews have been published, by Kapsalis and Moskowitz (1987), Bourne (1982), Szczesniak (1987), and Sherman (1988).

Due to the higher reproducibility and sensitivity of instrumental techniques, rheological methods may be advantageous for routine control operations, predicting consumer response, and for assessment of effects of ingredients, process variables and storage on texture. However, this is not an easy task. Texture is a multidimensional property and its evaluation by the consumer of the food product would be, necessarily, affected by the overall evaluation of the relevant textural attributes. The choice of the instrumental analysis for the particular attribute to be evaluated and the conditions used during the sensory evaluation are determinant factors not always under adequate control. In addition, in most cases, the consumer may not be able to give a detailed diagnosis regarding the various components of "texture." Even when texture is evaluated by trained panels, the degree of objectivity can be very low and the terminology used to describe texture is often specific to the food product under analysis. Here, the principles of rheology, followed by the principles of tribology, under the Application of Fluid Mechanics, are discussed for the sensory perception of foods.

M. A. Rao, *Rheology of Fluid, Semisolid, and Solid Foods,* Food Engineering Series,
DOI 10.1007/978-1-4614-9230-6_7,

Stimuli for Evaluation of Viscosity

Many studies have been devoted to the relationship between measured viscosity measured instrumentally and perceived fluid texture. In many cases, the systems that have been studied were model systems of hydrocolloid solutions. Therefore, the structures of the studied systems are far from those of real-fluid food products and the results obtained are of limited use. Szczesniak (1979) pointed out that to describe mouthfeel of beverages, 136 terms were generated from 5,350 responses. Of these, 30.7 % were viscosity-related terms. It was also pointed out that the term "mouthfeel" was better than "viscosity" that was applicable to liquid foods. The former has the same meaning as texture that has been defined by Szczesniak (1963) as "the composite of the structural elements of the food and the manner in which it registers with the physiological senses," and by Sherman (1970) as "the composite of those properties which arise from the physical structural elements and the manner in which it registers with the physiological senses." The high frequency and often the sole use of terms such as "thin," "thick," "viscous," "low viscosity," and "high viscosity" indicate that the sensory perception of viscosity is the most important mouthfeel sensation (Szczesniak 1979). Some of the terms for describing viscosity of fluid foods may be related to their shear rate versus shear stress data, often referred to as objective evaluation. For meaningful results, both the objective rheological data and sensory assessment data must be obtained and such studies are reviewed in this chapter.

Wood (1968) suggested that the stimulus associated with oral assessment was the apparent viscosity determined at 50 s^{-1}. Shama et al. (1973) and Shama and Sherman (1973) made an important contribution in identifying the stimuli associated with sensory assessment of foods by the nonoral methods: tilting a container and stirring a fluid food, and oral methods. They showed that instead of a single-shear rate as suggested by Wood (1968), the stimuli encompass a wide range of either shear rates or shear stresses. They matched the foods assessed sensorially with the shear rate versus shear stress data on the foods. Even after 33 years, the methodology described and the results reported have considerable merit. The results of Shama and Sherman (1973) on stimuli associated with sensory assessment of viscosity were confirmed in subsequent studies (Cutler et al. 1983; Kokini and Cussler 1984).

The key step in the procedure used was in the sensory assessment, especially in selecting a set of four fluid foods that could be arranged in a matrix shown in Fig. 7.1. The arrows were marked along the diagonals joining the corners of the matrix that indicated the direction of viscosity ratings in each pair examined. Each arrow pointed to the less viscous sample. It is important to note that for a wholly consistent evaluation of all four samples in a group, none of the triangles within the matrix should have all three arrows pointing in a clockwise or anticlockwise direction. Furthermore, one of the arrows in each triangle had to point in the direction opposite to the other two. Obviously, considerable preliminary work was done prior to choosing the four foods in a group. For the foods shown in Fig. 7.1, the panelist's sensory evaluation of viscosity is shown in Table 7.1.

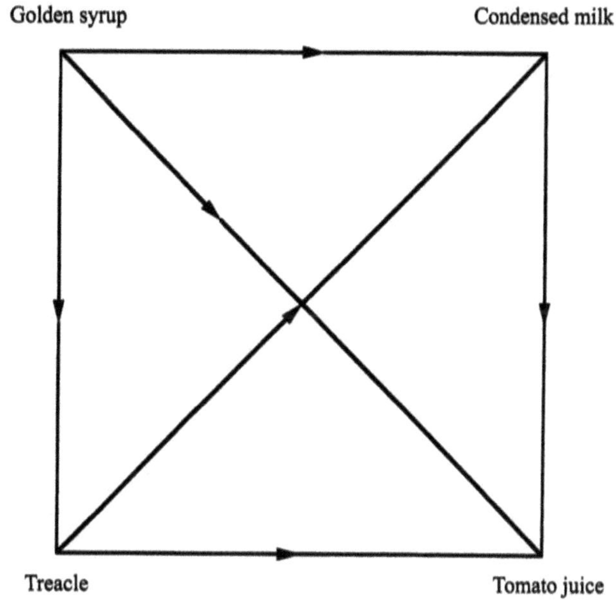

Fig. 7.1 A set of four fluid foods that could be arranged in a matrix for sensory assessment, based on Shama et al. (1973). Each arrow points to the less viscous sample

Table 7.1 Typical viscosity evaluation data of a sensory panelist (Shama et al. 1973)

Product	Code	Ranking	Code	Product
1.[a] Dil. golden syrup	W	>	B	Dil. treacle
2. Condensed milk	U	<	G	Dil. golden syrup
3. Condensed milk	H	<	R	Dil. treacle
4. Tomato juice	P	<	M	Golden syrup
5. Dil. treacle	V	>	L	Tomato juice
6. Tomato juice	D	<	0	Condensed milk

[a] Dil. Means diluted

Stimuli Associated with Tilting Container and Stirring

In Fig. 7.2, four sets of shear rate versus shear stress data are shown on a double logarithmic scale. The data of the soup intersected and crossed over the data of the three diluted treacles (molasses) at different points. According to the sensory evaluation, the soup was considered to have a viscosity similar to that of 80 % treacle and it was intermediate between the viscosities of the 77 and 84 % treacle samples. In the sensory assessment of viscosity by stirring, the stimulus that controls the subject's evaluation of viscosity is the shear stress ($\sim 10^1 - 10^3$ Pa) developed at a shear rate (90–100 s^{-1}) that does not alter substantially with the rheological characteristics of the food (Fig. 7.3). The subjects judged viscosity as the resistance offered by each sample to the stirring motion.

The torque measured during stirring is given by:

$$\text{Torque} = \text{surface area of rotating body} \times \text{radius of rotating body} \times \text{shear stress}$$

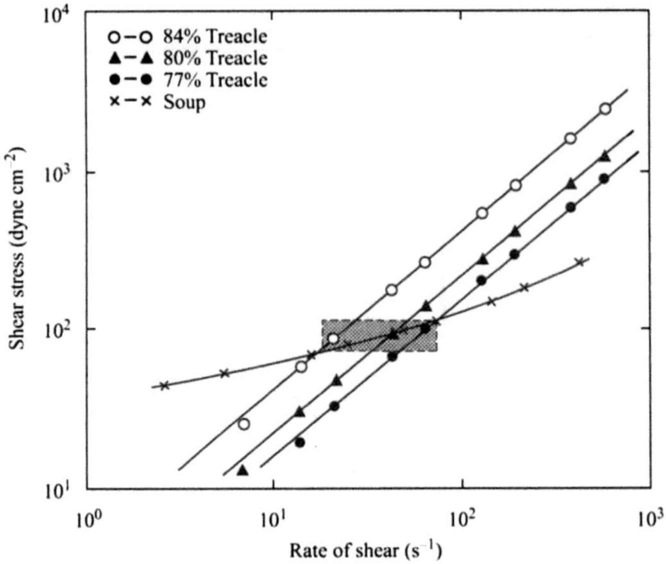

Fig. 7.2 The data of the soup intersected and crossed over the data of the three diluted treacles (molasses) at different points. According to the sensory evaluation, the soup was considered to have a viscosity similar to that of 80 % treacle and it was intermediate between the viscosities of the 77 and 84 % treacle samples (Shama et al. 1973)

Fig. 7.3 Stimuli controlling the sensory assessment of viscosity by the nonoral method stirring the fluid (Shama et al. 1973)

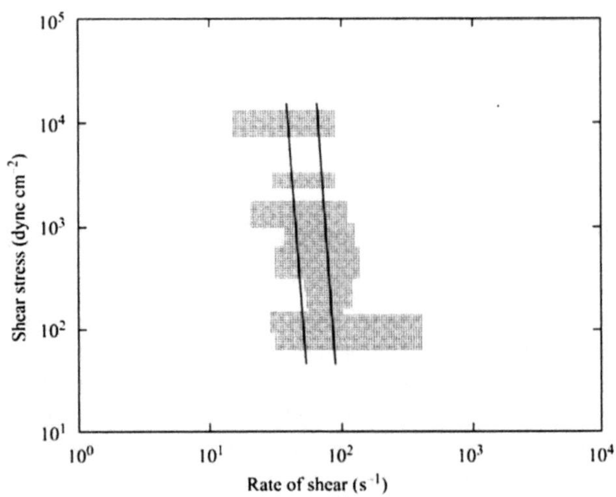

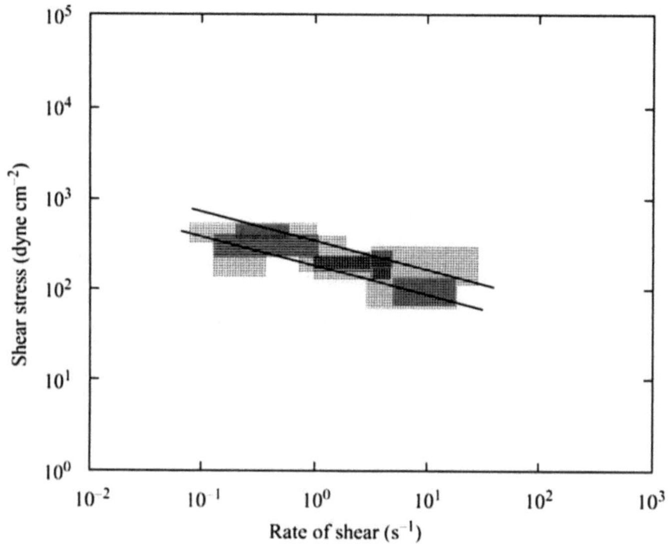

Fig. 7.4 Stimuli controlling the sensory assessment of viscosity by the nonoral method tilting a container (Shama et al. 1973)

In the evaluation of viscosity by tilting the container, panel members indicated that their judgments were based on the rate at which each sample flowed down the side of the beaker, that is, viscosity was judged from the behavior of the apex of the film which flowed down the side of the container when it was first tilted. The major variable was the thickness of the film flowing down the side and its rate of flow, that is, viscosity evaluation by tilting the container depends on the shear rate of (10^{-1}–100 s^{-1}) developed at a shear stress (6–60 Pa) (Fig. 7.4) which varies with the rheological properties of the sample.

Stimuli Associated with Oral Evaluation of Viscosity

Wood (1968) compared the sensory ratings provided by a panel of personnel with shear rate-shear stress data and concluded that the stimulus associated with the oral evaluation of viscosity was the shear stress developed in the mouth at a constant shear rate of 50 s^{-1}. However, Shama and Sherman (1973) showed that the stimulus depends on whether the food is a low-viscosity or a high-viscosity food (Fig. 7.5). For low-viscosity foods, the stimulus is the shear rate developed at an almost constant shear stress of 10 Pa. In contrast, for high-viscosity foods, the stimulus is the shear stress developed at an approximately constant shear rate of 10 s^{-1}. Shama and Sherman (1973) also suggested that for foods that contain oil as the continuous medium, the foods may not spread spontaneously over the tongue and hard palate, and

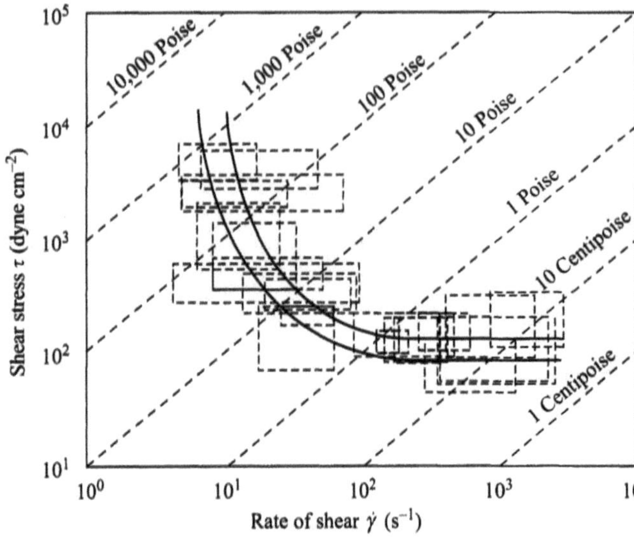

Fig. 7.5 Stimuli controlling the sensory assessment of viscosity by oral methods. Lines of constant viscosity are shown. (Shama and Sherman 1973)

the spreading coefficient may be associated with the oral evaluation of viscosity. While Fig. 7.5 contains constant viscosity lines, Fig. 7.6 has the rheological data of the several foods that were employed.

An additional factor that may play a significant role is the dilution of fluid foods with saliva. The amount of saliva is related to the rate of production that in turn depends on the composition of the fluid food. Therefore, matching relevant composition characteristics (e.g., pH) of the Newtonian and non-Newtonian foods used in the sensory evaluation may be desirable. The flow of low-viscosity Newtonian fluids in the mouth may be turbulent exhibiting higher sensory viscosity.

Comparison of Oral and Nonoral Assessment Techniques

Houska et al. (1998) determined the relationships for five sensory methods of oral and nonoral viscosity evaluation between viscosity scores and instrumentally measured dynamic viscosity for Newtonian fluid foods of low and medium viscosities. From those relationships, the effective shear rates for the five the sensory tests were estimated. Highest shear rates were predicted for viscosity perception by compression of samples between tongue and palate, and the lowest for pouring the fluid foods from a teaspoon. Mixing with a teaspoon, slurping and swallowing exhibited nearly the same dependencies of apparent shear rates on equivalent instrumental viscosity. All the relations were found to be of the hyperbolic type. The relationships between

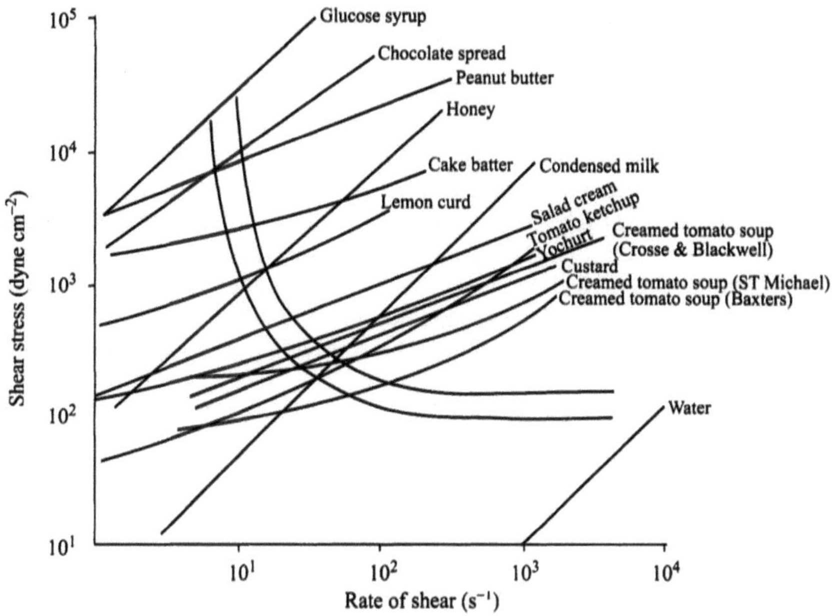

Fig. 7.6 Stimuli controlling the sensory assessment of viscosity by oral methods. Shear rate versus shear stress data of several foods used in the study are shown. (Shama and Sherman 1973)

apparent shear rates and equivalent instrumental viscosity were in good agreement with the relationship predicted by Shama and Sherman (1973) for oral perception.

Sensory Assessment of Viscosity of Gum Dispersions

Gums are used extensively in foods and their rheological behavior was discussed in Chap. 4. It will be recalled that in general, gum dispersions exhibit shear-thinning behavior. Based on the shear-thinning characteristics with a Brookfield viscometer and sensory assessment by a trained panel, gum dispersions could be divided into three groups (Szczesniak and Farkas 1962); however, true shear rates were not calculated in the objective measurement of flow behavior. Group A consisted of gums (e.g., 2 % cooked, 0.75 % phosphomannan, and 0.15 % xanthan gum) corn starch that exhibited a sharp decrease in viscosity with an increase in the shear rate; they were also judged by the trained panel to be nonslimy. Group C consisted of gums that showed a small to moderate change in viscosity and were considered by the trained panel to range from slimy to extremely slimy (e.g., 1 % locust bean gum, 1.3 % sodium alginate, 5 % low-methoxyl pectin, and 2.6 % methyl cellulose). Group B consisted of gums (e.g., 0.6 % guar gum, 1 % carrageenan, and 1 % gum karaya) showing an intermediate dependence on the shear rate and the panel's description that they were at least somewhat slimy.

Perceived Thickness of Gum Dispersions

The results of Szczesniak and Farkas (1962) do not provide any quantitative relationships between rheological behavior and sensory assessment of viscosity in part because the mathematical form of rheological behavior of the gum dispersions was not determined. Further, it was shown in Chap. 4 that the reduced variables: η/η_0 and $\dot{\gamma}/\dot{\gamma}_0$ can be used to describe rheological data of gum dispersions by a single master curve. Morris et al. (1984) correlated the perceived thickness (T) and stickiness (S) with the rheological parameters at 25 °C of concentrated gum dispersions with $\eta_0 > 10$ mPa s: lambda carrageenan, high-methoxyl pectin, guar gum, locust bean gum, and high guluronate-alginate. The correlating equations obtained were:

$$\log T = 0.25\log(\eta_0) + 0.10\log(\dot{\gamma}_{0.1}) + 1.31 \tag{7.1}$$

$$\log S = 0.25\log(\eta_0) + 0.12\log(\dot{\gamma}_{0.1}) + 1.25 \tag{7.2}$$

where, $\dot{\gamma}_{0.1}$ is the shear rate where $\eta_a = 0.1\ \eta_0$. To facilitate comparison with other studies on sensory assessment, Morris et al. (1983) used poise as the unit of viscosity.

The 2.77 w/v% pectin solution (log η_0 in poise = 1.64 and log $\dot{\gamma}_{0.1} = 2.30$) was the standard and the blind control, and was assigned a value of 100 for both thickness and stickiness, and a sample judged to be four times as "thick" as the standard was given a "thickness" score of $T = 400$. Thus, the ratio scaling approach of Stevens (1975) was used. Because the above equations were developed using the 2.77 w/v% pectin solution as the standard arbitrarily, panelists were also asked to assess the thickness of a range of Newtonian foods (mostly sugar syrups) against the standard pectin solution to obtain an equation for thickness:

$$\log T = 0.221\log(\eta_N) + 1.691 \tag{7.3}$$

The above equation was also applicable to stickiness, S, substituted in place of the thickness, T. Because Cutler et al. (1983) have shown that perceived thickness and viscosity were linearly related, the correlations for perceived thickness can be used for perceived viscosity. By combining Eqs. 7.1 and 7.3, one obtain an expression to calculate the viscosity η_N of a Newtonian liquid that would be perceived as having an identical thickness and stickiness to any polysaccharide dispersion with known η_0 and $\dot{\gamma}_{0.1}$:

$$\log \eta_N = 1.13\log \eta_0 + 0.45\log \dot{\gamma}_{0.1} - 1.72 \tag{7.4}$$

Cutler et al. (1983) extended the experimental techniques used by Morris et al. (1983) to fluid foods: syrups, chocolate spread, condensed milk, strawberry dessert sauce, lemon curd, tomato ketchup, rosehip syrup, milk, and sieved chicken soup. Again, the 2.77 w/v% pectin solution (log η_0 (poise) = 1.64 and log $\dot{\gamma}_{0.1} = 2.30$)

was the standard and the blind control, and was assigned a value of 100 for both thickness and stickiness; also, to facilitate comparison with earlier studies on sensory assessment, the viscosity of all samples was expressed in poise. Eq. 7.3 can be rewritten to obtain an expression for the perceived thickness of Newtonian foods:

$$T = 49(\eta_{\mathrm{N}})^{0.22} \tag{7.5}$$

The constant 49 is applicable to the selected standard fluid for scaling, but the exponent is independent of the standard chosen. Further, its low value means that to double the perceived thickness would require a 20-fold increase in the objective value of η_{N}. From Eq. 7.3, one can derive an expression for the perceived thickness of Newtonian foods that can be used to compare equivalent Newtonian viscosity with observed flow curves for non-Newtonian foods:

$$\log(\eta_{\mathrm{N}}) = 4.52 \log T - 7.65 \tag{7.6}$$

The effective oral shear rates at which the objective viscosity was equal to η_{N}, the equivalent Newtonian viscosity, calculated from the above equation were in good agreement with the results of Shama and Sherman (1973) (Fig. 7.7) (Cutler et al. 1983); the only exceptions to this observation were the highly shear-thinning xanthan gum dispersions that deviated slightly from the Shama and Sherman's results.

Spreadability: Using Force and Under Normal Gravity

A master curve of the rheological conditions applicable during spreading of lipophilic preparations with force on the skin using a method similar to that of Wood (1968) showed that the range of acceptable apparent viscosity was about 3.9 poise to 11.8 poise, with an optimum value of approximately 7.8 poise (Barry and Grace 1972). The preferred region was approximately bounded by shear rates 400–700 s^{-1} and shear stress 2,000–7,000 dyne cm^{-2} (200–700 Pa, respectively).

The ranges of values of shear stress and shear rate associated with pourability of two commercial salad dressings under natural conditions were 190–430 dyne cm^{-2} (19–43 Pa, respectively) and 0.9–7.3 s^{-1}, respectively (Kiosseoglou and Sherman 1983). The methodology used in this study was the same as that in Shama et al. (1973). These values agreed reasonably well with those prevailing when the viscosity of a food is judged by tilting the container and evaluating the rate of flow down the container wall. Also, spreadability evaluation was influenced by yield stress and shear-thinning characteristics; flow ceases when the operative shear stress falls below the yield stress.

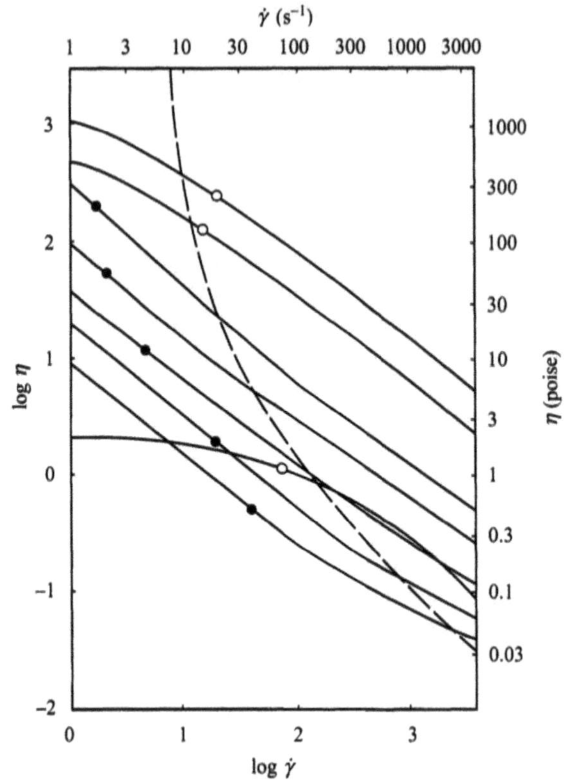

Fig. 7.7 Estimated oral shear rates by Cutler et al. (1983) are indicated by circles. Closed circles are of xanthan gum solutions; the dotted line represents the middle of the curve determined by Shama and Sherman (1973)

Application of Fluid Mechanics

There are a large number of terms that are used to describe food texture. However, only a few studies have attempted to relate the descriptive words to physical data (Jowitt 1974). Kokini et al. (1977) examined 12 textural terms and suggested that for liquid and semisolid foods the three terms: thick, smooth, and slippery, are adequate. Assuming that as a first approximation of the deformation process, a liquid food is squeezed between the tongue and the roof of the mouth, thickness (T) is assumed to be related to the shear stress on the tongue, that is:

$$T \propto (\text{shear stress on the tongue}, \sigma)^{\beta} \tag{7.7}$$

Assuming that $\beta = 1$, for a Newtonian fluid, the expression for the thickness (T) of a Newtonian fluid is:

$$T \propto \text{shear stress on the tongue}, \sigma = (\eta_N)^{1/2} F^{1/2} v \left(\frac{4t}{3\pi r_0^4} \right)^{1/2} \tag{7.8}$$

where, F is the normal force between the tongue and the roof of the mouths, r_0 is the radius of the tongue, v is the average velocity of the tongue, and t is time for assessment. From Eq. 7.8, to double the perceived thickness, the magnitude of the viscosity must be increased four-fold, and the perceived thickness is a function of time. The work of Cutler et al. (1983) (Eq. 7.5) shows that perceived thickness is independent of time and also predicts an increase in the perceived thickness with increase in viscosity, but a 20-fold increase in viscosity is suggested to double the perceived thickness.

For a non-Newtonian power law fluid ($\sigma = K\dot{\gamma}^n$) the expression for the perceived thickness that is proportional to shear stress on the tongue is $T \propto$ shear stress on the tongue,

$$\sigma = Kv^n \left[\frac{1}{h_0^{n+1/n}} + \left(\frac{F}{r_0^{n+3}} \frac{n+3}{2\pi K} \right)^{1/n} \left\{ \frac{(n+1)t}{2n+1} \right\} \right]^{(n^2/n+1)} \tag{7.9}$$

The shear stress felt on the tongue is assumed to be a function of flow within the mouth. That flow may be visualized as resulting from a combination of a lateral movement of the tongue (shear flow), given by the first part on the right hand side, and a compression movement of the tongue toward the palate, given by the second part on the right-hand side. Terpstra et al. (2005) also found linear relationships between calculated shear stress on the tongue and the orally perceived thickness, but that at high values of shear stresses the perceived thickness leveled off: >150 Pa for mayonnaise and >30 Pa for custard. For most of the studied mayonnaises, the contribution of the lateral movement of the tongue to the shear stress was orders of magnitude larger than the contribution of the squeezing movement of the tongue toward the palate.

Fluid Mechanics of Spreadability

For spreading foods, a torque is applied by the consumer during the spreading action that is equal to shears tress on knife × area of knife × length of knife (Kokini 1987). Therefore, spreadability is inversely proportional to the shear stress on the knife:

$$\text{spreadability} \propto (\text{shear stress on knife})^{-1} \tag{7.10}$$

Creaminess, Smoothness, and Slipperiness

Other sensory terms that can be related to various physical forces are smoothness and slipperiness:

$$\text{smoothness} \propto (\text{friction force})^{-1} = (fF)^{-1} \tag{7.11}$$

$$\text{slipperiness} \propto (\text{viscous force} + \text{friction force})^{-1} \propto \left[\frac{Kv^n}{h_s}\left(\pi r_0^2\right) + \eta F\right]^{-1} \tag{7.12}$$

where, f is a friction coefficient and h_s is the average height of aspertites on the tongue's surface. In addition, creaminess is related to thickness and smoothness:

$$\text{creaminess} \propto (\text{thickness})^{0.54} \times (\text{smoothness})^{0.84} \tag{7.13}$$

Creaminess in custards showed strong correlation with rheological parameters obtained from dynamic stress sweeps and the point where the food's structure mechanically broke up (critical stress) (de Wijk et al. 2006). These observations were interpreted as that high creaminess is related to high initial modulus (e.g., G' at 1 Pa) and a relatively low stress (or strain) where the product starts to flow (low critical stress or strain). The custards rated as very creamy had critical stress values of around 15 Pa, while those rated least creamy had values around 65 Pa. Thus, a structure that was harder to break up affected creaminess negatively. It was also suggested that rheological parameters obtained from other bulk rheological tests such as a shear rate versus shear stress flow curve, and dynamic stress sweep were less important. However, for prediction of creaminess of custard, it was suggested that in addition to the aforementioned rheological parameters, parameters related to friction should be included. Tribological methods can be used to characterize frictional characteristics; the study of Lee et al. (2002) is an example of such a study.

Mahmood et al. (2006) studied taste, thickness and creaminess of butter fat-in-water emulsions, stabilized by sodium caseinate, and with well-defined droplet-size distribution and rheological properties. The sensory ratings of creaminess and thickness were strongly correlated, and the higher values of both were attributed to samples of higher viscosity and higher oil volume fraction. Emulsions prepared with maltodextrin or xanthan having the same apparent viscosity were perceived to have significantly different levels of perceived creaminess. An important conclusion was that the apparent viscosity at 50 s^{-1} was insufficient to describe fully the perceived thickness or creaminess of the model emulsions.

Role of Size, Shape, and Hardness of Particles

Earlier, in Chaps. 2, 4, and 5, it was pointed out that many foods contain solid particles. Thus, the role of the size, shape, and hardness of particles in oral perception of texture is of interest. For example, in the confectionary industry the minimum particle size that can be comprehended by the palate is said to be about 25 µm. Further, particle sizes about 10–15 µm are considered to be the limit of diminishing effect. On the other hand, in tooth paste, the alumina trihydrate particles with an average diameter 5–20 μm are used and larger particles are known to contribute to gritty sensation in the mouth (Tyle 1993).

Tyle (1993) studied oral perception of grittiness and viscosity of synthetic particles (60 mg) suspended in fruit syrups (5 mL). The particles were of different size distributions and shapes: angular-shaped garnet 5.2–33.0, rounded micronized polyethylene 7.2–68.9, and flat mica platelets coated with titanium dioxide 28.1–79.6 μm. There was no effect on the thickness ratings of the studied particulate dispersions. Particles up to about 80 μm suspended in fruit syrups were not perceived to be gritty if they were soft and rounded (micronized polyethylene) or relatively hard and flat (mica). With the hard and angular garnet particles, grittiness was perceived above a particle size range of 11–22 μm. In this respect we note that cooked starch granules are soft, with most granules being no larger than about 80–100 μm, while many chocolate particles are about 40–45 μm, irregular in shape, but with rounded edges.

Role of Rheology in Perception of Flavor and Taste

The influence of the rheology of a particular food material on the perception of its taste or flavor can have two main origins. A physiological effect due to the proximity of the taste and olfactory receptors to the kinesthetic and thermal receptors in the mouth, since then an alteration of the physical state of the material may have an influence on its sensory perception, and an effect related to the bulk properties of the material (e.g., texture, viscosity), since the physical properties of the material may affect the rate and the extent with which the sensory stimulus reaches the gustatory receptors.

For fluid systems, mainly hydrocolloid solutions/dispersions have been used as model systems, it was demonstrated that, generally, an increase in viscosity of the system decreased the perception of sweetness. Different hydrocolloids were found to affect sweetness to differing extents (Vaisey et al. 1969; Pangborn et al. 1973). Vaisey et al. (1969) have also shown that hydrocolloid solutions with more pronounced shear thinning behavior tended to decrease in less extent the sweetness perception than those that are less pseudoplastic. Although sweetness was more difficult to detect when the viscosity was higher, for suprathreshold levels of sugar it was shown that the more viscous solutions were perceived as sweeter (Stone and Oliver 1966).

Moskowitz and Arabie (1970) found that the taste intensity (sweetness, sourness, saltiness, and bitterness) was related to the apparent viscosity of carboxymethyl cellulose solutions by a power function with a negative slope. Pangborn et al. (1973) observed that the influence of different hydrocolloids on the perception of some basic taste intensities (saltiness, bitterness, sourness) appeared to be more dependent on the nature of the hydrocolloid and the taste of the substance than on the viscosity level. In contrast, sweetness imparted by sucrose was found to be highly dependent on viscosity, that is, the hydrocolloid concentration; above a certain viscosity threshold, it was shown that the sweetness intensity of sucrose was significantly depressed. Saltiness was the taste attribute less affected, sourness, imparted by citric acid, was significantly reduced by all hydrocolloids tested, and for the other taste substances, the presence of a hydrocolloid generally enhanced the taste intensity of saccharin and depressed that of sucrose and caffeine (bitterness).

The effect of different kinds and levels of hydrocolloids on selected aromatic flavor compounds was also investigated (Pangborn and Szczesniak 1974). The hydrocolloids were found to decrease both the odor and the flavor intensities, and once again, the overall effects seemed much more dependent on the nature of the hydrocolloid and of the taste substance than on the viscosity level imparted by the biopolymer. Similar results were found in a subsequent study conducted by Pangborn et al. (1978), about the effect of the same hydrocolloids on the aroma, flavor, and basic tastes of three distinct beverage systems.

Launay and Pasquet (1982) studied the relationship between sweetness and solution viscosity using sucrose solutions thickened with guar gum prepared to two constant viscosity levels. The results showed that the sweetness intensity decreased in the presence of the gum, but this reduction was not dependent on the viscosity level, a result that seems to be in conflict with the results previously reported by Moskowitz and Arabie (1970).

Some results from the several studies mentioned above are difficult to compare and some apparent discrepancies are observed. The main causes for the differences are due to the different hydrocolloids and the taste substances that have been used, the different hydrocolloid concentration ranges that have been studied, and also, in some cases, from the inaccurate evaluation of the effect of the added taste substance, for example, sucrose, on the rheological behavior of the hydrocolloid.

Role of Hydrocolloid Concentration

The importance of the concentration regime of the hydrocolloid upon its effect on taste and flavor perception was clearly shown by Baines and Morris (1987). These authors studied the effect of guar gum concentration on flavor/taste perception in thickened systems, using solutions with constant concentration of sucrose and flavoring and a wide range of concentrations of three different samples of guar gum. The results obtained showed that the perceived intensity of both attributes was independent of polymer concentration up to c^* (the concentration threshold that marks the transition from a dilute solution of random-coil polymer molecules, to an entangled network where the molecules are not free to move independently, described in Chap. 4), but decreased steeply at higher degrees of space occupancy by the polymer $c[\eta]$, that is, above c^*. The interpretation that has been proposed for the observed behavior was that a restricted replenishment of surface depletion would occur with increasing coil overlap and entanglement of the hydrocolloid, thus, decreasing the perception of the sensory attributes.

Engineering Approach

An interesting engineering approach was proposed by Kokini and coworkers to model viscosity–taste interactions. Kokini et al. (1982) have studied the perception of sweetness of sucrose and fructose in solutions with various tomato solids contents, and with basis on the observed decreasing of sweetness intensity as the percentage of tomato solids increased, they have proposed a more complex but rather comprehensive physical model relating viscosity and taste intensity, based on the physics and chemistry in the mouth. This model was further successfully applied (Cussler et al. 1979) to explain the effect of the presence of hydrocolloids at different levels on the subjective assessment of sweetness and sourness, giving a new and more consistent interpretation of the results previously reported by Moskowitz and Arabie (1970), and also to explain the interaction of butyric acid in solutions of carboxymethylcellulose reported by Pangborn and Szczesniak (1974). The main assumptions are rather simple and logical: because the taste or odor attributes are assessed by very rapid chemical reactions at the human taste and olfactory receptors, the assessment process should be limited not by these chemical reactions but rather by a diffusion process in the mouth or in the nose. Thus, for example, variations in sweetness and sourness are proportional to the product of the solute concentration and the square root of the diffusion coefficient. The diffusion coefficients may be calculated from mass transfer considerations and are dependent on viscosity. The model seems to work quite well for those cases where mass transfer is the limiting factor. However, for attributes like saltiness and bitterness the above model does not work; the model also cannot explain why fructose and sucrose at the same concentration originate different taste intensities.

Role of Rheology in Swallowing

The swallowing process may be described in three phases: oral, pharyngeal, and esophageal. There has been much emphasis on the esophageal phase (Perlman 1999), but the pharynx serves a crucial role in the swallowing process by helping to propel the bolus down to the esophagus without regurgitation or aspiration (Langmore 2001). The purpose of the oral phase of swallowing is to prepare the food bolus for swallowing. When a solid food is masticated sufficiently, the base of the tongue pushes the bolus toward the back of the mouth into the pharynx (Kahrilas et al. 1993; Langmore 2001). As the bolus enters the pharynx, the pharyngeal muscle walls move in a squeezing action to further help propel the bolus downward. Meanwhile both the nasopharyngeal and laryngeal openings are shut off by neuromuscular actions to prevent entrance of fluid into these areas (Langmore 2001). For about 0.6 s after the start of the swallow the airway is open to any spillage from the bolus, which may be a very dangerous scenario for a person with a dysfunctional neuromuscular structure (Langmore 2001). At the end of the pharyngeal phase, the upper esophageal sphincter (UES) lets the bolus pass quickly into the esophagus (Langmore 2001).

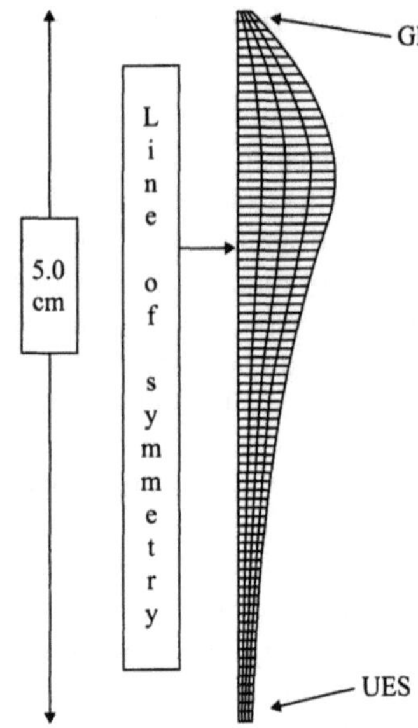

Fig. 7.8 Schematic diagram of the human throat as an axisymmetric pipe with moveable Walls. The inlet of that pipe is the glossopalatal junction (*GPJ*) and the outlet is the upper esophageal sphincter (*UES*)

Dysphagia, also known as deglutition, is defined as the "impaired ability to swallow" (Perlman 1999). An increase in bolus viscosity delays pharyngeal bolus transit and lengthens the duration of the opening of the UES (Dantas and Dodds 1990; Dantas et al. 1990; Reimers-Neils et al. 1994), the outlet of the throat. Pouderoux and Kahrilas (1995) report that the force of tongue propulsion increases when a bolus of greater viscosity is introduced. Kendall et al. (2001) suggested that smaller volumes of viscous liquids could also be key factor to a more effortless swallow. Here, the role of fluid rheology during swallowing is of interest.

A Model of the Swallowing Process

In basic terms, the swallowing process may be modeled (Chang et al. 1998) considering that the human throat is an axisymmetric pipe with moveable walls as shown schematically in Fig. 7.8. The inlet of that pipe is the glossopalatal junction (GPJ) and the outlet is UES. At the beginning of the flow process, the fluid is pushed into the inlet by the base of the tongue. The diameter of the pipe is widened at the same time to accommodate the passing liquid. Soon the outlet is opened and the wall of the pipe near the inlet begins to close and further pushes the fluid toward the outlet in a squeezing action. Toward the end of the flow process, the pipe collapses completely and all of the fluid is pushed through the outlet.

Using the above model, Meng et al. (2005) studied the pharyngeal transport of three fluids with different flow properties: a Newtonian fluid with $\eta = 0.001$ Pa s (representing water), another Newtonian fluid (above 3 s^{-1}) with $\eta = 0.150$ Pa s (representing 2.50% w/v barium sulfate mixture with a density of 1,800 kg m^{-3}),

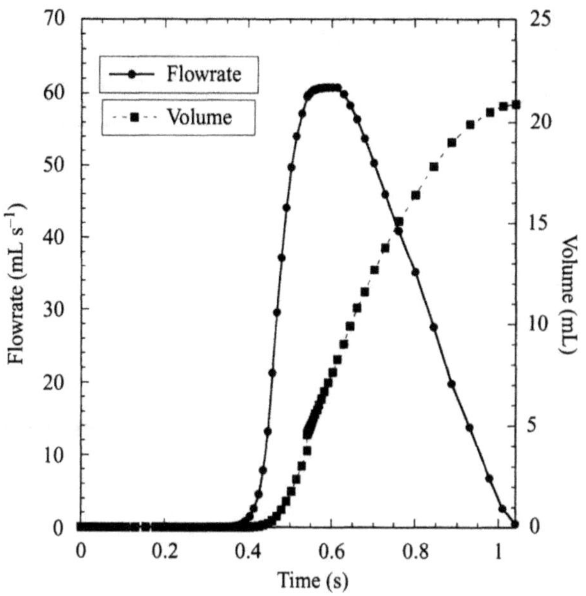

Fig. 7.9 Flow rate data through the upper esophageal sphincter showed a smooth curve

and a non-Newtonian, shear-thinning fluid with $K=2.0$ Pa s^n, $n=0.7$ (representing a starch-thickened beverage and with a density of 1,800 kg m^{-3}). The power law parameters for the starch-thickened beverage is typical of a starch-thickened beverage, such as apple juice (Meng and Rao 2005), and other foods served at nursing homes for dysphagic patients. The equations of continuity and motion were solved, using the initial and the boundary conditions of the pharyngeal wall movement that were based on clinical values published in the literature (Kahrilas et al. 1988, 1993; Cook et al. 1989). Much detailed information on the solution scheme and various results obtained can be found elsewhere (Meng et al. 2005). Here, the role of fluid properties is discussed.

Flow Rate and Cumulative Volume for a Newtonian Fluid

Flow rate data through the UES (Fig. 7.9) showed a smooth curve and the overall shape of the flow rate curve resembled that of Chang et al. (1998). Maximum flow rate values were lower than those reported in Chang et al. (1998), by about 20 mL s^{-1}, possibly due to the coarser nature of their mesh. However, the total volume transported through the UES with respect to time was also very similar to the study by Chang et al. (1998), showing approximately a total of 20 mL of fluid had passed into the esophagus by the end of the simulation.

Fig. 7.10 Flow rates of the three boluses: water, barium sulfate, and shear-thinning non-Newtonian fluid

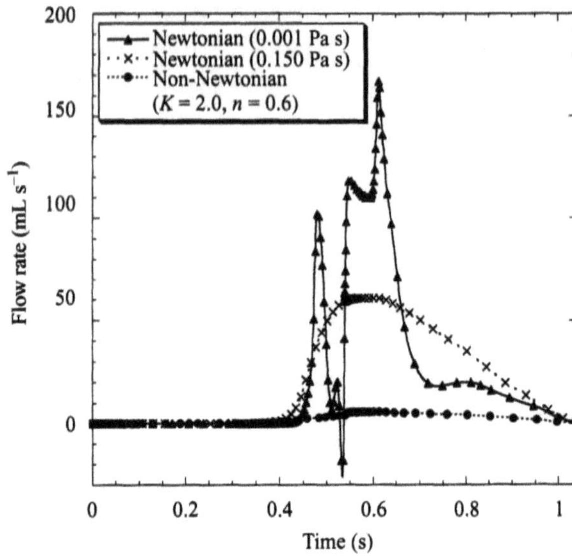

Effect of Fluid Rheology on the Swallowing Process

For a Newtonian water bolus with a viscosity of 0.001 Pa s and a density of 1,000 kg m^{-3}, the initial normal stress at the GPJ had to be reduced by 92 % (from 150 to 12 mmHg) and the duration at which the initial normal stress was held was reduced by 69 % (from 0.32 to 0.1 s) in order for the model to simulate a 20-mL swallow. Not using these corrections resulted in reverse flow (i.e., negative flow rate) at the GPJ. This was reasonable because the water bolus has a smaller viscosity than the other two liquids. Reverse flow may occur in the human throat if the pharyngeal walls are not sufficiently elastic to expand and accommodate the liquid. Such clinical data are not currently available in the literature, but would be valuable for the continuation of these studies. The computational changes may have varied the problem slightly, however, the effect of rheology on the swallowing time should still remain the same.

The flow rates of the three boluses (Fig. 7.10) showed significant differences from 0.54 to 1.04 s. The water bolus (Newtonian, 0.001 Pa s) showed a sharp increase at about 0.45 s, followed by a rapid decrease to negative values at about 0.53 s. At 0.53 s, the UES is widening, which may draw some of the fluid in the negative z direction in the case of a less viscous liquid such as water. A second maximum in flow rate was later seen at about 0.6 s followed initially by a rapid decrease to 20 mL s^{-1} at 0.7 s and then a more gradual decrease to 0 mL s^{-1} at 1.04 s (Fig. 7.10). The barium bolus (Newtonian, 0.150 Pa s) showed a moderately rapid increase to a maximum flow rate of 60 mL s^{-1} between 0.4 and 0.6 s, followed by a gradual decrease to zero flow rate at 1.04 s. The starch-thickened bolus (non-Newtonian) showed extremely low flow rates at all times compared to the Newtonian

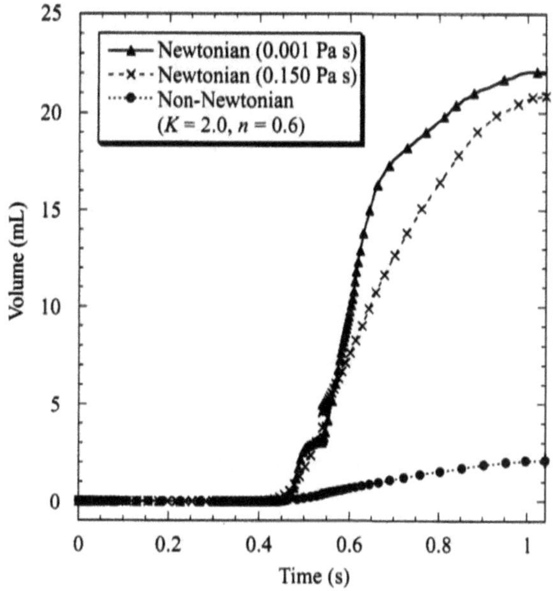

Fig. 7.11 Total fluid volume that passed through the UES for the three boluses: water, barium sulfate, and shear-thinning non-Newtonian Fluid

boluses. The difference in the shape of the flow rate curve for the water bolus indicated that there may be inertial effects due to its relatively low viscosity. The Reynolds number at the UES at the end of the simulation is 116 for the Newtonian water bolus, 4.0 for the Newtonian barium sulfate bolus, and 5.4 for the non-Newtonian bolus (data not shown).

Furthermore, the total fluid volume that passed through the UES for the shear-thinning non-Newtonian bolus was only around 2 mL (Fig. 7.11), about an order of magnitude lower than that of Newtonian boluses. This result can perhaps explain why people, especially dysphagic patients, often take more than one swallow to complete the same volume of liquid when the viscosity is higher. These data also emphasize the importance of the non-Newtonian, shear-thinning, nature of many starch-thickened foods in a dysphagia diet.

The shear rates at the UES for Newtonian and shear-thinning boluses were different. Given that the pharyngeal wall movements were identical in all three cases, much higher shear rates were found for the water bolus (Newtonian, 0.001 Pa s) even though the initial normal stress at the GPJ was much lower and was sustained for a shorter duration. The apparent shear rate at the UES at the end of simulation was 0.001 s^{-1} for the non-Newtonian bolus, compared to that at the same location of 0.009 s^{-1} and 27 s^{-1} for the barium and Newtonian water boluses, respectively. These apparent shear rates and the Reynolds numbers reported above clearly indicated that the consistency of the liquid influences pharyngeal bolus transport. More importantly, given the same boundary conditions (pharyngeal wall movement), the rate at which a fluid traversed through the UES was substantially greater for the water bolus than for either the barium sulfate or the non-Newtonian bolus.

Calculated values of the apparent viscosity (η_a) of the shear-thinning non-Newtonian bolus at the UES showed a rapid drop from an initial viscosity of 60 Pa s to about 1 Pa s between 0 and 0.2 s. The low viscosity was sustained from 0.2 to 0.5 s, during which both GPJ and UES re-opened. From 0.6 s to the end of the simulation at 1.04 s, the viscosity rose rapidly, showed a maximum at 0.6 s, followed by a gradual decrease to 40 Pa s at 1.04 s.

These data suggest that perhaps one reason shear-thinning non-Newtonian liquids are safer to swallow than thin Newtonian liquids is due to the reduced fluid flow during the second half of the swallowing process. The reduced flow in turn allows more time for air passages (e.g., entry to the trachea or the nasopharynx) to completely shut off prior to the arrival of food. As a result, the dysphagic patient does not aspirate as he or she would with a Newtonian bolus.

Effect of Rheology on Time to Swallow 1.0 mL

To further illustrate the differences in the swallowing processes of Newtonian and non-Newtonian fluids, t_{cv} was coined to represent the time to swallow a critical volume and was defined as the number of seconds needed to transport the first 1.0 mL of fluid into the esophagus. The greater the t_{cv} value, the safer is the swallow, as the muscles in the pharynx have more time to close off entryway to the air passages before food arrives. The parameter t_{cv} may be useful for characterizing the severity of deglutition in a particular patient. It may also be used as a benchmark for any improvement or deterioration in the patient. Because it would be difficult to obtain t_{cv} values clinically, computer simulations such as this one are beneficial to the understanding of dysphagia.

The effect of Newtonian viscosity (density assumed constant) on t_{cv} is shown in Fig. 7.12. In general, a Newtonian bolus with a higher viscosity results in a higher t_{cv}. A linear relation may be seen between t_{cv} and η; at viscosities higher than 1 Pa s (Fig. 7.12). The relationship between density and viscosity was not examined for the Newtonian bolus in our study because we only examined viscosities less than 15 Pa s and effect of density was important in very high viscosity fluids (Li et al. 1994).

Values of t_{cv} increase sharply with consistency coefficient K of shear-thinning non-Newtonian fluids for $K<0.5$ (Fig. 7.13). For K values between 0.5 and 1.0 there was a slight decrease in t_{cv}. For $K>2$ a linear dependence on K is seen for t_{cv} (Fig. 7.13). These data suggest that increasing the consistency coefficient of a shear-thinning fluid generally lengthens the swallowing time and may help in reducing the risk of aspiration. The effect of small values of K should be further examined for a better understanding of their effect on t_{cv}.

Values of t_{cv} showed a strong dependence on the flow behavior index, n, for both shear-thinning ($n<1$) and shear-thickening ($n>1$) fluids (Fig. 7.14). This strong dependence suggested that non-Newtonian fluids in general increase the critical swallowing time more effectively than Newtonian fluids, which underscores the importance of simulating pharyngeal swallows of non-Newtonian boluses.

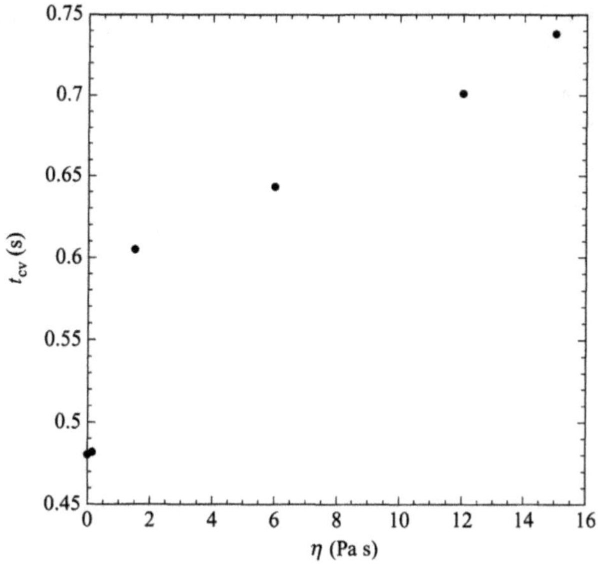

Fig. 7.12 Time to transport the first 1.0 mL of fluid foods with different viscosities into the esophagus

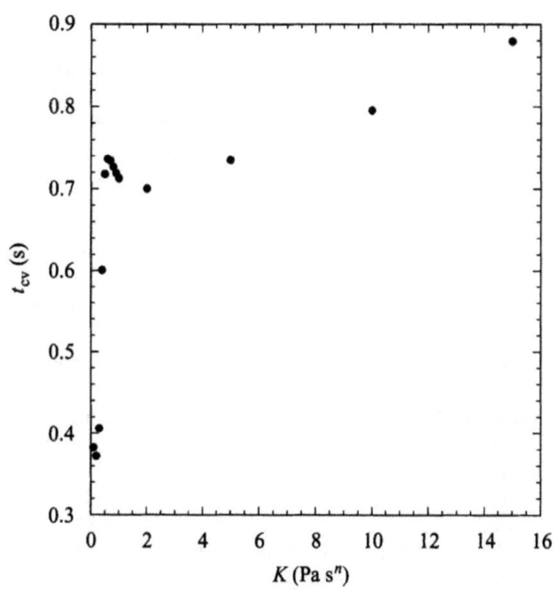

Fig. 7.13 Increasing the power law consistency coefficient of a shear-thinning fluid generally increased the swallowing time

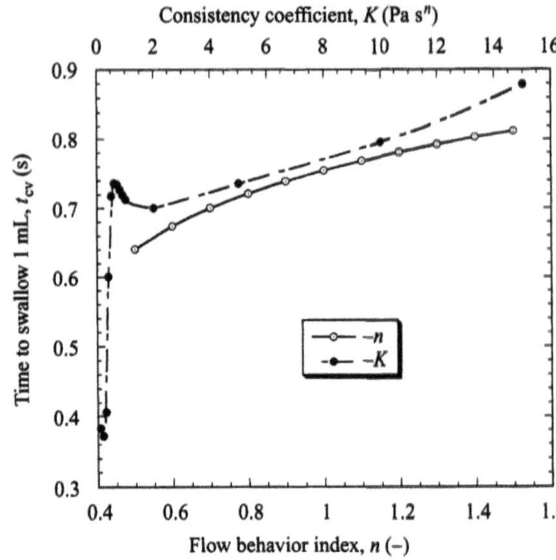

Fig. 7.14 In general, non-Newtonian fluids increased the critical swallowing time more effectively than Newtonian fluids

References

Baines, Z. V. and Morris, E. R. 1987. Flavour/taste perception in thickened systems: the effect of guar gum above and below c^*. *Food Hydrocolloids* 1: 197–205.

Barry, B. W. and Grace, A. J. 1972. Sensory testing of spreadability: investigation of rheological conditions operative during application of topical preparations. *J. Pharm. Sci.* 61: 335–341.

Bourne, M. C. 1982. *Food Texture and Viscosity*, Academic Press, New York.

Chang, M. W., Rosendall, B., and Finlayson, B. A. 1998. Mathematical modeling of normal pharyngeal bolus transport: a preliminary study. *J. Rehabil. Res. Dev.* 35(3): 327–334.

Cook, I. J., Dodds, W. J., Dantas, R. O., Massey, B., Kern, M. K., Lang, I. M., Brasseur, J. G., and Hogan, W. J. 1989. Opening mechanisms of the human upper esophageal sphincter. *Am. J. Physiol*, 257(5): G748–G789.

Cussler, E. L., Kokini, J. L., Weinheimer, R. L., and Moskowitz, H. R. 1979. Food texture in the mouth. *Food Technol.* 33(10): 89–92.

Cutler, A. N., Morris, E. R., and Taylor, L. J. 1983. Oral perception of viscosity in fluid foods and model systems. *J Texture Stud.* 14:377–395.

Dantas, R. O. and Dodds, W. J., 1990. Effect of bolus volume and consistency on swallow-induced submental and infrahyoid electromyographic activity. *Braz. J. Med. Biol. Res.* 23: 37–44.

Dantas, R. O., Kern, M. K., Massey, B., Dodds, W. J., Kahrilas, P. J., Brasseur, J. G., Cook, I. J., and Lang, I. M. 1990. Effect of swallowed bolus variables on oral and pharyngeal phases of swallowing. *Am. Physiol. Soc.* 258: G675–G681.

Houska, M., Valentova, H., Novotna, P., Strohalm, J., Sestak, J., and Pokorny, J. 1998. Shear rates during oral and nonoral perception of viscosity of fluid foods. *J. Texture Stud.* 29(6): 603–615.

Jowitt, R. 1974. The terminology of food texture. *J. Texture Stud.* 5: 351–358.

Kahrilas, P. J., Dodds, W. J., and Hogan, W. J. 1988. Effect of peristaltic dysfunction on esophageal volume clearance. *Gastroenterol.* 94(1): 73–80.

Kahrilas, P. J., Lin, S., Logemann, J. A., Ergun, G. A., and Facchini, F. 1993, Deglutitive tongue action: volume accommodation and bolus propulsion. *Gastroenterol.* 104: 152–162.

Kapsalis, J. G. and Moskowitz, H. R. 1978. Views on relating instrumental tests to sensory assessment of food texture. Application to product development and improvement. *J. Texture Stud.* 9(4): 371–393.

Kendall, K.A., Leonard, R.J., and McKenzie, S.W. 2001. Accommodation to changes in bolus viscosity in normal deglutition: a videofluoroscopic study. Annals of Otology, Rhinology & Laryngology (*Ann Oto Rhinol* Laryn), 110: 1059–1065.

Kiosseoglou, V. D. and Sherman, P. 1983. The rheological conditions associated with judgement of pourability and spreadability of salad dressings. *J. Texture Stud.* 14: 277–282.

Kokini, J. L. 1987. The physical basis of liquid food texture and texture-taste interactions. *J. Food Eng.* 6: 51–81.

Kokini, J. L. and Cussler, E. L. 1984. Predicting liquid food texture of liquid and melting semi-solid foods. *J. Food Sci.* 48: 1221–1225.

Kokini, J. L., Kadane, J., and Cussler, E. L. 1977. Liquid texture perceived in the mouth. *J. Texture Stud.* 8: 195–218.

Kokini, J. L., Bistany, K., Poole, M., and Stier, E. 1982. Use of mass transfer theory to predict viscosity-sweetness interactions of fructose and sucrose solutions containing tomato solids. *J. Texture Stud.* 13: 187–200.

Langmore, S. E. 2001. *Endoscopic Evaluation and Treatment of Swallowing Disorders*, Thieme Medical Publishers, Inc., New York, USA.

Launay, B. and Pasquet, E. 1982. Sucrose solutions with and without guar gum: rheological properties and relative sweetness intensity. *Prog. Food Nutri. Sci.* 6: 247–258.

Lee, S., Heuberger, M., Rousset, P., and Spencer, N. D. 2002. Chocolate at a sliding interface. J. Food *Sci.* 67(7): 2712–2717.

Li, M., Brasseur, J. G., and Dodds, W. J. 1994. Analyses of normal and abnormal esophageal transport using computer simulations. Am. J. Physiol. 266: G525–G543.

Mahmood, A., Murray, B. S., and Dickinson, E. 2006. Perception of creaminess of model oil-in-water dairy emulsions: influence of the shear-thinning nature of a viscosity-controlling hydrocolloid. Food-Hydrocolloids 20(6): 839–847.

Meng, Y. and Rao, M. A. 2005. Rheological and structural properties of cold-water-swelling and heated cross-linked waxy maize starch dispersions prepared in apple juice and water. Carbohydrate Polymers 60: 291–300.

Meng, Y., Rao, M. A., and Datta, A. K. 2005. Computer simulation of the pharyngeal bolus transport of Newtonian and non-Newtonian fluids. IChemE Trans. Part C—Food and Bioproducts Processing 83: 297–305.

Morris, E. R., Richardson, R. K., and Taylor, L. J. 1984. Correlation of the perceived texture of random coil polysaccharide solutions with objective parameters. Carbohydr. polym. 4: 175–191.

Moskowitz, H. R. and Arabie, P. 1970. Taste intensity as a function of stimulus concentration and solvent viscosity. J. Texture Stud. 1: 502–510.

Pangborn, R. M. and Szczesniak, A. 1974. Effect of hydrocolloids and viscosity on flavor and odor intensities of aromatic flavor compounds. J. Texture Stud. 4: 467–482.

Pangborn, R. M., Tabue, I. M., and Szczesniak, A. 1973. Effect of hydrocolloids on oral viscosity and basic taste intensities. J. Texture Stud. 4: 224–241.

Pangbom, R. M., Gibbs, Z. M. and Tassan, C. 1978. Effect of hydrocolloids on apparent viscosity and sensory properties of selected beverages. J. Texture Stud. 9: 415–436.

Perlman, A. L. 1999. Dysphagia: populations at risk and methods of diagnosis. Nutri.Clinical Practice 14(5): S2–S9.

Pouderoux, P. and Kahrilas, P. J. 1995. Deglutitive tongue force modulation by volition, volume, and viscosity in humans. Gastroenterol, 108: 1418–1426.

Reimers-Neils, L., Logemann, J. A., and Larson, C. 1994. Viscosity effects on EMG activity in normal swallow. Dysphagia 9: 101–106.

Shama, F. and Sherman, P. 1973. Identification of stimuli controlling the sensory evaluation of viscosity. II. Oral methods. J. Texture Stud. 4: 111–118.

Shama, F., Parkinson, C., and Sherman, P. 1973. Identification of stimuli controlling the sensory evaluation of viscosity. I. Non-oral methods. J. Texture Stud. 4: 102–110.

Sherman, P. 1970. Industrial Rheology, Academic Press, New York.
Sherman, P. 1988. The sensory-rheological interface, in Food Texture—Its Creation and Evaluation, eds. J. M. V. Blanshard and J. R. Mitchell, Butterworths, London.
Stevens, S. S. 1975. *Psychophysics—Introduction to Its Perceptual Neural and Social Prospects*, John Wiley, New York.
Stone, H. and Oliver, S. 1966. Effect of viscosity on the detection of relative sweetness intensity of sucrose solutions. *J. Food Sci.* 31: 129–134.
Szczesniak, A. S. 1963. Classification of textural characteristics. *J. Food Sci.* 28: 385–389.
Szczesniak, A. S. 1979. Classification of mouthfeel characteristics of beverages, in *Food Texture and Rheology*, ed. P. Sherman, Academic Press, New York.
Szczesniak, A. S. 1987. Correlating sensory with instrumental texture measurements—an overview of recent developments. *J. Texture Stud.* 18(1): 1–15.
Szczesniak, A.S. and Farkas, E. 1962. Objective characterization of the mouthfeel of gum solutions. *J. Food Sci.* 27: 381–385.
Terpstra, M. E. J., Janssen, A. M., Prinz, J. F., de Wijk, R. A., Weenen, H., and van der Linden, E. 2005. Modeling of thickness for semisolid foods. *J. Texture Stud.* 36(2): 213–233.
Tyle, P. 1993. Effect of size, shape and hardness of particles in oral texture and palatability. *Acta Psychological*: 111–118.
Vaisey, M., Brunon, R. and Cooper, J. 1969. Some sensory effects of hydrocolloid sols on swetness. *J. Food Sci.* 34: 397–400.
de Wijk, R. A., Terpstra, M. E. J., Janssen, A. M., and Prinz, J. F. 2006. Perceived creaminess of semi-solid foods. *Trends Food Sci.* Technol. 17:412–422
Wood, F. W. 1968. Psychophysical studies on the consistency of liquid foods, S.C.I. *Monograph: Rheology and Texture of Foodstuffs*s, pp. 40–49, Society of Chemical Industry, London.

Chapter 8
Application of Rheology to Fluid Food Handling and Processing

In this chapter, we consider application of rheology to handling and processing operations. However, it should be noted that there are many situations where rheology is applied. Earlier, sensory assessment and swallowing of foods were considered in Chap. 7. Table 8.1 contains some of the phenomena in which rheological behavior plays an important role and the typical shear rates encountered in them. The latter should also provide guidelines for obtaining the shear rate range over which rheological data should be obtained.

Velocity Profiles in Tubes

Equations describing velocity profiles can be used, among other applications, to study the effect of different rheological models on the distribution of velocities and to understand the concept of residence time distribution (RTD) across the cross-section of a pipe or a channel.

The velocity profile of a fluid flowing in a tube can be derived from the relationship

$$v_z = \int_r^{r_0} \left(\frac{\mathrm{d}v_z}{\mathrm{d}r} \right) \mathrm{d}r \tag{8.1}$$

where v_z is the velocity in the axial direction, r is the radial coordinate, and r_0 is the radius of the tube (Fig. 8.1). From a simple force balance for tube flow, one can obtain

$$\sigma_{rz} = \sigma_{\mathrm{w}} \frac{r}{r_0} = -\frac{r}{2} \frac{\mathrm{d}p}{\mathrm{d}z} \tag{8.2}$$

where σ_{rz} is the shear stress at any radius r, σ_{w} is the magnitude of the shear stress at the wall ($r = r_0$), p is the pressure, and z is the axial coordinate. Utilizing Eqs. 8.1 and 8.2 and noting that in the rheological equations described in Chap. 2, $\sigma = \sigma_{\mathrm{rz}}$ and

M. A. Rao, *Rheology of Fluid, Semisolid, and Solid Foods*, Food Engineering Series, DOI 10.1007/978-1-4614-9230-6_8, © Springer Science+Business Media New York 2014

Table 8.1 Typical shear rates of foods and pharmaceuticals in practice[a]

Phenomenon	Shear rates	Application
Sedimentation of particles	10^{-6}–10^{-3}	Spices in sauces and dressings
Draining under gravity	10^{-1}–10^{1}	Storage vessels
Extrusion	10^{0}–10^{3}	Snack foods
Calendaring	10^{1}–10^{2}	Dough sheeting
Pouring from a bottle	10^{1}–10^{2}	Fluid foods, medicines
Viscosity assessment by stirring	50–10^{2}	Sensory assessment by stirring
Viscosity assessment by tilting	10^{-1}–10^{2}	Sensory assessment by tilting
Oral viscosity assessment	10^{1}–10^{3}	Oral sensory assessment
Dip coating	10^{1}–10^{2}	Confectionery
Mixing and stirring	10^{1}–10^{3}	Food processing
Pipe flow	10^{0}–10^{3}	Food processing
Flow in upper esophageal sphincter		
Water, 1 mPa s	low–440	Swallowing
Newtonian, 150 mPa s	low–20	Swallowing

[a] Compiled from Sestak et al. (1983); Steffe (1996), and Chap. 7.

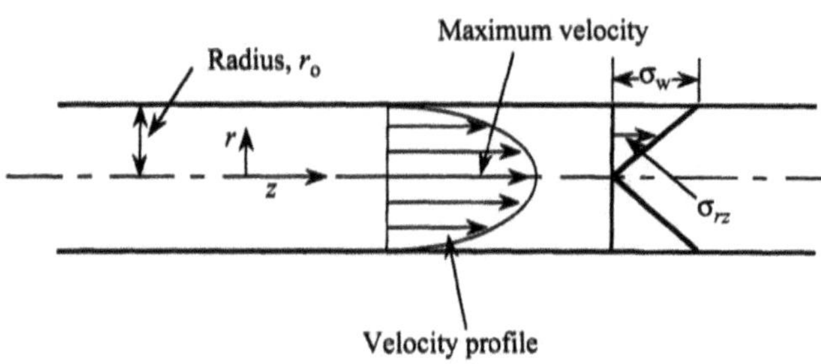

Fig. 8.1 Schematic diagram for analysis of laminar flow in a tube v_z velocity in the axial direction, r radial coordinate, r_0 radius of the tube

$\dot{\gamma} = (\mathrm{d}v_z/\mathrm{d}r)$, one can derive equations describing the velocity profiles for laminar flow in a tube.

The power law model $\sigma = K\dot{\gamma}^n$, Eq. 2.3, has been used extensively in handling applications. The relationship between the maximum velocity (v_{zm}) and the average velocity ($\bar{v}_z$) for the design of the length of a holding tube of a pasteurizing system is of special interest

$$\frac{v_{zm}}{\bar{v}_z} = \frac{3n+1}{n+1} \tag{8.3}$$

For Newtonian and pseudoplastic fluids in laminar flow, one can deduce from Eq. 8.3, the popular relationship that maximum velocity, at most, is equal to twice the average velocity. In contrast, in the case of shear-thickening (dilatant) fluids, the maximum velocity would be more than twice the average velocity.

Table 8.2 Velocity profile[a] and volumetric flow rate equations for power law, Herschel–Bulkley, and Bingham plastic fluids

Power law model: $\sigma = K\dot{\gamma}^n$

Velocity profile[b]: $v_z = \left(\frac{n}{n+1}\right)\left(\frac{\Delta P}{2KL}\right)^{(1/n)} [r_0^{(n/n+1)} - r^{(n/n+1)}]$

Volumetric flow rate: $\frac{Q}{\pi r_0^3} = \left(\frac{n}{n+1}\right)\left(\frac{\sigma_w}{K}\right)^{1/n}$

Herschel-Bulkley model: $\sigma - \sigma_0 = K_H(\dot{\gamma})^{nH}$

Velocity profile[c]: $v_z = \frac{2L}{\Delta P(m+1)K_H^m}\left[(\sigma_w - \sigma_0)^{(m+1)} - \left(\frac{r\Delta P}{2L} - \sigma_0\right)^{(m+1)}\right]$

when, $r_0 \le r \le r_0$

Volumetricflow rate: $\frac{Q}{\pi r_0^3} = \frac{(\sigma_W - \sigma_0)^{(m+1)}}{\sigma_w^3 K_H^m}\left[\frac{(\sigma_W - \sigma_0)^2}{m+3} + \frac{2\sigma_0(\sigma_W - \sigma_0)}{m+2}\frac{\sigma_0^2}{m+1}\right]$

Note that in the equations for the Herschel—Bulkley model $m = (1/n_H)$

Bingham plastic model: $\sigma - \sigma_0 = \eta'\dot{\gamma}$

Velocity profile: $v_z = \frac{1}{\eta'}\left[\frac{\Delta P}{L}(r_0^2 - r^2) - \sigma_0\ (r_0 - r)\right]$, when, $r_0 \le r \le r_0$

Volumetric flow rate: $\frac{4Q}{\pi r_0^3} = \frac{\sigma_W}{\eta'}\left[1 - \frac{4}{3}\left(\frac{\sigma_0}{\sigma_W}\right) + \frac{1}{3}\left(\frac{\sigma_0}{\sigma_W}\right)^4\right]$

[a] The average velocity can be obtained by dividing the equation for volumetric flow rate by the area of cross-section of the pipe.
[b] Maximum velocity occurs at the center line, $r = 0$.
[c] Velocity profiles are valid for $r_0 \le r \le r_0$, where the radius of plug $r_0 = (2\sigma_0 L/\Delta P)$; note the subscript of r is zero. The maximum velocity occurs when $0 \le r \le r_0$ and is obtained by substituting r_0 for r.

The volumetric flow rate Q is given by the equation

$$Q = \int_0^{r_0} 2\pi r v_z \mathrm{d}r = \pi \int_0^{r_0} v_z(r)\mathrm{d}(r^2) \tag{8.4}$$

Integrating by parts the second part of Eq. 8.4, and using the boundary condition $v = 0$ at $r = r_0$, one can obtain the general equation for the volumetric flow of a fluid in a tube

$$Q = \left(\frac{\pi r_0^3}{\sigma_W^3}\right)\int_0^{\sigma_W} \sigma_{rz}^2\left(\frac{\mathrm{d}v_z}{\mathrm{d}r}\right)\mathrm{d}\sigma_{rz} \tag{8.5}$$

Substituting for the shear rate appropriate expressions from different rheological models, one can derive equations relating Q and pressure drop Δp (Rao 1995).

Table 8.2 contains expressions for the velocity profiles and the volumetric flow rates of the three rheological models: power law, Herschel–Bulkley, and the Bingham plastic models.

In the case of fluids obeying the power law model, the pressure drop per unit length $\Delta p/L$ is related to Q and r_0 by the relationship

$$\frac{\Delta p}{L} \propto \frac{Q^n}{r_0^{3n+1}} \tag{8.6}$$

From this relationship, we see that for Newtonian foods ($n = 1$) the pressure gradient is proportional to the $(r_0)^{-4}$ power. Therefore, a small increase in the radius of the tube will result in a major reduction in the magnitude of the pressure gradient. In contrast, for a highly pseudoplastic fluid (e.g., $n = 0.2$), increasing the pipe radius does not have such a profound effect on the pressure gradient.

Energy Requirements for Pumping

Mechanical Energy Balance Equation

The energy required to pump a liquid food through a pipe line can be calculated from the mechanical energy balance (MEB) equation. The MEB equation can be used to analyze pipe flow systems. For the steady-state flow of an incompressible fluid, the MEB can be written as follows (Brodkey 1967):

$$gZ_1 = \frac{p_1}{\rho} + \frac{v_{z_1}^2}{\alpha} - W = gZ_2 + \frac{p_2}{\rho} + \frac{v_{z_2}^2}{\alpha} + E_f \tag{8.7}$$

where g is the acceleration due to gravity, Z is the height above a reference point, p is the pressure, v is the fluid velocity, W is the work output per unit mass, E_f is the energy loss per unit mass, α is the kinetic energy correction factor, and the subscripts 1 and 2 refer to two points in the pipe system (e.g., Fig. 8.2). In order to accurately estimate the energy required for pumping of a fluid food in a specific piping and equipment system, the term ($-W$; J Kg^{-1}) has to be estimated from Eq. 8.7 after the other terms have been evaluated as described next.

The velocities at the entrance and exit of the system can be calculated from the respective diameters of the tanks or pipes and the volumetric flow rate of the food. The energy loss term E_f consists of losses due to friction in pipe and that due to friction in valves and fittings

$$E_f = \frac{2 f v_z^2 L}{D} + \sum_1^b \frac{k_f v_z^2}{2} \tag{8.8}$$

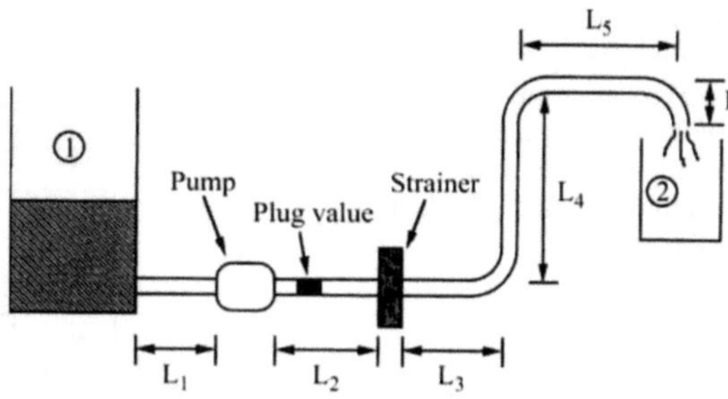

Fig. 8.2 Schematic diagram of flow system for application of mechanical energy balance (MEB) equation (Redrawn from Steffe and Morgan 1986)

where f is the friction factor, v_z is the velocity, L is the length of straight pipe of diameter D, k_f is the friction coefficient for a fitting, and b is the number of valves or fittings. It is emphasized that k_f is unique to a particular fitting and that different values of v_z, k_f, and f may be required when the system contains pipes of different diameters. Furthermore, losses due to special equipment, such as heat exchangers, must be added to E_f (Steffe and Morgan 1986).

Friction Losses in Pipes

Because many fluid foods are non-Newtonian in nature, estimation of friction losses (pressure drop) for these fluids in straight pipes and in fittings is of interest. One can estimate the friction losses in straight pipes and tubes from the magnitude of the Fanning friction factor, f, defined as

$$\frac{\Delta pf}{\rho} = \frac{2fLv_z^2}{D} \tag{8.9}$$

where f is the friction factor. For laminar flow conditions, Garcia and Steffe (1987) suggested that based on the work of Hanks (1978), the friction factor can be calculated from a single general relationship for fluids that can be described by Newtonian ($\sigma = K\dot{\gamma}^n$, Eq. 2.1), power law ($\sigma = K\dot{\gamma}^n$, Eq. 2.3), and Herschel–Bulkley ($\sigma - \sigma_{0H} = K_K\dot{\gamma}^{n_H}$, Eq. 2.5) fluid models.

For fluids that can be described by the power law (Eq. 2.3), the generalized Reynolds number (GRe) can be calculated from the equation

$$\text{GRe} = \frac{D^n v_z^{2-n}\rho}{8^{(n-1)}K}\left(\frac{4n}{3n+1}\right)^n \tag{8.10}$$

where D is the pipe diameter, v_z is the average velocity of the food, ρ is the density of the food, and K and n are the power law parameters of frozen concentrated orange juice (FCOJ). When $n=1$, Eq. 8.10 reduces to the Reynolds number for a Newtonian food with viscosity η

$$\text{Re} = \frac{Dv_z\rho}{\eta} \tag{8.11}$$

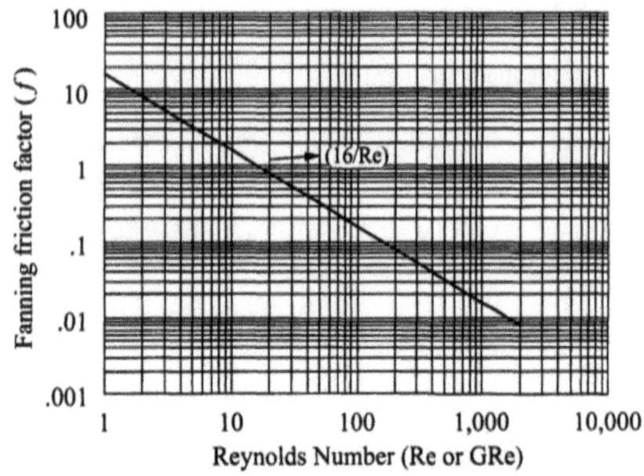

Fig. 8.3 The friction factor-generalized Reynolds number (GRe) relationship for power law fluids under laminar flow conditions. It can also be used for Newtonian fluids in laminar flow

In laminar tube flow, the Fanning friction factor can be calculated from the equation

$$f = \frac{16}{\mathrm{GRe}} \tag{8.12}$$

Figure 8.3 illustrates the friction factor versus GRe relationship for power law fluids under laminar flow conditions. It can also be used for Newtonian fluids in laminar flow with the Reynolds number being used in place of GRe. In fact, the Newtonian f versus Reynolds (Re) relationship was established much earlier than extension to non-Newtonian fluids. Once the magnitude of the friction factor is known, the pressure drop in a pipe can be estimated from Eq. 12.

For the laminar flow of a Herschel–Bulkley fluid ($\sigma - \sigma_{0H} = K\mathrm{K}\dot{\gamma}^{n_{\mathrm{H}}}$, Eq. 2.5), the friction factor can be written as

$$f = \frac{16}{\Psi \mathrm{GRe}} \tag{8.13}$$

where ψ is related to the yield stress σ_0 and the flow behavior index (n; Garcia and Steffe 1987)

$$\Psi = (3n+1)^n (1-\xi_0)^{1+n} \left[\frac{(1-\xi_0)^2}{(3n+1)} + \frac{2\xi_0(1-\xi_0)}{(2n+1)} + \frac{\xi_0^2}{(n+1)} \right] \tag{8.14}$$

where

$$\xi_0 = \frac{\sigma_0}{\sigma_{\mathrm{w}}} = \frac{\sigma_0}{(D\Delta p/4L)} \tag{8.15}$$

Because of the highly viscous nature of non-Newtonian foods, laminar flow conditions are likely to be encountered more often than turbulent conditions. Nevertheless, it is important to be aware of developments with respect to prediction of friction factors in turbulent flow of non-Newtonian foods. For turbulent flow, except for Newtonian fluids, the predicted magnitudes of friction factors for non-Newtonian fluids may differ greatly depending on the relationship employed (Garcia and Steffe 1987). However, for power law fluids, the relationships of Dodge and Metzner (1959) and Hanks and Ricks (1974) were found to predict similar magnitudes of the friction factors. For the Herschel–Bulkley model that is used when non-Newtonian foods exhibit yield stress, the analysis of Hanks (1978) was found to be the most comprehensive, but to use the derived relationship, it is necessary to perform a numerical integration and several iterations.

Kinetic Energy Losses

In the MEB equation, kinetic energy losses can be calculated easily provided that the kinetic energy correction factor α can be determined. In turbulent flow, often, the value of $\alpha=2$ is used in the MEB equation. When the flow is laminar and the fluid is Newtonian, the value of $\alpha=1$ is used. Osorio and Steffe (1984) showed that for fluids that follow the Herschel–Bulkley model, the value of an in laminar flow depends on both the flow behavior index *(n)* and the dimensionless yield stress (ξ_0) defined earlier. They developed an analytical expression and also presented their results in graphical form for α as a function of the flow behavior index *(n)* and the dimensionless yield stress (ξ_0). When possible, the values presented by Osorio and Steffe (1984) should be used. For FCOJ samples that do not exhibit yield stress and are mildly shear-thinning, it seems reasonable to use a value of $\alpha=1$.

Friction Loss Coefficients for Fittings

Steffe et al. (1984) determined magnitudes of the coefficient for a fully open plug valve, a tee with flow from line to branch, and a 90° short elbow as a function of GRe using apple sauce as the test fluid. They found that, as for Newtonian fluids, k_f increases with decreasing values of GRe. The regression equations for the three fittings were

$$\textit{three-way plug valve}\ k_f = 30.3\ \text{GRe}^{-0.492} \tag{8.16}$$

$$\textit{tee}\ k_f = 29.4\ \text{GRe}^{-0.504} \tag{8.17}$$

$$\textit{elbow}\ k_f = 191.0\ \text{GRe}^{-0.896} \tag{8.18}$$

In many instances, the practice is to employ values determined for Newtonian fluids, such as those in the Chemical Engineers' Handbook.

Pump Selection and Pipe Sizing

Steffe and Morgan (1986) discussed in detail the selection of pumps and the sizing of pipes for non-Newtonian fluids. Preliminary selection of a pump is based on the volumetric pumping capacity only from data provided by the manufacturers of pumps. Effective viscosity η_e was defined by Skelland (1967) as the viscosity that is obtained assuming that the Hagen–Poiseuille equation for laminar flow of Newtonian fluids is applicable

$$\eta_e = \left[\frac{(D\Delta p/4L)}{(32Q/\pi D^3)}\right] \tag{8.19}$$

An alternate form of Eq. 8.17 in terms of the mass flow rate, m, and the friction factor, f, is

$$\eta_e = \frac{fm}{4\pi D} \tag{8.20}$$

In calculating η_e from Eq. 8.19, either the port size of a pump or the dimensions of the assumed pipe size can be used. Based on the magnitude of η_e, the suitability of the pump volumetric size must be verified from plots of effective viscosity versus volumetric flow rate. It is emphasized that a pump size is assumed based on the volumetric pumping requirements and the assumption is verified by performing detailed calculations.

A comprehensive example for sizing a pump and piping for a non-Newtonian fluid whose rheological behavior can be described by the Herschel–Bulkley model (Eq. 2.5) was developed by Steffe and Morgan (1986) for the system shown in Fig. 8.2 and it is summarized in the following. The Herschel–Bulkley parameters were: yield stress = 157 Pa, flow behavior index = 0.45, consistency coefficient = 5.20 Pa s^n.

Pump Discharge Pressure

The discharge pressure of the pump can be calculated by applying the MEB equation between the pump discharge and the exit point of the system so that the upper seal pressure limits are not exceeded. The MEB equation for this purpose can be written as

$$p_1 = \left[g(Z_2 - Z_1) + \frac{p_2}{\rho} + E_f\right]\rho \tag{8.21}$$

The energy loss due to friction in the pipe, valve, and fittings was estimated to be 329.0 J kg^{-1}, and the discharge pressure of the pump, p_1, was estimated to be 4.42×10^5 Pa (Steffe and Morgan 1986).

Power Requirements for Pumping

The total power requirements for pumping are calculated by adding the hydraulic and the viscous power requirements. The former can be estimated from the MEB equation written for the work input, $-W$. We note that $p_1 = p_2$ and that E_f includes

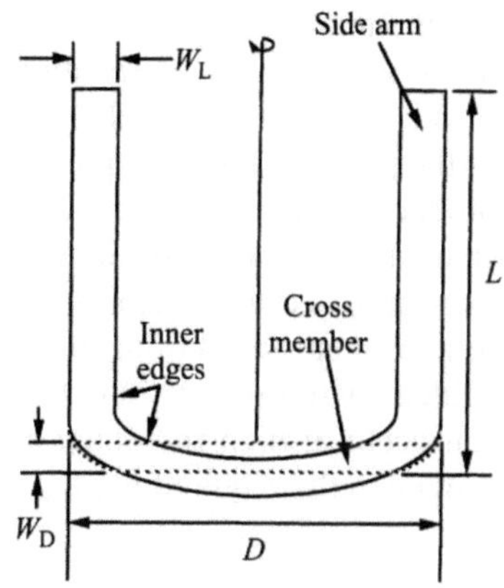

Fig. 8.4 Schematic diagram of an anchor agitator

not only the friction losses on the discharge section but also the inlet section. The former was, as stated earlier, estimated to be 329.0 J kg^{-1}, while the latter was estimated to be 24.7 J kg^{-1} by applying the MEB equation between the exit of the tank and the pump inlet. The work input ($-W$) was estimated to be 380.0 J kg^{-1} and because the mass flow rate was 1.97 kg s^{-1}, the hydraulic power input was estimated to be 749.0 J s^{-1}or 0.749 kW. For estimating the viscous power requirements due to energy losses in the pump due to friction, the operating speed and the effective viscosity of the fluid food must be calculated. The former can be calculated from the displacement volume per revolution of the pump and the required volumetric flow, while the latter can be calculated from Eq. 8.18. For a size 30 Waukesha pump, the volumetric displacement per revolution is 2.27×10^{-4} m^3 s^{-1} and hence the pump speed was 417 rpm, while the equivalent viscosity for the fluid food under consideration was 0.703 Pa s (Steffe and Morgan 1986). The energy losses in the pump were estimated from the manufacturer's data to be 0.835 kW. Therefore, the sum of the hydraulic and the viscous losses were 1.58 kW or 2.12 hp. These data allow selection of a suitable motor and drive system. In this example, pipe size was based on the pump port's diameter. It may also be based on plans for future expansion and ease of cleaning.

Power Consumption in Agitation

Mixing, also called agitation, of fluid foods is an important operation in food processing plants. The goals of a mixing operation include: homogenization, dispersion, suspension, blending, and heat exchange. Several types of agitators are used in the food industry and many, undoubtedly, are proprietary designs. Agitators used in the dairy industry were classified as (Kalkschmidt 1977): propellers: screw, edge, and ring; understirrers: disc, cross-bar, paddle, anchor, blade, gate-paddle, spiral, finger-paddle, and moving cutters. For the purpose of illustration, the commonly used anchor agitator is shown in Fig. 8.4. Sometimes, vertical baffles placed along the circumference of a mixing tank, are used to avoid vortex formation at high rotational speeds in low-viscosity foods.

It is important to match the agitator and agitation conditions to the characteristics of the product. For example, agitators for intact fruit must not shear or damage the product. Even in a low-viscosity fluid, like milk, the type of agitator, and its dimensions and rotational speed are important. After studying the mixing efficiency of two top-entering agitators: straight paddle and pitched blade impeller at various rotational speeds in milk storage tanks, Miller (1981) recommended a simple two-bladed paddle for use in cylindrical vessels. Because damage to milk fat globules can occur at high rotational speeds (e.g., 150 rpm), the lowest speed capable of providing the required mixing effect should be selected.

Role of Flow Behavior in Agitation

The energy consumption in agitation depends on the basic principles of fluid mechanics; however, the flow patterns in a mixing vessel are much too complex for their rigorous application. Therefore, empirical relationships based on dimensionless groups are used. Here, because most fluid foods are non-Newtonian in nature, the discussion emphasizes in-tank agitation of such fluids using top-entering agitators. A major complication is that the shear rate $(\dot{\gamma})$ is not uniform in an agitated vessel. For example, it has highest value at the point of highest fluid velocity; such a point occurs at the tip of the rotating agitator and decreases with increasing distance.

Henry Rushton and coworkers developed the concept of the power number (Po) for studying mixing of fluids that for Newtonian fluids is defined as

$$\mathrm{Po} = \frac{P}{D_\mathrm{a}^5 N_\mathrm{a}^3 \rho} \tag{8.22}$$

where P is power ($P=2\pi N_a \times T$; J s^{-1}), D_a is agitator diameter (m), T is torque on agitator (Nm), N_a is agitator speed (s^{-1}), and ρ is the density of orange juice (kg m^{-3}). Several types of agitators, some with proprietary designs, are used in the food industry.

In laminar mixing conditions of Newtonian fluids, Po is linearly related to agitator rotational Reynolds number, Re_a

$$\mathrm{Po} = \frac{A}{\mathrm{Re}_\mathrm{a}} \tag{8.23}$$

where A is a constant and

$$\mathrm{Re}_\mathrm{a} = \frac{D_\mathrm{a}^2 N_\mathrm{a} \rho}{\eta} \tag{8.24}$$

The value of the constant A depends on the type of agitator. Laminar mixing conditions are encountered as long as Re_a is less than about 10.

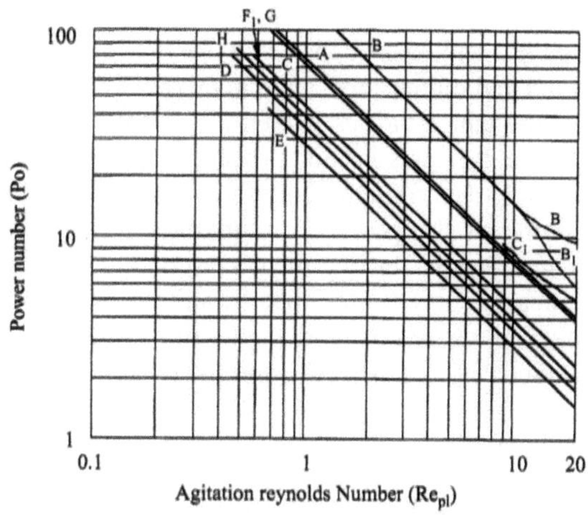

Fig. 8.5 Curves of power law agitation Reynolds number (Re_{pl}) versus power number (Po) for several agitators (Adapted from Skelland 1967) Curve A—single turbine with 6 flat blades, B—two turbines with 6 flat blades, C—a fan turbine with 6 blades at 45° C, D, and E—square-pitch marine propeller with 3 blades with shaft vertical and shaft 10 °C from vertical, respectively, F and G—square-pitch marine propeller with shaft vertical and with 3 and 4 blades, respectively, and H—anchor agitator

For a non-Newtonian food, the viscosity is not constant, but depends on the shear due to agitation. Therefore, for FCOJ and other non-Newtonian fluids, we would like to define an agitator Reynolds number that can be used in place of Re_a to estimate Po from data presented in Fig. 8.5. For fluids that follow the power law flow model, Metzner and Otto (1957) suggested a GRe for mixing

$$Re_{mo} = \frac{D_a^2 N^{2-n} \rho}{K} 8\left(\frac{n}{6n+2}\right)^n \tag{8.25}$$

However, Re_{mo} has not found wide acceptance because of deviations between predicted and experimental values of Po. A widely accepted procedure assumes that the average shear rate during mixing is directly proportional to the agitator rotational speed, N_a, that is

$$\dot{\gamma} = k_s N_a \tag{8.26}$$

and the apparent viscosity is

$$\eta_a = K(k_s N_a)^{n-1} \tag{8.27}$$

Substituting the expression for apparent viscosity in the rotational Reynolds number for Newtonian fluids, $Re_a = D_a^2 N_a \rho / \eta$, the power law agitation Reynolds number is

$$Re_{pl} = \frac{D_a^2 N_a^{2-n} \rho}{K(k_s)^{n-1}} \tag{8.28}$$

The curves of Re_{pl} versus Po for several agitators, adapted from Skelland (1967), are shown in Fig. 8.5. It is emphasized that each line in Fig. 8.5 is valid for a specific agitator, its orientation and dimensions, and the mixing tank dimensions, as well as the configuration of the tank's baffles. Given the strong likelihood of proprietary agitator designs and mixing tanks in food processing plants, it would be advisable to develop Re_{pl} versus Po data for agitation systems being used for a specific food.

Once the magnitude of the power law agitation Reynolds number (Re_{pl}) is known, assuming that it is equal to Re_a, the corresponding value (Po) can be determined from the applicable curve for the specific agitator, such as Fig. 8.5 or similar data. From the known values of Po, the diameter of the agitator (D_a), agitator rotational speed *(N_a)*, the density (ρ) of FCOJ, and the power required *(p)* for agitation can be calculated.

Estimation of the Constant k_s of an Agitator

Procedures for determining k_s of a specific agitator and mixing tank can be found in Rao and Cooley (1984). In one procedure (Rieger and Novak 1973; Rao and Cooley 1984), the constant k_s can be determined from a plot of log $P / KN_a^{n+1} D_a^3$ versus $(1-n)$; the slope of the line is equal to $-\log k_s$. For a given agitator, tests must be conducted such that the following data are obtained: P, the power ($P = 2.71 \pi Na \times T$; Js), D_a, agitator diameter (m), T is torque on agitator (Nm), N_a is agitator speed (s^{-1}), and the power law rheological parameters of test fluids so that a wide range of $(1-n)$ values are obtained. Typical values of the proportionality constant k_s for chemical industry impellers range from about 10 to 13 (Skelland 1967).

Scale-up of mixing vessels is an important task of engineers. Scale-up of mixing vessels requires prediction of the rotational speed, N_{a2}, in scale 2 that will duplicate the performance in scale 1 due to agitation at a speed of N_{a1}. An important assumption in scale-up is geometric similarity that is achieved when all corresponding linear dimensions in scales 1 and 2 have a constant ratio. One popular scale-up criterion is based on equal power per volume, $P\ V^{-1}$, because it is understandable and practical. Other scale-up criteria include (Wilkens et al. 2003): equal agitation Reynolds number, equal impeller tip speed, equal bulk fluid velocity, and equal blend time. An additional consideration is the high cost, both capital and operating, of high-speed mixing systems. Therefore, the highest rotational speed N_{a2} estimated using the above criteria may not be an economical option for implementation (Wilkens et al. 2003).

Residence Time Distribution in Aseptic Processing Systems

In flow systems, not all fluid and solid food particles remain for the same time periods, that is, the particles have a distribution of residence times. Danckwerts (1953) proposed the concept of RTD, and the theoretical and experimental principles of

RTD have been well reviewed by Levenspiel (1972) and Himmelblau and Bischoff (1968). The RTD functions and the role of RTD in continuous pasteurization systems was reviewed (Rao and Loncin 1974a, b). Here, only the necessary principles of RTD are discussed and for additional information the above references must be consulted.

First, we note that with few exceptions RTD studies represent lumped parameter analysis. Only in a few instances, such as that of fully developed laminar flow of fluids in tubes, one can attempt distributed parameter analysis in that the expressions for the RTD functions *C* and *F* can be derived from known velocity profiles.

Experimental RTD data on any equipment can be obtained by imposing at the inlet a pulse, step, or sinusoidal impulse and monitoring the response at the outlet. The tracer used must have physical properties such as density and viscosity similar to the test fluid food, and it should not in any manner alter the properties of the test fluid food. The sinusoidal impulse and response require more care and "sophisticated" equipment than the pulse and the step inputs. The response to a pulse impulse is called a *C*-curve, while the response to a step change is called an *F*-curve. For a closed vessel, that is, in which material passes in and out by bulk flow (i.e., no diffusion at inlet and outlet) the *C* and *F* functions are related as

$$C(\theta) = \frac{\mathrm{d}F}{\mathrm{d}\theta} \tag{8.29}$$

One measure of the distribution of residence times (ages) of the fluid elements within a reactor is the *E* function, defined so that $E\,\mathrm{d}\theta$ is the fraction of material in the exit stream with age between h and $h+d\theta$ (Levenspiel 1972). It can be shown (Levenspiel 1972) that the *C* and *E* functions are identical, and that for an isothermal process the ratio of the final (C) to initial (C_0) concentrations of either microorganisms or nutrients can be determined from the expression

$$\frac{C}{C_0} = \int_0^{\infty} \exp\left(-\frac{2.303t}{D(T)}\right) E(t)\mathrm{d}t \tag{8.30}$$

where $D(T)$ is the *D*-value of the microorganism or a nutrient at a temperature *T*. For a non-isothermal process, (C/C_0) can be determined by recognizing that the *D*-value is dependent on temperature and taking this dependence into consideration. At this point, we note that $E(t)=[E(\theta)/t]$ is used in estimating the number of survivors or percent nutrient retention according to Eq. 8.30.

Interpretation of RTD Data

Experimental RTD studies are usually interpreted in terms of the dispersion model (Eq. 8.31) or the equal-sized tanks in series model (Eq. 8.33)

$$\frac{\partial C}{\partial \theta} = \left(\frac{D}{vL}\right)\frac{\partial^2 C}{\partial X^2} - \frac{\partial C}{\partial X} \tag{8.31}$$

where

$$X = \frac{z}{L},\ \theta = \frac{t}{\bar{t}} = \frac{tv}{L},\ \text{and } D = \text{dispersion coefficient} \tag{8.32}$$

The former model is usually employed for flows in tubes and, thus, should be applicable to the interpretation of RTD in holding tubes. The theoretical RTD for laminar flow of Newtonian fluids can be derived as

$$C(\theta) = \frac{N(N\theta)^{N-1}}{(N-1)}e^{-N/\theta} \tag{8.33}$$

$$C(\theta) = E(\theta) = \frac{1}{2\theta^3} \tag{8.34}$$

For non-Newtonian fluids that can be described by the power law model, Lin (1979) showed that the RTD is given by the expression

$$E(\theta) = \frac{\mathrm{d}F(\theta)1}{\mathrm{d}\theta} = \frac{2n}{3n+1}\frac{1}{\theta^3}\left(1 - \frac{n+1}{3n+1}\frac{1}{\theta}\right)^{(n-1)/(n+1)} \tag{8.35}$$

This expression derived by Lin (1979) appears to be correct and needs to be tested with experimental data. There are a number of models for nonideal flows, that is, flows that fall between the ideal conditions of a perfect mixer and plug flow. Some of the models for nonideal flow were discussed in Levenspiel (1972) and Rao and Loncin (1974a).

Sizing Holding Tube Length

Holding tubes play an important role in thermal processing of fluid foods in HTST or aseptic flow systems. One simple but important relationship for determining the length of a holding tube (L) is based on the residence time of the fastest particle, that is, $L = v_{zm} \times$ holding time. Because the average velocity $(\bar{v}_z)$ of a fluid food can be readily calculated in many instances, a relationship between $\bar{v}_z$ and v_{zm} will be useful. Such a relationship applicable for the fully developed laminar flow of fluid foods obeying the power law model was given earlier

$$\frac{\bar{v}_z}{v_{zm}} = \left(\frac{n+1}{3n+1}\right) \tag{8.3}$$

For the special case of a Newtonian fluid $(n = 1)$, $v_{zm} = 2 \times (\bar{v}_z)$; this relationship provides a safe design criterion for Newtonian fluids and shear-thinning fluids in laminar and turbulent flow, but it will not provide a safe design for shear-thickening (dilatant) fluids. Even though there have been few reliable reports on shear-thickening (dilatant) behavior of fluid foods, the absence of such behavior must be confirmed by proper rheological tests in order to avoid underprocessing of fluid foods in holding tubes.

In practical terms, the Food Drug Administration requires that holding tubes have positive slope of at least 0.25 inch per foot to avoid dead spaces. Also, to conserve plant space holding tubes are coiled. Therefore, one can expect the flow conditions in holding tubes to be far from those in fully developed laminar flow in straight tubes with the result that the maximum velocities are less than twice the average velocities. The experimental data of Sancho and Rao (1992), to be discussed later, confirm the deviations from ideal fully developed laminar flow.

Experimental RTD Data

Most of the RTD data on thermal processing flow systems found in the literature were obtained using water as the test fluid. For a plate heat exchanger, Roig et al. (1976) found that the RTD data could be fitted with five tanks-in-series model. Veerkamp et al. (1974) also found that the RTD data in a holding tube could be fitted with the tanks-in-series model.

Heppell (1985) demonstrated that milk has a broader RTD than water in an infusion-type UHT sterilizer. Therefore, it is necessary to perform RTD studies on holding tubes and heat exchangers with either fluid foods that are to be processed or with model solutions that possess physical and rheological properties similar to the fluid foods. Sancho and Rao (1992) found that the maximum velocities in a holding tube were less than those estimated from fully developed flow assumption for both Newtonian and non-Newtonian fluids. Magnitudes of the dispersion number (D/vD_p) shown in Fig. 8.6 as a function of the Reynolds number for the Newtonian liquids and the GRe for the non-Newtonian liquids followed three distinct trends depending on the magnitude of Re or GRe (Re_{Gen}; Sancho and Rao 1992): (1) over the range 10–100, the dispersion number was nearly constant, (2) over the range 100–2,000, the dispersion number increased with increase in Reynolds number, and (3) over the range 2,000–10,000, the dispersion number decreased with increase in Reynolds number. Because the rheological behavior of the studied non-Newtonian guar gum solutions was known, interpretation of the data in terms of the GRe and grouping together with the Newtonian fluid data were possible.

Helical Flow

It is known that plug flow conditions are approached in laminar helical flow and for this reason Rao and Loncin (1974a) suggested that this type of flow could be used advantageously in continuous sterilization of foods. It is interesting to note that one continuous thermal processing system in which plug flow conditions are claimed to exist is based on helical flow (Anonymous, undated).

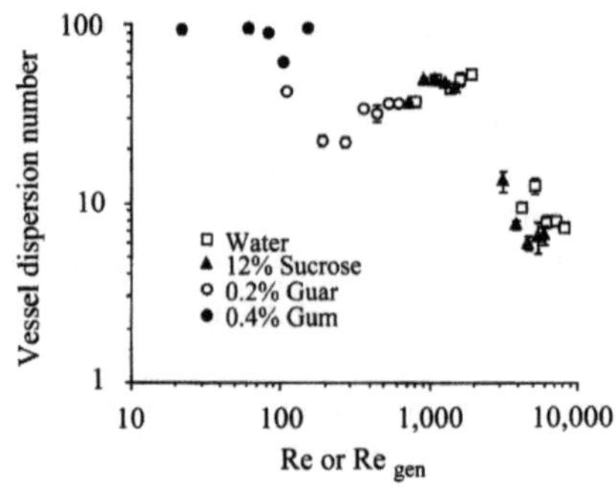

Fig. 8.6 Magnitudes of the dispersion number (*D/vd*) are shown as a function of the Reynolds number for the Newtonian liquids and the generalized Reynolds number for non-Newtonian fluids (Sancho and Rao 1992)

Heat Transfer to Fluid Foods

Thermorheological Models

In order to understand or study heat transfer phenomenon, the rheological behavior of a fluid food must be known as a function of both temperature and shear rate. For convenience in computations, the effect of shear and temperature may be combined into a single thermorheological (TR) model. A TR model may be defined as one that has been derived from rheological data obtained as a function of both shear rate and temperature. Such models can be used to calculate the apparent viscosity at different shear rates and temperatures in computer simulation and food engineering applications. For a simple Newtonian fluid, because the viscosity, η, is independent of shear rate, one may consider only the influence of temperature on the viscosity. For many foods, the Arrhenius equation (Eq. 2.43) is suitable for describing the effect of temperature on η

$$\eta = \eta_{\infty A} \exp(E_a/RT) \tag{2.43}$$

where $\eta_{\infty A}$ is the frequency factor, E_a is the activation energy (J mol^{-1}), R is the gas constant (J mol^{-1} K^{-1}), and T is the absolute temperature (K).

As stated previously, for non-Newtonian foods, the simple power law model (Eq. 2.3) can be used to describe shear rate ($\dot{\gamma}$) versus shear stress (σ) data at a fixed temperature

$$\sigma = K\dot{\gamma}^n \tag{2.3}$$

where n is the flow behavior index (dimensionless) and K (Pa s^n) is the consistency index. Two very similar TR equations have been obtained by combining the power law and the Arrhenius equations. The equation obtained by Christiansen and Craig (1962) was

$$\sigma = K_{TC}(\dot{\gamma}\exp\{E_{ac}/RT\})^{\hat{n}} \tag{8.36}$$

The second equation commonly encountered in the food engineering literature and that has been used in several studies (Harper and El-Sahrigi 1965; Vitali and Rao 1984a, b) is

$$\sigma = K_{\mathrm{TH}} \exp(E_{\mathrm{aH}} / RT)\dot{\gamma}^{\bar{n}} \tag{8.37}$$

In both Eqs. 8.36 and 8.37, $\bar{n}$ is the average value of the flow behavior index for data at all the studied temperatures. Vitali and Rao (1984a, b) showed that the activation energy terms in Eqs. 3 and 4 are related according to

$$E_{aH} = \bar{n}(E_{aC}) \tag{8.38}$$

Thermorheogy of Starch Dispersions

In many foods, such as soups, salad dressings, gravies, and sauces, starch is present in excess water conditions. Because of the drastic increase in magnitude of the apparent viscosity (η_a) after the initial temperature of gelatinization is achieved, the rheological behavior of starch in excess water during the transition from fluid-like to highly viscous behavior affects heat transfer during food thermal processing. Furthermore, in dispersions of native starches, the increase is followed by an irreversible decrease in viscosity. Thus, starch gelatinization contributes to unique and complex transitions in viscosity discussed extensively in Yang and Rao (1998a), Liao et al. (1999), and Tattiyakul and Rao (2000).

Most models of starch gelatinization were developed under isothermal conditions, based on apparent first-order kinetics and the Arrhenius equation to describe the effect of temperature on the gelatinization rate (Kubota et al. 1979; Dolan and Steffe 1990; Kokini et al. 1992; Okechukwu and Rao 1996). Because of the temperature history imposed during thermal processing, data obtained under isothermal conditions may not be suitable to describe changes in η_a during gelatinization. In addition to the temperature of the starch sample, shear rate or dynamic frequency has a significant effect on η_a or complex viscosity (η^*), respectively.

Model of Dolan et al. (1989)

A comprehensive TR model taking into consideration the effect of time t-temperature (T) history was developed by Dolan et al. (1989). The model for constant shear rate and starch concentration is

$$\eta_{\mathrm{dim}} = \frac{\eta_{\mathrm{a}} - \eta_{\mathrm{ug}}}{\eta_{\infty} - \eta_{\mathrm{ug}}} = \left[1 - \exp(-k\Psi)\right]^{\alpha} \tag{8.39}$$

where

$$\Psi = \int T(t) \exp\left(\frac{-E_{\mathrm{g}}}{RT(t)}\right) \mathrm{d}t \tag{8.40}$$

where η_{dim} is dimensionless apparent viscosity defined in Eq. 8.37, η_a is apparent viscosity at a specific time during heating, η_{ug} is apparent viscosity of the ungela-

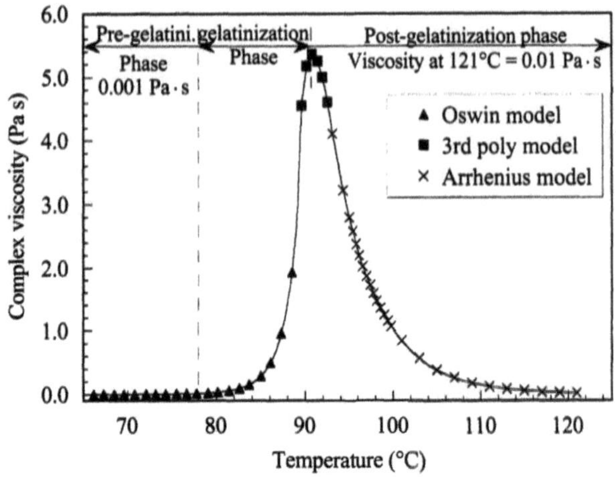

Fig. 8.7 Thermorheological data during gelatinization of a corn starch dispersion (Yang and Rao 1998b)

tinized dispersion, and η_∞ is the highest magnitude of η_a during gelatinization. The model was extended to include the influence of shear rate, temperature, concentration, and strain history (Dolan and Steffe 1990). Data obtained using back extrusion and mixer viscometers were used to evaluate the models. The activation energy of gelatinization (E_g) depended on the heating temperature (Dolan et al. 1989) and some of the factors affecting viscosity were negligible (Dolan and Steffe 1990). Applicability of Eq. 8.38 to η^* data during gelatinization of an 8% corn STD was demonstrated in Chap. 4.

Model of Yang and Rao (1998b)

Because η^* data are obtained at low strains with minimal alteration of the STD structure, they provide unique opportunities for studying applicable models. Furthermore, empirically obtained frequency shift factor (Ferry 1980) has been used successfully in time-temperature superposition studies on food polymer dispersions (Lopes da Silva et al. 1994), and the applicability of similar, if not identical, scaling of frequency was explored for STDs (Yang and Rao 1998a).

For fluid dynamics and heat transfer investigations related to food processing, the necessary η_a data may be obtained from models developed for η^* data using relationships based on the Cox–Merz rule (Rao 1992; Rao and Cooley 1992). These results were described in detail elsewhere (Yang 1997; Yang and Rao 1998a, b, c) and will be reviewed in brief here.

The averaged experimental TR data, shown in Fig. 8.7, were used to fit models for numerical simulation of heat transfer. Our first attempt was to fit equations to each segment in Fig. 8.7. The increasing segment (A–B) of the η^* curve that was sigmoid-shaped was fitted to a modified Oswin model (Eq. 8.41)

$$\eta^*\left(\frac{\omega}{\omega_r}\right) = 7.4 \times 10^{-6}\left(\frac{T}{100 - T}\right)^{6.208}; R^2 = 0.99 \qquad (8.41)$$

The segment (B–C) from 89.5 to 92.5 °C, that contained the end of increase in viscosity section, the peak viscosity value, and beginning of the decreasing part of the viscosity curve was well described by a third-order polynomial (Eq. 8.42)

$$\eta^*\left(\frac{\omega}{\omega_r}\right) = -69122.86 + 2244.36T - 24.28T^2 + 0.088T^3;\ R^2 = 0.99 \quad (8.42)$$

It is emphasized that rheological data described by Eqs. 8.41 and 8.42 of a STD cannot be predicted a priori and must be obtained experimentally. Because viscosities at ≥95 °C were severely affected by water evaporation in the rheometer (Yang and Rao 1998a), for temperatures from 95 to 121 °C (retort temperature; C–D), the decrease in magnitudes of the η^* with increase in temperature was assumed to follow an Arrhenius equation (Eq. 8.43).

$$\eta^*\left(\frac{\omega}{\omega_r}\right) = 4.11\exp\left[\frac{227.1\times10^3}{R}\left(\frac{1}{T}-\frac{1}{366.1}\right)\right];\ R^2 = 0.97 \quad (8.43)$$

Because of the availability of reliable TR data from 65 to 95 °C during gelatinization (Eqs. 8.42 and 8.43), as well as of both the experimental heat penetration data and computer simulation-based temperatures, it was considered safe to assume the temperature profile described by Eq. 8.43 from 95 to 121 °C. Comparison of the numerically calculated temperatures with the experimental profiles to be discussed strongly support the assumption of an Arrhenius model (Eq. 8.41) for the magnitudes of η_a from 95 to 121 °C.

In Eq. 8.43, the reference temperature 366° K was the mean value of temperatures of the decreasing segment of η^*. The value of E_a in Eq. 8.41 determined to be 227.1 kJ mol^{-1} is higher compared with those reported for gelatinized corn starch at lower temperatures. Given the large decrease in viscosity, the value of E_a is reasonable and it is in good agreement with the value of E_{ag}=240 kJ mol^{-1} in the model of Dolan et al. (1989) for the increasing viscosity segment (Yang and Rao 1998a). The decrease in viscosity with increasing temperature that occurs just after the peak viscosity has been reached is due to the decrease in starch granule size as a result of granule disruption and with E_{ag}=240 kJ mol^{-1}. In contrast, for the same STD sample after it was heated up to 95 °C and cooled to 25 °C, that is, after granule rupture has been completed, the value of E_a determined from the viscosities measured at 35 and 75 °C was 14.5 kJ mol^{-1}.

Tattiyakul (2001) derived a single equation to describe the data in Fig. 8.7

$$\begin{aligned}\eta^*(T) = &\left(7.4\times10^{-6}\left(\frac{T}{100-T}\right)^{6.208}\right)[H(T-65.0)-H(T-89.5)]\\ &+(-69122.86+2244.36T-24.28T^2+0.088T^3)\\ &[H(T-89.5)-H(T-95.0)]\\ &+\left(4.11\exp\left[\frac{227.1\times10^3}{R}\left(\frac{1}{T}-\frac{1}{366.1}\right)\right]\right)\\ &[H(T-95.0)-H(T-121.0)]\end{aligned} \quad (8.44)$$

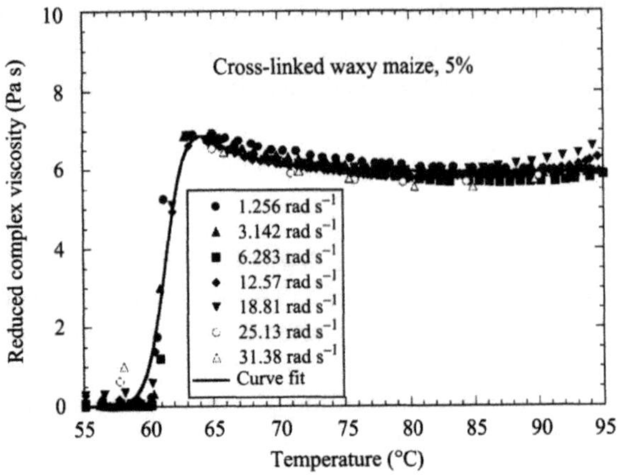

Fig. 8.8 The η^*-T profiles of a 5 % CWM STD obtained at values of a) from 1.26 to 31.38 rad s^{-1} (Tattiyakul 2000)

where $H(x)$ is the Heaviside step function that equals to zero when x is less than zero and one when x is greater or equals to zero.

In general, other native starch dispersions will exhibit similar viscosity versus temperature profiles as in Fig. 8.7, while cross-linked starch dispersions, due to limited granule rupture, will not exhibit a sharp decrease in viscosity of the segment C–D. The η^* versus T profiles of a 5 % CWM STD obtained at values of ω from 1.26 to 31.38 rad s^{-1}, 3 % strain, and a heating rate of 2.1 °C min^{-1} (Fig. 8.8) followed the equation (Tattiyakul and Rao 2000)

$$\eta^*\left(\frac{\omega}{\omega_R}\right)^\beta = \eta^*_{\text{peak}} \times \left(1 - \frac{1}{1+\left\{m_1 \times e^{m_2\times(T-T_o)}\right\}}\right) \times \frac{1}{\left[1+(T-T_p)^2\right]^{m3/T}} \tag{8.45}$$

where η^*_{peak} is the average peak complex viscosity at temperature T_p, which was equal to 6.9 Pa s at T_p = 64 °C in this experiment, T_0 is the gelatinization onset temperature; T_0 = 60 °C, m_1, m_2, and m_3, are constants with values: 0.0674, 2.332, and 2.1, respectively. Equation 8.45 had R^2 = 0.992 and chi-square value 2.57.

Yang and Rao (1998a) used a modified Cox–Merz rule (Eq. 8.46) to determine the parameters relating the dynamic and steady shear data, and a TR model for apparent viscosity was derived, Eq. 8.47.

$$\eta^*(\omega) = C\left[\eta(\dot{\gamma})\right]^{\alpha'} |_{\omega=\dot{\gamma}} \tag{8.46}$$

$$\eta_a = \left[C'\eta^*(T)\left(\frac{\dot{\gamma}_r}{\dot{\gamma}}\right)\right]^{\alpha'} \tag{8.47}$$

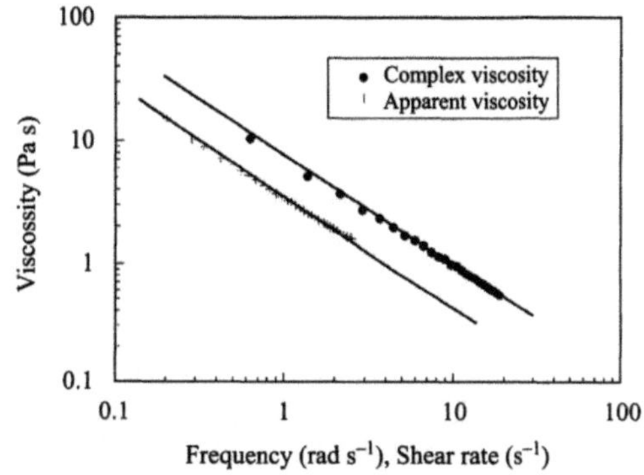

Fig. 8.9 Modified Cox–Merz rule for a gelatinized starch dispersion (Yang and Rao 1998b)

where η^* is complex viscosity (Pa s) at frequency ω (rad s^{-1}), η_a, is apparent viscosity (Pa s) at $\dot{\gamma}$ the shear rate (s^{-1}), C is a constant, α is the shift factor, C' *is* C^{-1}, α' *is* α^{-1}, $\dot{\gamma}_r$ the reference shear rate, and η^* *(T)* is experimental complex viscosity data expressed as either a polynomial or a modified Oswin function, and temperature *(T)* is the independent variable. Equation 8.47 is a convenient TR model to numerically simulate thermal and other food processing problems. Based on data at 25 °C (Fig. 8.9), the constant C and shift factor α of the modified Cox–Merz rule were found to be 2.07 and 1.01, respectively.

Continuous Flow Sterilization

Heat Transfer to Shear-Thinning Fluids

Many fluid foods are sterilized during flow in tubular heat exchangers. Pioneering theoretical studies were conducted on heating or cooling at constant wall temperature along the length of heat exchanger whose results are of interest to continuous flow sterilization of Newtonian and non-Newtonian fluid foods (Graetz 1883; Lyche and Bird 1956; Wissler and Schechter 1959). To obtain theoretical solutions, those studies were based on the assumption of temperature-independent physical properties. Two TR expressions have been introduced (Kwant et al. 1973): isoviscous and non-isoviscous flows. Isoviscous flow is defined as flow of a fluid with a temperature-independent physical properties so that the velocity fields remain constant throughout the heat transfer section of a heat exchanger; non-isoviscous flow is defined as flow of a fluid with temperature-dependent rheological behavior. Using numerical procedures, Christiansen and Craig (1962) obtained more realistic velocity profiles by using a combined power law and Arrhenius equation, Eq. 8.36, in which the effect of temperature upon the rheological properties of the fluid being heated was incorporated.

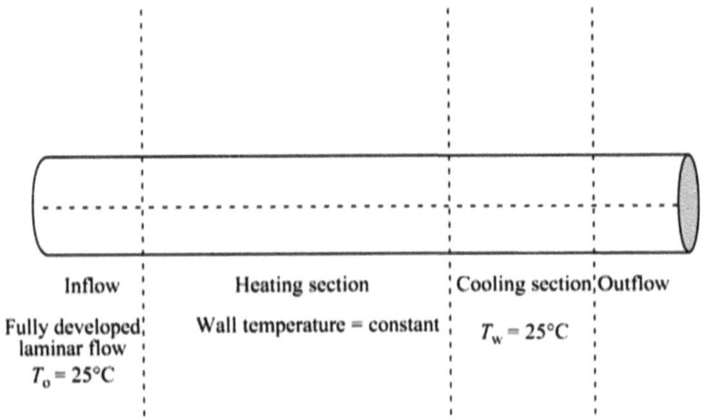

Fig. 8.10 Tubular sterilizer with only heating and cooling sections (Liao 1998)

Physical Model of Non-Isothermal Laminar Flow Tube Sterilizer

Simpson and Williams (1974) studied a continuous sterilizer without a holding section, consisting of just a heating and a cooling section, illustrated in Fig. 8.10 (Liao 1998). The fluid food was pumped at an initial temperature (T_0, arbitrarily selected as 25 °C) with a fully developed velocity profile throughout the cross-section at the entrance of the tube is shown in Fig. 8.9. The heating section was held at a constant wall temperature (T_w) and was terminated when the accumulated lethality value on the centerline reached 5.0 min, which is close to the practical thermal process design criteria in the food industry. At the end of the heating section, the heated fluid food was immediately cooled in a cooling section also held at another constant wall temperature of 25 °C, also arbitrarily selected. The analyses of this and other studies on continuous sterilization (Kumar and Bhattacharaya 1991; Liao et al. 2000) were based on the assumptions: The flow is at a steady state; The flow is assumed to be rectilinear and axisymmetric about the centerline which makes the problem two-dimensional instead of three-dimensional; The fluid thermal conductivity, specific heat, and density are independent of temperature and pressure; At the entrance to the heat transfer section, the flow has a uniform temperature and is fully developed; Thermal energy generation in the fluid by viscous dissipation or by any reaction is negligible; The fluid is heated (or cooled) at constant wall temperature; Natural convection effects are negligible; The fluid is considered to be homogeneous and there is no slip at the wall.

Simpson and Williams (1974) pointed out that the most important effect of the non-isothermal nature of the process is that radial temperature gradients will cause radial viscosity gradients leading to considerable distortion of the velocity profiles. They developed methods for the design of continuous sterilization equipment for pseudoplastic fluids with and without temperature-dependent rheology and recom-

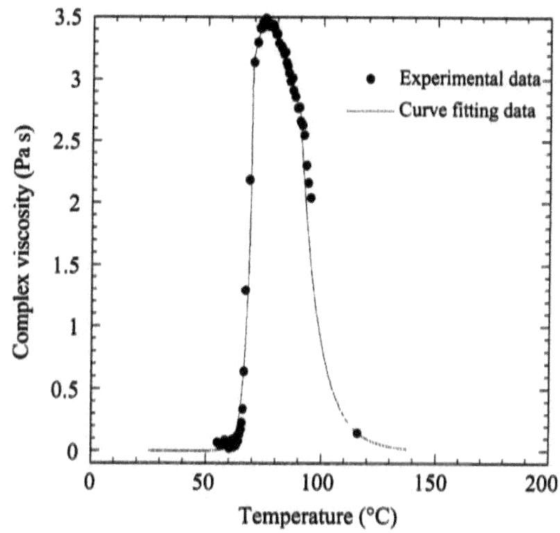

Fig. 8.11 Thermorheological data during gelatinization of a waxy rice starch dispersion (Liao 1998)

mended a total dimensionless length (ξ) of 1.2 with 0.8 for the heating section and the remaining 0.4 for the cooling section

$$\xi = \left(2\alpha / r_0^2\ \bar{v}_z\right) \tag{8.48}$$

This dimensionless length is valid within ±2 % for $0.3\,n \leq 1.0$ and $-4 < \psi < 4$, and the strength of coupling of the momentum and energy transport equations was characterized by the equation

$$\psi = \frac{E_a}{R}\left(\frac{1}{T_0} - \frac{1}{T_w}\right) \tag{8.49}$$

Results similar to those of Simpson and Williams were obtained by Kumar and Bhattacharaya (1991).

Continuous Sterilization of a Fluid Food Containing Starch

During thermal processing, the viscosity of a starch-containing fluid can increase drastically with temperature because of the amylose released and starch granule swelling after the initial temperature of gelatinization has been reached. After the dispersion has attained its maximum viscosity, the granules start to rupture so that the viscosity decreases with temperature (Okechukwu and Rao 1995; Yang and Rao 1998a). Therefore, the TR model obtained by combining the power law model with the Arrhenius equation (Eqs. 8.36 and 8.37) completely fails to describe the rheological behavior of starch-containing fluids during the course of gelatinization.

Liao (1998) studied continuous sterilization of a 4 % waxy rice starch dispersion (Fig. 8.11). Based on experimental rheological data, the TR data (Fig. 8.10) applicable to the fluid flowing in the heating section were obtained using the methodology of Yang and Rao (1998a). A single straight tubular heat exchanger (Fig. 8.9) in which the starch-containing fluid was first heated and then cooled over an additional length of 10 m. Since starch gelatinization is irreversible and there was no other phase transition of the gelatinized starch paste after it entered the cooling section, the Arrhenius equation describing the effect of temperature and a power law type of equation were applied to the fluid in the cooling section.

$$\eta_a = \eta_0 \exp\left(\frac{E_a}{RT}\right)\left(\frac{\partial v_z}{\partial r}\right)^{n-1} \tag{8.50}$$

The starch dispersion was considered to exhibit Newtonian behavior before gelatinization because of high water content of the starch dispersion. Computations were carried out for two tube diameters: 0.018 and 0.024 m, flow rates: 1.0 and 1.5 L min^{-1}, and two heating wall temperatures: 139 and 145 °C. The changes in the other physical properties with temperature were neglected since the variation was not significant compared with the drastic change of viscosity with temperature during gelatinization. Density (971.83 kg m^{-3}), specific heat (4,199 J kg^{-1} K^{-1}), and thermal conductivity (0.668 W m^{-1} K^{-1}) were assumed to be constant and to be the same as those of water at 80 °C.

The governing conservation equations and the boundary conditions in non-isothermal laminar tube flow can be formulated as follows:

Continuity equation

$$\frac{\partial v_z}{\partial z} + \frac{1}{r}\frac{\partial}{\partial r}(r v_r) = 0 \tag{8.51}$$

Momentum equation

$$\rho\left(v_z \frac{\partial v_z}{\partial z} + v_r \frac{\partial v_z}{\partial r}\right) = \frac{1}{r}\frac{\partial}{\partial r}(r\sigma_{rz}) + \frac{\partial \sigma_{zz}}{\partial z} \tag{8.52}$$

$$\rho\left(v_z \frac{\partial v_r}{\partial z} + v_r \frac{\partial v_r}{\partial r}\right) = \frac{1}{r}\frac{\partial}{\partial r}(r\sigma_{rr}) + \frac{\partial \sigma_{rz}}{\partial z} \tag{8.53}$$

The stress components in Eqs. 8.52 and 8.53 are given by

$$\sigma_{rz} = \eta\left(\frac{\partial v_z}{\partial r} + \frac{\partial v_r}{\partial z}\right) \tag{8.54}$$

$$\sigma_{zz} = -p + 2\eta\eta\frac{\partial v_z}{\partial z} \tag{8.55}$$

$$\sigma_{rr} = -p + 2\eta\frac{\partial v_r}{\partial r} \tag{8.56}$$

The energy equation is

$$\rho c_p \left(v_z \frac{\partial T}{\partial z} + v_r \frac{\partial T}{\partial r} \right) = k \left[\frac{1}{r} \frac{\partial}{\partial r} \left(r \frac{\partial T}{\partial r} \right) + \frac{\partial^2 T}{\partial z^2} \right] \tag{8.57}$$

The boundary conditions for the equations are

$$\text{At } z = 0; \quad 0 < r < r_0, \quad v_z = \bar{v}_z \left(\frac{3n+1}{n+1} \right) \left[1 - \left(\frac{r}{r_0} \right)^{(n+1)/n} \right],$$

$$v_r = 0 \text{ and } T = T_0$$

$$v_r = 0 \text{ and } T = T_0$$

At $z \geq 0$, $r = r_0$, $v_z = 0$, $v_r = 0$, and $T = T_w$, and at $z \geq 0$, $r = 0$, $v_r = 0$.

A commercial computational fluid dynamics package (FIDAP Version 7.6, Fluid Dynamics International, Evanston, IL) based on the finite element method was used to solve the governing continuity, momentum and heat transport equations. A mesh was defined with more nodes near the wall and the entrance of the tubular heat exchanger to resolve the larger variations of temperature and velocities near the wall and the entrance.

Analysis of the Heating Section

Some of the results from Liao (1998) are presented and more details can be found in the thesis and in Liao et al. (2000). As the fluid flowed in the heating section, the fluid near the wall was heated first and later the thermal energy front traveled in the radial direction. The radial viscosity gradient, thus, caused the flow profiles to distort considerably. Figure 8.12 shows the centerline velocities and temperatures in dimensionless form along the length of the tube; the sharp increase in velocity at the beginning of heating and cooling sections is to be noted.

The maximum velocity was always on the centerline as expected and was higher than the average velocity by as much as 4.2 times the average value at the beginning of the heating section. As the fluid continued traveling along the tube, the centerline velocity decreased to almost the same magnitude as the average velocity. The profile became much flatter and tended toward plug-flow. The portion of the flow blunted is relevant to the TR behavior caused by gelatinization of starch and the bluntness increased as the fluid traveled further in the heating section. As the temperature on the centerline reached the initial temperature of starch gelatinization, which is about 62 °C, and the viscosity of the fluid traveled on the centerline began to be dependent on temperature, the plug-flow was completely developed ($\zeta = 0.24$). At this axial location, the starch-containing fluid traveled along the centerline and began to gelatinize.

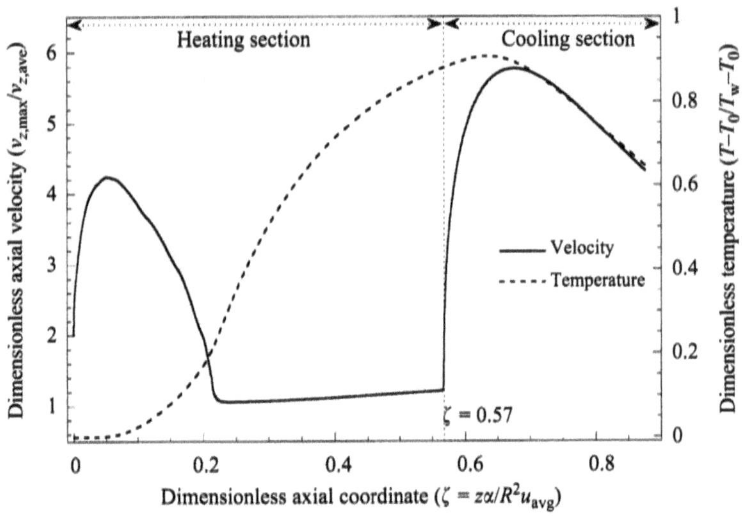

Fig. 8.12 Dimensionless axial velocity and temperature on the centerline along the length of the tube, r_0=0.012 m; T=145 °C; Q=1 L min^{-1}. Note the sharp increase in velocity at the beginning of heating and cooling sections

As pointed out by Simpson and Williams (1974), the blunting effect caused by the pseudoplasticity of the fluid increased the total probability of sterility by some 20 orders of magnitude with the incorrect assumption that the food was Newtonian.

Analysis of Cooling Section

When the heated fluid entered the cooling section, the temperature of the fluid traveling near the tube wall decreased rapidly causing the viscosity of the fluid to increase. Again, this spatial variation in viscosity forced the fluid traveling away from the wall to move faster; consequently, the maximum centerline velocity was approximately five–eight times the average value (Fig. 8.12). The temperature on the centerline continued to increase in the initial part of the cooling section and it was high enough for effective sterilization. The additional lethality gained in the cooling section could contribute toward additional food safety which makes the design more conservative.

Length of Heating Section

Simpson and Williams (1974) assumed that the heating section was terminated when the temperature gain on the centerline was 95 % and an F_0 of 2.45 min was attained at the end of heating section. In this work, an accumulated F_0 of 5 min on the centerline was specified before the fluid exited the heating section. For the non-isothermal continuous tube flow, F_0 was calculated as

Fig. 8.13 Cumulative lethality on the centerline along the heating section for two flow rates: 1.5 and 1.0 L min^{-1} (B), heating wall temperatures: 139 (D) and 145 °C, and tube diameters: 0.018 and 0.024 m (C)

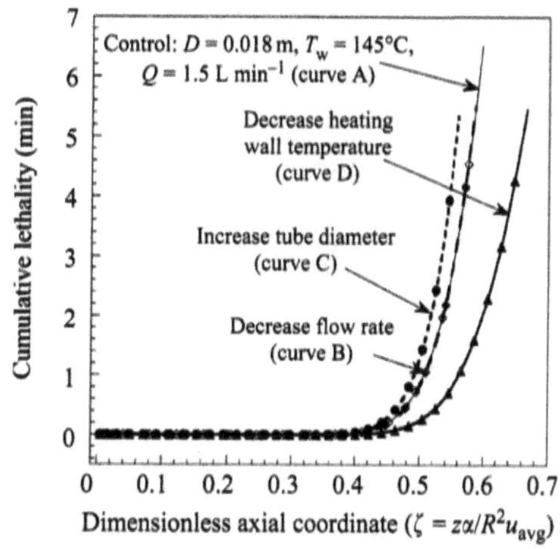

$$F_0(L,r) = \int_0^t \frac{\dfrac{T - T_R}{10z - \text{value}}}{v_z(z,r)}\, dz \qquad (8.58)$$

where z−value $= 2.303\ \text{RT}_\text{R}^2/E_a$, and F_0 is the equivalent heating time at a reference temperature T_R for a process whose actual temperature T varies with time t. For the nutrients, an average concentration which provides a measure of the nutritional quality of the processed food can be calculated. Thiamine retention was also evaluated, using a z-value of 35 °C and D_{121} °C value of 158 min (Feliciotti and Esselen 1957), as an indicator for product nutrient retention

$$\tilde{C}_f(z) = \frac{\int_0^{r_0} C_f(z,r) v_z(z,r) 2\pi r dr}{\int_0^{r_0} v_z(z,r) 2\pi r dr} \qquad (8.59)$$

It was assumed that bacterial spores and nutrients travel at a constant distance from the wall throughout the tube.

The dimensionless length of heating section was about 0.56–0.60 when the heating wall temperature was 145 °C. Tube wall temperature (139 °C) resulted in an increase in the dimensionless length of heating section to $\zeta_H = 0.67$. For higher heating wall temperature, the effective contribution to food lethality started earlier in the tube so that shorter length was required to achieve sterility (Fig. 8.13). In contrast to changes in the heating wall temperature, increasing the flow rate resulted in the increase of the actual length of heating section (other parameters held constant) because of the change in the resident time required in the tube to achieve the sterility. Increasing the flow rate by 150 % caused a 50 % increase in the actual length

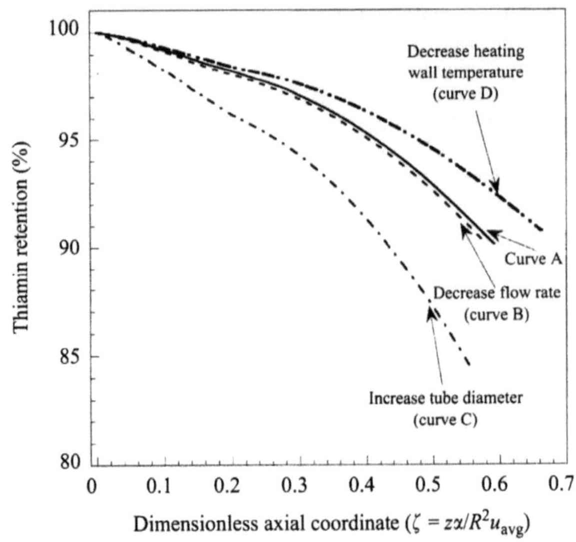

Fig. 8.14 Thiamine retention along the length of the heating section for two flow rates: 1.5 and 1.0 L min^{-1} (B), heating wall temperature: 139 (D) and 145 °C, and tube diameters: 0.018 and 0.024 m(C)

of heating section, but as can be seen in Fig. 8.14, the average thiamine retention remained unchanged. Decreasing the heating wall temperature produced higher nutrient retention along the tube but the final nutrient content at the end of heating section was almost unchanged because longer length of tube needed to achieve sterility. Increasing the tube diameter resulted in more destruction of nutrient.

Influence of Rheological Behavior on the Heating Length

Fluids with different rheological behavior required different heating tube lengths to achieve the target value of 5.0 min of lethality at the SHZ for the specific set of variables: r_0 = 0.009 m; T_w = 139 °C; Q = 1.5 L min^{-1}. A dimensionless heating length of 0.69 was needed for a fluid without any starch that obeyed the temperature-dependent power law model (Eq. 8.36). A dimensionless length of 0.85 was found if the fluid was assumed to be Newtonian. The shorter dimensionless heating tube length of 0.67 for starch-containing fluid is because of the occurrence of the plug-flow profile in the heating tube (r_0 = 0.009 m; T_W = 139 °C; Q = 1.5 L min^{-1}).

Role of Rheology in Thermal Processing of Canned Foods

Heat Penetration Parameters

Many foods, such as those containing large-size solids (e.g., soups, cut vegetables), are sterilized in sealed cans. In thermal processing of canned foods, the formula methods provide a good approximation of the lethality when the heat penetration data can be described by a single straight heating curve (Eq. 8.60). The parameters of Eq. 8.60, particularly f_h, permit a formula method to evaluate processes under conditions different from those at which experimental data were obtained

$$\log \frac{RT - T(t)}{j(Rt - IT)} = -\frac{t}{f_h} \tag{8.60}$$

$$j = \frac{RT - T(t)}{RT - IT} \tag{8.61}$$

The parameter j is a measure of the lag to achieve a uniform heating rate and is associated with the position of the cold spot or slowest heating point, the can size, and the IT (Ball and Olson 1957); basically, these three factors determine the time to achieve a uniform heating rate. Although f_h is usually defined as the time required to traverse one logarithmic cycle on the temperature scale, the physical meaning of f_h is more complex and is associated with the mode of heat transfer. Ball and Olson (1957) derived analytical solutions for f_h, in both ideal thermal convection (Eq. 8.62) and conduction (Eq. 8.63), respectively

$$f_h = 2.303 \frac{c_p \rho V}{UA} \tag{8.62}$$

$$f_h = \frac{2.303}{\left[\frac{(2.405)^2}{R^2} + \frac{\pi^2}{L^2}\right]\alpha} \tag{8.63}$$

Generally, Eq. 8.60 is applicable to foods where the heat transfer mode is well-defined, such as clarified fruit juices and thin soups that heat by convection mode, and corned beef and chicken loaf that heat by conduction mode. Broken heating (BH) heat penetration data can usually be fitted by two straight lines with different slopes on a semilog plot; in some cases, the heating curve is represented by more than two straight lines. Since there is a significant change in the heating rate, one can interpret that the break is the result of an increased resistance to heat transfer.

Ball and Olson (1957) associated viscosity data of corn starch dispersions with BH heat penetration data, but a direct correlation with starch gelatinization was not established because the data were obtained at a constant 60 °C (140°F) with a Scott orifice viscometer. Stevens (1972) investigated BH phenomena, but due to limited computer capabilities and the lack of suitable TR data complete numerical simulation was not conducted.

Ball (1923, 1927, 1928) studied BH curves and proposed Eqs. 8.52 and 8.53 as extensions of the single heating curve method

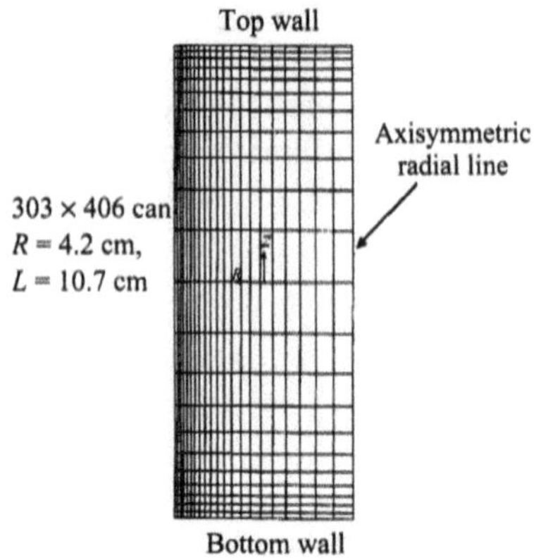

Fig. 8.15 Finite element grid for studying heat transfer to a canned (303 × 406 can) starch dispersion (Yang 1997; Yang and Rao 1998b, c)

$$\log(g_{\text{bh}}) = \log(jI) - \frac{x_{\text{bh}}}{f_h} \tag{8.64}$$

$$\log(g_{\text{h2}}) = \frac{f_{\text{h}}\log(jI) + (f_2 - f_{\text{h}})\log(g_{\text{bh}}) - B_{\text{b}}}{f_2} \tag{8.65}$$

Although Eqs. 8.64 and 8.65 are known to be not very precise (Ball and Olson 1957), they are still used in the food industry. The break point (x_{bh}) is determined as the intersection of the two curves.

Numerical Solution of Transport Equations

The governing partial differential equations for heat transfer to an SD in a 303 × 406 can with: $R=4.2$ cm and $L=10.7$ cm, and no head space were solved and the details are given in Yang (1997) and Yang and Rao (1998a, b). Because the SD exhibited non-Newtonian behavior after the initial gelatinization temperature was reached, the problem could not be approximated to simple heat transfer by conduction.

The boundary conditions for solving the equations were (see Fig. 8.15): (1) at the wall on the left, $r=R$, $0\leq z\leq L$, $T=RT$, $v_z=0$, and $v_r=0$; (2) top wall at $z=L$, $0\leq r\leq R$, $T=RT$, $v_z=0$ and $v_r=0$; (3) bottom wall at $z=0$, $0\leq r\leq R$, $T=RT$, $v_z=0$ and $v_r=0$; and (4) axisymmetric line at $r=0$, $0\leq z\leq L$, $\partial T/\partial r=0$, $\partial v_z/\partial r=0$, and $v_r=0$. The initial conditions were: $0\leq r\leq R$, $0\leq z\leq L$, $T=IT$, $v_z=0$, and $v_r=0$. Since the magnitude of η_a during gelatinization could be from 10 to 2,000 times higher than that of ungelatinized SD, the changes in the other fluid properties with the temperature were neglected, and were assumed to be equal to those of water: $\rho_0=1{,}000$ kg m^{-3}, $k=0.66$ J m^{-1}K^{-1} s^{-1}, $c_p=4{,}180$ J kg^{-1} K^{-1}, and $\beta=0.00053$ K^{-1} and 0.0005 KT^{-1} at 70 °C. The governing equations with boundary and initial conditions described above were solved with FIDAP fluid dynamics analysis program. A mesh made up of 2,115 nine-node isoparametric elements, 45 in the radial direction and 47 in the axial direction was used (Yang 1997; Yang and Rao 1998a, b).

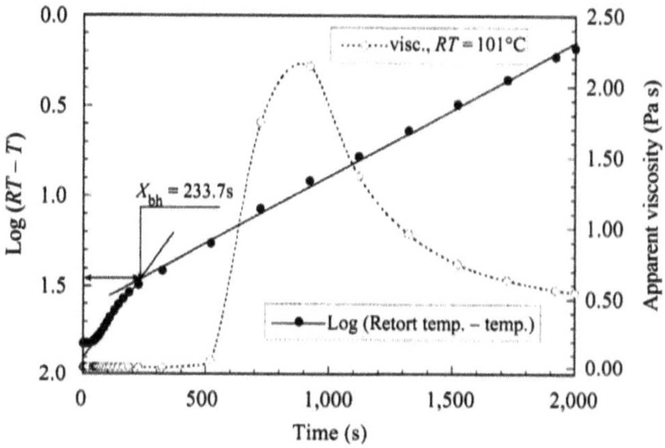

Fig. 8.16 Broken heating curve from simulation of heat transfer to a canned starch dispersion at a retort temperature *(RT)* 101 °C. The log *(RT – T)* versus time at the assumed slowest heating point and the corresponding fluid apparent viscosity (η_a) versus time profiles are shown (Yang 1997)

The temperatures at a point one-third the height of the container on the axis were examined because traditionally it is considered to be the slowest heating point in natural convection-heating products. The log *(RT – T)* versus time at the assumed slowest heating point and the corresponding η_a versus time profiles are shown in Figs. 8.16–8.19 for different heating conditions.

Estimated Heat Penetration Parameters and Broken Heating Phenomena

The following salient features of the calculated heat penetration parameters were noted: (1) the *j*-values did not change significantly because of the short lag phase and use of the corrected zero time; (2) the f_h values were nearly the same except for the simulations at *RT* 131 °C and can size 211 × 400; (3) the x_{bh} values could be separated into two distinct groups, one for simulations at *RTs* 101 and 111 °C, and the other for simulations at *RTs* 121 and 131 °C; (4) the f_2 values decreased as the *RT* increased. A considerable decrease in f_2 values occurred when the *RT* was increased from 101 to 111 °C; and (5) the temperatures at which x_{bh} occurred were approximately 70 °C for simulations with *RTs* at 101 and 111 °C, and close to the *RT for* simulations with *RTs* at 121 and 131 °C.

The two process conditions: *RT* = 101 and 111 °C, *IT* = 35 °C that had the shortest x_{bh} values, 3.9 and 3.2 min, respectively, were also characterized by the longest heating times during which the magnitudes of η_a were greater than 0.55 Pa s and also the highest magnitudes of η_{final}, 0.55 and 0.06 Pa s, respectively; in addition, the temperature at which the x_{bh} occurred was close to the initial temperature of gelatinization. The significant difference between the f_2 value (f_2 = 22.7 min) at *RT* = 101 °C and at *RT* = 111 °C (14.8 min) was due to the large differences in the magnitudes of η_a over the entire heating process after initiation of gelatinization (Figs. 8.16 and 8.17). At *RTs* 121 and 131 °C, the x_{bh} s were approximately four times higher than those at *RTs* 101 and 111 °C, and the temperatures at which the x_{bh} s occurred were close to the corresponding retort temperatures. The x_{bh} s for these

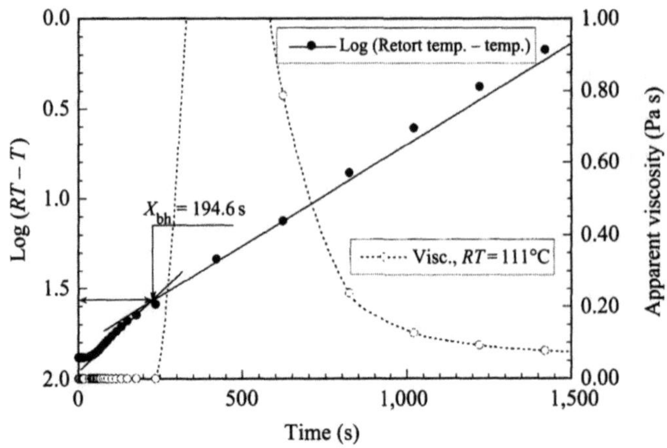

Fig. 8.17 Broken heating curve from simulation of heat transfer to a canned starch dispersion at a retort temperature *(RT)* 111 °C. The log *(RT–T)* versus time at the assumed slowest heating point and the corresponding fluid apparent viscosity (η_a) versus time profiles are shown (Yang 1997)

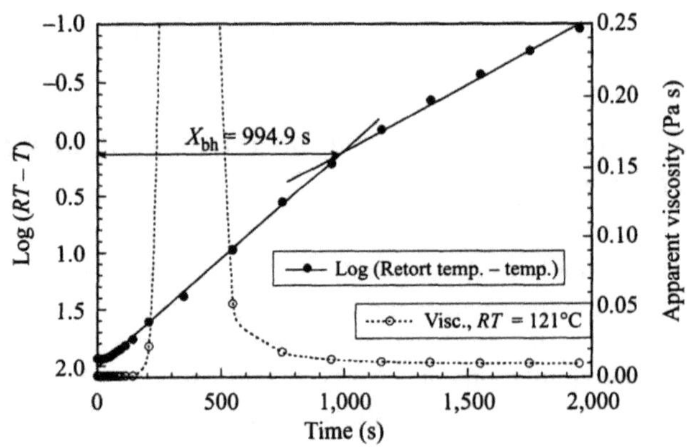

Fig. 8.18 Broken heating curve from simulation of heat transfer to a canned starch dispersion at a retort temperature *(RT)* 121 °C. The log *(RT–T)* versus time at the assumed slowest heating point and the corresponding fluid apparent viscosity (η_a) versus time profiles are shown (Yang 1997)

conditions were mostly due to the significant decrease in the heat transfer driving force, that is, *(RT–T)*.

Heat Transfer to a Starch Dispersion in an Intermittently Rotated Can

Agitation during thermal processing is an effective means of providing induced convection, which results in a higher heating rate and more uniform heating. Agitation also prevents separation of different ingredients in the food product during thermal processing (Rao and Anantheswaran 1988). Among the many factors

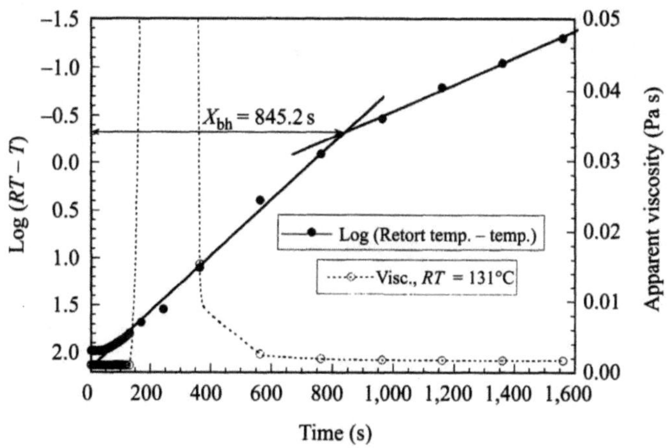

Fig. 8.19 Broken heating curve from simulation of heat transfer to a canned starch dispersion at a retort temperature (*RT*) 131 °C. The log *(RT−T)* versus time at the assumed slowest heating point and the corresponding fluid apparent viscosity (η_a) versus time profiles are shown (Yang 1997)

that influence heat transfer rate to a can of food under sterilization listed in Lopez (1987), the mode of container agitation, the agitation speed, the fluid viscosity, the radius of agitation, and the headspace volume and motion are top-ranked. One of the most widely used can agitating methods in the food industry is axial rotation provided by a spiral-continuous retort; the Sterilmatic, whose pilot-plant scale unit is the Steritort.

Tattiyakul et al. (2002a) examined heat transfer to a canned 3.5 % corn starch dispersion (303 × 406) undergoing axial rotation at an intermittent angular velocity Ω_t between 0 and 146 rpm. The starch dispersion had an initial temperature of T_0 and was heated at a wall temperature, T_w. Several assumptions were made to simplify the problem: (1) constant thermal conductivity (k), specific heat (c_p), and volume expansion coefficient (β); (2) constant fluid density (ρ_0 = 1,000 kg m^{-3}) in the governing equations except in the buoyancy term, where density (ρ) is approximated by Boussinesq approximation $\rho=\rho_0(1-\beta\ (T-T_0))$; (3) homogeneous fluid; (4) negligible heat generated by viscous dissipation; (5) constant can wall temperature of 121 °C after 2 min; (6) no-slip boundary condition; (7) no headspace; and (8) negligible centrifugal force due to the reel (primary) rotation.

The governing conservation equations in the inertial frame of reference were solved with the boundary conditions of temperature and velocity

$$r = R;\quad 0^\circ \le \theta \le 360^\circ : T_w = RT,\quad v = \Omega_t R$$

where r is the distance from the can center (m), R is the can radius (R = 0.04048 m), θ is the angular position in the can (°), RT is the retort temperature (°C), and Ω_t is the can angular velocity ($\Omega_t = 2\pi N/60$ rad s^{-1}), and N is the can's rotational speed in revolution per minute (rpm).

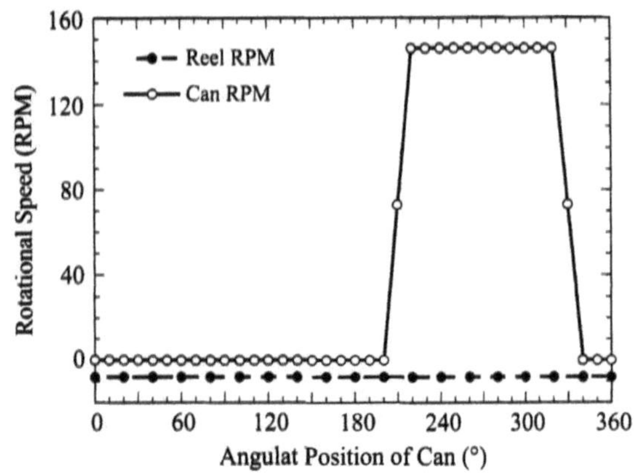

Fig. 8.20 Idealized can rotation profile in an Steritort that was used in simulation

Intermittent Rotation Profile

For the intermittent rotation of a can, an idealized rotation profile based on that provided in an Steritort was employed; the can's rotational profile, adapted from that in Ladeinde (1988), is shown in Fig. 8.20. The simplified mathematical formula for the can rpm in an Steritort whose reel rotates at 8 rpm can be expressed as

$$\begin{aligned} N = {} & \frac{N_c}{20}(\Theta - 200)[H(\Theta - 200) - H(\Theta - 220)] \\ & + N_c[H(\Theta - 220) - H(\Theta - 320)] \\ & + \left(N_c - \frac{N_c}{20}\right)(\Theta - 320)[H(\Theta - 320) - H(\Theta - 340)] \end{aligned} \tag{8.66}$$

where Θ is the angular location of the can in an Steritort, *H(x)* is the Heaviside step function that equals to zero, when *x* is less than zero, and one, when *x* is greater or equals to zero, and N_c is the maximum can rotational speed that can be calculated from the reel rotational speed, N_r; the shell diameter, D_r, and the can diameter; D_c. Assuming a no-slip condition, the linear velocity of the can surface equals the linear velocity of the reel

$$\pi D_c N_c = \pi D_r N_r \tag{8.67}$$

and

$$N_c = D_r N_r / D_c \tag{8.68}$$

The initial temperature and velocity conditions simulated were: $0 \le r \le R$; $0° \le \theta \le 360°$, $T_0 = 30°C$, $u = 0$, $v = 0$.

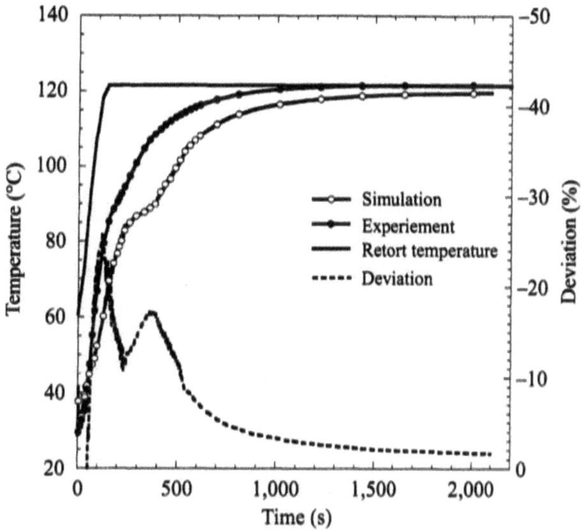

Fig. 8.21 Numerically predicted time-temperature profile at the center of a canned 3.5 % corn starch dispersion intermittently rotating at 0 and 146 rpm, and experimental heat penetration data

Thermorheological Behavior of the Model Food

The model fluid food studied was 3.5 % corn SDs whose thermorheological properties were described by the model developed by Yang and Rao (1998a)

$$\eta_a = \left[\frac{1}{2.07}\eta^*(T)\left(\frac{\dot{\gamma}_r}{\dot{\gamma}}\right)\right]^{1/1.01} \tag{8.69}$$

where $\dot{\gamma}_r$ is the reference shear rate (s^{-1}), $\dot{\gamma}$ is shear rate (s^{-1}), and T is temperature (°C), η_a is apparent viscosity (Pa s), and η^* (T) is complex viscosity (Pa s) of the 3.5 %corn starch dispersion expressed by Eq. 8.44. Again, the other physical properties of the starch dispersion were assumed to be constant (ρ_0 = 1,000 kg m^{-3}, k=0.66 J m^{-1} K^{-1} s^{-1}, c_p=4,180 J kg^{-1} K^{-}1, and β=0.00053 K^{-1}).

From the numerical simulation, much useful information on the time and spatial profiles of fluid velocity, viscosity, and temperature were obtained (Tattiyakul 2001; Tattiyakul et al. 2002a). For brevity, only the numerically predicted time-temperature profile at the center of a canned 3.5 % corn starch dispersion intermittently rotating at 0 and 146 rpm, and the experimental heat penetration data are shown in Fig. 8.21. The major discrepancy (10–20 %) between the predicted center temperature and the experimental data obtained using an Steritort was found in the heating time range of 100–400 s, that coincided with the time during which the starch gelatinized. There may be two reasons for the discrepancy between the predicted and experimental results.

First, there was a large difference between the heating rate at which the TR data were obtained (2.1 °C min^{-1}) and the heating rate to which the SD was actually subjected to in an Steritort (~30 °C min^{-1} for the first 120 s). According to Yang (1997), the peak viscosity of an 8 % corn starch dispersion heated at 6.0 °C min^{-1} heating rate was found to be about 75 % of that heated at 2.1 °C heating rate.

Second, it should be noted that the movements a can undergoes in one cycle of an Steritort include: (1) sliding in the slot that holds the can at about 90 °C and 270°; (2) falling of the can onto the baffled portion of the shell at about 0° (or 360°) before rotation; (3) rotation of the can on the baffled portion of the shell as it is pushed from about 340 and 200°; and (4) falling of the can back into its slot at the end of the baffled portion of the shell (about 180°). Even though each additional movement occurs for a short time, its effect on heat transfer, especially during the early stage of heating (100–400 s), is significant. The temperatures showed good agreement, that is, deviation less than 5 %, between the predicted and the experimental values during the temperature range in which accumulated bacterial lethality is significant ($T>100$ °C). Therefore, the predicted temperature profile is considered to be conservative for thermal processing application.

The effect of different flow properties on heat transfer to canned dispersions undergoing intermittent axial rotation was studied by Tattiyakul et al. (2002b). In addition to the 3.5 % corn starch dispersion discussed earlier, a 5 % CWM dispersion whose rheological data are shown in Fig. 8.9 and a tomato concentrate that followed a simpler thermorheological behavior were studied. Because of the high apparent viscosity over a wide temperature range of the CWM dispersion, it had the slowest time-temperature profile (not shown here). Results on the effect of continuous axial rotation on heat transfer to a canned starch dispersion can be found in Tattiyakul (2001).

Empirical Correlations for Heat Transfer to Fluids Flowing in Tubes

Empirical correlations are based on dimensionless groups so that they are applicable for a wide range of physical and thermal properties and process conditions. In turn, the dimensionless groups are obtained from dimensional analysis. Dimensional analysis is based on the assumption that the only fundamental units are those of mass, length, time, temperature, and heat, and that the quantity under consideration is a function of several known variables. As an example, the heat transfer coefficient in convection heat transfer in containers can be expected to depend on the characteristic dimension of the container, the physical and thermal properties of the fluid food, and rotational speed of can agitation. The pertinent dimensionless groups for an application can be determined by the application of Buckingham's Pi theorem, which states that the relationship among q quantities whose units may be expressed in terms of $u$fundamental units may be written as a function of $q-u$ dimensionless groups (Bird et al. 1960).

There are many well-known dimensionless groups that are used in transport phenomena. Earlier, the Reynolds number was used to correlate data on pressure drop in pipe flow. For correlating data on heat transfer, often the dimensionless groups

Nusselt (Nu), Reynolds (Re), Prandtl (Pr), and Grashof (Gr) are used. They are defined as

$$\text{Nusselt number, Nu} = \frac{h_c D}{k} \tag{8.70}$$

$$\text{Pandtl number, Pr} = \frac{c_p \eta}{k} \tag{8.71}$$

$$\text{Reynolds number, Re} = \frac{D v \rho}{\eta} \tag{8.72}$$

$$\text{Grashof number, Gr} = \frac{g L^3 \rho^2 \beta \Delta T}{\eta^2}. \tag{8.73}$$

For non-Newtonian fluids, the viscosity in the above equations must be replaced by an appropriate apparent viscosity. In pipe-flow applications, as shown in Chap. 3, the pipe-flow apparent viscosity (η_{ap}) based on the diameter D and the average axial velocity in the tube ($\bar{v}_z$) can be used in deriving the appropriate dimensionless numbers

$$\eta_{ap} = K' \left(\frac{8 \bar{v}_z}{D} \right)^{n'-1} \tag{8.74}$$

For example, one can obtain the GRe for pipe flow

$$\text{GRe} = \frac{D^{n'-1} \bar{v}^{2-n'} \rho}{K' 8^{n'-1}} \tag{8.75}$$

In terms of the consistency coefficient and flow behavior index of the

$$\text{GRe} = \frac{D^n v^{2-n} \rho}{8^{n-1} K} \left(\frac{4n}{3n+1} \right)^n$$

In addition to the dimensionless groups mentioned above, the ratio of the viscosity at the wall, η_w, to that of the bulk fluid, η_b, is used to account for significant temperature gradients between the wall and the bulk fluid. It is helpful to think of dimensionless groups as ratios of forces or effects in a system (Bird et al. 1960; Foust et al. 1980). For example, the Reynolds number can be written as the ratio of inertial forces to viscous forces; thus, a low value of the Reynolds number means that the viscous forces are large in comparison to the inertial forces. The Prandtl number, that is used extensively in studies on convective heat transfer, is the ratio of molecular diffusivity of momentum to molecular diffusivity of heat.

The Nusselt number contains the heat transfer coefficient, h_c (Wm^{-2} K), defined as

$$q = h_c A(\Delta T) \tag{8.76}$$

where q is the rate of heat transfer (W), A is the area (m^2) of heat transfer, and ΔT is the temperature difference (K) and is the driving force to heat transfer. Equation 8.76 is often referred to as Newton's law of cooling.

There have been several empirical heat transfer relationships for laminar flow in tubes proposed which attempt to allow for variation in rheological properties. Empirical equations developed by several authors for the fluids with different rheological behaviors have been well reviewed by Skelland (1967). Here, a few equations containing dimensionless groups are given to illustrate such relationships.

Newtonian Fluid

An empirical expression was presented by Sieder and Tate (1936) for heat transfer to highly viscous Newtonian fluids in laminar flow through tubes that has found extensive application

$$\frac{h_m D}{k} = 2.0\left(\frac{wC_p}{kL}\right)^{1/3}\left(\frac{\eta_b}{\eta_w}\right)^{0.14} \tag{8.77}$$

where η_b is the viscosity at the bulk temperature and η_w is the viscosity at wall.

Power Law Fluid

The equation proposed by Charm and Merrill (Charm 1971) for heat transfer to power law fluid is

$$\frac{h_m D}{k} = 2.0\left(\frac{wC_p}{kL}\right)^{1/3}\left[\frac{K_b(3n+1)}{K_w 2(3n+1)}\right]^{0.14} \tag{8.78}$$

where h_m is the mean heat transfer coefficient (Wm^{-2} K), K_b is the consistency index (Pa s^n) at the bulk average temperature, and K_w is the consistency index (Pa s^n) at the wall temperature.

Empirical Correlations for Heat Transfer to Canned Fluids

Rao and Anantheswaran (1988) reviewed studies on convective heat transfer to canned fluids in detail. Here, the dimensionless groups and relationships applicable to both Newtonian and non-Newtonian fluids are reviewed in brief.

The rotational Reynolds Re_a number is defined as

$$\mathrm{Re_a} = \frac{D_a^2 N \rho}{\eta} \tag{8.79}$$

where D is the characteristic dimension, for example, the diameter of the can, and N is the rotational speed. For a non-Newtonian fluid being heated in a can, Anantheswaran and Rao (1985b) suggested that the pipe apparent viscosity can be used in the various dimensionless groups. The expressions for the generalized Reynolds ($\mathrm{GRe_a}$), Prandtl (GPr), and Grashof (GGr) numbers are

$$\mathrm{GRe} = \frac{D^n v_z^{2-n} \rho}{8^{(n-1)} K} \left(\frac{4n}{3n+1} \right)^n \tag{8.80}$$

$$\mathrm{GRr} = \frac{c_p K 8^{n-1}}{k N^{1-n}} \left(\frac{3n+1}{4n} \right)^n \tag{8.81}$$

$$\mathrm{GGr} = \frac{g L^3 \Delta T \rho^2 N^{2-2n}}{K^2 8^{2n-2}} \left(\frac{4n}{3n+1} \right)^{2n} \tag{8.82}$$

Anantheswaran and Rao (1985a, b) developed the following respective correlations for heat transfer to cans under end-over-end agitation for Newtonian and non-Newtonian fluids

$$\mathrm{Nu} = 2.9\,\mathrm{Re_a^{0.436}}\,\mathrm{Pr^{0.287}} \tag{8.83}$$

$$\mathrm{Nu} = 1.3(\mathrm{GRe_a})^{0.485}(\mathrm{GPr})^{0.367} \tag{8.84}$$

The correlations were for data obtained with model fluids in copper cans with headspace volume in the range 3–9 % and the rate of heat transfer was independent of the radius of rotation in the range 0–14.9 cm.

Heat transfer to canned liquids being heated in an Steritort was examined by Rao et al. (1985). Both Newtonian (water, and 30, 40, 50, and 60 % w/v, aqueous sucrose solutions) and non-Newtonian fluids (0.3, 0.4, 0.5, and 0.75 % aqueous guar gum solutions) were used. The heat transfer coefficient for both Newtonian and non-Newtonian fluids was correlated by an equation that was the sum of the contributions due to natural and forced convections

$$\mathrm{Nu} = A[(\mathrm{Gr})(\mathrm{Pr})]^B + C[(\mathrm{Re_a})(\mathrm{Pr})(D/L)]^D \tag{8.85}$$

The contribution of natural convection heat transfer was significant with the Newtonian fluids. The coefficients A, B, C, and D for Newtonian fluids were: 0.135, 0.323, 3.91×10^{-3}, and 1.369, respectively. For the non-Newtonian dispersions, the generalized dimensionless numbers: $\mathrm{GRe_a}$, GPr, and GGr were used in place of $\mathrm{Re_a}$, Pr, and Gr, respectively. The magnitudes of the coefficients A, B, C, and D for the non-Newtonian guar gum dispersions were: 2.319, 0.218, 4.10×10^{-7}, and 1.836, respectively.

References

Anantheswaran, R. C. and Rao, M. A. 1985a. Heat transfer to model Newtonian liquid foods in cans during end-over-end agitation. *J. Food Eng.* 4: 1–19.

Anantheswaran, R. C. and Rao, M. A. 1985b. Heat transfer to non-Newtonian liquid foods in cans during end-over-end rotation. *J. Food Eng.* 4: 21–35.

Ball, C. 0. 1923. Thermal process time for canned food. Bull. 37. National Research Council, Washington, DC.

Ball, C. O. 1927. Theory and practice in processing. *The Canner* 64: 27–32.

Ball, C. O. 1928. Mathematical solution of problems on thermal processing of canned food. *Univ. of Calif (Berkeley) Publications in Public Health.* 1(2): 15–245.

Ball, C. O. and Olson, F. C. W. 1957. *Sterilization in Food Technology*, 1st ed., McGraw Hill Book Co., New York.

Bird, R. B., Stewart, W. E. and Lightfoot, E.N. 1960. *Transport Phenomena*, John Wiley & Sons, New York.

Brodkey, R. S. 1967. *The Phenomena of Fluid Motions*, Addison-Wesley, Reading, MA.

Charm, S. E. 1971. *Fundamentals of Food Engineering*, 2nd ed., AVI Publishing Co., Westport, CT.

Christiansen, E. B. and Craig, S. E. 1962. Heat transfer to pseudoplastic fluids in laminar flow. *AIChE J.* 8: 154–160.

Danckwerts, P. V. 1953. Continuous flow systems. *Chem. Eng. Sci.* 2: 1–13.

Dodge, D. W. and Metzner, A. B. 1959. Turbulent flow of non-Newtonian systems. *AIChE J.* 5: 189–204.

Dolan, K. D. and Steffe, J. F. 1990. Modeling rheological behavior of gelatinizing starch solutions using mixer viscometry data. *J. Texture Stud.* 21: 265–294.

Dolan, K. D., Steffe, J. F. and Morgan, R. G. 1989. Back extrusion and simulation of viscosity development during starch gelatinization. *J. Food Process Eng.* 11: 79–101.

Feliciotti, E. and Esselen, W.B. 1957. Thermal destruction rates of thiamine in pureed meats and vegetables. *Food Technol.* 11:77–84.

Ferry, J. D. 1980. *Viscoelastic Properties of Polymers*, 3rd ed., John Wiley, New York.

Foust, A. S., Wenxel, L. A., Clump, C. W., Maus, L. and Andersen, B. 1980. *Principles of Unit Operations*, 2nd ed., John Wiley, New York.

Garcia, E. J. and Steffe, J. F. 1987. Comparison of friction factor equations for non-Newtonian fluids in pipe flow. *J. Food Process Eng.* 9: 93–120.

Graetz, L. 1883. Uber die Warmeleitungsfahigkeit von Flussigkeiten (On the heat transfer in liquids) *Annalen der Physik und Chemie*, part. 1,18: 79–94.

Hanks, R. W. 1978. Low Reynolds number turbulent pipeline flow of pseudohomogeneous slurries, in *Proceedings of the Fifth International Conference on the Hydraulic Transport of Solids in Pipes Hydro- transport.* May 8–11. Paper C2, p. C2–23 to C2–34, Hanover, West Germany, cited in Garcia and Steffe 1987.

Hanks, R. W. and Ricks, B.L. 1974. Laminar-turbulent transition in flow of pseudoplastic fluids with yield stress. *J. Hydronautics* 8: 163–166.

Harper, J. C. and El-Sahrigi, A. F. 1965. Viscometric behavior of tomato concentrates. *J. Food Sci.* 30: 470–476.

Heppell, N. J. 1985. Comparison of the residence time distributions of water and milk in an experimental UHT sterilizer. *J. Food Eng. 4*: 71–84.

Himmelblau, D. M. and Bischoff, K. B. 1968. *Process Analysis and Simulation*, John Wiley and Sons, New York.

Kalkschmidt-J. 1977. Ruehr- und Mischeinrichtungen—unter besonderer Beruecksichtigung der Milchwirtschaft. [Stirring and mixing equipment for the dairy industry.]. *Fette,-Seifen,-Anstrichmittel* 77(9): 357–359. *Food Science Technol.* Abstract 76-04–P0664.

Kokini, J. L., Lai, L.-S., and Chedid, L. L. 1992. Effect of starch structure on starch rheological properties. *Food Technol.* 46 (6) 124–139.

Kubota, K., Hosakawa, Y., Suzuki, K., and Hosaka, H. 1979. Studies on the gelatinization rate of rice and potato starches. *J. Food Sei.* 44: 1394–1397.

Kumar, A. and Bhattacharya, M. 1991. Numerical analysis of aseptic processing of a non-Newtonian liquid food in a tubular heat exchanger. *Chem. Eng. Comm.* 103: 27–51.

Kwant, P. B., Fierens, R. H. E., and Van Der Lee, A. 1973. Non-isothermal laminar pipe flow—I. Theoretical. *Chem. Eng. Sci.* 28: 1303–1316.

Ladeinde, F. 1988. Studies on thermal convection in self-gravitating and rotating horizontal cylinders in a vertical external gravity field. Ph.D. dissertation, Cornell University, Ithaca, New York.

Levenspiel, O. 1972. *Chemical Reaction Engineering*, 2nd ed. John Wiley and Sons, New York.

Liao, H.-J. 1998. Simulation of continuous sterilization of fluid food products: the role of thermorheological behavior of starch dispersion and process. Ph.D. thesis, Cornell University, Ithaca, NY.

Liao, H.-J., Tattiyakul, J., and Rao, M. A. 1999. Superposition of complex viscosity curves during gelatinization of starch dispersion and dough. *J. Food Process Eng.* 22: 215–234.

Liao, H.-J., Rao, M. A., and Datta, A. K. 2000. Role of thermorheological behavior in simulation of continuous sterilization of a starch dispersion. *IChemE Trans. Part C—Food and Bioproducts Process.* 78(C1): 48–56.

Lin, S. H. 1979. Residence time distribution in continuous sterilization process. *Process Biochem.* 14(7): 23–25.

Lopes da Silva, J. A., Gonçalves, M. P., and Rao, M. A. 1994. Influence of temperature on dynamic and steady shear rheology of pectin dispersions. *Carbohydr. Polym.* 23: 77–87.

Lopez, A. 1987. *A Complete Courses in Canning and Related Processes: Book II—Packaging-Aseptic Processing Ingredients*, The Canning Trade Inc., Baltimore, Maryland.

Lyche, B. C. and Bird, R. B. 1956. The Graetz-Nusselt problem for a power-law non-Newtonian fluid. *Chem. Eng. Sei.* 6: 35–41.

Metzner, A. B. and Otto, R. E. 1957. Agitation of non-Newtonian fluids. *AIChE J.* 3: 3–10.

Miller, E. J. 1981. The design and operation of agitators for use in whole milk storage vessels. *N.Z. J. Dairy Sei. Technol.* 16(3): 221–229. *Food Sei. Technol.* Abstract 80-10–P1720.

Okechukwu, P. E. and Rao, M. A. 1995. Influence of granule size on viscosity of corn starch suspension. *J. Texture Stud.* 26: 501–516.

Okechukwu, P. E. and Rao, M. A. 1996. Kinetics of cornstarch granule swelling in excess water, in Gums & Stabilisers for The Food Industry 8 (ed. G. 0. Phillips, P. A. Williams, and D. J. Wedlock), pp. 49–57, The Oxford University Press, Oxford, U.K.

Osorio, F. A. and Steffe, J. F. 1984. Kinetic energy calculations for non-Newtonian fluids in circular tubes. *J. Food Sei.* 49: 1295–1296 and 1315.

Rao, M. A. 1992. Measurement of viscoelastic properties of fluid and semisolid foods, in *Viscoelastic Properties of Food*, eds. M. A. Rao and J. F. Steffe, pp. 207–232, Elsevier Applied Science Publishers, London.

Rao, M. A. 1995. Rheological properties of fluid foods, in *Engineering Properties of Foods*, eds. M. A. Rao and S. S. H. Rizvi, 2nd ed, pp. 1–53, Marcel Dekker, Inc., New York.

Rao, M. A. and Loncin, M. 1974a. Residence time distribution and its role in continuous pasteurization Part I. *Journal Lebensmittel Wissenschaft und Technologie*, 7: 5–13.

Rao, M. A. and Loncin, M. 1974b. Residence time distribution and its role in continuous pasteurization Part II. *Journal Lebensmittel Wissenschaft und Technologie*, 7: 14–17.

Rao, M. A. and Cooley, H. J. 1984. Determination of effective shear rates of complex geometries. *J. Texture Stud.* 15: 327–335.

Rao, M. A. and Anantheswaran, R. C. 1988. Convective heat transfer to fluid foods in cans. *Adv. Food Res.* 32: 39–84.

Rao, M. A. and Cooley, H. J. 1992. Rheology of tomato pastes in steady and dynamic shear. *J. Texture Stud.* 23:415–425.

Rao, M. A., Cooley, H. J., Anantheswaran, R. C. and Ennis, R. W. 1985. Convective heat transfer to canned liquid foods in a Steritort. *J. Food Sci.* 50: 150–154.

Rieger, F. and Novak, V. 1973. Power consumption of agitators in highly viscous non-Newtonian liquids. *Trans. IChem. E.* 51: 105–111.
Roig, S. M., Vitali, A. A., Ortega Rodriguez, E. and Rao, M. A. 1976. Residence time distribution in the holding section of a plate heat exchanger. *Journal Lebensmittel Wissenschaft und Technologie*, 9: 255–256.
Sancho, M. F., and Rao, M. A. 1992. Residence time distribution in a holding tube. *J. Food Eng.* 15: 1–19.
Sestak, J., Zitny, R., and Houska, M. 1983. Simple rheological models of food liquids for process design and quality assessment. *J. Food Eng.* 2(1): 35–49.
Sieder, E. N., and Tate, G. E. 1936. Heat transfer and pressure drop of liquids in tubes. *Ind. Eng. Chem.* 28: 1429–1435.
Simpson, S. G. and Williams, M. C. 1974. An analysis of high temperature/short time sterilization during laminar flow. J. *Food Sei.* 39: 1047–1054.
Skelland, A. H. P. 1967. *Non-Newtonian Flow and Heat Transfer*, John Wiley, New York.
Steffe, J. F. 1996. *Rheological Methods in Food Process Engineering, 2nd ed.*. Freeman Press. East Lansing, MI, USA.
Steffe, J. F. and Morgan, R. G. 1986. Pipeline design and pump selection for non-Newtonian fluid foods. *Food Technol.* 40(12): 78–85.
Steffe, J. F., Mohamed, I. O., and Ford, E. W. 1984. Pressure drop across valves and fittings for pseudoplastic fluids in laminar flow. *Trans. Am. Soc. Agric. Engrs.* 27: 616–619.
Stevens, P. M. 1972. Lethality calculations, including effects of product movement, for convection heating and broken heating foods in still-cook retorts. Ph.D. dissertation, Univ. of Massachusetts, Amherst, MA.
Tattiyakul, J. 2001. Heat transfer to a canned starch dispersion under agitation: Numerical simulation and experiment. Ph.D. Thesis, Cornell University, Ithaca, NY.
Tattiyakul, J. and Rao, M. A. 2000. Rheological behavior of cross-linked waxy maize starch dispersions during and after heating. *Carbohydr. Polym.* 43: 215–222.
Tattiyakul, J., Rao, M. A., and Datta, A. K. 2002a. Heat transfer to a canned corn starch dispersion under intermittent agitation. *J. Food Eng.* 54(4): 321–329.
Tattiyakul, J., Rao, M. A., and Datta, A. K. 2002b. Heat transfer to three canned fluids of different thermo-rheological behavior under intermittent agitation. *IChemE Trans. Part C—Food and Bioproducts Process* 80: 20–27.
Veerkamp, C. H., Romijn, A. J. M., and Pol, J. C. 1974. Influence of varying residence time distribution on inactivation of microorganisms during pasteurization of egg products. *Lebensm.-Wiss. u. -Technol.* 1: 306–310.
Vitali, A. A. and Rao, M. A. 1984a. Flow properties of low-pulp concentrated orange juice: serum viscosity and effect of pulp content. J. *Food Sci.* 49: 876–881.
Vitali, A. A. and Rao, M. A. 1984b. Flow properties of low-pulp concentrated orange juice: effect of temperature and concentration. *J. Food Sci.* 49: 882–888.
Wilkens, R. J., Henry, C., and Gates, L. E. 2003. How to scale-up mixing processes in non-Newtonian fluids. *Chem. Eng. Progress* 99(5): 44–52.
Wissler, E. H. and Schechter, R. S. 1959. The Graetz-Nusselt problem with extension for a Bingham plastic. *Chem. Eng. Prog. Symp. Ser.* 29–34.
Yang, W. H. 1997. Rheological behavior and heat transfer to a canned starch dispersion: computer simulation and experiment. Ph.D. thesis, Cornell University, Ithaca, NY.
Yang, W. H. and Rao, M. A. 1998a. Complex viscosity-temperature master curve of cornstarch dispersion during gelatinization. *J. Food Proc. Eng.* 21: 191–207.
Yang, W. H. and Rao, M. A. 1998b. Transient natural convection heat transfer to starch dispersion in a cylindrical container: numerical solution and experiment *J Food Eng.* 36: 395–415.
Yang, W. H. and Rao, M. A. 1998c. Numerical study of parameters affecting broken heating curve. *J. Food Eng.* 36–37: 43–61.

Index

M. A. Rao, *Rheology of Fluid, Semisolid, and Solid Foods,* Food Engineering Series,
DOI 10.1007/978-1-4614-9230-6,

MIX
Papier aus verantwortungsvollen Quellen
Paper from responsible sources
FSC® C105338

If you have any concerns about our products,
you can contact us on
ProductSafety@springernature.com

In case Publisher is established outside the EU,
the EU authorized representative is:
Springer Nature Customer Service Center GmbH
Europaplatz 3, 69115 Heidelberg, Germany

Printed by Libri Plureos GmbH
in Hamburg, Germany